EARTHCARE: Global Protection of Natural Areas

Other Titles in This Series

The Changing Economics of World Energy, edited by Bernhard J. Abrahamsson

Indonesia's Oil, Sevinc Carlson

Desertification: Environmental Degradation in and around Arid Lands, edited by Michael H. Glantz

The Nuclear Impact: A Case Study of the Plowshare Program to Produce Natural Gas by Underground Nuclear Stimulation in the Rocky Mountains, Frank Kreith and Catherine B. Wrenn

Natural Resources for a Democratic Society: Public Participation in Decision-Making, edited by Albert Utton, W. R. Derrick Sewell, and Timothy O'Riordan

Social Assessment Manual: A Guide to the Preparation of the Social Well-Being Account for Planning Water Resource Projects, Stephen J. Fitzsimmons, Lorrie I. Stuart, and Peter C. Wolff

Water Needs for the Future: Political, Economic, Legal, and Technological Issues in a National and International Framework, edited by Ved P. Nanda

Environmental Effects of Complex River Development: International Experience, edited by Gilbert F. White

Food from the Sea, Frederick W. Bell

The Geopolitics of Energy, Melvin A. Conant and Fern Racine Gold

Westview Special Studies in Natural Resources and Energy Management

EARTHCARE: Global Protection of Natural Areas
edited by Edmund A. Schofield

The Sierra Club's Fourteenth Biennial Wilderness Conference was cosponsored by the National Audubon Society, with over 100 organizations participating. It was among the largest assemblies ever held of statesmen, scientists, lawyers, conservationists, and government and UN officials concerned with the preservation, protection, and restoration of natural areas and wildlife. This volume contains the major papers and statements from panel sessions. In coverage and comprehensiveness, it is nothing less than a compendium and handbook on protection and conservation of the Earth's natural areas.

Edmund A. Schofield specializes in research on polar ecosystems and is interested in the adverse impacts of pollutants on organisms and ecosystems. Dr. Schofield holds a Ph.D. in botany from Ohio State University. He was formerly director of research for the Sierra Club and is currently affiliated with The Institute of Ecology at the Holcomb Research Institute, Butler University, Indianapolis, Indiana.

Earth as Photographed from *Apollo 17*.

Do not dishonour the earth lest you dishonour the spirit of man. Hold your hands out over the earth as over a flame. To all who love her, who open to her the doors of their veins, she gives of her strength, sustaining them with her own measureless tremor of dark life. Touch the earth, love the earth, her plains, her valleys, her hills, and her seas: Rest your spirit in her solitary places. For the gifts of life are the earth's and they are given to all. . . .

—Henry Beston
"The Outermost House: A Year of Life
on the Great Beach of Cape Cod"

EARTHCARE: Global Protection of Natural Areas

Proceedings of the Fourteenth Biennial Wilderness Conference

edited by Edmund A. Schofield

with the assistance of
Paul M. Glassner and Allan B. Novick

published in cooperation with the Sierra Club
and the National Audubon Society

LONDON AND NEW YORK

First published 1978 by John J. Dziak and Raymond G. Rocca

Published 2018 by Routledge
52 Vanderbilt Avenue, New York, NY 10017
2 Park Square, Milton Park, Abingdon, Oxon OX14 4RN

Routledge is an imprint of the Taylor & Francis Group, an informa business

Copyright © 1978 by Taylor & Francis

All rights reserved. No part of this book may be reprinted or reproduced or utilised in any form or by any electronic, mechanical, or other means, now known or hereafter invented, including photocopying and recording, or in any information storage or retrieval system, without p ermission in writing from the publishers.

Notice:
Product or corporate names may be trademarks or registered trademarks, and are used only for identification and explanation without intent to infringe.

Library of Congress Cataloging in Publication Data
Wilderness Conference, 14th, New York, 1975.
    EARTHCARE: Global protection of natural areas.
    (Westview special studies on natural resources management)
    Sponsored by the National Audubon Society and the Sierra Club.
    Includes bibliographies.
    1. Nature conservation–Congresses. 2. Environmental protection–Congresses. 3. Conservation of natural resources-Congresses. I. Schofield, Edmund. II. National Audubon Society. III. Sierra Club. IV. Title.
QJl75.A1W54 1975a   333.7   76-15569
ISBN 0-89158.034-4

ISBN 13: 978-0-367-02248-8 (hbk)

CONTENTS

Foreword, *Chaplin B. Barnes and Nicholas A. Robinson*   xv

Preface, *Edmund A. Schofield*   xix

EARTHCARE Conference International Advisory Committee   xxi

PART I
CEREMONIES AND MAJOR ADDRESSES

OPENING CEREMONY

1. Welcome, *Nicholas A. Robinson*   3

2. World Environment Day Message from the Secretary-General of the United Nations, *Ismat T. Kittani*   4

3. World Environment Day Message from the Executive Director of the United Nations Environment Programme, *Kai Curry-Lindahl*   7

4. Introduction to EARTHCARE, *René J. Dubos*   8

5. EARTHCARE: A Petition, *read by Marian S. Heiskell*   14

6. Presentation of the EARTHCARE Petition to the United Nations, *Nicholas A. Robinson, Maurice F. Strong, and Marc Schreiber*   16

OPENING PLENARY SESSION

7. Welcome, *Raymond J. Sherwin* — 23

8. Welcome, *Elvis J. Stahr* — 26

9. Keynote Address: The Evolution of EARTHCARE—From Wilderness to Global Protection, *Russell W. Peterson* — 29

BANQUET

10. Presentation of the EARTHCARE Award to His Excellency Carlos Andrés Pérez, President of the Republic of Venezuela — 41

11. Presentation of the John Muir Award to the Honorable William O. Douglas, Associate Justice of the Supreme Court of the United States — 47

12. EARTHCARE Address: Alaska—Northern Outpost on Center Stage, *Jay S. Hammond* — 51

CLOSING CEREMONY

13. A Call to Action, *David Brower* — 61

PART II
FORMAL CONFERENCE SESSIONS

NATURAL SYSTEMS

14. Introduction: The Primordial Interdependence, *Edmund A. Schofield* — 75

The Atmosphere

15. Introduction, *Chaplin B. Barnes* — 103

16. Atmospheric Pollution and Energy Production, *Thomas B. Johansson* — 105

Contents

17. The Impact of Urban Air Pollution on Remote
    Regions, *John W. Winchester*   119

The Hydrologic Cycle: The Oceans

18. The Third United Nations Conference on the Law
    of the Sea and the Protection of the Marine
    Environment, *Frank X. Njenga*   131

19. The Great Ocean Debate: A Clash of Realities,
    Arthur G. Bourne   143

The Hydrologic Cycle: Coastal Marine and Island Environments

20. Coastal Marine Environments, 1975, *Rimmon C. Fay*   153

21. Conservation of the Coastal Marine Environment:
    A North European Viewpoint, *Lars Emmelin*   164

22. The Coastal-Zone Development Dilemma of
    Island Systems, *Edward L. Towle*   169

The Hydrologic Cycle: Rivers

23. Water and Life: River Resources and Human
    Concerns, *David H. Stansbery*   187

24. Conservation Problems in the Amazon Basin,
    Ghillean T. Prance   191

25. Development Projects in the Nile Basin, with
    Special Reference to Disease Repercussions,
    Emile A. Malek   207

The Hydrologic Cycle: Freshwater Wetlands

26. Freshwater Wetlands, *Arthur R. Marshall*   267

27. Wetlands: Assessment and Access, *G.V.T. Matthews*   278

28. Wetland Conservation and Management Problems,
    Eskandar Firouz   291

Terrestrial Ecosystems: Temperate Forests

29. Uses and Problems of the Temperate Forest in the United States, *Eleanor C. J. Horwitz*   303

30. EARTHCARE and Temperate Forests, *F. Herbert Bormann*   309

31. Wilderness Fire Management in Yosemite National Park, *Jan W. van Wagtendonk*   324

Terrestrial Ecosystems: Tropical Forests

32. Emphasis on Tropical Rain Forests: An Introduction, *Lawrence S. Hamilton*   337

33. Regeneration Problems of the Tropical Rain Forest, *Arturo Gómez-Pompa*   341

34. Scientific, Aesthetic, and Cultural Values of Rain-Forest Wildlife, *Edgardo Mondolfi*   347

Terrestrial Ecosystems: Savannahs and Other Grasslands

35. Conservation Problems of Savannahs and Other Grasslands, *Kai Curry-Lindahl*   359

36. Wildlife of Savannahs and Grasslands: A Common Heritage of the Global Community, *Norman Myers*   385

Terrestrial Ecosystems: Deserts

37. The Perplexity of Desert Preservation in a Threatening World, *Roy E. Cameron*   411

38. The Nature of Desert Ecosystems, *Avraham Yoffe*   443

39. Mareotis: A Productive Coastal Desert in Egypt, *M. Imam*   451

The Polar Regions

40. The Polar Regions and Human Welfare: Regimes for

Environmental Protection, *Gerald S. Schatz* 465

41. Oil and Gas in the Canadian Arctic: Exploitation, Issues, and Perspective, *Douglas H. Pimlott* 479

HUMAN SYSTEMS AND INSTITUTIONS

42. Environmental Laws and Conventions: Toward Societal Compacts with Nature, *Nicholas A. Robinson* 513

Human Settlement and Natural Areas

43. EARTHCARE in France, *François Gros* 547
44. Nature and Human Settlement, *Roland C. Clement* 554

Interdependence: The Interaction of Societies and Ecosystems

45. Interdependence: Society's Interaction with Ecosystems, *Gerardo Budowski* 563
46. The Interaction of Ecological and Social Systems in the Third World, *M. Taghi Farvar* 567
47. Interdependence: The Technological and Economic Aspects, *Robert Muller* 583
48. The Exporting and Importing of Nature, *Roderick Nash* 599

Economic Development and Environmental Protection

49. The Urgency of Preserving Natural Ecosystems, *Roger Tory Peterson* 617
50. The Developed Nations in an Interdependent World, *Victor Ferkiss* 623
51. Ecologically Sound Development in a Developed Country: Canada, *Jeanne Sauvé* 635

52. Environmentally Sound Development: The Search for Principles and Guidelines, *Noel J. Brown*   643

53. Opportunities for Environmental Protection in the Developing World, *William L. Finger*   650

Institutional Measures for the Preservation of Ecosystems

54. Institutional Approaches to Global Protection of the Environment, *J. Michael McCloskey*   655

55. Environmental Impact Analysis: The First Five Years of the National Environmental Policy Act in the U.S.A., *Oliver Thorold*   660

56. Habitat Management: Problems and Prospects, *Margaret M. Stewart*   678

57. National Parks: Atonement for Environmental Sins, *Nathaniel P. Reed*   695

58. Individual Initiatives, *F. Wayne King*   703

59. Biotic Impoverishment, *Thomas E. Lovejoy*   711

60. Private Action for the Global Protection of Natural Areas, *R. Michael Wright*   716

61. United Nations Environment Programme, *Kai Curry-Lindahl*   740

62. Zapovedniks: National Preserves of the Soviet Union, *Philip R. Pryde*   754

63. Zapovedniks, *Vladimir A. Borissoff*   758

Postscript

64. The Invasion of Cyprus and the Huge Environmental Price, *Renos Solomides*   765

APPENDIXES

APPENDIX A

Declaration of the United Nations Conference on the Human Environment — 771

APPENDIX B

Convention on Wetlands of International Importance, Especially as Waterfowl Habitat — 779

APPENDIX C

1. The Antarctic Treaty — 787
2. Agreed Measures for the Conservation of Antarctic Fauna and Flora — 796

APPENDIX D

Resolution Adopted by the Tropical Forests Panel — 805

APPENDIX E

Zapovedniks in the Soviet Union, *Philip R. Pryde* — 807

APPENDIX F

Biographical Information — 811

APPENDIX G

Conference Participants — 831

CREDITS — 837

FOREWORD

Chaplin B. Barnes
Nicholas A. Robinson
Vice-Chairmen and Coordinators,
EARTHCARE Conference

A senior official of the United Nations Environment Programme (UNEP) in attendance at "EARTHCARE," the Fourteenth Biennial Wilderness Conference, remarked, "It is so refreshing to be at a meeting where natural areas are the focus of attention!" His comment reflects the uphill struggle that confronts conservationists, naturalists, anthropologists, and thoughtful developers, as well as the many other people who are concerned with maintaining the health and stability of Earth's ecosystems. Sadly, most of the official's time had to be spent on pollution and politics, little of it on nature.

The so-called "environmental crisis," which was first perceived clearly by large numbers of people in 1970 and by increasing numbers of people since then, is due to the excesses of technology, the sprawl of cities, the pressures of population growth, and the competition among commercial interests to carve up and exploit natural resources. EARTHCARE, which we define as "the global protection of natural areas," was our attempt to redress those imbalances in more than a token way.

Like the superb conferences of the 1960s, whose proceedings have appeared as *The Careless Technology* and *Future Environments of North America,* the EARTHCARE Conference provided a forum for careful and well-documented assessments by established experts. The present volume of the conference's proceedings offers a global perspective on the state of Earth's natural systems and, like the proceedings of prior wilderness conferences, offers the reflective analysis that is too often lost in day-to-day conservation efforts. It establishes the rationale and the need for a vigorous international effort to preserve and protect natural areas and ecosystems.

Leaders of government too often attend to short-term goals under the guise of environmentalism. They patch up immediate crises but rarely seek to treat the underlying disease. They largely ignore the dangers to human life and health implicit in the current destruction and deterioration of natural systems. The limits of Earth's "carrying capacity" for the human species, not understood or respected, are stretched and exceeded. Only when a knowledge of and respect for these limits guide our decisions on development, urban redesign, industrialization, agriculture, trade, resource harvest, and even warfare, will the future of the world's peoples be truly secure.

Preparations for the EARTHCARE Conference required two years. They took place during a global economic recession, in the wake of an energy crisis, and at a time when many in government had begun to divert their attention from the decisions that environmental protection requires. The euphoria of the United Nations Conference on the Human Environment, held in Stockholm in 1972, had evaporated: the voices that once had called for a change in values and behavior in order to preserve nature were muted amidst the babble of vested interests opposed to making the trade-offs such a change would require. Thus, it was not easy to organize or to conduct the EARTHCARE Conference; yet, in spite of the problems, the conference was a success. It was a lot of work. Success came only through the herculean efforts of many individuals, most of them volunteers. Their contributions of time, services, funds, materials, advice, and encouragement were acknowledged in the *EARTHCARE Program/Journal*, so ably edited by Vivien Fauerbach. Her volume was the primer for those who attended the conference.

For Nick Robinson, as convenor, the EARTHCARE Conference fulfilled a personal aspiration of several years' standing. This record of its proceedings is intended to encourage others to join us in the teaching of EARTHCARE. If it even begins to provide the common coin that will enable conservationists, educators, scientists, and statesmen in all countries jointly to espouse with us a policy of caring for the Earth, it will have served a wholesome purpose.

The EARTHCARE Conference could not have been possible without the selfless and expert diligence of Ed Schofield, whose

Foreword

guiding hand assembled these papers, or the helping hand that the omnipresent and ever-cheerful Burt Butler lent to speakers, participants, and organizers alike. There would have been no conference had Ann Singer not nurtured, cajoled, and lobbied every bit of the organizing labor. Had Ann Kreizel and Joan Stanley never developed their most professional funding structure, there would not have been the financial wherewithal for these meetings. Patricia Rambach's counsel and tireless wanderings among United Nations byways assembled contacts, ideas, and contributors that enlivened the meetings in an essential way. That the press and media carried word of the conference deliberations around the world is the contribution of Deb Appel, without whose work far fewer would have benefitted from the collective wisdom of the conferees. Dozens of others, from our registrar, Betty Mulder, and our field trips coordinator, Bob Parnes, to our organizer of student volunteers, Arne Youngerman, and scores more made the Fourteenth Biennial Wilderness Conference a success.

We know we speak for the distinguished conference co-chairmen, Judge Raymond Sherwin and Dr. Elvis Stahr, and for all conference attendees, in thanking all of the individuals who created EARTHCARE and in applauding especially the leaders just named.

The conference constitutes, above all, testimony of the National Audubon Society, the Sierra Club, six honorary sponsors, and more than one hundred other sponsoring organizations of the fact that care of the Earth must be among society's first-rank priorities. These groups, numerous as they are, plus the more than 200,000 signatories to the EARTHCARE Petition, have called for a policy of respecting nature in every human act. Their call is still largely unheeded, but they will not be Cassandras. Rather, their task will be to teach wise use of natural areas, not merely to cry out against the impending peril.

Their call is universal; it knows no barriers of politics, national boundary, economic system, language, or culture. The very word "EARTHCARE," fashioned of whole cloth, bespeaks this unity.

United Nations, New York
April 1978

PREFACE

Edmund A. Schofield
The Institute of Ecology, Butler University, U.S.A.

In editing these *Proceedings* for publication, my primary aim has been to present as complete, accurate, and useful a record of the First EARTHCARE Conference as possible. A secondary aim has been to provide a few supplementary and explanatory materials (such as the appendixes) that enhance and complement the record. By doing so, I hoped to place the conference in the broad context it deserves, for its value transcends the particular time or place in which it occurred: a permanent record of what happened in New York City during a few days early in June 1975 deserves to be broadcast far and wide.

The process of editing took far longer and was more tedious than I had expected. There are many reasons why that was so (and perhaps I should have expected it to be so), but principal among them is the fact that the participants in the conference represented many professions, disciplines, and cultural traditions. They were specialists in law, science, government, and conservation—some of them spanning two and even more fields of endeavor—and they came to the conference from some fifteen different countries of Africa, Asia, Europe, and the Americas.

Their charge was to assess from their individual perspectives the health of Earth's natural systems and the prospects for preserving or restoring them. Their styles and approaches were as diverse as their backgrounds and as broad as the charge they were given. Thus, the very source of the conference's value—namely, its extraordinarily broad diversity of approach and outlook—was also a cause of delay.

Yet, it is precisely because the conference was such a broadly based event, and of potential value to people everywhere and for years to come, that the delay was both necessary and justified. I hope the result speaks for itself.

Many people helped me, and in various ways, to prepare the manuscript. It would be impossible to acknowledge each and every one of them, but I do want specifically to thank Paul M. Glassner for the invaluable help he gave at a critical stage in the process. His help was voluntary and both dedicated and most competent.

Even more important was the thorough, careful, and painstaking work of Allan Novick of Westview Press, who spent many weeks scrutinizing the edited manuscript for errors, inconsistencies, and other editorial blunders. His superbly capable work has improved this volume's quality many fold. I wish to thank him for his concern and patient consideration.

Further, I want to thank the originators and organizers of the conference (among them Nicholas Robinson, Chaplin Barnes, and Ann Singer) for their generous support and encouragement during my more than two years of association with them at various stages of the conference. I want also to thank Westview Press for its support and patience during the year and a half it took me to edit this volume. May its forbearance be amply repaid!

I am grateful to Dr. William G. Mattox for translating the passage from *Adam Homo* by Frederik Paludan-Müller and to Mr. Charles Chang Chao for assisting me in translating the passage from *Tao Teh Ching* by Lao Tzu and for drafting the passage in old-style Chinese characters.

Finally, I dedicate my own contributions to the entire EARTHCARE project to one for whom I have the profoundest respect: former Associate Justice of the United States Supreme Court, William O. Douglas. Justice Douglas did so much for so many for so long during his enlightened tenure on the court, and embodies the principles to which the Sierra Club and EARTHCARE are dedicated, that it was only fitting he should receive the John Muir Award at the conference. We owe him so much.

EARTHCARE Conference
International Advisory Committee

Gerardo Budowski
Director General, International Union for Conservation
of Nature and Natural Resources

Norman Cousins
Editor, *Saturday Review*

René J. Dubos
Professor Emeritus, The Rockefeller University

Thor Heyerdahl
Anthropologist and Explorer

Perez M. Olindo
Director, Kenya National Parks

Roger Tory Peterson
Ornithologist

Maurice F. Strong
Executive Director, United Nations Environment Programme

Barbara Ward
(Lady Jackson, D.B.E.)
President, International Institute for
Environment and Development

EARTHCARE: Global Protection of Natural Areas

PART I

CEREMONIES AND MAJOR ADDRESSES

OPENING CEREMONY

1. Welcome

Nicholas A. Robinson
Sierra Club, U.S.A.

As convenor of this conference, it is my pleasure, on this third World Environment Day, to call to order the first session of the First EARTHCARE Conference on the global protection of natural areas. We are privileged and honored indeed to be meeting in the Economic and Social Council chamber at United Nations headquarters, a chamber in which many of our concerns have been debated and will be debated in the future. We are pleased especially to give recognition to World Environment Day and to begin these deliberations on a day which is being observed around the world, this being just one of a great many observances.

Those of you who are familiar with the charter of the United Nations know that the words "environment," "natural science," and "nature" do not appear in the charter, which was written in 1945. Nevertheless, broad determinations, principles, and ends are embodied in it. The charter states:

> We, the peoples of the United Nations, are determined . . . to promote social progress and better standards of life in larger freedom; to practice tolerance and live together in peace with one another as good neighbors; to unite our strength to maintain international peace and security; and to employ international machinery for the promotion of the economic and social advancement of all peoples.

Certainly, in maintaining the quality of the global environment upon which we and all other species depend, we are fulfilling some of the highest principles upon which the United Nations was created, principles which are themselves debated in this very chamber.

We are privileged to have with us today some of the leaders of conservation organizations from around the world, as well as the scientists and other people from many walks of life who are concerned with the protection of natural areas and who will speak as members of the conference's panels in the course of the next three days.

It is my privilege, now, to turn these proceedings over to Mr. Kittani, who will speak on behalf of the secretary-general of the United Nations, Dr. Kurt Waldheim, and who will also have some remarks of his own.

2. World Environment Day Message from the Secretary-General of the United Nations

Ismat T. Kittani
United Nations Secretariat

I am pleased and honored indeed to represent the secretary-general at this important conference. The secretary-general greatly regrets that his official engagements abroad in connection with the delicate negotiations on the question of Cyprus, which opened in Vienna today, prevent him from being present personally to greet you.

I shall first read the message from the secretary-general on this occasion, and shall then say a few words of my own.

Message of the Secretary-General
Dr. Kurt Waldheim

World Environment Day, 5 June, was established by the General Assembly to mark the opening of the 1972 United Nations Conference on the

Human Environment in Stockholm, Sweden. The conference proclaimed that both aspects of man's environment—the natural and the man-made—were essential to his well-being and to the enjoyment of basic human rights, even the right to life itself.

World Environment Day gives the international community the opportunity to renew its pledge to further global cooperation in the care and use of the Earth's resources. This year's World Environment Day theme is "Human Settlements, Our Man-made Environment."

Man's habitat, under ever-growing pressure, faces today a host of crucial issues, the solution of which is essential if we are to ensure the well-being of all the people in the very places where they live and work. The United Nations Conference on Human Settlements, to be held in Vancouver next year, will mark a major step in the endeavor to resolve these problems.

I should like, on this third World Environment Day, to call upon governments, nongovernmental organizations, and all concerned citizens throughout the world, to intensify their joint efforts for the preservation and improvement of the human environment for the benefit of all people, and for generations yet unborn.

I should like briefly to add, and to emphasize to you, certain points that at least some of us in the United Nations regard as being of particular significance in this regard.

Above all, I should like to stress the crucial role that has been played in the past, and that must be played in the future, by nongovernmental organizations, citizens' groups, and concerned individuals in maintaining pressure upon governments to ensure that irreplaceable natural areas are not further destroyed. Your contribution to this in the past has been invaluable; we shall require it even more in the future.

Natural areas are important, not only for the knowledge they give us about basic life systems, but also because—with the relentless increase in the world's population and urbanization—they are essential sanctuaries for mankind's physical and spiritual sustenance.

With your help and encouragement, the United Nations has been a major initiator of global environmental protection programs that could, if they are fully supported and adequately financed, make an immense and unique contribution to not only the preservation but also the actual enhancement of the human environment. This is the vital work upon which the United Nations Environment Programme is engaged.

In our interdependent world, we know that no single issue or problem can be resolved in isolation. We know that the problems of human environment embrace virtually all aspects of human life on this planet. We also know that, although particular problems may vary from area to area, the essential problems are global and common to all. It is therefore crucial that we meet problems with a comprehensive and global approach and with a shared dedication to solve them. All nations, whether they be rich or poor, developed or developing, must come to share a common concern for the human environment and be persuaded to take urgent actions to arrest and reverse the wanton destruction we continue to inflict upon it.

The fundamental problem before us may be simply stated: How can we meet the legitimate needs and aspirations of the present generation without jeopardizing those who will follow? The solution to this problem is infinitely more complex. No one can claim to have devised a formula that will reconcile these often divergent interests. But it is through the process of debates in conferences such as this that we may grope our way together, as fellow-citizens of this planet, toward solutions that meet the essential interests of all.

Our immediate task is to ensure that we do not knowingly permit harmful practices to continue, but resolve to set aside areas of outstanding beauty and wilderness for the use of future generations. It is our responsibility to pass on to them, untarnished and undefiled, the heritage we received.

William Pitt the Elder once described London's parks as "the lungs of London." So should we view open places, parks, and natural wildernesses as the lungs of mankind—not as luxuries, but as vital, life-sustaining organs.

3. World Environment Day Message from the Executive Director of the United Nations Environment Programme

Kai Curry-Lindahl
United Nations Environment Programme

Mr. Maurice F. Strong, executive director of the United Nations Environment Programme, regrets very much that he is unable to be here today in person because of other commitments. I have a message from him, which I am going to read to you.

Message of the Executive Director
Maurice F. Strong

On the occasion of the Fourteenth Biennial Wilderness Conference on global protection of natural areas, I take the opportunity to extend to you and to the participants of the conference my congratulations and best wishes for success in this most significant endeavor. Your meeting has a special meaning to the United Nations Environment Programme, not only because it underscores, dramatically and profoundly, the meaning of World Environment Day, but, more especially, because it is based on a theme which was one of the fundamental premises of the Stockholm Conference in 1972, namely, the care and maintenance of a small planet.

Moreover, your initiative is clearly in keeping with the objectives of the Declaration on the Human Environment, which, in effect, has outlined certain EARTHCARE principles and guidelines to which we all subscribe—namely, that the natural resources of the Earth—including air, water, land, flora, and fauna, and especially representative samples of natural ecosystems—must be safeguarded for the benefit of present and future generations through careful planning and management.

The capacity of the Earth to produce vital renewable resources must be maintained and, wherever practicable, restored or improved. Man has a special responsibility to safeguard and wisely manage the heritage of wildlife and its habitat, which are now gravely imperiled by a combination of adverse factors.

It is our hope that your EARTHCARE Conference will give scope and meaning to these principles and will encourage governments to translate them into positive policies and practical action. Your traditional regard for prudent management of the Earth's natural resources and your commitment to an ethic of conservation have been a profound source of encouragement to us in the United Nations Environment Programme, and a demonstration of what is possible for dedicated citizens in the environmental community. We hope that your work will continue to inspire and encourage our efforts, and that your meeting this week will reinforce our own activities in the field of conservation education and citizen involvement.

As should be clear from the message of this day, attitudes and values of behavior of the people will be decisive determinants of the success of the global effort to protect the environment. The real catalyst for change is the citizen who can express his concern by his own actions, by organizing collective action for creating sound environmental lifestyles, and by making local and national governments yet more aware of the urgency of the need.

Citizen action can provide the vital impetus to create change for the better, and there is no better opportunity to practice citizen action than on World Environment Day.

Your meeting today is a splendid example of the creative use of World Environment Day opportunities.

Thank you, and heartiest congratulations.

4. Introduction to EARTHCARE

René J. Dubos
The Rockefeller University

Civilization and Wild Nature

EARTHCARE—such a lovely word, so symbolic of the quality of relationships that should link mankind to the Earth. It conveys better than an eloquent essay the spirit of this conference on the "global protection of natural areas." We must take *care* of the *Earth*, for the simple reasons that we care for it and it cares for mankind.

Civilized people have long found it difficult to cultivate a rational and loving attitude toward high mountains, deep forests, marshes, swamps, deserts, and other wild places. Our ancestors were not concerned with the protection of such natural areas; rather, they looked upon them with a sense of fear and alienation.

Everywhere, until modern times, the word *wilderness* has had connotations of insecurity and even of terror. The Puritans referred to the huge forest that covered most of New England as a "heidious and desolate wilderness full of wilde beastes and wilde men." Practically all the immigrants who settled in the New World during the past three centuries also regarded mountains and primeval forests with fear and contempt. The same attitude can be found throughout the Greco-Roman and Judeo-Christian civilizations until the eighteenth century, and also, despite what is commonly believed, in non-Judeo-Christian civilizations, including the Oriental ones.

Wherever they have settled, people of all cultures and all religions have tamed wild nature and transformed it so as to create from it environments better adapted to their immediate needs. What we call "nature" is in most cases a humanized nature.

Mastering the wilderness has thus been for millennia one of the great enterprises of mankind, but now that this mastery has been achieved in many places, the spirit of struggle between mankind and nature is progressively giving way to an awareness of man's fundamental kinship with wild nature. This new attitude first took the form of an emotional and spiritual admiration for the wilderness, richly expressed in poetry and in painting. It was strengthened by the theory of evolution, which helped man to comprehend that, biologically, he is still linked to the rest of creation by countless organic bonds. Enlightened public attitude toward the wilderness has thus come to incorporate a wide spectrum of emotional, spiritual, and philosophical values that are symbolized by Thoreau's phrase, "In wildness is the preservation of the world." These values set the conservation movement on its way, and their influence on modern life is increasing.

In addition to these subjective values, there is now emerging a scientific recognition of the fundamental role played by wilderness in the economy of the Earth. Ecological studies have revealed that each of the diverse components of wilderness—of the diverse natural areas—makes a special contribution to the maintenance of life on Earth. It is this diversity of contributions that I wish to discuss, because it is at the heart of the present conference.

The Contributions of Natural Areas

Scientific studies have made it clear that destroying natural areas means more than impoverishing the emotional and spiritual heritage of human life. It also reduces the ecological diversity and the productivity of the Earth. It renders ecosystems less resistant to pests and to climatic stresses, less able to support a great variety of plant and animal species, less efficient in trapping solar energy by photosynthesis. The protection of natural areas can thus be justified by the role they play in the economy of the Earth, and by the practical contributions they indirectly make to human life.

Forests, prairies, swamps, marshes, deserts, and all other types of natural areas are each characterized by a special flora and fauna that has evolved over immense periods of time and that cannot possibly be duplicated; each natural area constitutes a unique reservoir of biological species. The multiplicity of these reservoirs is the best and indeed the only insurance we have against the dangers of ecological catastrophes inherent in the oversimplified ecosystems resulting from industrialization, urbanization, and—perhaps even more—modern agriculture.

While it is possible to preserve a few wild species of animals, plants, and microorganisms in botanical and zoological gardens or in biological laboratories, it is impossible in practice to save the largest part of the living things of the wilderness once their natural habitats have disappeared. Yet, any loss of biological species can be significant, not only for the balance of life on the globe, but also for human welfare, and the importance of the loss cannot be predicted. For example, no one can predict what species of insect might prove useful for the future development of biological control methods.

In brief, the maintenance of ecological diversity is imperative on several counts. To mention only two: (1) it minimizes the chances of ecological catastrophes; and (2) it assures the existence of repositories of genetic types from which new species can be drawn to modify or diversify the hereditary endowment of domesticated animals and plants.

Energy from Natural Areas

Surprising as it may seem, natural areas are the greatest "producers" of energy on Earth. In any given year, the amount of energy captured by all vegetation on Earth through photosynthesis vastly exceeds the total amount of energy used by mankind for its daily life and for driving even its most extravagant technologies. The precise figures cannot be ascertained, but this is not important because what matters is the magnitude of the difference.

The green plants that live on the surface of the Earth capture, or fix, approximately 840 trillion kilowatt hours of energy per year through their photosynthetic activities. Of this grand total, about two-thirds are fixed by land vegetation, especially forest, and one-third is fixed by the vegetation growing in water, especially in marine estuaries, in various wetlands, and in ocean plankton areas. Ultimately, all forms of life and the very characteristics of the terrestrial atmosphere depend on the energy derived from the sun by photosynthesis, chiefly in natural areas.

In 1973, in contrast to the 840 trillion kilowatt hours of energy captured by the green vegetation of the Earth, mankind consumed only 70 trillion kilowatt hours. The twelvefold difference between these two amounts of energy is so great that it seems to justify a carefree attitude. Indeed, complacency was justifiable in the past, but the situation is rapidly changing, and for three reasons: (1) the increase in the world's population; (2) the greater per capita use of energy; and (3) the introduction of new technologies that threaten the productivity of natural areas.

At present rates of growth, little more than ten years will be required for a doubling of energy consumption in the world at large. If this rate of increase were to be maintained, the world's consumption of energy would overtake the production of new energy by green plants within a century. Such a change in the energy balance would unquestionably cause profound disturbances in natural systems and, of course, in human life. In fact, it would probably mean the destruction of life.

The present ratio of the energy produced by natural areas to the energy consumed by mankind is further threatened by the fact that many pollutants of air and water (including heat pollution) damage natural systems and thereby decrease the amount of energy produced by photosynthesis. One example will suffice to illustrate the potential magnitude of these effects.

Acid Rain

Rainwater has been rendered very acidic in many parts of the world by nitrogen oxides from internal-combustion engines and by sulfur oxides from power plants that burn fossil fuels. These gases are dispersed through the atmosphere and carried over the globe by air currents before they return to Earth in rain.

Acid rain causes profound ecological disturbances both on land masses and in water bodies. It has been calculated that if the concentration of acidic substances now reaching New England in contaminated rains were to be maintained for ten years, there might be a 10 percent decrease in photosynthesis and therefore in energy production; for this region alone, the decrease would mean a loss of energy equivalent to that produced by fifteen 1,000-megawatt power plants.

It might be thought that agriculture compensates for the loss of productivity of natural areas by providing energy in the form of cultivated plants. In fact, however, the energy contributed by agriculture is less than that contributed by natural areas: cultivated lands yield the equivalent of approximately 50 trillion kilowatt hours of energy per year, much less than 10 percent of that produced by natural areas. Furthermore, the significance of this contribution is decreased by the fact that scientific agriculture cannot function without the expenditure of enormous amounts of energy from fossil fuels.

Forests, wetlands, estuaries, and oceans are thus the great producers of the energy that keeps all forms of life—including human life—going. Therefore, from the point of view of the Earth's energy balance, the protection of natural areas has a crucial significance for global ecology.

Man-made Natural Areas

In theory, the expressions *wilderness* and *natural areas* should refer only to environments that have not been disturbed by human intervention. In practice, however, there are very few such places left in the world. For this reason, it is more useful to define natural areas by their characteristics than by their histories.

The ecological health of the Earth will depend increasingly upon the ecological quality of what could be called, paradoxically, "man-made natural areas." We must therefore try to understand the conditions that made areas of native wilderness ecologically successful and that endowed them with emotional and spiritual qualities. By using this knowledge we can, perhaps, reintroduce into many parts of the world the diversity, the local suitability, and other qualities implied by the word *wilderness*.

Management of Wilderness

In fact, mankind is rapidly moving everywhere toward the management of wilderness—odd as the juxtaposition of these two words may sound. We make ourselves judges of how many deer or elephants should be allowed to survive—for the good of their habitats and for their own good!—in the wild regions of America, Africa, or Asia. There were even plans to protect Niagara Falls against the impact of natural forces. For several thousand years, Niagara Falls has been receding because of erosion and rock falls. A joint Canadian-American "fallscape" committee, now defunct, had been formed to develop methods for controlling this process that disturbs the ideal public image of the falls—the purpose having been to preserve their visual purity against the attack of natural forces. There could be no more telling proof than this of how the quality called wilderness is now almost as much a state of mind as a creation of nature.

We can thus envisage for the future several kinds of management of natural areas.

One kind of management is symbolized by the plan contemplated for Niagara Falls; it would have continued the falls' existence as one of nature's masterpieces, even though their natural evolution would have been interrupted by human efforts.

A second kind of management, probably more interesting, would be to help natural forces reestablish the original natural ecosystems of certain areas—for example, woodlands in the Mediterranean basin and prairie vegetation on the North American continent.

A third kind of management would be to maintain certain kinds of man-made ecosystems that have emerged more or less empirically and yet have proven highly successful—for example, the humanized landscapes of East Anglia.

A fourth kind of management would be to create entirely new man-made ecosystems on the basis of scientific knowledge by introducing into areas that have been biologically ruined or impoverished the complex ecological values that spontaneously exist in various wildernesses of the globe.

Depending upon the place and the economic situation, one or the other of these approaches is justified. But for any of them to have a lasting success, it is essential that we preserve areas of native wilderness to the extent that this is possible. It is only by studying wild nature that we can learn to properly manage humanized nature.

5. EARTHCARE: A Petition

Read by Marian S. Heiskell
U.S.A.

Three years ago, acting upon the recommendations of the United Nations Stockholm Conference on the Human Environment, the General Assembly declared that safeguarding the environment is a prerequisite to "the enjoyment of basic human rights—even the right to life itself." Each government is responsible for securing such rights.

For the first time in Earth's history, our species has the capacity to violate the environment on a scale that endangers the existence of all species. The heedless exploitation of nature and the careless use of resources already threaten our inheritors with a world physically and spiritually impoverished. We must act now to renounce such a perilous course and to conduct our affairs in harmony with nature.

We have little time to reshape our global community. The next thirty years will produce a doubling of world population and an even greater explosion of expectations. We shall not cope with the consequent stresses on social organization without first reordering our values in recognition that human life is a part of nature. Our decisions must be grounded in ecological principles.

The biosphere comprises myriad interrelations among plants and animals, land, air, and water. We can modify some ecosystems, even enhance them, but we may not abuse them. Given our elementary knowledge, we cannot press the limits and capacities of natural systems without incurring the gravest danger. We are not immune to the hazard of ecological breakdown resulting from irreparable acts.

Regardless of our political, economic, or social organization, the lessons of natural science must guide all human affairs. Despoliation of natural environments denies science keys to better understanding and enriching all life. Without environmental protection, short-term economic gain impairs economic stability in the longer span of time. No new development should proceed until its environmental impact has been appraised. The drive to exploit, regardless of consequences, must yield to rational use of natural resources. Human needs must be met, excessive demands rejected.

The nations gathered at the 1972 Stockholm Conference on the Human Environment proclaimed that protection of the human environment is the "duty of all Governments." They also agreed that their responsibility to secure natural systems from disruption is a necessary condition for achieving fundamental human rights. Too seldom have these commitments been honored.

Daily, governments violate our human rights by ignoring the intimacy between the natural and human environments. Such behavior threatens the security of nations, individually and collectively. After the Stockholm conference, we hoped that governments would move quickly to avert harm to nature and adopt a policy of EARTHCARE which creates the conditions for assuring human rights. Too little has been done.

Members of the United Nations have welcomed petitions seeking to assure fundamental rights. We add our own, as basic as any received before. Our right to receive protection of our common global environment must be honored.

6. Presentation of the EARTHCARE Petition to the United Nations

The EARTHCARE Petition was presented by Nicholas A. Robinson, convenor of the EARTHCARE Conference and chairman of the Sierra Club's International Committee, to Maurice F. Strong, executive director of the United Nations Environment Programme, at a press briefing held on 23 October 1975 at United Nations headquarters in New York. Dr. Marc Schreiber, director of the United Nations' Division of Human Rights, also attended the briefing. Abridged and edited versions of their comments are presented below. Parts of Mr. Strong's and Dr. Schreiber's remarks consist of replies to questions from the press.

Presentation of the Petition

Nicholas A. Robinson

Mr. Executive Director, as convenor of the EARTHCARE Conference, it is indeed my pleasure to present to you this EARTHCARE Petition, which calls upon the United Nations to examine the denials of human rights that occur daily with respect to environmental protection. As the members of the United Nations stated in the General Assembly resolution and the Stockholm Declaration, protection of the environment is prerequisite to all of the other rights specified in the Universal Declaration of Human Rights.

Three years ago, at Stockholm, the nations made it quite clear that they intended to do something about this. Now, some 200,000 citizens and, we suspect, several thousand more whose signatures (because of the time lag) are still in the mails to us, think that their governments have not lived up to the pledge they made three years ago at Stockholm—that governments accepted at Stockholm the duty to do much more than they are presently doing.

The petitioners range from Mr. Justice Douglas of the United States Supreme Court to individuals who happened to come into a gasoline service station and sign the attendant's rather greasy copy of the petition—signatures of those who use the environment daily in a most basic way! There are many signatures from Sri Lanka and from Sweden, giving a whole range and spectrum of signatories. All of them agreed with the petition's thesis, namely, that they have a right to life, and that, daily, their right to life is not being honored.

The EARTHCARE Conference was our own "mini-Earthwatch," if you will. The scientists who gathered there from around the world presented a "State of the World" message. They restated the case that had been made at least as early as 1972—namely, that unless we stop preying upon natural ecosystems we will cause the ruination of man and of civilization as we have come to know it. What is more, if we do not start changing our lifestyles, the time draws near in which our predatory activity will take a serious toll.

We believe that conflicts involving the natural environment will pose a security threat to all members of the United Nations. In the course of the Law of the Sea negotiations, we see an attempt to solve some of the security problems, but the negotiations have been very slow on this point, and the environment has been a secondary factor in negotiations. The nations signatory to the Antarctic Treaty are questioning whether they ought to divide up the natural resources of Antarctica and exploit them, rather than reserving that area for science.

We hope, in submitting the EARTHCARE Petition to you and, through you, to the secretary-general—and in the secretary-general's subsequent submission of the petition to the Human Rights Commission of the Economic and Social Council—that we will have the kind of review of these promises that the members of the United Nations made in 1972, and that the Economic and Social Council itself eventually will debate what it means to say, "Protecting the environment is a necessary prerequisite to protecting human rights."

Since it is difficult to deal with the general proposition that the denial of environmental protection is a denial of human rights, we will supplement this petition with three specific, well-documented cases of the denial of human rights.

The three documented cases that we will present to the Human Rights Commission deal with (1) mercury poisoning in several countries; (2) the rights of peoples indigenous to tropical rain forests; and (3) the depletion of the stratospheric ozone layer.

Mercury poisoning is an example of toxic discharge. Principle 6 of the Stockholm Declaration explicitly states that "the discharge of toxic substances... must be halted." The lifestyles of tropical rain-forest peoples are very different from our own. Those peoples are, uniformly, minorities within their own territories. In harmony with nature, they make perfectly good livelihoods. But change is going on around them—change that takes away their livelihoods and the guarantees of the Universal Declaration of Human Rights, without offering them anything in return. The depletion of the stratospheric ozone layer as a consequence of activities carried out in the most industrialized states will affect both their own citizens and those of states that do not engage in such activities. Scientists have estimated that some 300,000 people would die of skin cancer every year if the ozone layer were eliminated or reduced.

I recall that, at the beginning of this decade, Secretary-General U Thant said we had ten years before the damage would be done. But individuals who have been poisoned by mercury do not have ten years: they are suffering now, or dying, or may already be dead. If the Earth's ozone layer is being depleted by the release of man-made chlorofluorocarbons as some investigators think it is, then we do not have ten years in which to change our lifestyles: by the time the ozone layer has been decimated, it will be too late to change. Our lifestyle will have changed anyway—involuntarily and for the worse.

Mr. Strong, it is with honor and some humility that I pass this burden, literally and figuratively, to you.

Acceptance of the Petition

Maurice F. Strong

Thank you very, very much, indeed. I am delighted to have this opportunity of receiving the EARTHCARE Petition. In thanking you, Mr. Robinson, may I also thank the organizers and sponsors of the EARTHCARE Conference for confirming and extending the commitments of Stockholm, as well as the more than 200,000 citizens for their continuing commitment to the cause of protecting the environment on which the future of our "Only One Earth" depends. Thank you very, very much.

That some 200,000 people are prepared to attach their names and signify thereby their commitment to this petition confirms my belief that the environment issue is alive and well, and that the number of committed people who continue to see it as one of the most important issues—if not the most important issue—that confronts the world community is growing. I hope they will be joined by not only tens of thousands, but by millions of others, because the right to life must surely be the most basic of human rights.

Those who endanger the environment on which life depends commit an act of aggression against human rights, an act of aggression against the right to life itself. The United Nations Conference on the Human Environment at Stockholm helped make this clear in 1972: the Stockholm Declaration started the process of writing this concept into standards of behavior amongst nations; it will, I hope, be followed by the evolution of international law in this field. But none of this will be possible if it is not undergirded by the most basic commitment of people, of people who *do* understand that offenses against the environment—particularly offenses against that aspect of the environment on which life depends—are, and should be considered, offenses against human rights.

It is quite clear, I think, that the environment issue, as it has progressed in the last five years, is a mixture of pluses and minuses, of accomplishments and disappointments. I will not try to catalogue them for you here because I have already done so in my statement to the Governing Council of the United Nations Environment Programme (UNEP). However, I think that, if one were to look at the environment issue since we began preparations for the Stockholm Conference five years ago, one would have to say that there has been a tremendous amount of progress on the whole. There has been some slippage in the high degree of interest that some industrialized countries used to have in environmental issues because of the energy and economic crises; on the other hand, the environment issue is today a global issue, in the real sense that countries all over the world are concerned about it. They have not all been converted, by any means, and the problems have not all been solved, but there is now a truly global commitment to solving these problems, and a group of committed people is working on them in virtually every country in the world. So, we have made very significant progress.

I came to the United Nations Environment Programme on

leave of absence from my government for what was supposed to have been two years. I was able to get the leave of absence extended a number of times because of the importance that I and the government of Canada attach to assuring that the decisions of Stockholm were properly implemented in the setting up of UNEP. We now have a very good staff—I would say one of the best groups of people I have ever had the pleasure of working with. Our Environment Fund is over the top; our program is well launched; we are working at the edge of our present political mandate. The task is now one of getting on with the job of assuring that what we have begun will continue. The first cycle of the task I took on in the United Nations has come to a logical end. The larger task has just begun.

The team we have in Nairobi is a first-rate team fully able to carry on. Indeed, I have been shifting responsibilities over the last year or so more and more to that team as it has been built up. We have very, very able people there. Dr. Mostafa K. Tolba, my deputy, who has a great breadth of experience, is a top-notch manager whose commitment to environment well preceded the Stockholm conference. He was minister in his own government, was president of the academy of sciences, and has a world reputation in the scientific field. He is one of the ablest administrators I have ever worked with. R. Bruce Stedman, assistant executive director for the Bureau of the Environment Fund and Management, is one of the United Nations' ablest administrators. If you knew some of the rest of our professional team, I think you would agree with me that we have assembled in Nairobi a professional team that is second to none.

It takes more time when such a team is international; it takes more time when one is working in a place that has no pre-existing infrastructure of international life. Nevertheless, I can tell you that I am very proud of the people on the team and have no qualms at all about their ability to carry on. However, as I pointed out to the Second Committee in my statement, they are operating far from the other organizations that they must work with, and they need the continued commitment and support of the governments of their colleagues in the system if they are to continue to perform their job effectively. They are certainly capable of doing so. I know they will do so, but we cannot forget them up there in Nairobi: they need continued support, and I am sure that they will get it.

I am delighted to be given the opportunity of presenting to the secretary-general the EARTHCARE Petition, which calls specifically for the issue of environmental protection to be looked at in terms of human rights. I certainly intend, in communicating it to him, to make sure that it carries my own full, personal commitment also.

Thank you.

Concluding Remarks

Marc Schreiber
United Nations, Division of Human Rights, U.S.A.

During the last few years, the human-rights bodies of the United Nations have been concerned with introducing the human element, the human factor, into a number of activities of the organization that, at first sight, did not seem to be concerned with the protection of human rights in the literal meaning of that term. For instance, an effort was made to introduce the principle of respect for the individual in devising solutions to such worldwide problems as population control and economic development. That effort was also made with respect to the environment, under Mr. Strong's powerful leadership, when the problem loomed upon us rather dramatically. Some of the delegates at the Stockholm conference and we in the Secretariat helped him find a formula that emphasizes the human element; it appears in Principle 1 of the declaration.

As you know, the Universal Declaration of Human Rights has provisions about right to life, liberty, and security of the person. Now we probably must go further and find the appropriate formula to elaborate those provisions. Thus, since it is quite feasible to discuss environmental protection in terms of human rights, we will try to give some volume, impact, and scope to this particular approach.

[The Declaration of the United Nations Conference on the Human Environment constitutes Appendix A of the present volume. —Editor]

OPENING PLENARY SESSION

7. Welcome

Raymond J. Sherwin
Sierra Club, U.S.A.

The New Yorker magazine once published a cartoon that depicted Manhattan as constituting virtually all of the United States, with several of the larger states appearing as microscopic satellites. As a native of one such appendage, I enjoy welcoming you, from so many even more remote places, to this great city.

It is an immense pleasure to me to have the chance to greet you, and to meet the many talented speakers. Because of their generosity in giving their time and knowledge, I have great hopes that this conference will achieve several goals: that it will teach, that it will bind us together in common purpose, and that it will inspire us to go back home and work harder.

The tutorial function is clear: this conference marks an essential step in the evolution of conservation philosophy. You will recall that the Sierra Club was created to explore, enjoy, and preserve California's Sierra Nevada and other scenic resources of the United States. Over the years, men like John Muir, William Colby, Francis Farquhar, Richard Leonard, Will Siri, Edgar Wayburn, David Brower, and many others have fought for the Yellowstones, the Yosemites, the Grand Canyons, and the other natural wonders of North America. Then, in a natural progression, there came a time when the air, water, and land pollution that accompanied an exploding population and technology supervened. These uninvited by-products loosed insidiously destructive influences that threatened not only the dwindling natural resources, but life itself.

So, in 1969, the Sierra Club addressed itself to these subjects of wider scope, and, in our subsequent efforts to cope with them, we have encountered additional substrata of difficulties. For example, as we studied pollution we were led to the question of energy. Now, as the function, supply, and distribution of external sources of energy have become increasingly controversial, one problem persistently gets in the way of solutions that otherwise seem so logical: while our confidence in the boundless inventiveness of our technologists may be well justified, the same cannot be said for our human institutions. One illustration should suffice: our experience in attempting to control the distribution and abusive use of any contraband—whether alcohol, dangerous drugs, narcotics, or the relatively innocuous marijuana—has been dismal. The record cannot give us confidence that we can avoid thefts and sabotage or even just careless losses in handling deadly plutonium. Yet presently we are confronted with the necessity of extending such controls to the international scene, where the difficulties are compounded by many orders of magnitude, and where the temptations are magnified in similar proportion. Thus, the decisive role of the human relations factor and the importance of achieving cooperation are manifest.

Now we face another lesson we have been slow to absorb. We seem often to give only lip service to the concept of the universal interdependence of living creatures and their habitats. It is not that the interrelationships of ecosystems have been unnoticed; the literature is extensive. What happens is that we are appalled at the magnitude of the problems posed by the logical consequences of that fact. To be candid, it is a logic that even highly competent activists among the Sierra Club membership are still hesitant to recognize. Although the fate of gene pools, or even the health of our forests, estuaries, and inland waterways may depend on international cooperation, we face the formidable tasks of achieving an understanding that will lead to conventions, treaties, or other compacts and will overcome a reluctance to commit the costs and effort.

I suppose that one principal reason for rejecting the total implications of interdependence is that each of us is pressed by urgent neighborhood problems. Of what moment does the peril of acid rain seem when the bulldozer threatens the immediate destruction of our forests—even though the trees may wither later from the wastes in the atmosphere? Or, why should we bother to worry about international tanker construction codes designed to minimize oil spills if our shoreline faces more immediate damage from onshore refinery construction?

These special difficulties are attended by other, more normal obstacles. Any conservation campaign has in it the potential to be abrasive. To the lumberman, the mere fact that we choose to discuss tropical and temperate forests looks as if we are singling him out for adverse criticism. The industrialist, manufacturer, or utility executive is already hypersensitive to the mention of pollution of any kind. So, if we internationalize our fields of study, we step on more and more toes. Nevertheless, leaders of these very target industries have helped make it possible to produce a conference directed toward uncovering and publishing the facts that underlie the most difficult problems. This is something of an act of faith, but it is justified by history. Undoubtedly, previous wilderness conferences have helped achieve a community of understanding of the importance of nature and of the steps necessary for the protection of natural areas. Here, we can help achieve a similar consensus with respect to the international mutuality of our concern. We can also hope to follow precept in the establishment of permanent bonds of friendship among us all.

On the matter of welding common bonds, I shall say no more. Like sharing a concert, a drama, or an exhibit, the experience we shall have here together will serve as the catalyst.

Finally, we come to the necessity of taking the product of our participation home. It does little good if we rest content with each other's wisdom and company. Here, we speak to the converted. But the time is ripe to gain a larger audience, for, at least in the United States, people are hungry for information on subjects they know are of immense importance but about which they feel desperately uninformed. The population of the United States may be unique in that regard, though I doubt it.

In contemplating the means of taking our concerns home, again we can take a cue from Sierra Club history. Long ago, we learned of the power of the facts and often personally went out and gathered them. Assembled here are the people who know the facts about the areas of most critical concern around the globe. Seldom will there be such an opportunity for us to learn—to have our eyes opened so that we, too, may speak with authority. Who knows but what our proceedings at EARTHCARE may strengthen everywhere the efforts to save the deserts and their myriad secrets, the savannahs and their lush assemblies of wildlife, the forests with their birds and butterflies, the oceans and their life-support systems and underwater "gardens"? If so, how much the richer we shall have made the whole world!

This is the fourteenth biennial wilderness conference sponsored by the Sierra Club. We are delighted and grateful for the participation of the National Audubon Society, without whose energetic and knowledgeable cooperation it could not have happened. To their president, Dr. Elvis Stahr, and to a considerable number of their staff and associates, we offer our profound thanks. Beyond mentioning Audubon, I fear to be specific lest I offend a host of others to whom we are indebted for support in terms of work and financial support. They include a number of conservation, business, and industrial organizations, as well as many individuals. For them, I hope that the quality of the program will be our thanks.

8. Welcome

Elvis J. Stahr
National Audubon Society, U.S.A.

This is not the first, but the fourth biennial wilderness conference I have had the privilege of attending, but it is the first in the distinguished history of these conferences that the Sierra Club has planned and carried out jointly with another great, nationwide conservation organization. As president of that cosponsoring organization, the National Audubon Society, and as cochairman of the EARTHCARE Conference with Judge Sherwin, I find it gratifying, and a genuine honor, to join him in welcoming you to this, the Fourteenth Biennial

Wilderness Conference, in this, the headquarters city of the National Audubon Society. And a very special wilderness conference this is, for EARTHCARE is the first to be primarily international and thus to endeavor to contribute to the global protection of natural areas, an historical concern of the National Audubon Society and a very special concern of the Sierra Club as well.

For years, the National Audubon Society has worked, from time to time, with the Sierra Club on matters of mutual interest and concern, of which there were and are many. But, since the great United Nations Conference on the Human Environment, held in Stockholm in 1972, Audubon has been working especially closely with the Sierra Club on a whole spectrum of international issues. So the almost daily association of a number of the most talented young staff members of both organizations in preparing for this conference in recent months has been not only productive, but pleasurable and positive.

Since the Stockholm conference, both Audubon and the Sierra Club have established ongoing, working relationships with the United Nations Environment Programme (UNEP), which came into being as the principal practical outcome of the Stockholm conference. UNEP's executive director, Maurice Strong, an incomparable person indeed—who, among his other duties, has served as an advisor to this conference—is committed to be in Kenya, the host country of UNEP, for World Environment Day, but I am very pleased to see that the distinguished director of Kenya's national parks, Dr. Perez Olindo, has come here for World Environment Day, and for this conference.

I mention UNEP because I want to get on record the exciting point that Maurice Strong and UNEP have given nongovernmental organizations (NGOs, as they are called in United Nations parlance) an unprecedented opportunity to work closely with an official agency of the United Nations, which, of course, is itself an organization of governments. In Mr. Strong's greetings to the EARTHCARE Conference, both the effectiveness of those working relationships and the relevance of EARTHCARE to UNEP's global program are stressed.

Is EARTHCARE important? In other words, is it worth your time and mine to be deeply concerned with the wise use, intelligent preservation, and responsible stewardship—in short, with the care—

of basic resources of this little planet, all of which, except the light and heat of the sun, are finite? Your very presence here provides your own answer to that question, and I congratulate you on it. Moreover, the respected and expert speakers and panelists with whom we will be interacting would not have come here—many of them from great distances—if they had not thought that the care of this Earth is important.

In addition, we are proud and grateful to count among the sponsors of EARTHCARE over one hundred citizen conservation organizations from around the world, representatives of many of which are present.

All such involvement, I submit, is proof positive that many, many people have become aware of and have determined to deal realistically with, global environmental problems. Many—maybe all of those problems—have the potential to become global crises: population, energy, the great whales, to mention just three diverse and excellent examples. We are convened here these few days to ponder such problems in the context of the entire biosphere, the very life-support system of our planet—in short, to bring both intellect and commitment to the care of this Earth, this "Only One Earth."

Many of you have come a long way to join us here, but none of us has really left home to do it. These few tons of rock and water wrapped in this thin wisp of atmosphere are home to us all—and our only home. Let us do nothing less than all we can to ensure that our home is well cared for, and that its housekeeping is neither abandoned to the selfish nor left wholly to chance.

9. Keynote Address: The Evolution of EARTHCARE— From Wilderness to Global Protection

Russell W. Peterson
Council on Environmental Quality, U.S.A.

The Advocates and Critics of EARTHCARE

Three weeks ago, the budget appropriation for the Council on Environmental Quality (CEQ) was sent to the floor of the United States House of Representatives for a vote. The appropriation passed, but I was considerably distressed by the opposition of one congressman, who referred to the council and its staff as "posy-pluckers." Since this is an international conference, some of you may not be familiar with this term. Let me explain. In essence, "posy-pluckers" are people who pick flowers. Beyond its basic meaning, however, the term carries a contemptuous connotation: in the context in which it was used, it suggests that CEQ is engaged in frivolous work of little value to the republic and certainly not deserving of tax dollars.

Any environmentalist quickly learns to develop a thick skin, and I have been called worse things than a "posy-plucker." What depressed me was the congressman's summation of his argument against approval of CEQ's budget. He concluded as follows:

> Mr. Speaker, I say again [that] the posy-picking, environmental extremists will put this country in the dark ages of no gas, no heat, no lights, if we in Congress do not wake up to the way they are contributing to the destruction of our free society. The combination of environmental extremists, zealous regulators, and funny-money politicians is simply more than our economy can take.

I was pleased to be invited to address this conference, and am deeply honored to deliver the keynote address. It may seem odd for me to begin with the words of one critic of a small United States agency—we all have much larger concerns than the size of CEQ's budget or even its existence. Yet this critic's words have direct relevance to our objective here. The terms of the criticism depressed me, for they suggest that the environmental movement in the United States has failed to explain its concerns properly to at least one of my government's policymakers.

I know that this congressman is not alone in considering environmentalists extreme in their positions; I could cite a long litany of quotations to the same effect from editorial pages, advertisements, and newspaper articles. Judging from my personal experiences with other countries, the United States is not the only nation in which environmental protection is viewed by some as an impractical, even trivial interference with the serious business of economic growth. This situation has an all too direct significance to EARTH-CARE: it indicates a major obstacle we must overcome if we are to succeed with the objectives of this meeting.

The National Parks Idea

Just over 100 years ago, our nation's national park system was born. As is the case with any new concept, acceptance of the parks idea took time; it was many years before national parks were fully accepted and a comprehensive national park system had become a reality. Internationally, too, acceptance of the national park idea started slowly; fifty years passed after the founding of Yellowstone National Park before national parks became an accepted idea in a few countries. Then, finally, the idea caught hold. Parks proliferated, particularly after the Second World War. Today, there are around 1,500 parks and equivalent reserves in well over 100 nations.

With this spread of the parks concept in recent years has come an increasing recognition of the international importance of national parks. The International Union for Conservation of Nature and Natural Resources was started in 1948 and its International Commission on National Parks shortly thereafter. In 1962, the First World Conference on National Parks was held, with the theme, "National Parks Are of International Significance." Ten years later came the United Nations Conference on the Human Environment at Stockholm, which gave significant attention to parks. And three months later, the Second World Conference on National Parks was held.

After more than a century, parks are now accepted virtually worldwide. But parks are only one form of protection for natural areas, as you well know. Wilderness is a more recent concept, and it, too, has started slowly.

Wilderness and Global Protection

The thirteen biennial wilderness conferences preceding this one focused on protecting natural areas in North America. This fourteenth conference—the First EARTHCARE Conference—marks a significant departure from past formats in two particularly important ways. First, it invites delegates from North America and other continents to pool what they have learned at home, and to determine whether ideas developed in one place are applicable to environmental problems in another. Second, and more significant, however, is the fact that in this meeting the concept of protecting natural areas is linked with the protection of our global life-support system. Wilderness is considered here in the context of critical global environmental needs, not as a separate objective.

Mankind needs wilderness, but he also needs to protect the full range of the world's ecosystems. Professor Dubos spoke eloquently yesterday of protecting ecosystems, and in yesterday's opening ceremony, as well as in past wilderness conferences, the many values of natural areas have been described. I do not intend to cover the same ground. I simply wish to emphasize that the protection of ecosystems is integral to the maintenance of our total life-support system. Maintaining the integrity of our global life-support system is what today's environmental concern is all about.

Interdependence

The theme of this conference, therefore, marks an important shift in perspective—an encouraging shift, in that it reflects an understanding of our common interdependence on a single planet. Yet such agreement will be of little value unless we can convince policymakers in our various nations that our environment is, indeed, something that we ought to be extreme about. After all, we have only one environment—and if and when this one ever goes, man goes with it. The problem, of course, is to convince those who must weigh economic and ecological decisions that the destruction of the planet is a distinct possibility.

Such talk will sound apocalyptic. Various prophets and crackpots have been preaching the end of the world since the beginning of recorded history—and we're still here. The human species is at least 3 million years old, so it is perfectly understandable that many of our critics should—and do—dismiss the possibility of man's disappearance as nonsense. Ultimate pessimism is nothing new. Yet there *is* something new about man's relationship to his environment these days, something that makes our times unique in our species' long residence on this planet: man's recently developed ability to grossly injure the Earth.

This ability, I would estimate, did not fully surface until after World War II—a few years before the Sierra Club sponsored the first wilderness conference. Before then, man could not do much lasting, extensive damage to the Earth. His impact on the Earth was less severe than it is now, and much less apparent.

Threats to the Earth's Resilience

As we know, the Earth has a prodigious capacity to absorb man's garbage, break it down into life-sustaining components, and recycle them back into the ecosystem. But this natural resilience of the Earth is not infinite. It depends on three conditions: first, that the human population remain relatively small and slow to grow; second, that man's products be relatively simple in composition; and third, that man's tools be relatively small and limited in scope. As long as these three conditions held, the Earth could take all the punishment man could hand out. None of the three conditions holds true any longer.

Population

As to human population, there are now 4 billion of us, a number that may have more significance to you if you realize that it took man over 3 million years to build his population to 1 billion. That occurred in the year 1830. It then took one century to add the second billion, thirty years to add the third billion, and only fifteen years to add the fourth billion. That is where we are today. At our current rate of growth, our children will see 12 billion more individuals added during their lifetimes.

When one considers all of the problems that affect the quality of the human environment, one certainly must put population growth at the top of the list. Those of you who have studied in so much depth the rise and fall of populations of other species certainly appreciate the fact that man is not exempt from those forces that control population. Our current rate of increase in population is not a natural rate; on the contrary, it is highly unnatural, resulting from man's invention of medicine and widespread adoption of public health measures, which markedly lowered the death rate without proportionately decreasing the birth rate.

Products and Wastes

The second condition for preserving the ecosystem's resilience is that man's wastes be relatively simple in composition. But man's products are now incredibly varied and strange. About 2 million chemical substances are known, and hundreds more are developed each year. We have created synthetic compounds that nature cannot break down, and we pour thousands of troublesome chemicals into our skies, our water, and our soils, with little understanding of their long-term effects.

Currently, CEQ is concerned about chlorofluorocarbons—a group of supposedly inert gases. Americans are most familiar with them under the trade name "Freon." A number of highly qualified chemists and other scientists fear that Freons may not be inert at all, but may be drifting up into the stratosphere, diminishing the ozone shield that envelopes the globe, and exposing us to dangerous levels of ultraviolet solar radiation. At the moment, this proposition is theoretical—that is, the possibility of a reaction between Freons and ozone in the stratosphere makes chemical sense, but we cannot be sure such a reaction is occurring until we have more adequately sampled the stratosphere.

We do know, however, that the production of aerosols with Freons has increased almost exponentially since Freons were introduced to commerce in the early 1950s. In 1954, according to one estimate, 188 million aerosol-spray cans were produced in the

United States; by 1974, production in the United States had jumped to above 3 billion cans—equivalent to fourteen for every citizen. Even this total, staggering as it is, represents only about half of world production.

Man's Tools

Finally, the impact of our works and our tools has become massive. Modern technology so multiplies the effort of one man that he can perform more work—and do more damage—than a hundred or even a thousand men could a century ago.

At the end of World War II, the largest oil tankers had a deadweight capacity of 18,000 tons; today, oil tankers of 250,000-ton capacity are commonplace, and several tankers of 540,000-ton capacity are under construction. The loss of just one of these tankers, fully loaded, would dump as much oil into the sea as thirty of the largest tankers operating in 1945.

In January 1975, there were four supertanker spills. According to current projections, the amount of oil moving around the world will double every ten years. By the year 2000, therefore, we will have six times as much traffic as we do now. We must expect that tanker accidents, groundings, and spills will increase.

It is true that the oceans and estuaries can reduce much of the oil to harmless materials, but this takes time; huge, sudden discharges far exceed the environment's ability to repair itself. What will happen to that self-repairing capacity if the January 1975 rate of spill were to continue—if by January 2000 we have twenty-four spills a month instead of only four?

Through spills and normal refinery operations, we are pouring more than 1 million tons of petroleum into our oceans, rivers, and estuaries annually. Some of the costs—bird kills and disfigured beaches—are known to us because they are so obvious. We have good reason to suspect others—for example, potentially irreversible damage to food webs that have sustained the seas for eons—but we cannot easily measure them.

These three factors—population growth, the complexity of man's products, and the massive impact of man's tools—represent a significant and rapid change in man's ability to injure the Earth. Taken together, they not only add to each other's impact, but multiply it into unprecedented hammer blows at the Earth's resilience.

The Profligate Depletion of Resources

In light of this genuinely new development in man's long dependence on the Earth, I think the charge of extremism takes on a new character. Who are the extremists? Are they the environmentalists, who argue that economic and technological development can outstrip man's ecological budget, or are they the growth advocates, who seem to believe that there is no end to the largess of nature?

In the early 1950s, an American geologist, M. King Hubbert, warned that oil production in this country would peak by 1970. Had we heeded his warning, my nation would have begun trimming back its consumption then, but we did not. Our oil production did peak in 1970, and has been going downhill ever since—and will continue in that direction no matter how many more holes we drill. Like spendthrifts who had inherited a large but nonetheless finite fortune, we went on spending our bank deposit of oil as if there were no end to it—and the "free society" of which our congressman-critic spoke became indentured to the economic servitude of oil.

I hope the United States and the rest of the world will learn something from this experience, but it does not appear likely that they will. Currently, we are all rushing, rushing ahead in using the world's supply of oil as fast as we can.

Hubbert says the world will peak out in the production of oil around the year 2000, twenty-five years from now. His arguments are convincing to me. We had better be ready by that time with alternative sources of energy or with a way of life that uses much less energy, or else the world will experience a crisis such as it has never seen before.

Depletion of other kinds of resources has been proceeding at a spendthrift rate. For example, water consumption in the United States doubled between 1950 and 1975. According to the United States Geological Survey, it will double again by the year 2000—not because of population growth, but to meet new demands from the minerals and manufacturing industries.

During the entire decade from 1960 to 1970, 800 hectares of rural land were converted to urban use every day in the United

States. This is land permanently removed from the possibility of agricultural production—in a world that is short of food and in a nation that depends on crop sales overseas to maintain a balance of trade. Similarly, in 1950, the United States consumed 2 billion tons of new materials and minerals, or about 12,000 kilograms per citizen. By 1972, consumption was up more than 100 percent, to 4 billion tons, some 18,000 kilograms per capita.

I hope, too, that all nations will learn in time that the resources of the Earth are finite, and that environmental degradation respects no national boundaries. We know that the annual catch of fish is declining, even though the fishing nations are employing more equipment and more sophisticated techniques than ever before. We know that pollution can be carried in the skies and in the waters far from the place of its origin, and that the citizens of one nation may have to pay for the abuse of the Earth by the citizens of another. We know, finally, from dozens of unanticipated environmental mishaps (for example, the Aswan Dam, the Saint Lawrence Seaway, and the careless slaughter of predators that kept other predators in check), that the various parts of our ecosystem are interrelated and sensitive in often surprising ways. As the ecologists phrase it, "Everything depends upon everything else."

It is this point, this point of view, this essential interdependence of man and a common environment, that we must strive to convey to the policymakers of our respective nations. If they criticize our efforts as extremist, it is not necessarily because of their bad will or their shortsightedness, but because we have failed to express our concerns with sufficient skill. We must learn to state the case for our ecosystem as forcefully as others state the case for our economies. Unless we do so, we will almost certainly fail to provide the EARTHCARE, the global protection of our natural areas, that this conference is all about.

Clashes at the Frontiers

This past year, my travels have brought home forcefully to me how important it is for all of us to work together to take care of our Earth. I just returned last weekend from ten days of touring Alaska. I visited, among other sites, the Clarence Rhode National Refuge in the Yukon Delta, and there saw tens of thousands of birds nesting:

Whistling Swans, Sandhill Cranes, Canada Geese, many varieties of ducks, shorebirds, and—circling everywhere—the Arctic Tern: birds that came from Antarctica, from South and Central America, even some that were migrating through on their way to nest in Siberia. In one small area, there in Alaska, concentrated during this critical part of their life cycle, was this fantastic variety of bird life, so important to people in many distant places and so vital to many other aspects of our life-support system.

In that same small area, a few hundred natives lived, practicing the subsistence culture their forefathers had practiced for many centuries, living in harmony with that nesting area of so many species of bird life. I was thrilled by their expression of the great spiritual value of being able to live in harmony with nature. To hear them describe this value made me stop feeling sorry for their lack of some of the "conveniences" of our culture. But there, in Alaska today, the forces of development are starting to collide with some of those great natural assets.

I visited the vast national parks in East Africa last year—another treasure for all of the people on Earth. Those natural areas are now being protected, but there, too, the forces of development will impinge with increasing severity as the months and years go by.

I went also to Lake Nakuru, in Kenya, which Roger Tory Peterson says provides the world's greatest bird spectacle. It is there that the World Wildlife Fund and others have worked so harmoniously with the local people and with President Kenyatta in bringing worldwide resources to bear to save that great treasure.

Carrying Capacity and "The Tragedy of the Commons"

I went to Niger, center of the great drought, and talked to President Kountché about the work they are doing to seed the clouds and to drill more wells. I told him about "the tragedy of the commons": about how one sheep farmer decided to graze his sheep on the village common; how the sheep prospered and multiplied in that luscious grass; how a second farmer decided to do likewise, and a third, and a fourth; how, before long, the sheep were overgrazing, causing the grass and then the sheep to die and thus the farmers to fight. I told him, "They had exceeded the carrying capacity of the village common just as you in Niger, President Kountché, have exceeded the carrying capacity of your land."

Immediately, President Kountché began to discuss this relationship. He said that if they hadn't had so many domestic animals in Niger, several million wouldn't have died. Obviously, those animals that died had been eating grass that would have been available to others. If there hadn't been a marked increase in human population in recent years, there wouldn't have been so many animals.

President Kountché hadn't planned to send anybody to the World Population Conference in Bucharest, but at that juncture he decided to do so. When I was at the conference in Bucharest, a person tapped me on the shoulder one day to say that he was from Niger and that President Kountché had told him to be sure to say hello. The point I want to make here is that we too infrequently recognize the fundamental importance of the population of human beings to the many other problems that we are focusing on.

Population and the "Green Revolution"

The food problem is a symptom, not a basic cause. If it weren't for the population problem, there wouldn't be any food problem.

I went to Mexico and met with Dr. Norman E. Borlaug—who, as you know, won the Nobel Peace Prize for the work he did in stimulating the "green revolution"—and also with leaders in agricultural work from many developing countries. I was impressed by the work they are doing on integrated pest management—breeding more resistant strains of wheat, for example, so that farmers need not rely so heavily on chemicals, with the threats they bring to many other aspects of our ecosystem.

I said to Dr. Borlaug, "You've been quoted as saying some years ago that all the 'green revolution' could do was give us twenty to thirty years more to solve the population problem. Do you still think that's true?"

"Absolutely," he said. "What's more, we've already wasted eight of those years."

I asked people who were there from other countries if they agreed with that assessment. Every one of them said yes. These are the people who are working to increase food production throughout the world. They say, "Without a solution to the population problem, there is no solution to the food problem."

So, when we talk about EARTHCARE, about the quality of the human environment, obviously it is fundamental that we talk about the number of human beings for whom we are working to provide a quality environment.

A Declaration of Interdependence

Yes, human beings have the power to destroy the Earth. If we recognize our interdependence, however, we can arrest man's accelerated degradation of our planet and set about restoring its capacity to support future generations.

Let us, then, express the hope that this EARTHCARE Conference will set a pattern for continuing international cooperation in an increasingly interdependent world. Accordingly, allow me to conclude my remarks this morning by proposing, for people everywhere, a "Declaration of Interdependence," a recognition of our need to care for our Earth and to care for each other:

Declaration of Interdependence

We, the people of the planet Earth, with respect for the dignity of each human life, with concern for future generations, with growing appreciation of our relation to our environment, with recognition of limits to our resources, and with need for adequate food, air, water, shelter, health protection, justice, and self-fulfillment, hereby declare our interdependence and resolve to work together in brotherhood and in harmony with our environment to enhance the quality of life everywhere.

BANQUET

10. Presentation of the EARTHCARE Award to His Excellency Carlos Andrés Pérez, President of the Republic of Venezuela

The EARTHCARE Award was presented to President Carlos Andrés Pérez of Venezuela on behalf of the National Audubon Society and the Sierra Club by Mr. Kent Gill, president of the Sierra Club. The award was made during the banquet ceremonies at the New York Hilton Hotel. Dr. Ramón Escovar Salom, minister of foreign affairs of the Republic of Venezuela, accepted the award on behalf of President Pérez.

Presentation

Kent Gill
Sierra Club, U.S.A.

It is our pleasure tonight to select from among the current heads of the world's nations one who has a demonstrated record as an outstanding conservationist and leader for environmental quality in his land: His Excellency Carlos Andrés Pérez, President of the Republic of Venezuela. On behalf of the National Audubon Society and the Sierra Club, it is our pleasure to recognize his contributions with a special EARTHCARE Award.

A national leader who insists on conservation as a condition for progress, who has activated a program that expresses the principles of the Stockholm conference, President Pérez made environment a major issue in his campaign for election. Soon after he assumed the office of president, he enunciated a far-reaching environmental program to his people by emphasizing a new value system toward nature and calling upon Venezuelans to declare peace with their surroundings. Then, in a series of decrees and proclamations which I shall detail in the citation itself, he set in motion the program. A recent proposal of that program has been the addition of 2 million hectares of land surrounding Angel Falls as an addition to Gran Sabana (Cainama) National Park. [The decree for this action was promulgated in September 1975.—Editor] Here this evening to receive the award on behalf of President Pérez is the distinguished minister of foreign affairs of that country, The Honorable Dr. Ramón Escovar Salom.

Dr. Escovar, with this award go our best wishes to President Pérez and to your people. If I might read the award citation in English:

EARTHCARE Award
Presented to
His Excellency, Carlos Andrés Pérez
President of the Republic of Venezuela

The United Nations Declaration on the Human Environment proclaims that "Man is both creature and moulder of his environment, which gives him physical sustenance and affords him the opportunity for intellectual, moral, social, and spiritual growth." The Declaration calls on all nations and all peoples "to exert common efforts for the preservation and improvement for the benefit of all the people and for their posterity."

Since these historic principles were pronounced at Stockholm, Venezuela has demonstrated by its example that protection of the environment and preservation of the nation's heritage are consistent with development. President Carlos Andrés Pérez has called for public understanding of the essential worth and importance of an equilibrium between Man and his environment. "Conservation," he has said, "is not a barrier to economic and social development, but a precondition of its achievement."

Proclaiming a continuing study of the ecological recovery of Venezuela, the President issued a remarkable series of decrees which placed a two-year moratorium on the killing of any wild animal; created a national park and enlarged another; prohibited the destruction of mangrove trees and the discharge of sewage into mangrove swamps; established protected zones in several water basins and fragile coastal areas in which development will be restricted; ordered the Ministry of Education to lay special stress on the teaching of conservation and to train teachers and prepare teaching materials accordingly; halted the granting of building permits in major urban areas until a city-planning strategy is developed; prohibited new industry within the metropolitan area of Caracas or any of the major arteries leading to it; and arranged for the decentralization of certain existing industries.

Mr. President, in recognition of your distinguished leadership in promoting the rational use and conservation of natural areas, and in advocating environmental protection as an integral part of development, the convenors of this Conference present you with this special "EARTHCARE Award." You have truly heeded the words of the Stockholm Declaration and are fulfilling the obligation to EARTHCARE.

Presented at the EARTHCARE Conference on Global Protection of Natural Areas this sixth day of June 1975, in New York City.

Acceptance

Dr. Ramón Escovar Salom
Minister of Foreign Affairs, Venezuela

Mr. President, Mr. Chairman, and distinguished guests; I wish to take this opportunity to express to you not only the gratitude of the president of Venezuela, but also very briefly to give you an account of what Venezuela wants to do and what she is doing for the care and conservation of her natural resources.

The president of Venezuela is most grateful for this award, this great distinction. He receives it, and I receive it on his behalf, as an encouragement to continue in our present undertaking as a commitment to the future. As the award citation said, Venezuela and its government have taken important decisions in order to protect her fauna, her trees, her water, and her air.

[Dr. Escovar spoke in Spanish. The remarks published here are the edited transcript of the simultaneous English translation.—Editor]

The measures we have taken are not only a matter of natural conservation. We have also adopted them for "political conservation," because we consider that the water, the air, the trees, and the space are not the privilege of the few: they belong to the citizens of the country as a whole and, on a planetary level, to the entire world.

The air belongs to no one in particular, and green areas do not belong to anyone in particular. They belong to the inhabitants of the entire country, and to all those who wish to enjoy them. This means that the value of natural resources is a profoundly democratic value. It is a good thing to say this, and to repeat it time and time again in democratic countries, because democracy and nature are interlinked. One cannot defend nature without at the same time defending democracy and the intrinsically political nature of nature itself. Nature has a profoundly democratic value. This is what Venezuela, as a small country, wishes everybody to become aware of.

We want it to be made known to all countries that we in Venezuela are defending our natural resources. In so doing, this is no aggression against anyone at all. All we are trying to do is actively defend what we have and to participate in what is going on throughout the entire world. Venezuela is a small country of 1 million square kilometers and 14 million people, situated in northern South America. In this era, together with the great countries and with all peoples of the world, we wish to defend natural resources.

Venezuela does not aspire to direct the world, it aspires to participate in it, to participate in it with the right we all have to breathe and to enjoy the trees, the grass, and the waters. This we learned nearly 200 years ago in the United States Constitution, although we were then so young. In our participation, we continue to follow the basic principles of western civilization, which is based upon the defense of man and of citizens, the right of people to participate in the culture of the country and the society in which they live, and their right to participate in the life of the society as well. We have carried this on to ecology.

We are here precisely to recall that man has removed himself from nature and that we must get back together with nature. This effort is a most important one; it is universal. It does not represent a single party nor a single political interest; it represents the entire human species. If there is one effort that can unite all men, it is this.

Before, we tried to protect citizens from the powers of Earth;

today, we must ally the citizens' rights on Earth with the need to defend the natural resources that will sustain us. This means, too, that nature must be made a part of democracy, and we must give nature a certificate of membership.

Together, we must liberate nature. This is the new, universal unity, the new attitude toward the world and toward life, so that in the twenty-first century it will be said of this generation that both men and women understood their destiny and defended it. This is all. It seems simple, but the fact is it is very difficult to achieve.

Venezuela, for example, is one of the major oil producers and exporters in the world. Our policy is to conserve oil because it does not belong to one generation but to many future generations and because we wish to defend the dignity of petroleum as a natural, nonrenewable resource.

At this time, there is great confusion in the world. Many believe that the developing countries who are protecting their own nonrenewable resources are creating an economic imbalance in the world. I wish to say to you, in the spirit of good faith and goodwill that prevails this evening, that it is not in any way our intent or desire to create an imbalance. What we want is a new balance, a balance that will enable us to protect our resources as others have protected theirs.

The president of Venezuela is a leader of conservationism in Latin America. He is not the only one, but he is an enthusiastic advocate of the idea of defending nature and of the need to do so. If the award that has been conferred on the president of Venezuela in New York this evening has any meaning at all, it must surely be that it is a distinction conferred on the president of a Latin American country, a president who has emphasized and attached great importance to the defense of the natural resources of his country. I feel sure that anyone who examines this policy will agree with it— even if he is neither Venezuelan nor even Latin American.

I know that very shortly you will celebrate the bicentennial of the United States. I wish to remind you that Venezuela, like all Latin American countries, was born within the same historical and spiritual context and has the same ethical values as the United States. I sincerely believe that one of the reasons the world has so many grave shortcomings at present is that it has lost its old relationship with ethical values. It was a sincere conviction and belief in ethical values that created the United States, and it is ethical values that have, throughout history, made people successful and in accord with the destiny of history. Neither the destiny of a country nor human destiny can be forged unless ethical values underlie it.

Now the task is a more complex one. We must combine trees together with animal life so that we can live in harmony. We must combine the water, the air, and the earth in a harmonious context. This is the task we face at present, and this is what is needed for the future. This may be called "ecological balance"; you may call it "justice" or "democracy," as you wish.

Some years ago, we all believed that progress consisted in the development of industrial civilization. Now we realize that we must defend the Earth. We must defend and attach proper value to the natural resources we have been endowed with. The progress of a country can no longer be measured by the predominance of industry over agriculture, but must be measured by the manner in which it succeeds in achieving a harmonious and balanced development.

The proof that agriculture does no harm to any country is that the United States is the most powerful industrialized country in the world and is also the most powerful agricultural country in the world. Hence, we must correct our past conceptions of what progress means. Progress actually means development with the concurrent fulfillment of man, his happiness and well-being, and his development as a balanced human being. And this is the great effort we shall have to make from now on. Therefore, on behalf of the president of Venezuela, I wish to express my most heartfelt gratitude for this magnificent distinction.

I should like to say to you in ending that the world is a non-renewable resource, and that it therefore should be used sparingly!

Thank you very much.

11. Presentation of the John Muir Award to the Honorable William O. Douglas, Associate Justice of the Supreme Court of the United States

The Sierra Club's John Muir Award was presented at the EARTHCARE banquet to an outstanding champion of conservation in the United States, the Honorable William O. Douglas, who at the time of the conference was an associate justice of the United States Supreme Court. The Honorable Raymond J. Sherwin, judge of Superior Court in Solano County, California, vice president for the Sierra Club's international program, and (with Dr. Elvis J. Stahr, president of the National Audubon Society) cochairman of the EARTHCARE Conference, presented the award to Justice Douglas on behalf of the club. In accordance with longstanding tradition, the recipient's identity was not disclosed until the banquet itself was under way.

At the time of the conference, Justice Douglas had been in New York University's Institute for Rehabilitation Medicine for several weeks, recuperating from the effects of a stroke he had suffered some six months before. In spite of his physical handicap, he appeared at the banquet to receive the award in person. Presentation of the award to Justice Douglas at the EARTHCARE Conference has proven to have been especially timely because his health forced him to retire from the court in November 1975. He had served on the court longer than any other justice in history.

Presentation of the Award

Hon. Raymond J. Sherwin
Sierra Club, U.S.A.

At its biennial wilderness conferences, the Sierra Club awards its highest honor, the John Muir Award. That is my pleasure this evening.
In the past, the John Muir Award has been bestowed upon such people as William Colby, Olaus Murie, Ansel Adams, William Starr, Francis Farquhar, Harold Bradley, and Sigurd Olson. This

evening's recipient is no stranger to honors of the highest kind. For example, less than two years ago, the leaders of the legal profession in this country convened in Washington, D.C., to honor this man for what he had achieved in his professional life. At that convocation, former Chief Justice Earl Warren called him an "enemy of monopoly, the champion of free enterprise and of individual rights." Chief Justice Warren Burger referred to him as follows:

> [A] strong, articulate individualist, with a restless, questing mind and spirit, who finds respite from arduous work in climbing mountains and visiting strange lands and peoples, he is learned in the social, economic, and political scene, yet devoted to nature with a prescience and concern for the environment far ahead of his time.

It is on this theme that I should like to recount a personal experience.

We first met on the shores of Garnet Lake, of which you may read in his book, *My Wilderness.* The first evening before the campfire can't be said to have been typical, but it revealed the diversity of his interests and was a clue to what was to follow in the next several days and nights. Serious discussions about the economy of the country and about intricate, scientific aspects of the Sierra Nevada were mixed with informal discussions on many less formidable subjects and even songs—such as his inimitable performance (if you will pardon me) of "Egyptian Ella." If you have ever heard him sing it, you will know what I mean!

That experience signifies the element of his spirit that has led to this evening's ceremony: namely, an abiding love of the land and of all living creatures that inhabit it, a love that infuses his entire work and pleasure in life. The powerful mind, spirit of independence, and dogged determination that have put him at the top of his chosen profession, coupled with his lifelong identification with nature and natural resources, have also made him our most effective advocate for conservation. Whenever the rest of us falter, we have only to turn to his books and other writings, his judicial decisions, and his speeches for renewed inspiration.

I should like to present to him the John Muir Award, which reads as follows:

The John Muir Award

The John Muir Award is presented to William O. Douglas, who has given so much for so long for the cause of wilderness and preservation of the land, and who by his words and deeds has spoken eloquently for that cause in the great tradition of John Muir.

*Signed by Kent Gill, president of the Sierra Club,
for all the members of the Sierra Club
June 6, 1975*

Acceptance

*Hon. William O. Douglas
Associate Justice, Supreme Court of the United States*

Mr. Chairman, I thank you for the great honor paid me this evening. When I was advised that this honor was coming my way, I had only one question: Would it disqualify me from sitting in any Sierra Club cases coming before the Court? I answered that question as follows: If this would disqualify me, then I am already disqualified because of my love and admiration for John Muir, who played a very powerful influence in my early life. Receiving an award in his name merely confirms the historic fact of a warm influence and a warm and abiding relationship. I am greatly honored.

Other men greatly influenced me in my early years. One of them was William Borah of Idaho; another was Clarence Darrow—whom you all know, at least by reputation. It never would have occurred to me, or to any other member of our court, to step out of a case because Borah's or Darrow's name was on a brief. Those who practice before us realize that a counsel with a friend on the court is at somewhat of a disadvantage: he has to do a better job on the merits than another fellow might have to do, since the judge is conscious of his affection and therefore leans over backwards to scrutinize his friend's case rather critically.

Disqualification of a judge is an important problem, especially when the absence of a judge leaves only eight, creating the chance of a four-to-four decision. Mrs. Douglas and I have protested many acts of public agencies and private enterprise, have appeared on the scene and picketed them, and have marched in protest. In those cases, I suppose I might have to be disqualified to sit, but so far as I know none of the cases the Sierra Club has in the courts involve any activity of that nature, or of any other nature, on my part.

I suppose my education to John Muir has filled me with certain prejudices. If so, they are prejudices I'm proud of because I think the waters of our rivers and wonderful lakes were given to all the people, not just to a privileged few to fill up with garbage and industrial waste. If we are to keep America eternally beautiful, we must be busy on a thousand different fronts every week of every year, fighting all these little invasions, including the invasions of the air, such as one sees here in New York City. (A native recently commented that he didn't like air unless he could see it. Well, you can see it here!)

The hike the chairman mentioned was down the John Muir Trail in Tuolumne Meadows, South. It is grand and beautiful country! John Muir was afraid its grandeur would be lost and that people would never see it; now we're beginning to wonder if it will be lost because too many people see it! We are at the stage where we are going to vacation more and more in the outdoors.

My only worry about the Sierra Club is not that it's too active or too busy, but that some of its local chapters may be captured by some timber man and other special interests. I think one chapter in the South has suffered that fate already. This is a form of subversion we don't know very much about.

Once more, let me thank you all for this beautiful and touching award. I'll always cherish it.

12. EARTHCARE Address:
Alaska—Northern Outpost on Center Stage

Jay S. Hammond
Governor of Alaska, U.S.A.

For almost thirty years, the fabric of my life has been interwoven with Alaska's land and people. I've made my living trapping fur, hunting game, catching fish, and I have coursed more than 2 million miles (3.2 million kilometers) in Alaska—first by dogs and then by boat and small bush aircraft—mostly over the uninhabited expanses of its huge interior and over its almost unending shoreline. I have built log cabins and homesteaded on a remote Alaskan lake, and there watched the annual cycle spin—at first slowly from congealing cold and then in frantic short-lived, flamboyant summer haste, as if the land and all its creatures must renew themselves double-time to keep pace with their allotted destinies.

With my wife, who is Alaska-born, I've made a home and raised a family in a small fishing village of but 200 neighbors. I've attended Alaska's university, worked in local, state, and federal public agencies, and when Alaska achieved statehood I was privileged to represent in the legislature a sprawling rural district ten times the size of New York State with less than one one-thousandth of its population.

Now, I am privileged to speak to you as the new governor of Alaska and to share with you something of how Alaskans hope to immerse themselves in the stream of human history—at least in that small meander which is given to a single generation to course and to try to comprehend.

Alaska and Its People

Once, Alaska was set apart by its remoteness; now, we Alaskans must tap other sources to distinguish ourselves or to dramatize our differences from the other states; distance will no longer do. One distinction is, of course, the fact that—unlike most people in the other states—many in Alaska elected to go there and are not simply there through the accident of birth. As well, we feel a sense of social youthfulness, rejuvenated by newly won political responsibilities and economic options and by intimate contact with a fresh and powerfully beautiful land.

Alaskans feel a special closeness to each other, a shared destiny, a partnership against common adversaries, real and imagined. We Alaskans cherish, I must admit, that fancied independence felt by people in all frontier lands.

Nevertheless, like a side channel of a glacial stream, Alaska can seek its own course only within the confines of the central river valley, for we are, of course, political partners with forty-nine other states. By and large, we are reasonably satisfied with that arrangement, though we do have a small third party clamoring for independence.

After all, we have accepted economic transfusions from the nation for a century, and now that we are nurtured to the brink of economic health, we cannot spurn our donors, nor can we turn our nation's fellow citizens back at Alaska's borders. We can, however, ask that they be made aware of current circumstances and not come north if it is but to cast themselves, penniless, on the steps of our exhausted public welfare offices.

For, contrary to some suppositions, Alaska—from an economic point of view—is already vastly overcrowded. We have the nation's highest unemployment rate and a degree of poverty in our villages which makes Appalachia seem affluent by contrast: the cost of sustenance for one Alaskan is roughly twice the amount of income he produces. Thus, those who have no resources or employment assured well in advance should be aware that our virtually bankrupt state simply cannot now accommodate more bodies.

Alaska's Contributions

Of course, each of us shares a common global problem: people must eat to survive. They must build and trade to live a civil life for which the human spirit strives. But now awareness grows that there are so many people here on Earth and so many more each minute who are eating, building, and trading, that they threaten discord in that finely tuned, finitely resilient life-support system we call nature.

Alaska already has contributed substantially to mankind's survival and civility. Through 10,000 years, that great land has nurtured the biological and cultural needs of Eskimo, Indian, and Aleut societies. For almost a century, our rich and unpolluted waters have produced huge numbers of fish and shellfish for export around the world. In 1972, for example, 2.4 million tons of fish were caught in the waters off Alaska. The once virgin forests of southeastern Alaska have produced vast amounts of lumber and other timber products for export. The beauty, wildness, and challenge of Alaska have been recreational and spiritual resources for countless thousands.

Petroleum and Change in Alaska

Now the surging development of Alaska's oil and gas resources has hurled us onto center stage, where the spotlight of national attention has served to shadow the continuing value of renewable and amenity resources.

For some decades, oil and gas will dominate our regional economy as well as our worldwide relationships. I say this with no little apprehension, apprehension shared by most Alaskans who have not yet determined whether each new-found oil province should be viewed as holy grail or cup of bitter vetch—his decision depending upon his own sampling to date. And, of course, important though petroleum is in our industrialized society, it is inert and soulless stuff. We must consciously address the real affairs of people, who are of life and spirit.

Though we Alaskans, like our sprawling land, are richly diverse in our lifestyles and our attitudes, there is among us strong agreement about some essential elements. We want to maintain the health, the beauty, the productivity of a land that is strong but vulnerable. We want to freely enjoy and use—not abuse—our natural resources. We want the opportunity for economic and social satisfaction. We want an open, accessible, responsive system of self-government. We want to create a different kind of society in our North—modern in every sense, but in harmony with nature. To define this kind of new relationship and to discuss how it may be achieved is, of course, the very purpose of this conference.

I would like to take a few moments to describe some aspects of the Alaskan situation, because not only is the outcome of our efforts important in environmental terms for that small global acreage we hold in trust, but more, it is significant as both an experiment and, perhaps, as a precedent for people facing similar challenges throughout the world.

Coping with Change

Regardless of what goals a society sets, it cannot possibly achieve them unless it is able to control the rate and thrust of change. No part of Earth today is so isolated, no government so powerful, that its society has change under control—certainly not Alaska. Like a canoeist in white water, we are confronted with a need to sweep past rocks and, at times, to backpaddle to avoid capsizing. For, unfortunately, unlike that canoeist, we cannot haul a spluttering society up from the river for a second try. Alaska, it has been said, may well be the last chance to do things right the first time.

In Alaska today, change explodes on every hand. That brought by oil and gas development is calibrated in megatons. Each oil and gas discovery seems to some a golden gusher, but to others a gumbo of new-found griefs. Such development, of course, brings money; money is spent; these expenditures bring jobs; jobs entice more people to the state; the influx of people changes communities, lifestyles, land uses, and the quality of surroundings.

Who controls petroleum development within Alaska? This is by no means a question one can quickly answer. One might point to the Organization of Petroleum Exporting Countries, to key members of our congress, to the president, to the secretary of the interior, or to monolithic oil corporations. All have their influence, as do the major landowners of Alaska: the federal government, the state, and Alaska native corporations. As landowners, they can decide when and where prospective oil lands are leased. They can, as well, establish the conditions under which exploration and development will take place and thereby provide protection of the environment through the planned siting of facilities. They can demand shares of profits, both to raise revenue and to modify rates of development.

The Outer Continental Shelf

Each of the three types of landowners has approached petroleum development differently. The federal government of the United States has embarked upon a widespread effort to increase domestic fossil fuel production. In Alaska, federal agencies have begun accelerating exploration of our outer continental shelf. The government is exploring intensively as well in arctic Alaska, in Naval Petroleum Reserve Number Four.

From the national standpoint, this federal activity may be justified as a reaction to threats of energy shortages imposed by foreign nations. To Alaskans, already reeling under the impact of the trans-Alaska pipeline, these federal proposals threaten a clout of community and environmental change far beyond our ability to cushion or control. We are concerned that exploration and development in storm-struck waters of the Gulf of Alaska and of the southern Bering Sea will result in oil spills that could devastate our fisheries. Moreover, we do not want gluttonous onshore facilities serving offshore fields to gobble up prime lands or to consume areas that are critical for wildlife. People from the tiny coastal villages already have come to us and asked for help in fending off the juggernauts of such potential impact.

Our position on development of the outer continental shelf is positive and constructive: we would put a rein upon development simply to assure that we can steer, not strangle it. We need time for necessary research, to build up environmental protection capabilities, and to plan for onshore impact. Moreover, we want agreement not to explore for oil in those areas of the outer continental shelf that are of outstanding value for the food they produce.

In partnership with the federal government, we would carefully guide and control exploration and development in those areas where such activities are permitted. We believe that any oil or gas discovered on our outer continental shelf can be brought to consumers just as soon—and at far lower cost on any scale—under these conditions as by pursuing that most hectic pace set forth so arbitrarily by the federal government.

The Role of Alaska Natives

Alaska natives have a major role in oil and gas development. Through profit-seeking corporations established by the Alaska Native Claims Settlement Act of 1971, the native people own over 40 million acres (16.2 million hectares) in fee simple, including subsurface mineral rights in many areas. Native corporations have entered into joint ventures with firms involved in petroleum exploration and development. Thus, natives now have a hand on both the wheel and throttle of change, especially in the western and northern portions of Alaska. As landowners in almost every coastal and riverside community, natives will also influence the siting of docks, ports, tank farms, pipelines, airfields, warehouses, and other facilities necessary for petroleum resource development.

State Lands

Finally, the Alaska state government also has its hands on controls of change; we can exert substantial influence upon its course through oil and gas leasing policies, taxation, and general police powers applied to both land and water use.

The Alaska Statehood Act allows the state to select 104 million acres (42.1 million hectares). To date, we have selected about 70 million acres (28.3 million hectares). The state, of course, selected lands with oil and gas potential—among them the now famous Prudhoe Bay. In addition, the state received title to tide and submerged land up to three miles (4.8 kilometers) out from mean high tide, around our 34,000-mile (54,700-kilometer) shoreline. These lands are extremely important, not only for their own petroleum potential, but for the opportunity their ownership provides to control all coastal development, including development of the outer continental shelf.

For many years after Alaska achieved statehood in 1959, its petroleum resource leasing policy was geared almost solely to short-term needs for revenue. The times are changing rapidly. Within two or possibly three years, revenues from Prudhoe oil should relieve the budgetary pangs and pressures we have been beset by. Thus, in the future, state oil and gas leasing policies should prove one of the most effective ways to steer those juggernauts of growth. Moreover, I predict that Alaskans soon will urge their leaders not to gun the engines, but to ride the brakes, so that we may better plot our course while it is yet before us. The costs of growth rampaging uncontrolled are sensed more sharply and borne with more reluctance by us all.

Land-use Planning

Those of you who know Alaska realize that petroleum is by no means the only issue demanding public decision now. Another is, of course, land use, one aspect of which I will discuss with you because of its significance to all.

The nation will own 60 percent of Alaska's lands even after native and state land selections are complete. There has never been a coherent policy, much less a comprehensive plan, for federal lands within our state, one reason being the fragmented holding of diverse—oftentimes competing—federal agencies. Just as the state is moving toward a land and water planning process, we fervently desire that federal policies will soon provide both guidelines and a mandate for coordinated planning on federal holdings in Alaska. We shudder when we see one federal bureau proposing a new park within an area the agency next door proposes to carve up with road and pipeline corridors. We cannot help but wonder if any viable land plan could possibly survive the resultant scar tissues and contusions.

Lands of National Interest

By far the most important opportunity for government expression on management direction for Alaska's public lands is the decision now before Congress regarding the so-called "Lands of National Interest." As many of you know, these lands are select parts of public domain identified by the Department of the Interior as having outstanding scenic, wildlife, or other prime values. The secretary has submitted recommendations to the Congress regarding permanent assignment of 82 million acres (33.2 million hectares)— to federal agencies. Within one or two years, Congress is expected to act upon these proposals.

When I hear people speak enthusiastically about this last-chance opportunity to preserve natural environments for all time, I often feel that they perceive these lands as they would gigantic postcards or centerfolds in outdoor magazines: beautiful, but flat, dimensionless, and essentially without meaning save as spectacle. The perception of Alaskans is much different. We do not view these lands from without, but rather, from within. We have umbilical connections to the land, which are severed only with pain and protest. We hunt them and we fish them, both for pleasure and for food. We span them casually by boat, snowmobile, aircraft, and—even yet—by dogs to get from one place to another. We prospect in them, cut cabin logs, and many yet trap fur. These lands are neither empty, nor are they just utilitarian. They bind us all together, providing us some degree of harmony both with each other and with natural realities. They are the mortar of our society and the matrix of our character.

Land Management and Natural Values

We will urge Congress to reflect these dynamic and organic dimensions of national-interest lands with actions in the coming months. While acknowledging the need to create new parks, refuges, and other traditional conservation modes, we will, as well, ask Congress to establish permanent and innovative mechanisms for cooperative public-land management.

Natural values, unfortunately, have no respect for boundaries set by bureaucrats—especially those square-cornered boundaries we insist upon imposing for the convenience of lawyers and surveyors. Land managers must have the freedom to ignore these lines in moving toward the larger purpose: that of matching values and the capabilities of the countryside with changing human needs. Through thoughtful designations and the creation of a system for cooperative land management and planning, Congress can assure that national-interest lands serve in equity those local, regional, national, and international publics who have a stake in the way Alaska's land and water are used.

Growth and Equity

The development of outer continental shelf oil and gas poses two of mankind's major modern dilemmas relating to land and water management. One is the question of resource sharing: when lands and ocean resources cannot provide for all the wants of everyone, who should get what share? Even in our uncrowded state, we face this problem every day. In fisheries management, for example, we wrestle with foreign fleets and the problems of too many United States fishermen seeking a share of far too small a catch. We are asking the same question in land-use decisions, knowing that wilderness lovers, hunters, miners, and all other land users cannot each have all they want. To sustain anything at all, all must relinquish something.

The other question is just as fundamental: what human and environmental costs should properly be borne in the name of those twin gods, progress and development? Should we pay them blind obeisance? Economic development is the road by which much of mankind hopes to reach a more abundant life. But we are coming now to realize that the road is not without potholes and blind alleys. The old presumption that all growth is good ignored the fact that unbridled growth may not necessarily be healthy; accordingly, henceforth we will demand that socioeconomic biopsies be performed well in advance, to determine whether that growth be malignant or benign.

Fueling the engines of America with Alaskan oil and gas may well be necessary, but it is not necessary to risk our fisheries unduly, to sunder our communities with the impact, to squander scenic values, or to sacrifice our wildlife. Too often, in the sunlit euphoria of burgeoning economic development, we are prone to lose sight of what might be termed "the dark side of the boom." We must address economic development as we should all issues: with human perspectives and priorities held paramount.

Prospect

It is extraordinarily humbling to be asked to lead Alaska in these challenging, fast-changing times. Our problems are not at all unique; they are shared by every state and nation the world over. However, in Alaska, "future shock" comes in its highest amperage, and—as the plastic world chafes hard at the primordial—insulation cracks and aggravates the jolt. Yet our social and environmental setting is unique and our opportunities are unparalleled.

They are not opportunities for Alaskans alone, but for the nation and even for the world at large. America and Alaska must forge policies in partnership for all critical issues of energy and land or water use.

We know the task will not be easy. Sometimes it seems impossibly complex, akin not only to counting those proverbial pinhead-dancing angels, but to choreographing them as well. I have pledged my administration to an orderly course of social development toward goals that Alaskans have themselves identified as being of high priority. With your help, these goals can be harmonious with the swelling theme of global stewardship of the natural environment.

CLOSING CEREMONY

13. A Call to Action

David R. Brower
Friends of the Earth International, U.S.A.

> *What we need is a band of angels organized along the lines of the Mafia.*
>
> —Anonymous

People concerned about humanity, equity, and peace the world over are learning that the overdrain of resources and the degradation of people are part of a single, all-encompassing, accelerating crisis. It is beginning to become universally apparent that no one will buy or sell, work or play, love or hate, or be any color at all on a dead planet. The Four Horsemen are saddling up.

Mindless Growth

The threat to the Earth of mindless growth is real, despite the editors who deny it, the admen who weave rugs to sweep it under, the in-house scientists who allege that all is well, or the political and corporate and labor leaders who have decided that the environment annoys them and must not interfere with their careers.

The day-by-day diminuendo of Earth quality barely disturbs us, but if you think back no more than fifteen years, you know the loss as you remember how enjoyable breathing used to be, how enjoyable used to be the taste of water, the sound of birdsong, or the renewal that could come safely from a walk in the city evening with stars in it or on a lonely trail into wildness the ages have made perfect.

Renewal and Relapse

As the sixties ended, an alarm was sounded. Perhaps looking at ourselves from the moon told us something: we were on an island, with no place else to go, and we were caring for it miserably. Earth Day. A new environmental consciousness. Recycling. The National Environmental Policy Act. The pleasure and sanity of slowing down. Could it last? Yes, it could, and must.

But each day's news tells of the relapse, of the slipping back into the rush to use tomorrow's resources for today's convenience, to take the easy trip and charge it to the kids. Early in his administration, Mr. Nixon said it was "now or never" for rescuing the environment—then voted for the never. He and the big contributors resumed the attack, and President Ford, alas, carries it on. A "Strength through Exhaustion Program" is labeled "Project Independence." It is a massive effort to speed the using up of oil—even our own security reserves—to free us from the Organization of Petroleum Exporting Countries. We are told to remain calm, no one has *proved* that increasing health hazards at work or at home, that nuclear proliferation and extra radiation everywhere, will hurt us— and they *do* mean jobs and more gross national product. So, honor the first man to split the atom, and let's race to see who will be the last!

Few of the traditional conservation organizations are much concerned about that severest environmental threat—nuclear proliferation; most of them ignore the inflation, unemployment, and inequity produced by environmental abuse. They are backing away from requiring the logging industry to be as responsible to the environment and society, and as careful of wilderness, as the rest of us must be. And, in Alaska, "the last chance to do it right the first time"— the opportunity to preserve the last great wilderness—is evaporating. Battles against highway trust funds, the effluents and influence of Detroit, offshore drilling, supertankers and superports, and reactor and radiation hazards pre-empt our attention, as they must.

Summing it all up: we are confronted with crises. They can depress and overwhelm us, or we can call them opportunities, seize them, and organize and delegate well enough to overcome them well.

Decisions to Come

Great effort is needed to prepare the public for its next major decision date, 1976—which coincides with the year in which we celebrate what the old American Dream accomplished and prepare for the New Dream.

By the end of the decade, we must turn the United States away from the brink it is speeding toward and thus do our best to set the kind of example for other nations that used to be expected of us.

There are vacuums to fill or to help others fill. There are seminars and conferences to be held, jobs to be revised and staffed, and enough rethinking indulged in to let humanity build a conservation conscience in all its fields of activity. Now is the time for all men to come to the aid of their sanity. It could be fun.

Many groups are working to discern and move toward the alternatives to the anthill civilization that world leaders are hurtling toward. We join, and intend to speed, a major effort to bring about the transition that must be made if the Earth and its living passengers are to flourish.

Realization of the crisis is now widespread enough to generate the necessary power for change. The only real energy crisis would be the failure to recognize that power and to use it. The good parts of existing systems can be used to beat the bad parts in steps that are orderly, just, and swift. We are concentrating on solutions feasible now and fixable later, if they need fixing. We concede that the researcher's cause is good and must continue, well-funded; but unless we act now on what they already know or reasonably suspect, no one will be around to complete his studies.

The Movement

Unfortunately, the environmental movement can command but a fraction of what industry and government spend on environmental disruption. We make up for our deficit as well as we do because our members and staff possess diverse skills and resources, and because the public is willing to listen carefully to people it knows are not out to feather their own nests. Our members—whether from the sciences, industry, professions, student groups, or working people, urban or rural—are distinguished both by the depth of their concern for nature and by their understanding of the human processes necessary to effect change in a complex society.

To do more we need more members—perceptive, motivated, and committed. We know, with William H. Murray, that "the moment one definitely commits oneself, then Providence moves too.... A whole stream of events issues from the decision ... which no man could have dreamt would have come his way."

We ask that you join the movement—those of you who haven't yet—in the most generous category you can. You contribute heavily, on demand, to government, the Pentagon, oil companies, utilities, and Detroit (and who among them is in command?). We hope you will remember organizations like our own in your budgeting—considering how full our agenda is kept, willy-nilly, by these other people you support.

When you join and participate in our programs, you catalyze and support many good things; you also keep informed. You provide moral support. You have an opportunity to lead and to enjoy it, and you also get some benefits. You help us speak for you, and you for us and for yourself. You count!

Lobbying

A strong conservation lobby is vital. In the United States, some of the organizations lobby substantially in Congress and in state capitols, as people and funds are available. There are environmental lobbyists in Washington, D.C., who are experts in the fields of energy, wildlife, nuclear power, air and water pollution, and wilderness and public-lands issues; in addition to lobbying, they stimulate grassroots pressure on Washington. Many of the organizations spend a major part of their funds and energy advocating strong environmental legislation and urging public support for it. They inform, and try to persuade, congressmen and their staffs to enact the necessary laws, and the executive branch to enforce them. When all else fails, they sue. We in Friends of the Earth (FOE) have had to sue oftener than we like, and hope the government will soon take its laws more seriously.

The lobbying activities of conservation organizations—unlike those of industry—are not tax deductible. Small contributions in large numbers let your voice be heard.

Like many organizations, FOE has field representatives—in New York, San Francisco, Seattle, Fairbanks, Kansas City, Los

Angeles, and in the Northern Plains states. We also have a European representative in Paris, two staff people in the United Kingdom, and one in Nairobi. They add to FOE's effectiveness, to our publishing, to international meetings, and to conservation efforts where various countries can help each other, with special emphasis on energy and the limits to growth. We also publish a lot—a separate story.

The Individual

Conservationists are learning how to be the Ralph Naders and Rachel Carsons of their blocks. Neither of these people was overcome with a sense of futility. Each made a difference, and indeed, each *is* making one, even though Rachel Carson died many years ago. It was said of her that "she did her homework, minded her English, and cared." The expert knowledge of our members, thus applied, accomplishes what money can't. And this knowledge is applied again and again—in appearances at conferences, on television and radio, at schools, at neighborhood gatherings, and in letters to legislators, editors, and friends, and all the president's men.

The absence of dams in the Grand Canyon, of Boeing supersonic transports in the eardrums, and of whale-killing in all the seas (by all but two nations), is evidence of how the power of a conservation member is focused and multiplied through organizations. We look to that power to supersede nuclear power (peaceful and warring) and its transcendent threat to the planet.

Attitudes and Alternatives

We need to change some attitudes, to evolve new positions, especially on energy. Critical battles over energy concern us all. A quick look at changing attitudes shows the need for further change.

By the late sixties, further acceptance of damming places like the Grand Canyon stopped: too much was lost for too little gain. Better alternatives, it seemed, were coal-fired steam plants. Then came the highly destructive Four Corners strip-mining scheme: a disaster visited upon the Navajo and Hopi so that Los Angeles, Phoenix, and Las Vegas could have more light than they needed.

We fight such schemes on many fronts. But if dams and strip-mined coal are bad, then what? Substitute an increased use of oil, and coax the public to buy more? That gave us Santa Barbara, breakaways in the Gulf of Mexico, breakups of tankers by the score, and the imminence of far greater disaster, as described in Noël Mostert's book, *Supership* (1974). It also gives us the severing of wilderness and culture by the several trans-Alaska pipelines now being promoted to exhaust our oil as rapidly as possible to make us independent of Arab oil.

Natural gas is clean, but it is poisonous, explosive, and scarce, and deregulation will create neither a jot nor a tittle more of it—clean or not. "Clean" atomic power is the dirtiest. High-level radioactive waste, generated to aid our present comfort, must be isolated from the environment for half a million years or more because of its slow rate of decay. Our failure to do this will perpetrate a hazard for generations five times as distant in our future as the Neanderthal is distant in our past. The old, now defunct United States Atomic Energy Commission did not tell the public much truth about this, or about the danger from mining, transporting, burning, reprocessing, and hiding atomic fuels. Half a million years is a long, long time to keep gene-deforming poisons out of the hands of small children and big adults. There is no way now; there is not likely to be one.

Geysers, magnetohydrodynamics, space platforms? Fusion perhaps? These dreams are more likely to exacerbate than to help the problem. Overuse of all forms of energy has seriously disrupted the environmental balance essential to our own lives. Yet the compulsion to find and use still more energy is absorbing so much time, money, and even energy that we are about to emulate King Midas, except that everything we touch becomes not gold, but inedible, unbreatheable, unwearable joules.

The Remedy

There is a remedy, and we stress it. Obvious though it is, the route to sanity is barely heeded in ruling circles: Use *less!*

Everyone used less before, and we can do it again, perhaps using 4 percent less energy each year, rather than 4 percent more. If we matched Sweden's skill, the United States could have the same per capita gross national product with half the United States' present, per capita use of energy—a fair goal for 1985. When we drop to one-third of our 1974 energy use per capita, we will use as much as

France. (When last we checked, the French were not living in caves or freezing in the dark!) On the French standard, solar energy and wind (indirect solar) energy could by themselves meet United States needs by the year 2000. No reactors. No superships. Just sense. Such a goal is the logical one, and quite probably the only feasible one. So we are on sound ground in resisting the Exxon-Chase-Mobil-Detroit-*et alii* urge for a surfeit of energy growth—and the consequence of irreversible, Earth-wide tragedy.

We welcome and promote ideas for energy conservation in the search for energy, and in construction, choice of materials, recycling, space heating, mass transit, agriculture, packaging; in lifestyles; and especially in Detroit. The question remains, "What should organizations and industries do?"

A List of Things to Do

An attenuated preamble should by now lead to a "things-to-do" list for conservation organizations and for industry. The list should urge organizations and industry to step forward, in mutual support, to persuade the various publics and their governments to initiate crash programs toward tenable, sustainable goals that would allow the various cultures to survive. This kind of effort is required around the world, to turn mankind's course in time and to prevent the population crash and the culture crash that seem otherwise to be inevitable.

The United States, with its superlative appetite for exploiting natural resources, has been the primary contributor to the threat to the planet's life-support systems. Yet the United States has an important role to play in leading a vigorous turnabout, for within its bounds the environmental attack has been so swift that we can still see the dust swirling from beneath the feet of those who launched it. In other nations, many of them eager to follow the United States' mistake, the attack has not been swift enough to make its source unmistakably clear.

This is not an effort to denigrate my own country, but rather an attempt to explain some of the events of the last few decades; nor is it an attempt to assign guilt. Let us instead realize that no leading system of government on Earth has yet realized the full intensity of the environmental threat confronting it. None has realized the ultimate tragedy that will come if we insist on throwing a geometric curve at the Earth in our demands upon its capital of organic and inorganic resources and in dissipating them beyond all hope of recovery. There was once enough of the world and of time. There is no longer enough of either, given our numbers, multiplied by our desires.

General Amnesty Needed

We can, to begin with, declare an amnesty for all of us who desist from what we have done heretofore. We might as well do so, for we cannot undo our doings. But we can, so to speak, redo our intentions: we can begin to agree that guilt must really attend those who insist on carrying on with old, indefensible habits which a finite planet cannot sustain. We are stranded together on a ship that is drifting through space, and there is no possibility of rescue except through that which we ourselves discern, define, and deliver.

Let that be a summary of my own view from the United States, assisted by the travels I have enjoyed in Europe. But let a man of authority speak on the same subject—Sir Frank Fraser Darling, in his foreword to the British edition of Paul R. Ehrlich's *The Population Bomb* (1971):

> The one side of us, our heart, subscribes to the doctrine that a world-wide raising of the standard of living to something near that which we enjoy ourselves is necessary on grounds of common humanity. But the other side, our head, should realize this is impossible, and that if it did take place by some immediate miracle we should be in a condition of disaster. The resources are not there and we have not achieved the means of circulating what is needed. Furthermore, it is quite ridiculous to assume that the western standard of living is the most desirable. I say this, having known and enjoyed much simpler days. Our hope is in reducing world population (our own very much included) and maintaining reduction until we can evolve a new economics—one not based on a constant growth-rate of that figment, gross national product. Man as a species is capable of a joyous flowering, but until population is in control, the species as a whole is going to be continually degraded. We are using the planet's finite resources prodigally and asking more from its rehabilitative power than its somewhat debilitated condition can provide.

Standard of Living

Sir Frank's appraisal thus modifies the one made by Adlai Stevenson in 1965. When Stevenson alluded to a "liberation of resources undreamed of until this day," he hoped that this liberation could be extended to all. But in the following 3,600 days, almost 700 million people were added to the number on the Earth—all of them with growing appetites, few of these assuaged, and an ever-decreasing proportion of them likely to be so in the future. We knew the old dream could not be extended. We could sense that an earlier formula, "Standard of Living *equals* Resources *times* Technology *divided by* Population," was not quite working out. We could perceive that not only was population thinning out the standard, but that technology was thinning it, too. Technology was accelerating the liberation of resources, yes, but also accelerating their decimation and their scattering. It was not creating; it was finding, moving, using up, then looking for the energy to repeat the process with progressively poorer materials, moving them faster, making them into smaller, less recoverable fragments for a diminishing proportion of the Earth's growing masses of people. The formula had changed: "Standard of Living *equals* Resources *divided by* Technology, All *divided by* Population." Technology itself was not doing this, of course—its managers were. They had never needed to learn otherwise. Now they need to.

I do not think that many conservation organizations have been aware of this change; certainly, awareness has come only recently to those I worked with for almost forty years. Nor do I think many industries are aware of the change. At this moment, I cannot name any that are. Nor, to judge from what we have been hearing, are the United Nations agencies really in the picture—except for the United Nations Environment Programme.

Reshaping Perspectives

What we need to do is to seek out and endow with leadership those organizations, nonprofit or industrial, that are flexible enough to get us through this decade, and on into the next with a fair chance of flourishing again. We may then be on the way toward reshaping man's perspective enough to cool the unconscionable drain on resources. If the overdeveloped countries can renounce, the normal countries will have less to catch up with, and may in due course conclude from their own observations that catching up is not all that desirable.

If human pressures begin so to alter, there is a fair chance that humanity can contemplate its span in thousand-year periods—rather than merely hope that it can survive for the next thousand days or so. And what is so great about mere survival? You can survive in jail!

Thirteen Tenable Goals

To sum up, I hope we can agree to look on the following ends as desirable and, having agreed, bring them to pass:

1. The first thing to ask concerning any proposed new development, public or private, in our own country or abroad, is: What would it cost the Earth, organically as well as inorganically?
2. The next question should be a searching out of the least-studied alternative—suppose we simply did not undertake the development? Would not the array of ultimate benefits be impressive?
3. We must realize that an expanding population is a threat and a diminishing population a boon—diminished ultimately, and by willingly accepted means, to the true, long-term carrying capacity of the Earth.
4. Population limitation should begin in the affluent countries, where numbers multiplied by desire cost the Earth most; to halve the numbers and double the desire will get us nowhere.
5. The search for new places to exploit must ultimately cease; we should increasingly occupy ourselves, our science, our technology, and our good sense, in going back over where we have been, restoring and healing the injuries caused by our depredations.
6. We must realize that wildness, and large areas of wildness which we call wilderness, indeed hold answers to questions that man has not yet learned how to ask—answers about

the life-force that worked quite well before man arrived, the force that man can and should work *with* instead of against.

7. High on our list of priorities should be ways to use our genius so as to diminish our requirements for energy, realizing how great has been the harm to the environment resulting from man's inability to manage energy wisely enough.

8. We should understand that the so-called "green revolution" is fraught with danger: that it is indeed using up in a flash of time the organic wealth, energy, and diversity that only eons could assemble, and that it builds a hope which can be dashed in the cruellest of ways by predictable genetic or other disasters.

9. We must agree that the poisons which make the chemicals of life unavailable to life are not to be buried or broadcast on land, in the sea, in the air, or in space. Instead, they are to be kept under man's control until he can disassemble them and recycle them usefully.

10. We must discover that war and pollution are now outmoded, even though they have produced jobs and profits in the past. Satisfaction will come from abating the untenable, not from continuing it. Personal, corporate, or national profit should ensue from a recycling revolution.

11. We must conclude that the way to change course is not to construct a wall that is crashed into head-on, but to fashion a curve that can absorb safely the momentum which our cultures have developed, and redirect that momentum on a safe course.

12. We shall not require unanimity, because that gives too much power to the veto and takes too much from the essential human diversity which makes us interesting. We will learn to walk with each other and to agree with each other as far as we can; otherwise, there can be no real conversation and no understanding.

13. It will become clear in all man's activities that a finite Earth imposes limits—and is likely to impose them harshly if man does not himself impose them rationally first. We must be rational enough to impose limits outselves, and to

impose them before our numbers and desires have obliterated the heritage of organic wealth that made our existence possible and which is essential to keep it from being impossible.

There is an idealism about the foregoing list that suggests it could prevail only in a world of dreamers. But then, man does dream; each of us has a poet inside, and music—and love for people near and far that we have been too embarrassed to display. Idealism will prevail, if only because the alternatives are prohibitively grim. The idealism will be driven into focus and into use by the most important driving power we have—self-interest—and by our comprehension, in this time of unprecedented crisis, that *self* cannot exist alone. There must be the *other* forms of living things, inhabiting the ecosphere, with which we were always interrelated and upon which we were always dependent, but which were never before in such diminishing supply as to frighten us into reason.

When you've arrived at a brink, progress lies behind you. It is better to turn around before taking another step.

References

Ehrlich, Paul R. 1971. *The Population Bomb*. London: Friends of the Earth/Ballantine.

Mostert, Noël. 1974. *Supership*. New York: Alfred A. Knopf.

PART II
FORMAL CONFERENCE SESSIONS

Earth as Photographed from Near the Moon (Courtesy NASA).

O vast Rondure, swimming in space,
Cover'd all over with visible power and beauty,
Alternate light and day and the teeming spiritual darkness,
Unspeakable high processions of sun and moon and countless stars above,
Below, the manifold grass and waters, animals, mountains, trees,
With inscrutable purpose, some hidden prophetic intention,
Now it seems my thought begins to span thee.

—Walt Whitman
"Passage to India"

Of all the accomplishments of technology, perhaps the most significant one was the picture of the Earth over the lunar horizon. If nothing else, it should impress our fellow man with the absolute fact that our environment is bounded, that our resources are limited, and that our life support system is a closed cycle.

—Frank Borman

NATURAL SYSTEMS

14. Introduction: The Primordial Interdependence

Edmund A. Schofield

The Institute of Ecology, Butler University, U.S.A.

A Living Earth

"Our Earth is a vast machine, insatiably reconsuming itself" (Elder 1976). This simple and startling statement implies some fundamental facts about the Earth: that there is, for example, a driving force, and a cycle of continual change as well—that Earth's natural systems are not static, but, rather, dynamic. It implies also that, whatever and however many its component parts, the Earth is one—a unity melded by a vast network of meshing interconnections.

It was around this theme of all-pervading interconnectedness that the EARTHCARE Conference was organized, in recognition that the best models for human systems are natural systems. For the natural systems of Earth—the multifarious ecosystems, the grand biogeochemical cycles, the spectacular array of biomes and species—are the primordial interdependence.

The Driving Forces

A vast force that moves entire continents and opens up seas where once there were no seas resides deep within the Earth. It drives an immensely slow and relentless process of continual decay and renewal through which continents are created, cleaved, merged, and perhaps even consumed. As Whipple (1968) has stated, "The fragile crust of the Earth, floating on the heated and deformed rocks underneath, is not the stable and permanent layer that it appears from everyday experience. Not only is it clearly drifting about the main body of the Earth, but it is certainly cracking and buckling through geologic ages."

Another force, solar energy, drives most of the other, quicker, cycles—epicycles—that circulate on the surface of the Earth: the hydrologic cycle, the carbon cycle, the nitrogen cycle, and the other grand cycles of nature. Yet another force, gravity, is immanent in all of the grand cycles. It keeps the cycles in mesh with one another: without it, the "vast machine" would fly apart.

Every year, 1.73 billion megawatts of solar energy reach the Earth; the sun is thus dominant in the dynamics of the surface of the Earth. The immense flow of solar energy propels the atmosphere in a complex pattern of winds, determining weather and climate and driving the ocean's circulation in patterns linked with those of the atmosphere, and it propels all of life as well.

The water and gases in the oceans and atmosphere react chemically with the solid surface of the continents and transport material from one place to another. Matter continually changes state from solid to gas to liquid to solid again, moving from place to place, absorbing and releasing energy. And matter and energy continuously move into and out of the living components of the ecosphere, blurring the basically artificial distinction between living and nonliving. In Thoreau's words, "There is nothing inorganic." The product of this ceaseless exchange and transformation is Earth's glorious array of biomes, ecosystems, and species.

The components and processes of the ecosphere are in a state of virtual dynamic equilibrium, so that while matter and energy continuously circulate, the cycles themselves remain essentially unchanged. Yet, over geological spans of time, apparently even the cycles have changed.

Thus the processes that operate on the surface of the Earth—including all biological processes—are due to interactions between the cycles propelled by solar energy and manifestations of the internal heat engine (the mountains, volcanoes, and rock forced up from the interior), all under the organizing and cohesive aegis of gravity.

Genealogy of the Planet

Earth is the third planet from the sun, which is situated 150 million kilometers away. A planet of ordinary size and mass (6 septillion tons), it has but one satellite, the moon.

The visible universe (as seen by astronomers) has been developing as a system for more than 10 billion years. Some 5 billion years ago, Earth's sun coalesced and its nuclear furnace started. The oldest rocks so far found on Earth are about 4 billion years old; since they were derived from the sediments of older, preexisting rocks, Earth's granitic crust must be more than 4 billion years old. In fact, its age is usually given as 4.5 billion or 5 billion years. Only a few hundred million years after Earth came into being, life appeared.

The Richness and Variety of Life

Life on Earth (the biosphere) exists within a thin film of soil, water, and air on the surface of the planet (the ecosphere). The interplay between life and the ambience that sustains it is the currency of the modern science of ecology. Ecologists, the practitioners of ecology, have divided the biosphere into spatial units of varying sizes called ecosystems, which consist of plants, animals, and microorganisms together with their physical environment. Large ecosystems with similar configurations are combined into major biogeographic provinces called biomes—forests, grasslands, deserts, tundra, and so forth.

The continual flow of energy and matter has assumed a more and more central place in the science of ecology: the flow of energy and the cycling of matter link together all of the seemingly disparate and unrelated elements of the ecosphere. Recognizing this flow and circulation, the late A. G. Tansley of England coined the term "ecosystem" in 1935 and stimulated the birth of an entire scientific subdiscipline, ecosystems analysis. Ecologists who write in the Germanic and Slavic languages use the equivalent term, "biogeocoenosis." Coupled with Darwin's elegant theory of organic evolution, the concept of ecosystems has yielded profound insights into the intimate relationship of organisms to each other and to their physical environment.

Most of the following brief descriptions of Earth's natural

systems draw heavily upon the recent insights of ecology. They are meant to be introductory to the papers that follow, not exhaustive nor even authoritative treatises. Whatever errors of fact or interpretation they contain only prove the folly of presuming too much—not to mention the infinite complexity of nature.

The Atmosphere: Earth's Ocean of Air

The atmosphere is a thin and fragile membrane, a ubiquitous mediator between sea and land, tropics and poles, the organic and the inorganic, animals and plants, one ecosystem and another. Without it, life as we know it would be out of the question. Yet, without life, the atmosphere itself would be quite different from what it is, for the atmosphere and life have evolved in tandem: step by step, eon after eon, they have helped to create one another.

The Earth's atmosphere was not always what it is now. At first, billions of years ago, the atmosphere consisted of such gases as ammonia, hydrogen, methane, and water vapor, spewed from the Earth's innards through volcanoes. The first-generation atmosphere—constantly riddled by lightning and continually bombarded by intense sunlight for hundreds of millions of years—changed slowly and begot the first, simple forms of life, forms of life that could survive only in the sea because the harsh, intense, primordial sunlight destroyed all unshielded cells. The sunlight's effect on the atmosphere was now abetted by the processes of life.

As eons passed, the atmosphere continued to change; life assumed more varied and complex forms. Slowly, protective layers in the upper atmosphere, created by the impact of sunlight on the molecules of gas, shielded the Earth's surface from sunlight's more intense rays—especially after green plants appeared and began to emit molecular oxygen, which previously had been rare.

Once there were significant amounts of molecular oxygen in the atmosphere, the appearance of an ozone layer, which shielded living things from the sun's ultraviolet rays, was inevitable. Released from the constraints imposed by ultraviolet light, life assumed a prominent place in the softened sunlight. It spread rapidly, becoming even more rich and diverse.

At present, the atmosphere consists of molecular nitrogen (78.08 percent), not ammonia as it once did; molecular oxygen

(29.95 percent) in addition to the ever-present water vapor; argon (0.93 percent); and carbon dioxide (0.34 percent in 1970), not the large amounts of methane as at first. And except for the carbon dioxide—which varies locally, primarily in response to biological processes, and which has been increasing by some 2 or 3 percent per decade since 1900—and the water vapor—which varies between close to 0 and 2 or 3 percent—the composition of the atmosphere is remarkably uniform. This is so because the atmosphere is heated from below, not from above as the ocean is. It is, therefore, ceaselessly homogenized in its lower 100 kilometers by a continuous mixing motion driven by the solar energy absorbed by, and then reradiated from, the surface of the Earth.

The atmosphere is densest at sea level, thinning rapidly upward. While it extends outward some 10,000 kilometers from the surface of the Earth, about 99 percent of it is held within the lowest 30 kilometers by gravity. Most of the water is trapped within the troposphere, which is the lowermost 15 to 25 kilometers of the atmosphere. The troposphere is the arena of weather, the turbulent mixing caused by the atmosphere's absorption of energy radiated from the sun-warmed surface of the Earth. It is within this thin surface layer of air that nearly all life occurs.

In the troposphere, the atmosphere's temperature drops with increasing height up to the tropopause, some 13 to 30 kilometers above the surface. In the stratosphere, which extends upward from the tropopause another 32 or so kilometers, the atmosphere's temperature remains constant in the lowest few kilometers, but then increases upward to the stratopause, at a height of about 48 kilometers. In the mesosphere, which is situated between 48 and 88 kilometers above the Earth's surface, the atmosphere's temperature again drops with increasing height. At an altitude of 88 kilometers, the mesopause occurs, and, above it—beyond 88 kilometers—the thermosphere.

It is the water vapor in the troposphere that makes it so turbulent. Water vapor, which is measured as "humidity," strongly absorbs certain wavelengths of sunlight (e.g., the near infrared) and the heat emitted by the Earth—the emitted heat being solar energy in a "degraded" form. Thus, that portion of the sun's energy which passes unimpeded to the ground does not immediately return to outer space, but is temporarily detained within the ecosphere. The water vapor, which evaporates from the seas and other bodies of water and is given off by plants, rises when it is heated, condensing and forming clouds, which further impede the spaceward flow of heat from the lower regions of the atmosphere. Were it not for the water vapor in the Earth's atmosphere, the planet would be far colder than it is.

Carbon dioxide, which is present in minute amounts in the atmosphere, also absorbs and retains some of the heat emitted by the sun-bathed surface of the Earth. Locally, its concentration may fluctuate under the influence of natural biological processes (primarily respiration), and every year human activities (the burning of fossil fuels, forest fires, and so on) contribute some 15 billion tons of additional carbon dioxide to the atmosphere. Yet all of the carbon dioxide in the atmosphere would disappear within twenty years if the decay of once-living creatures did not replenish it—a measure both of carbon dioxide's scarcity in the atmosphere and of its intimate connection to living systems.

In concert with Earth's other great ocean, the ocean of air plays a central role in redistributing the energy of the sunlight that continually flows into the ecosphere: the sun's energy virtually picks up the ocean's waters, and the atmosphere—also powered by sunlight—broadcasts the waters over the entire face of the land. Thus, with the oceans, and propelled by the power of sunlight, the atmosphere creates and perpetuates the Earth's varied ecosystems. Persistent patterns of climate, set up by the lay of land and the spin of a tilted, orbiting Earth, sustain a resplendent mosaic of forest, grassland, and desert, fringed by tundra and great ice sheets and drained by countless rivers.

The Hydrologic Cycle

Water is by far the most abundant single substance in the ecosphere. Inorganic itself, it is the medium of all life processes, flowing through living matter mainly in the stream of transpiration—from the roots of plants and out through their leaves—but also supplying their hydrogen in the course of photosynthesis. This global scale flow of water is called the hydrologic cycle. The atmosphere and the oceans, Earth's two great interlocking "commons," are its principal components.

They are constantly in motion, driven by the energy from the sun and the rotation of the Earth. Together, they redistribute solar energy, much of which has been absorbed by water. In the process, they scatter moisture over the entire face of the globe, determining the distribution of living things. The moisture in the atmosphere represents latent energy derived from the sun, which is released in storms. Each day, more energy enters and propels the hydrologic cycle than has been utilized to date by all civilizations.

The hydrologic "cycle" is not a single simple cycle, but rather, is a series of pathways by which the water in the ecosphere passes from one physical state to another, from one part of the Earth to another, and into and out of the living and nonliving components of ecosystems. Propelled by solar energy and gravity, it permeates the entire hydrosphere, extending upwards to some 15 kilometers in the atmosphere and downward into the crust of the Earth about 1 kilometer. Living systems follow in its wake. It is one of the grand cycles of the Earth, and if it can be said to begin or end anywhere, it is in the vast world ocean, which contains nearly 1.5 billion cubic kilometers of water, or 97.3 percent of all water on Earth.

While at any one instant a mere 0.005 percent of Earth's water is in transit, some 390,000 cubic kilometers of water enter the atmosphere every second as water vapor from the oceans and land, and from transpiration by plants. Yet the atmosphere contains only 0.001 percent of Earth's water. The amount in watercourses (0.0001 percent) is even smaller.

Something slightly more than 2 percent of all water on Earth is locked in icecaps and glaciers, 85 percent of it in the Antarctic icecap. This enormous mass of ice is equivalent to the flow of all rivers in the United States for 17,000 years, and to the flow of the Mississippi River for some 50,000 years.

Earth's vast quantities of water—in the oceans, in icecaps, and in the atmosphere—in tandem with sunlight and gravity, sustain and control the pattern of living communities virtually everywhere on Earth. It is the *sine qua non* of life. The oceans, coastal marine environments, rivers, and freshwater wetlands—even the atmosphere and dry-land terrestrial ecosystems—are integral components of the hydrologic cycle. Water is continuously coursing through all of them, albeit in different amounts, at various rates, and along diverse routes.

The World Ocean

There are almost 1.5 billion cubic kilometers of water in the "world ocean," some 97 percent of all the water on Earth. Situated mostly in the southern hemisphere, it covers 71 percent of the Earth's surface to an average depth of 3.8 kilometers. The Pacific Ocean, which occupies just over half of the Earth's surface (more than is occupied by all of the dry land combined) is the deepest ocean; the Arctic Ocean, the shallowest.

Like all other components of Earth's natural systems, not even the vast oceans (nor even the continents, for that matter) have always been situated where they are now. Sea-floor spreading and continental drift—consequences of what is known as plate tectonics—have slowly opened up broad oceans where once there was *terra firma,* driving continents against each other and thrusting up huge mountain ranges.

On the floors of the oceans, the separation of the plates caused deep rifts in the abyssal plain. Elsewhere, in compensation, molten rock poured forth, feeding a planet-girdling network of largely submarine ridges. The biosphere—life—has followed, or tried to follow, these upward and downward extensions of the ecosphere. Weather patterns and hence entire climates have been profoundly altered by the drifting of the continents as well—altering in turn the character and distribution of living things. Thus is the Earth continually replenished.

The inhabitants of Iceland know well the network of rifts and ridges. They call their strand of it—the part that gave rise to Iceland—the Midgard Serpent. It writhes "in giant fury," they say, "trying to come ashore." The emergence in 1963 of the new island of Surtsey, as well as the violent eruption in 1973 on the nearby island of Heimaey, were both results of the serpent's continual "writhing" off the southern coast of Iceland. Since they settled Iceland eleven centuries ago, Icelanders have witnessed 110 onshore and 14 offshore eruptions of the "serpent." They could scarcely have ignored their furiously restless bedfellow, "which surrounds the whole world, and . . . now lies in the middle of the ocean round the Earth, biting its own tail. . . ." Thus, it appears that, under the name Midgard Serpent, Earth's system of midoceanic rifts and ridges forms a cornerstone of Iceland's national culture and of a wider Germanic culture as well.

The world ocean is continuously exchanging energy (heat) and matter (oxygen, carbon dioxide, water, and so on) with the atmosphere. Through this interplay with the atmospheric "sea," the world ocean permeates the entire biosphere, playing a key role in the interaction of all ecosystems. Its influence is all the more pervasive because the ocean is a continuous expanse, not a series of discrete subunits like the continents.

The ocean circulates unceasingly because polar and tropical seas heat unequally—partly because of the Earth's tilt, partly because the polar ice reflects a large portion of the solar energy that reaches the surface. This induces winds, which set up ocean currents that redistribute the original solar energy and, thereby, the supply of moisture to land ecosystems. Thus, the circulation patterns of the world ocean actually determine the pattern of life on the land. Without the continual redistribution of heat by the ocean, in fact, no present form of life could exist on Earth.

Most of the energy that reaches the world ocean from the sun is absorbed by the top kilometer of water. This leads to stability in the vertical structure of the ocean. Also, because of water's unique combination of physical properties, marine environments fluctuate more slowly and to a far smaller degree than do terrestrial or freshwater environments, which are subject to the rapid and erratic fluctuations of the less stable troposphere.

The ecological structure of marine biotic communities resembles that of grassland ecosystems: a few rapidly photosynthesizing plants are cropped continually by numerous herbivorous animals. But, in spite of its enormous size and the rapid photosynthesis, the world ocean sustains no more than one-third of the Earth's total biological production. This is so because it is only in the upper, lighted kilometer of the ocean that photosynthesis by tiny, free-floating algae (phytoplankton) takes place—when it can: a lack of certain critical nutrients in the surface waters of the ocean (due to the ocean's vertical stability) inhibits photosynthesis, with the result that biological productivity in 90 percent of the world ocean—nearly two-thirds of the Earth's surface—is comparable to that of deserts.

It is only in offshore and coastal areas that receive runoff from the land, or where persistent winds and currents induce the upwelling of nutrient-rich bottom waters, that there are sufficient nutrients to sustain full-tilt photosynthesis. Even coastal areas, however, are less productive than some land ecosystems. Thus, the great world ocean, while absolutely essential to all life on Earth, works its life-sustaining miracles mostly from afar, indirectly through the mediation of the atmosphere and with the assist of land, freshwater, and river ecosystems.

Coastal Marine and Island Environments

Coastal Marine Environments

Coastal marine environments are unlike either open-ocean or terrestrial environments, yet they are strongly influenced by both; in fact, they might be considered extensions of both. Certainly, most continental shelves are, properly speaking, neither terrestrial nor truly oceanic, but share important characteristics of both. Further, rocky beaches differ markedly from sandy beaches and headlands from bays, just as estuaries and salt marshes differ from coastal situations where the transition from water to land or from salt to fresh water is abrupt. Barrier reefs and mangrove swamps are still different coastal marine environments that are confined to the tropics. Mangrove swamp forests are excellent examples of what are called "ecotones"—transition zones between much larger ecological formations (in this case, rain forest and oceanic ecosystems). The richness and variety of life are often greater in ecotones than in either of the two dominant formations, in part because ecotones share species with both of them.

While coastal environments differ from each other in fundamental ways, the organisms that inhabit them do have this in common: they must be able to endure daily exposure to profoundly different conditions, for they live delicately balanced between the extremes of oceanic and terrestrial, or of marine and freshwater, environments. On the other hand, they benefit because water's high specific heat ameliorates maritime environments and because nutrient-laden runoff from land, coupled with the mixing action of tides and storms, stimulates biological productivity. Many coastal ecosystems are, in fact, among the most productive on Earth.

Estuaries are "nutrient traps," primarily of the nutrients in runoff from terrestrial ecosystems, that usually are scarce in the open ocean. Biological activity is, therefore, very high in estuaries, and many species of the open ocean rely completely on these lush habitats during the earliest stages of their life cycles.

Islands

There are many kinds and many sizes of islands. Islands occur in all climatic zones, and they are abundant. In the Pacific and Indian Oceans alone, for example, there are more than 30,000 of them. Some islands are associated with nearby continents (and are, in fact, really part of the continents), while others are associated with vast submarine rift and ridge systems on the ocean floor. Many of the former islands consist of ancient, "continental" rocks; most of the latter are of more recent, volcanic origin.

Islands range in size from that of Greenland (1.8 million square kilometers) to that of mere rocks that are scarcely above water at high tide. New Guinea and Borneo, in the western Pacific Ocean, are the world's second- and third-largest islands, respectively. Other well-known or notable islands are Madagascar, the islands of Japan, Sri Lanka, the islands of New Zealand, the Galápagos Islands, the Hawaiian Islands, the Aleutian Islands, and the Malay Archipelago. There are many, many others.

The wide range of sizes among islands, as well as their various distances from continents and from each other, have inspired important ecological discoveries about the relationship between biotic diversity and key characteristics of the habitat. In fact, the entire science of ecology has been immeasurably enriched by the subdiscipline of island biogeography, the most important proponent of which was the late Robert H. MacArthur of Princeton University. From the initial realization that there is a simple relationship between the number of species that inhabit an island and the island's size, for example, have flowed invaluable insights into the structure and functioning of ecosystems everywhere. The theory of island biogeography has important practical applications in conservation as well, since it aids one in establishing the most appropriate size for natural ecological preserves.

Since islands are surrounded by ocean water, their climates are usually more moderate and stable than those of continents: at least there are fewer wide or rapid fluctuations in temperature over the course of a year. On the other hand, island environments are not necessarily monotonous: the higher islands, especially, support a wide variety of habitats because the differences in elevation lead to a zonation of microclimatic conditions. Thus, the habitats on an island may range from littoral to alpine or desertic.

Islands are ephemeral and fragile patrons of biological experimentation: their unique mixture of characteristics encourages and accelerates many processes of evolution—especially geographically isolated, geologically young, or sparsely colonized islands, on which beached "waifs" encounter such unusual ecological conditions as vacant niches and a lack of competition. Faster and more nearly random speciation is the result. In fact, it was in the Galápagos Islands that Charles Darwin gained some of his most profound insights into the process of speciation. In the Malay Archipelago, Alfred Russel Wallace, with Darwin a codiscoverer of biological evolution, gained equally crucial ecological insights.

The ecological "freedom" of isolated oceanic islands encourages genetic innovation: new and often bizarre forms of life that would succumb in the ecologically more rigorous environments of continents originate and persist on islands. Island species are thus very vulnerable to the depredations of aggressive newcomers. Island animals, some of them flightless birds, often are totally unafraid of potential predators and hence suffer great losses at the hand of newcomers that evolved in the competitive ecological arenas of continents. Island floras exhibit a similar vulnerability to alien animals and plants. Thus, while islands are centers of biological innovation, they also have disproportionately high numbers of threatened, endangered, and extinct species.

Rivers

Rivers and streams have no independent existence of their own: they reflect regional climate and the landscapes through which they flow, and often are hardly more than aboveground extensions of the water table. The topography, bedrock, soils, vegetation, and climate in a stream's watershed determine the character of the stream; at the same time, streams mold the landscapes through which they flow. In Aldo Leopold's words, "Soil and water are not two organic systems, but one. Both are organs of a single landscape."

Flowing water has indelibly etched its presence into the surface of all landscapes: ancient river valleys run through even the driest deserts, their flow terminated by imperceptibly slow changes in regional climate. Others lie buried hundreds of meters under gravel and sand deposited by the great Pleistocene glaciers, while yet others extend on continental shelves as submarine canyons many kilometers seaward from the land. They occur everywhere on Earth, even—though to a very limited degree—in Antarctica. The Nile, the Amazon, the Yangtze, the Ob-Irtysh, the Huang, the Congo, the Mississippi-Missouri, and the Amur are the world's longest river systems.

Streams are integral parts of the grand cycles of the biosphere—of the hydrologic cycle, which also encompasses the seas, the atmosphere, and groundwater and is driven by sunlight and gravity; and of the geomorphic cycle of uplift and erosion through which much of the Earth's landscape has been produced. The hydrologic cycle and the biogeochemical cycles operating within a watershed ecosystem are inseparable. Aldo Leopold epitomized the situation again when he stated that "waters, like soil, are part of the energy circuit."

Energy and matter are continually flowing from the land into stream ecosystems, which literally feed on the produce of the land—on the litter from streamside vegetation and on other terrestrial organic detritus. Without a continuous influx of nutrients from the land, stream ecosystems would be far less fertile than they are. As it is, biological processes in streams are subordinate to geological phenomena, and life's hold is more tenuous overall because oxygen and light often are limiting.

Each year, rivers transport millions of tons of sediment, depositing them in their floodplains, in lakes, and in the ocean. This is the source of the well-known fertility of floodplains. Dissolved nutrients and other mineral elements, extracted from solid bedrock by the action of physical and chemical weathering, are also carried downstream by rivers. There is thus a continual interchange between river and landscape, the river eventually relinquishing what it removes.

Rivers reflect the geographic zones and biomes through which they flow. Tropical rivers support a greater diversity of life than do temperate rivers, while temperate rivers support more species than do Arctic rivers. The water level of tropical rivers usually fluctuates more widely during the course of a year than does that of temperate rivers. The Nile, which by some accounts is the longest river on

Earth, flows from grassland savannah through desert to the sea, while the Amazon, which is nearly as long as the Nile (some say longer), flows primarily through tropical rain forests to the ocean. The Arctic rivers of North America (the Colville and the Mackenzie) and Eurasia (the Volga, the Lena, the Ob-Irtysh, the Yenisey, and so on), flow northward across tundra—in some cases from headwaters in the boreal forest— into the Arctic Ocean. Since their headwaters begin to melt each spring long before their mouths do, they become choked with ice and water that cannot flow northward to sea level until the ice has melted.

Where rivers flow into the ocean or a sea, the commingling of the waters, abetted by tidal action and the abundance of critical nutrients, supports high biological productivity. Without the contribution of the rivers (and, ultimately, of the land), coastal marine ecosystems would be far less productive than they are.

Freshwater Wetlands

Freshwater wetlands—marshes, swamps, fens, mires, pools, sloughs, potholes, bayous, wet meadows, pocosins, muskegs, and so forth—assume many forms and differ from each other in many fundamental ways. They are as varied as the landscapes within which they occur. Hence their many names. In essence, they are but aboveground extensions of the water table, sustained by special characteristics of the surrounding topography and underlying geology. Hence, any terrain having the water table at or near the surface is a wetland; for most, water supply and topography are the most important factors governing their occurrence.

Some freshwater wetlands are large and persistent features of the landscape, others are small or only intermittent and seasonal features. All wetlands, however, are ephemeral on a geological time scale. Some of them are totally independent of aboveground freshwater systems such as lakes and rivers, while others are closely linked to nearby rivers and lakes, among them the so-called "river-overflow lands." In freshwater wetlands, although there is no daily tidal ebb and flow as there is in coastal marine environments, the water level does fluctuate—but on a seasonal or annual schedule, in response to variations in local and regional rainfall.

The biological productivity of freshwater wetlands varies from type to type. While most types tend naturally to be fertile (for example, freshwater marshes and swamps are among the most productive of ecosystems), others—such as bogs—are quite infertile. In fact, bogs, or at least their aquatic component, are so lacking in certain essential nutrients that many plants—such as sundews, pitcher plants, bladderworts, butterworts, and other insectivorous plants—have tapped an alternative source of nutrients: the corpses of their insect prey. Nonetheless, vast quantities of organic matter—peat—have accumulated in bogs, primarily because anaerobic conditions predominate.

Bogs are most common in regions that were glaciated during the Pleistocene Epoch (10,000 to 2.5 million years ago). They support a spongy cover of sphagnum mosses, heath shrubs, sedges, and so on. Pocosins are bog-like habitats found in the southern United States; the muskeg is a far-northern variation of the bog habitat, with trees. Swamps and marshes, wetlands with mineral soils vegetated principally with trees or grasses, respectively, are found throughout the world.

Glaciated areas in which drainage is impeded, such as the "prairie-pothole" region of the northern United States and southern Canada (570,000 or more hectares in area), have large expanses of wetlands. Probably more than half of North America's duck population breeds in the prairie-pothole region.

Many parts of the Arctic tundra qualify as wetland because they are saturated with water for most or all of the growing season—a condition due primarily to the impermeable permafrost just under the insulative mat of tundra vegetation. On the other hand, since the amount of precipitation is low, tundra is more properly classified as polar desert.

Innumerable migratory waterfowl nest in the Arctic wetlands during the long polar "day" (summer) and then fly south in the fall to open water for the winter. These birds are as dependent upon their breeding grounds in the north as they are upon their temperate or tropical wintering grounds, since a key element in sustaining large populations of waterfowl is sufficient suitable nesting area.

Because organic matter and plant nutrients accumulate in many types of wetlands, there is great temptation to drain them for agriculture. This, of course, displaces many species of plants and wildlife that are totally dependent on wetland habitats.

There are more than 55 million hectares of freshwater wetland in the continental United States alone; in Alaska, there are some 14 million hectares of wet tundra, 26.5 million hectares of moist tundra, and over 4 million hectares of low-brush muskeg-bog. One of the largest wetlands in the world is the Sudd of the Sudan (1.3 to 1.4 million hectares of permanent swamp, more in the rainy season), which lies adjacent to the White Nile. There, and in the marshes of Mesopotamia in southern Iraq (estimated to cover as much as 5 million hectares), human beings have lived for millennia in harmony with large freshwater wetlands. The largest wetland in Europe, the Pripet Marches, occupies 4.6 million hectares in the extreme western part of the Soviet Union.

Wetlands are habitat for many species of mosses, grasses, sedges, rushes, reeds, and bulrushes. Most have a great capacity for absorbing water during wet periods, for filtering out silt and dissolved compounds, and for gradually releasing clear water during dry periods: rivers that originate in or that pass through marshes and other wetlands are clearer and have more constant flow rates than rivers that depend solely upon surface runoff—especially runoff from cultivated landscapes. Nevertheless, despite their contributions to other ecosystems and their role in molding the landscape, most wetlands—Florida's magnificent Everglades (nearly 1.8 million hectares), for example—are very dependent upon the specific features of the landscapes in which they have developed. Hence, any significant alteration or disruption of the hydrologic and topographic systems that sustain wetlands may cause their destruction.

Tropical Forests

There are several types of tropical forest. Among the most conspicuous and noteworthy of them are the equatorial or tropical rain forest, the monsoon forest, and the tropical scrub forest.

The tropical rain forest occurs where the rainfall is between 125 and 1,250 centimeters per year. One type, the "equatorial type," receives frequent torrential thunderstorms, while the other, the "tradewind type," receives steady, almost daily, rains. Cloud forests, too, are supplied with moisture from condensation as well as rain. There is no dry season, scant variation in temperature, and no cold period. Soils are lateritic.

There are tropical rain forests on the east coast of Central America, in the Amazon basin, on the west coast of Africa, in the Congo basin, Malaya, the East Indies, the Philippines, on New Guinea, on the west coast of India, the east coast of Madagascar, and on the Pacific islands.

Tropical rain forests are among the most diverse ecosystems on Earth, their diversity being due to the equable climate, long-term climatic stability (for example, a lack of Pleistocene glaciation), and a steady influx of species from the subtropical and temperate regions. Pockets of endemism ("ecological islands") of plant and animal species are common. In spite of their great diversity, however, they are very vulnerable to any kind of disruption, for, in the absence of the protective vegetation, nutrients are rapidly washed out of the soil by heavy and frequent rains.

Occupying large areas on the peripheries of the tropical rain forests are moist forests that lose their leaves during the dry season. In some instances, bamboos, which are present elsewhere only in second-growth stands, are abundant in the climax of the "monsoon" or tropical deciduous forest. In the monsoon forests of Burma, teakwood is a typical species.

There are monsoon, or tropical deciduous, forests along the coast of Brazil, in Venezuela, Guyana, Ecuador, India, Bangladesh, Indochina, southern China, and northeastern Australia.

Lianas (woody vines) and epiphytes (smaller plants that grow on trees, shrubs, or other larger plants) are locally abundant in the monsoon forest, but they usually are fewer and smaller than in the tropical rain forest. As in the tropical rain forest, the soils of the tropical deciduous forest are, for the most part, lateritic (that is, they have been leached of certain elements by the abundant rainfall). As in the tropical rain forest also, because of the abundant light, heat, and moisture, life processes are so rapid in the litter on the forest floor that organic debris is quickly degraded.

Tropical scrub forests occur in southwestern Mexico, northeastern Brazil, eastern Africa, southwestern Africa, northwestern India, and northern Australia—tropical areas where rainfall is lower than in monsoon forests but greater than in grasslands.

The amazing diversity of plants in tropical rain forests also characterizes their animal life. While there are many species, however, there are few individuals of any one species, which means that species are easily extirpated from an area when their habitats are destroyed or when they are overexploited: the smaller and less dense a population, the greater the chance it will be entirely destroyed by some catastrophe and the less chance there will be individuals to recolonize a devastated area afterwards. Among the species of animals in tropical rain forests are the Jaguar, the Ocelot, tapirs, many monkeys, sloths, and a wide array of bird species.

Temperate Forests

Forests, areas with a more or less continuous cover of trees, usually occur where there are relatively high amounts of precipitation in conjunction with an underlying soil and an absence of extreme temperatures, although some do occur in quite dry situations. In most forests there are several layers of different types of vegetation: a dominant layer of tall trees, often a layer of shorter trees, a layer of shrubs, a layer of herbs, and often a layer of simple, low-growing plants called the moss layer. In forests composed of evergreen species, the light that reaches the forest floor is always attenuated by leaves; in temperate forests composed primarily of deciduous trees, bright light reaches the forest floor during winter and early spring. It is during the early spring that many wildflowers in the herb layer burst into bloom.

The undisturbed forest has a self-regulating, self-protecting function: the forest cover itself prevents erosion of the soils that sustain it. The forest ecosystem holds nutrients within the soil and within the trees themselves, thereby minimizing their loss. Precipitation that falls on the forest penetrates the soil and leaves it slowly and continually through its uptake and transpiration by the trees themselves, and by flowing through the soil; under the influence of gravity, it empties into streams and rivers.

Forests in the temperate zones—between the tropics of Cancer and Capricorn and the Arctic and Antarctic circles—assume various forms, which for convenience's sake have been given formal names: the boreal coniferous ("needle-leaf") forest of northern North America and northern Eurasia (the "taiga"); the temperate deciduous forest of eastern North America, southern South America, eastern Asia, and western Europe; the temperate rain forests of southeastern and western North America, eastern Asia, eastern and southern

South America, southeastern Africa, Tasmania, eastern Australia, and New Zealand; and the evergreen-hardwood forest (the "sclerophyll forest")—including open woodlands and shrubs ("scrub" or "chapparal")—in western and southern North America, western South America, northern and southern Africa, southern Europe, and Australia.

Very little remains of the mixed forests of coniferous and broad-leaved deciduous trees that originally covered much of central Europe, eastern Asia, and northeastern North America, or of the mixed forests of coniferous and broad-leaved evergreen trees that once covered much of the Mediterranean lands. Vast tracts of the taiga are still in the natural climax state, however.

Temperate deciduous forests occur where precipitation, which is evenly distributed through the year, amounts to between 65 and 230 centimeters per year and where drought is rare. Some snow falls during the winter. Temperate rain forests receive from 125 to as many as 890 centimeters of precipitation per year, some of it snow. Again, this is distributed evenly through the course of the year. In the taiga, 40 to 100 centimeters of precipitation fall each year, much of it in the form of snow.

Through its intricate structure, the forest provides a variety of habitats for plants and animals. Animals of the temperate forests include bears, wild boars, deer, squirrels, numerous insectivores, and rodents. Predators include foxes, wolves, and wildcats, although the latter two are on the decline.

Savannahs and Other Grasslands

Grasslands, whether they be savannahs, prairies, steppes, *velds, llanos, campos, paramos,* or *pampas,* occur where available moisture is insufficient to support the growth of more than scattered individual trees but is sufficient to support an uninterrupted cover of vegetation. That is, grasslands develop where the precipitation regime prevents the establishment of closed-canopy forests, but where the perennial vegetation (predominantly grasses) is not discontinuous like that of deserts. In temperate latitudes, grasslands occur in the rain shadows of mountain ranges; in the tropics, they occur where wind patterns cause air passing over land to dry out and where precipitation is thus less than necessary for trees. Grasslands may be hot or cold: as with deserts, the key to their occurrence is precipitation, not temperature. More precisely, grasslands occur where the *availability* of water—which is determined by the interaction of precipitation, temperature, and air movement—is sufficient for herbaceous, but not for woody, perennials. Some kinds of tundra qualify as grassland on the basis of their vegetative cover, though on the basis of precipitation they are deserts.

The eastern parts of the North American grassland—the tallgrass or true prairie—were originally a mixture of tall grasses, some of them up to 2 meters in height. The short-grass plains, which extended westward to the Rockies, probably differed from the tall-grass prairies because of lower precipitation. The steppes of Asia are dry, generally treeless, grasslands similar to the short-grass plains of North America.

Savannahs are subtropical or tropical grasslands with scattered trees and other vegetation able to withstand dryness. They are transitional between strict grasslands and tropical forests. Like at least some parts of the North American prairie and, undoubtedly, other grasslands, savannahs may owe their overall character—and in some cases their very presence—to intermittent fires, whether or not they are caused by human beings.

Grasslands—even the tropical savannahs of eastern Africa, which must endure irregular rainfall and which do not produce a thick layer of organic litter—support enormous herds of hoofed herbivores; carnivores; and "top carnivores." The wild herds of the world's grasslands do not overgraze their ranges and yet exist in greater numbers than domesticated cattle, which often do overgraze their ranges, cause erosion, and thereby reduce the fertility of the grassland. The large herds of wild herbivores migrate long distances, timing their feeding to coincide with rainfall and plant growth. They graze selectively, returning nutrients to the grassland in their wastes, and their grazing keeps the vegetation young and productive. Rather than depleting their forage, they leave sufficient tissue for the plants they eat to recover.

In North America, the American Bison, rodents in large numbers ("prairie dogs," mice, ground squirrels, and pocket gophers), rabbits, snakes, hawks, weasels, kit foxes, coyotes, ground-nesting birds such as quails, and the like, feed on the grasses and other vegetation, on the herbivores, or on other carnivores.

Saiga antelope, wild horses, and wild asses roam—or used to

roam—in large numbers on the steppes of Asia. The natural grazers of the South American *pampas* are the guanacos, and in Australia kangaroos fill this role. All of these grazers have been largely displaced by man's domesticated grazers.

Temperate grasslands occur in central North America, eastern Europe, central western Asia, Argentina, and New Zealand. Precipitation, which is evenly distributed throughout the year, or is greatest in the summer, varies between 30 and 200 centimeters per year. Their soils are black (the so-called "black prairie soils"), chestnut, or brown in color, and almost all of them have a layer of lime.

Tropical grasslands or savannahs cover a wide belt on either side of the Equator between the Tropic of Cancer and the Tropic of Capricorn. The grasses of the tropical savannah are taller than those of the temperate grasslands, often reaching heights of over 3 meters. The climate is always very warm, and there is a long dry season; therefore, savannah plants must be able to withstand drought.

The principal locations of tropical savannahs are the Pacific coast of Central America, the Orinoco basin, Brazil south of the Amazon basin, north-central, eastern, and south-central Africa, Madagascar, India, southeastern Asia, and northern Australia. Rainfall in these areas varies between 25 and 190 centimeters per year, with almost no rain falling during the cool season.

Deserts

Deserts, like grasslands, develop where precipitation is too sparse to support the growth of forests—in the rain shadows of mountains, for example, or on the west coasts of continents in the tropics. They are arid areas of sparse to absent vegetation, receive 25 centimeters or less of precipitation per year, and occur in all latitudes. They may be very hot or very cold: low precipitation, not high temperature, is their distinguishing characteristic. Technically, even Antarctica and much of the Arctic tundra are deserts—"polar deserts" and "cold deserts"—according to some classifications. But for most purposes, the term desert is reserved for temperate and tropical areas and is not applied to high-latitude ecosystems.

If semi-arid areas are included, about 30 percent of the Earth's dry-land surface (some 32 million square kilometers) may be classified as desert; approximately 5 percent is extremely arid. In warm or hot deserts, rainfall ranges between essentially 0 and 25 centimeters per year, in cold deserts, between 5 and 20 centimeters per year. What rain does fall in deserts comes at very irregular, unpredictable intervals. At Iquique, Chile, in the Atacama Desert, the mean annual rainfall over a 25-year period was 0.13 centimeter. At Calama, Chile, also in the Atacama, there was no rain at all for one 13-year period. Typically, the drainage of deserts is "internal"; that is, it does not empty into the sea. Notable exceptions are the Nile and Colorado river systems.

Warm deserts occur in southwestern North America, in Peru and northern Chile, northern Africa, Arabia, southwestern Asia, eastern Africa, southwestern Africa, and central Australia. Cold deserts exist in Patagonia, trans-Caucasus Asia, central Asia, and intermontane western North America.

In warm deserts, the diurnal range of temperature is great, due primarily to the clarity and dryness of the air. In cold deserts, most precipitation falls in winter, some of it as snow. Frosts are common, occurring during one-third to one-half of the year.

Desert soils are virtually devoid of humus. Those of warm deserts are usually reddish and often sandy or rocky, those of cold deserts gray and also often sandy or rocky. Saline soils develop in both warm and cold deserts because the salts produced by weathering are not leached away.

Perennial desert plants grow with wide spaces between them, and the drier the habitat the more widely spaced they are. This suggests that the plant biomass is roughly proportional to the amount of precipitation, which apparently is so because the roots of the plants are in contact with each other, preventing the establishment of competitors by efficiently absorbing all moisture that penetrates the surface of the soil. Perennial desert plants are, of course, drought resistant, and in many cases can photosynthesize all year because of the chlorophyll in their stems. They are adapted to drought in various ways: some have drought-resistant seeds, some have small, thick leaves that they shed during dry spells, and some—the New World cacti, for example—are succulent and capable of storing water in their stems. Some desert plants have specialized photosynthetic mechanisms that conserve water, which along with sunlight and carbon dioxide is an essential component of photosynthesis.

Many desert animals are small enough to hide under stones or in burrows during the day, when the heat is most intense. Some desert rodents, which live in cool burrows, are largely nocturnal in their habits.

Despite the heat stress that might be induced by a dark coloration (dark colors absorb more sunlight than do lighter colors), the inhabitants of deserts tend to be either black, or else buff, sandy, or reddish-gray in color, so that they resemble the terrain on which they live. Although paler colors would tend to reduce heat stress, the darker colors are common because they have camouflage value—i.e., their ecological function is said to be "cryptic."

The Polar Regions

Arctic and Anti-Arctic: Shallowest Sea, Highest Continent

Because the Earth's axis of rotation is not perpendicular to the ecliptic, but is tilted some 23°27', two very large regions—the Arctic in the north and the Antarctic in the south, which together occupy some 15 percent of the globe—undergo continuous darkness and daylight for weeks or even months each year. But aside from this and a few other shared characteristics, the Arctic and Antarctic are for the most part utter opposites. They are opposites in more than simply their names or geographic positions, for the Arctic is dominated by a central sea, the Antarctic by a central continent.

A shallow, ice-covered sea centered almost symmetrically on the North Pole, the Arctic Ocean covers some 14 million square kilometers and is surrounded by North America (including Greenland) and Eurasia, the largest land masses on Earth. It is Earth's shallowest ocean, and is situated in the center of the so-called "land hemisphere."

By contrast, Antarctica is a high, ice-covered assemblage of land masses, also about 14 million square kilometers in area. Centered almost symmetrically on the South Pole, it is surrounded by the greatest expanses of water in the world—Earth's highest continent, set in the midst of the "water hemisphere."

These contrasting characteristics have led to profound differences between the animals and plants of the Arctic and those of the Antarctic.

The Polar Biotas

For example, there are no amphibians, flying insects, land mammals, ferns, or trees in Antarctica, and only two species of flowering plants, both of them confined to the northernmost tip of the Antarctic Peninsula. There are no indigenous peoples in Antarctica and only a few thousand human inhabitants at the present time—all of them transient.

In the Arctic, on the other hand, there are at least two species of amphibians, one (perhaps two) species of reptiles, several species of ferns, many species of flying insects and land mammals, and hundreds of species of flowering plants, some of them low, prostrate "trees." Further, more than a million people reside permanently in the Arctic; but while it is densely populated by Antarctic standards, the Arctic supports considerably less than 1 percent of the Earth's total population. Truly, the life stream of the Arctic flows in a lean mixture.

Geography and Climate

The contrasts in the animals and plants of the two polar regions can be explained in two ways. First, the continents that extend into the Arctic provide terrestrial boreal species with easy access to high northern latitudes, while the 1,000-kilometer and greater gaps between southern temperate land areas and the continent of Antarctica very effectively block the migration of land-based species. Second, and at least as important, despite the fact that the Earth is actually almost 5 million kilometers farther from the sun during the Northern Hemisphere's winter than it is during the Southern Hemisphere's winter, the Antarctic is much the colder. (The lowest temperature ever recorded on Earth, −88 C, was recorded on the Antarctic ice cap.) This is because Antarctica, 96 percent of it permanently covered with ice, is the highest of all the continents and because winds there flow predominantly northwards, radiating outward from the center of Antarctica and virtually insulating it from the warmth of lower southern latitudes. In the far north, on the other hand, the summer's warmth retained by the waters of the Arctic Ocean (which is covered by only a thin layer of ice) moderates temperatures during the winter.

The polar regions—especially the Antarctic—have a profound influence on climate throughout the globe. Their vast expanses of polar ice, which reflect much more sunlight than do bare soil, vegetation, or open water and therefore induce low temperatures in those regions, significantly affect global patterns of atmospheric circulation.

The Polar Oceans

The Antarctic, or Southern, Ocean, which surrounds Antartica, is no mere extension of the Atlantic, Pacific, and Indian Oceans. In basic ways, it is a virtually separate body of water, its surface waters driven eastwardly around Antarctica by the prevailing winds. Its northern boundary, called the Antarctic Convergence, is a zone of significant ecological change marked by a sharp, 5- or 6-degree difference in surface temperature, and by similarly sharp differences in salinity and related physical characteristics. It occupies an irregular belt between 50° and 60°S latitude.

Some authorities claim that the great Southern Ocean extends to the southern coasts of Australia and Africa, which is the northern limit of icebergs. In any case, Antarctic waters influence the ocean depths at least as far north as 40°N latitude.

While most of the Arctic Ocean is not highly productive as oceans go, the Antarctic Ocean is richer in life than any other oceanic area, primarily because nutrient- and oxygen-rich bottom waters well upward around Antarctica. In both polar regions, but especially in the Antarctic, marine ecosystems and their associated faunas have important influences on the less fertile coastal land ecosystems nearby, primarily in the form of nutrients transported ashore by breeding birds.

Tundra: The Prairies of the North

The terrestrial areas of the Arctic are covered with the vast, virtually treeless tundra, a relatively simple biome dominated by sedges, grasses, mosses, lichens, and minute prostrate "shrubs," all of which grow close to the ground. Only the uppermost few centimeters of soil beneath the tundra's insulating mat of vegetation ever thaw, even at the height of the Arctic summer. Below the thin, thawed layer under the mat lies the permafrost—perennially frozen ground. Permafrost with a high proportion of ice in it is referred to as ice-cemented permafrost.

Large expanses of tundra are covered with standing water during the summer, and there are many shallow lakes and small ponds. But the apparent abundance of water is illusory, for precipitation is quite sparse. On the basis of precipitation, it is a true desert. The standing water is due to the existence of permafrost, which impedes drainage, and to the short, cool summer season, which inhibits evaporation.

The tundra is one of the largest biomes and the only essentially continuous land biome on Earth, girdling the Earth across the northern parts of North America, Greenland, and Eurasia. It supports such animals as caribou and reindeer, Arctic Foxes, lemmings, ground squirrels, and the nearly exterminated Musk Oxen. It supports countless birds as well, most of them migratory species. Waterfowl are especially abundant. Many, particularly the wading birds, spend most of the year in the tropics, but breed in the Arctic.

Biotically, the tundra is quite homogeneous in composition—that is, tundra species are essentially the same, or at least closely related, throughout, whether they inhabit North America, Greenland, or Eurasia. In contrast, the biotas of the tropical rain forests of South America and Africa, for instance, are quite distinct, though the climates are similar. Since so many tundra species have pan-Arctic distributions, ranging across lands and seas around the North Pole in both the Old and New Worlds, they usually are less vulnerable, as species, to human activities than would be species confined to discontinuous or smaller ranges such as islands. The vulnerabilities of tundra organisms derive from other characteristics of the tundra environment.

References

Elder, John. 1976. *The Bowels of the Earth*. London, New York: Oxford University Press.

Whipple, Fred L. 1968. *Earth, Moon, and Planets*, 3rd ed. Cambridge, Mass.: Harvard University Press.

* * *

It is interesting to contemplate a tangled bank, clothed with many plants of many kinds, with birds singing on the bushes, with various insects flitting about, and with worms crawling through the damp earth, and to reflect that these elaborately constructed forms, so different from each other, and dependent upon each other in so complex a manner, have all been produced by laws acting around us. . . . There is grandeur in this view of life with its several powers, having been originally breathed by the Creator into a few forms or into one; and that, while this planet has gone circling on according to the fixed law of gravity, from so simple a beginning endless forms most beautiful and most wonderful have been, and are being evolved.

—Charles Darwin

*I thought of the wandering air—its pureness, which is its beauty;
the air touched me and gave me something of itself.*

—Richard Jefferies
"The Story of My Heart"

The Atmosphere

15. Introduction

Chaplin B. Barnes
National Audubon Society, U.S.A.

In discussing the great global commons, the oceans and the atmosphere, we are concerned with portions of the Earth (or, in the case of the atmosphere, with a thin layer of life-giving gases that envelop the Earth) that in large measure have been historically beyond the jurisdiction of individual nations—areas that are, in fact, the common property of all nations and individuals. To a considerable degree, this attribute of the oceans is rapidly being modified, both by the unilateral action of individual nations and by negotiations at the Law of the Sea Conference; it remains true, however, of the atmosphere.

Not only politically, but ecologically, too, we are dealing in the case of these commons with issues that are by their very nature international. Not only are significant disruptive actions taken anywhere on the Earth likely to have adverse consequences elsewhere, but, equally, action to protect the environment taken by one government may be to no avail if similar action is not taken by other governments, or, in some cases, by all governments.

As Barbara Ward and René Dubos (both advisors to this conference) noted more than three years ago, these commons are a shared resource of humanity, one that can be protected only by a shared caring and a shared commitment.

In the chapter entitled "The Shared Biosphere" of their wonderfully perceptive, unofficial report to the secretary-general of the United Nations Conference on the Human Environment, *Only One Earth*, they wrote of this aspect of the atmosphere in the following terms:

The global interdependence of man's airs and climates is such that local decisions are simply inadequate. Even the sum of all local separate decisions, wisely made, may not be a sufficient safeguard and it would take a bold optimist to assume such general wisdom. Man's global interdependence begins to require, in these fields, a new capacity for global decision-making and global care. It requires coordinating powers for monitoring and research. It means new conventions to draw up ground rules to control emissions from aircraft and to assess supersonic experiments. It requires a new commitment to global responsibilities. Equally, it needs effective action among the nations to make responsibility a fact.

In the brief time allowed us, we shall be considering the life-supporting role of the atmosphere, its characteristics and properties, the threats to it that we can identify—the transport of pollutants, the modification of weather, and the depletion of the stratospheric ozone shield, among others—and, finally, the possibilities for common action to protect the atmosphere.

In considering protective action, I should like to suggest three basic principles to be kept in mind:

1. The burden of proof as to whether an action with potential impact on the atmosphere should be taken should be, not on those who advocate caution, but on those who propose the action.
2. Because the atmosphere is more easily disrupted than many other systems, we should exercise special care to assess and eliminate any adverse impacts upon it.
3. The extent to which substances released into the atmosphere are transported around the globe dictates that there be international, cooperative research in the setting of standards and in the assessment of environmental impact.

I am happy to join with you in benefiting from the experience and wisdom of two distinguished scientists, both of whom are experts in the field of the atmosphere and who are long-standing friends and colleagues. My own role as a lawyer and a layman will be only to moderate the discussion.

16. Atmospheric Pollution and Energy Production

Thomas B. Johansson
University of Lund, Sweden

Introduction

For millions of years, the atmosphere has contributed to the evolution of our present Earth, with its highly developed ecosystems and human life. The atmosphere is a great resource of vital importance for life. It serves as a transport medium and reservoir of oxygen and carbon dioxide; it filters radiation from the sun to allow biological life; it helps to balance the temperature of the Earth; and it is an important link of the hydrologic cycle—to mention just a few of its functions.

Human activity has now reached a point where it is inadvertently affecting sensitive atmospheric functions. A major activity is energy production. Burning of fossil fuels injects sulphur oxides, carbon dioxide, and particulate matter into the atmosphere. This paper will discuss these pollutants, with emphasis on atmospheric particles and their long-range transport by the atmosphere.

Sources, Residence Times, and Sinks of Particulate Matter in the Atmosphere

Our interest in atmospheric particles is related to two types of possible effects: the influence of particles on important processes in the atmosphere, and their effects on a site where they settle after atmospheric transport. Let us start with a brief review of some properties of atmospheric particles.

The life of particles in the atmosphere is shown schematically in Figure 1. Produced by any of a number of sources, they are transported and may be transformed during transport by chemical reactions, evaporation or condensation of water (or other volatile substances), or coagulation with other particles, and finally deposited somewhere.

Particles in the atmosphere originate from a number of natural and man-made processes. Table 1 gives a rough estimation of particu-

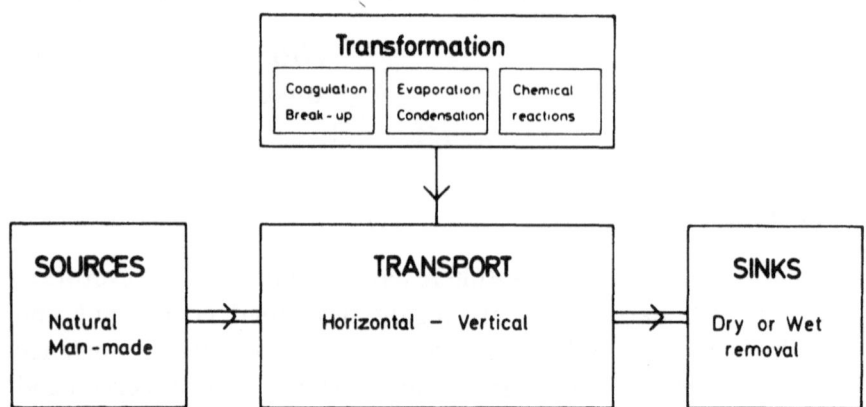

Figure 1. Flow Chart of Particles in the Atmosphere.

late sources having emissions in millions of tons per year. The human contribution is approximately 10 percent, according to this estimate.

These estimates represent sources of a global scale. This is not necessarily the best or only measure of the effects of the particles. For different problems, several parameters are of prime interest: chemical composition, particle size, and injection height, to mention a few.

The geographic distribution of man-made sources is important. Excluding oceans (70 percent of the Earth's surface) and polar regions, and considering man's tendency to build urban agglomerates, it can be easily shown that the man-made portion of the atmospheric aerosol may be very significant on a local or regional scale. On a global scale, effects of deposited aerosols have to be considered.

Two basic processes control the occurrence of particulate matter suspended in air. Mechanical processes acting on matter may inject particles. Examples of such processes are erosion, creating wind-blown dust, and the bursting of small air bubbles at the sea surface, which injects material into the air. The diameter of mechanically made particles usually exceeds several micrometers. The second basic process is vapor condensation. Vapors may be produced by forest fires, volcanoes, or anthropogenic activities such as combustion. Condensation particles are very small, but by coagulation and other processes, they approach a size of from 0.1 to 1 micrometer.

The burning of fossil fuels is a large source of particles less than 1 micrometer in diameter. As they are very hard to remove by pollution

Table 1. Particulate Matter Released Each Year into the Atmosphere by Natural and Anthropogenic Sources

Sources	Quantity of Particulate Matter Released (megatons per year)	
	Natural Sources	Anthropogenic Sources
Primary particle production:		
Fly ash from coal	–	36
Iron and steel industry emissions	–	9
Non-fossil fuels (wood, mill wastes)	–	8
Petroleum combustion	–	2
Incineration	–	4
Agricultural emission	–	10
Cement manufacture	–	7
Miscellaneous	–	16
Sea salt	1,000	–
Soil dust	200	–
Volcanic particles	4	–
Forest fires	3	–
Subtotal	1,207	92
Gas-to-particle conversion:		
Sulphate from H_2S	204	–
Sulphate from SO_2	–	147
Nitrate from NO_x	432	30
Ammonium from NH_3	269	–
Organic aerosol from terpenes, hydrocarbons, etc.	200	27
Subtotal	1,105	204
Total	2,312	296

Source: Butcher and Charlson (1972), p. 163.

control devices, a very substantial fraction of them is emitted.

The lifetime of a particle in the air is of primary importance to its transport. Lifetimes vary with meteorological conditions but range from 10 to 100 hours for particles in the 1 to 10-micrometer range and from 100 to 1,000 hours for particles in the 0.1 to 1-micrometer range (Esmen and Corn 1971). This allows for long transport distances, especially for the smaller sizes where pollution particles may dominate.

What goes up must come down. Suspended particles are removed from the atmosphere by dry deposition (gravitational settling and turbulent transfer) and/or wet deposition (rainout or washout—the former term refers to processes within the cloud and the latter to processes below the cloud base). The efficiency of these processes is dependent upon the size and chemical composition of the particles.

Particles in the Atmosphere

Concentrations of particulate matter in the atmosphere vary depending on meteorological parameters that affect both production and transportation of particles. Much effort has gone into attempts to establish baseline levels of total particulate matter as well as certain trace elements. Measurements have been made at remote continental locations, oceanic areas, and polar regions. The results of this work are of basic importance also for urban air pollution work, since all man-made atmospheric particles are imposed on the natural environment and an understanding of their relative importance is necessary to the design of efficient air pollution control strategies.

Although most work has been done using filter samplers, total mass may not be the best parameter to use for evaluating effects of particles in the atmosphere. There is evidence from several continental locations (mostly urban) that the size distribution of particulate matter, studied on a mass scale, is bimodal (Whitby 1973). One mode contains particles between 0.1 and 1 micrometer in diameter, and the other mode contains particles between 1 and 10 micrometers in diameter. Natural sources may contribute the major fraction of the coarse particles, while anthropogenic sources may contribute a major fraction of the fine particles. Although the bimodal distribution is supported on a mass basis, single compounds or elements may not adhere to it. More work is required to build a solid base for the design of monitoring programs. As human activities predominantly generate "submicrometer" particles, this size range may be a much better index of particulate pollution levels than the total amount. (See Table 2.)

Small particles are of prime importance as air pollutants (Friedlander 1973). From the health standpoint, these particles may penetrate the deeper parts of the respiratory system, where a large number of them are deposited and taken into the blood stream. Existing pollution control devices can efficiently remove particles larger than several micrometers, but submicrometer particles are quite hard to remove. The submicrometer particles tend to be enriched in toxic materials such as heavy metals present in fossil fuels, creating a possible public health hazard. These particles may also affect the climate by interfering with the sun's radiation and causing a net cooling effect which would reduce the heat given off by carbon dioxide, a process that will be discussed below. Fine particulates can also cause cloud nucleation and visibility reduction because of their long residence time in the atmosphere.

More emphasis should be placed on the significance of particle sizes. As pointed out, the effects of atmospheric aerosols are largely related to small particles. In this size range, we also find most of the pollution, at least outside the local area of aerosol-producing activities.

Only with a basic understanding of the natural processes affecting particles in the atmosphere can efficient programs be designed to study important developments over time. Otherwise, costly and inefficient monitoring systems are likely to result.

Long-Range Transport

As indicated above, particle residence times in the atmosphere range from days to weeks. During this time the particles follow the air mass and may move over great distances in a relatively short time. Consider, for example, the results of a series of air mass trajectories in Figure 2. In another study, the movement of an air mass was followed from a point in Europe every third day for a one-year period. After twenty-four hours, more than half of the air parcels had moved outside a circle with a 600-kilometer radius, and after sixty hours, outside a circle with a 1,300-kilometer radius. These observations

Table: The Concentration of Particles in the Atmosphere over Various Types of Areas

Type of area	Concentrations of All Widths of Particles in the Air (cubic micrometers per cubic centimeter)	Concentrations of Particles Less than 1 Micrometer Wide in the Air (cubic micrometers per cubic centimeter)	Number of Condensation Nuclei per Cubic Centimeter of Air
Clean continental	2 to 10	0.5 to 3	50 to 500
Average continental (west of United States)	15 to 40	4 to 7	2,000
Downwind of urbanized area	20 to 50	10 to 25	5,000 to 10,000
Polluted urban area	100 to 250	50 to 80	1,000 to 10,000

14-18 January 1974.

The Atmosphere

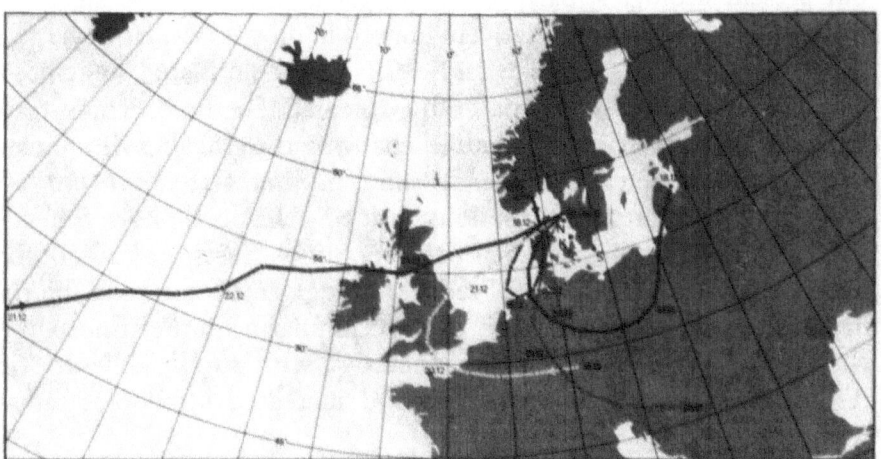

Figure 2. Air Mass Trajectory. The movements of an air mass may be followed by studying wind directions and speeds at various altitudes, retrogressing in time. Such an air mass trajectory analysis is illustrated with the episode recorded on the Swedish west coast from 22 to 24 December 1972. The positions of the air masses are indicated for each day. The dots correspond to positions at six-hour intervals. *Source:* C. Brosset in *Ambio* 2: 2-8 (1973), page 7. Based on the data of Anne-Beate Henrikson, Swedish Meteorological and Hydrological Institute, Stockholm.

demonstrate the opportunities for long-range transport of atmospheric aerosols.

The fact that long-range aerosol transport actually occurs has been observed on many occasions. Among the best-known examples are the finding of Sahara dust in the West Indies and the finding of dust from the Black Sea region in Scandinavia. Other examples include transport from volcanoes and forest fires. Nuclear weapons tests have offered an opportunity to follow radiation clouds during long periods of time. This type of transport is, of course, directly linked to the large-scale atmospheric circulation pattern.

Effects of the Long-Range Transport of Aerosols

Man's activities release many substances into the atmosphere, which dilutes and transports pollutants, and these are finally deposited somewhere. In general, these emissions pose no immediate threat to our present ecological balance on a regional or global scale, since the amount of pollutants is relatively limited in most cases. However, we have detected substances like DDT, PCBs (polychlorinated biphenyls), and mercury in the Arctic and in the Greenland ice sheet, demonstrating that man's activities are indeed changing the global environment.

This slow deterioration may show no dramatic, short-term effects. Its effects are more insidious, reducing the possibilities for ecosystems to remain in continual balance. Such effects are caused, for example, by the emission of Freons (a trade name for certain fluoromethanes) and carbon dioxide, both accumulating in the atmosphere with no apparent short-term impact but with such likely drastic results as higher air temperatures and increased ultraviolet radiation.

An important principle to remember when evaluating the possible effects of worldwide dispersion of pollutants is illustrated in the studies of nuclear fallout. As mentioned above, the clouds of radioactive material yielded by nuclear weapons tests have been followed over extreme distances in the atmosphere. One of the active elements present is cesium. Although the total amount of cesium released into the atmosphere from weapons tests is as low as 270 kilograms, it can be detected worldwide. The cesium content in lichens, reindeer, and man were measured in northern Scandinavia. Higher levels of cesium were found with greater frequency in this food chain. The average radiation dose absorbed by the population over a period of one generation corresponded to a 2 to 3 percent increase in the intake of naturally occurring radiation (Lindén et al. 1974). This clearly demonstrates the influence of distant aerosol sources and the possibility of significant effects in cases where the particle deposition is extremely small.

Air pollution studies are often limited to urban areas, where it is an obvious problem with clear public health implications. The object of this paper is to stress that air pollution also has a regional and a global dimension. The transport of pollutants through the atmosphere is worldwide, and some pollutants may be approaching levels of significant interference with the various functions of important ecosystems.

The deposition of sulphur compounds emitted from burning of fossil fuels is an example of an air pollution problem which has been

The Atmosphere

extensively studied (Bolin 1971). The effects of acid sulphur deposition include:

1. Acidification of rivers and lakes.
2. Increased corrosion rates.
3. Increased leaching of nutrients from soils.
4. Reduced forest growth rates.
5. A time-lagged, big change of pH in soils expected to occur when the buffering capacity of certain soils has been consumed.

Here we face situations where the dramatic effects are rather slow and where we can do very little to correct them once they appear.

Transport of Sulphur in Northern Europe

Sulphur occurs naturally in water, soils, and biological material, and a continuous interplay between these media takes place. The atmosphere is a link in this process. If we want to evaluate the effects of human activities, we must study them against this natural background.

A few hundred million tons of sulphur annually pass through the atmosphere as part of the sulphur biogeochemical cycle. Most of the sulphur comes from natural sources, but a significant fraction—approximately 40 million tons (360 million kilos) or about 10 percent—is man-made. As this anthropogenic contribution is concentrated in densely populated areas, it is clear that the human fraction may be very substantial, and it is estimated that 75 percent of the sulphur over western Europe is caused by man.

Sulphur is injected into the atmosphere by man as sulphur dioxide (SO_2) through combustion processes. A small fraction (2 to 3 percent) of the SO_2 is oxidized to sulphur trioxide (SO_3) in the smokestack. These pollutants react with water adhering to the surface of particles to form sulphuric acid (H_2SO_4).

In the late 1960s scientists observed a systematic change of precipitation pH over northern Europe. They determined that the most acidic precipitation falls in Belgium and Holland, where pH in 1966 reached values below 4. In southern Sweden, the 1966 pH was approximately 4, down from a measurement of about 6 in 1956. The variation of the deposition of excess acid through precipitation is shown in Figure 3. Excess acid is the total amount of acid less the total amount of base, measured over a period of one year.

It is important to find the sources of this increased acidity and especially to distinguish between distant and local contribution. In the 1972 Swedish case study for the United Nations Conference on the Human Environment, the sulphur problem was reviewed. As anthropogenic sulphur emissions in Europe have a mean atmospheric residence time of two to four days, they may very well be transported over large distances. To investigate sulphur and soot transport in Sweden, concentrations of soot and sulphur were recorded at three locations in the south of Sweden. By air mass trajectory analysis, the data were divided into groups corresponding to transport sectors. Figure 4 shows clearly increased levels in the southwest sector, thus supporting the transport hypothesis. Time analyses have also been done. These data revealed episodes of unusually high levels of soot and sulphate occurring simultaneously at several locations. Figure 2 is a trajectory analysis of such an episode showing the possible transport route on this occasion. The report concluded that about 50 percent of the sulphur deposited in southern Sweden originated in continental Europe or England.

The 1972 Swedish case study also made estimates of the effects of the acid deposition for several different prognoses of sulphur emissions for the rest of the century. Increased emission levels, as well as reduced levels, were analyzed. It was found that serious economic

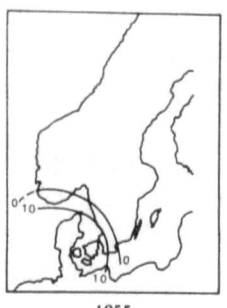
1955

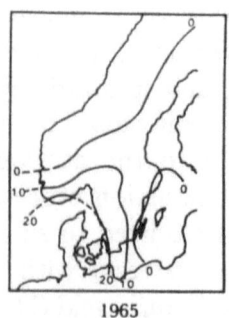
1965

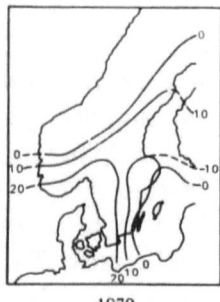
1970

Figure 3. Annual Deposition of Excess Acid in Precipitation Over a Fifteen-Year Period in Sweden (measured in milligrams of hydrogen ions per square meter). *Source:* Bolin 1971.

The Atmosphere

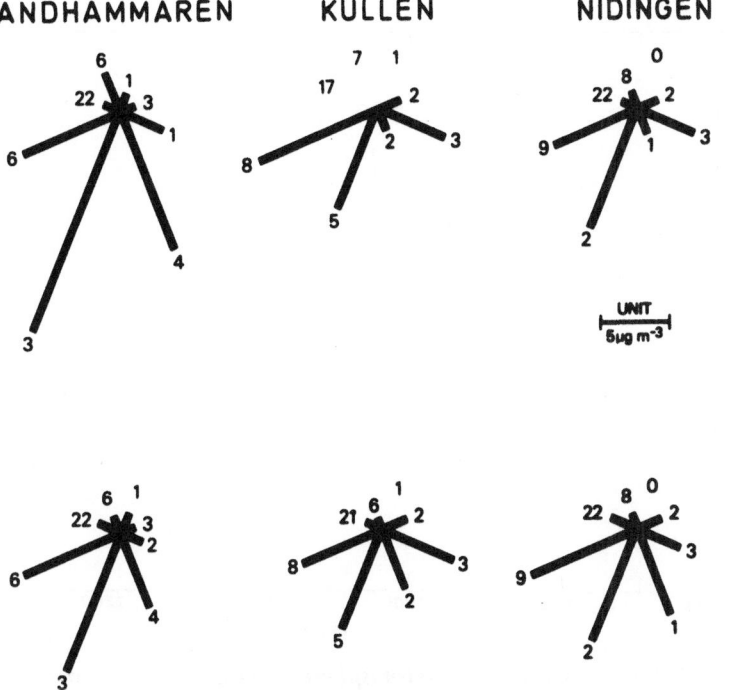

Figure 4. Soot (Upper Row) and Sulphate (Lower Row) Concentrations at Three Coastal Stations in Sweden. The stations are indicated in Figure 2. The vectors correspond to different 45° sectors of transport directions, with the length of the vector indicating the mean value of the concentration during the number of days indicated. *Source:* Rodhe et al. 1972.

consequences, due to the effects discussed above (nos. 1–5), would follow, with losses up to several hundred million dollars per year by the year 2000 in the case of doubled emissions. The costs of measures aiming to reduce emission levels were found to be high but not excessive compared to the costs of emission effects. It should be observed that the cost estimates of the effects included only direct costs and made no allowance for long-term environmental losses.

Carbon Dioxide

Another major emission from the burning of fossil fuels is carbon dioxide (CO_2). Unlike carbon monoxide, which is emitted by automobiles, for example, carbon dioxide is not toxic. It may affect our

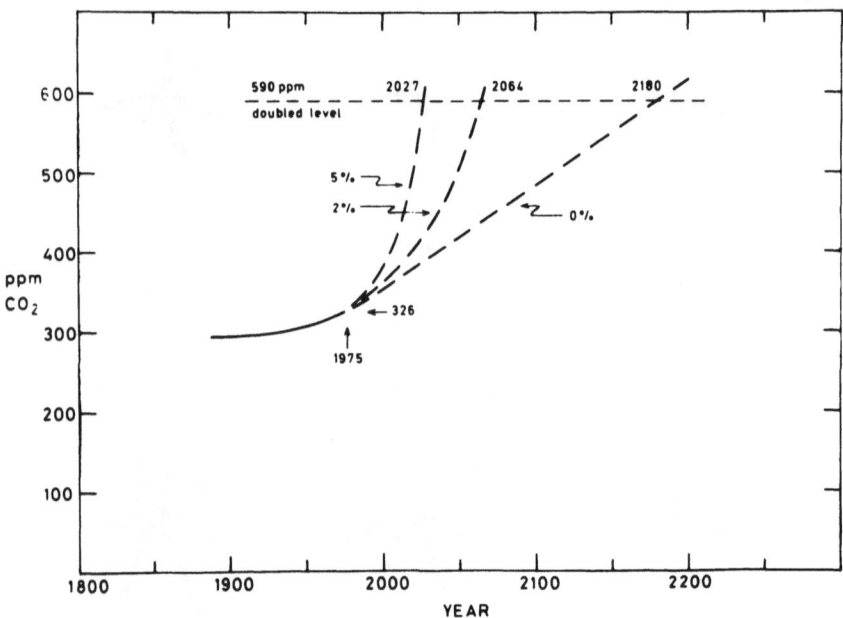

Figure 5. Levels of CO_2 in the Atmosphere. Measured Levels of CO_2 in the atmosphere up to 1975, and projections of CO_2 concentrations during three assumptions of annual percent increase of fossil fuel combustion. Calculations assumed unchanged importance of the oceans. Atmospheric CO_2 would double by the year 2064, given an annual increase in combustion of 2 percent. Zero percent indicates combustion at today's rate. *Source:* Adapted from Bolin 1975.

climate, however (Bolin 1975), and impose an environmental limit on fossil fuel burning.

The average temperature on Earth is determined by a delicate balance between incoming radiation from the sun and the Earth's outgoing heat radiation. The sun's radiation is partly scattered back into space by clouds and other reflecting materials. By changing these reflecting surfaces we may affect the radiation balance. Such a change might result, for example, from large oil spills or clear-cutting of forest lands. Carbon dioxide, a naturally occurring compound, absorbs the outgoing heat radiation from the Earth and thus heats the atmosphere. It is also an important link in ecological cycles. The concentration of CO_2 in the atmosphere was approximately 295 parts per million (ppm) before the industrial revolution (late 1700s) and has now reached 325 ppm. So far, this has had only a small impact on climate, possibly due to counteracting effects of particulate matter emissions. As shown in Figure 5, however, with projected large increases of fossil fuel consumption, concentrations in the natural level of atmospheric CO_2 may double shortly after the year 2000. It is important to observe that the residence time of carbon dioxide in the atmosphere is very long (approximately 1,000 years). This forms the basis for the increasing concentration.

Our knowledge of the climatic system is still too limited to make a firm prediction of the effects we can expect from such an increase of atmospheric carbon dioxide. Calculations have been made, however, that indicate the magnitude of changes we may experience. With a doubled CO_2 level, we can expect an average global temperature increase of 3° C, and much more in the polar regions. This may lead to a rather rapid (approximately ten years) melting of polar sea ice, which would affect the climatic system because much less radiation would be scattered back into the atmosphere by the dark seawater than by the ice. Thus, once melted, the polar ice would probably not refreeze and man will have introduced an irreversible change.

The resultant heating of the polar regions could affect climate in several ways. Evaporation from the oceans might increase and lead to more cloudiness. The slow melting of the large inland ice on Greenland and even Antarctica might occur. A melting of this ice would, unlike the floating polar ice, elevate the water level of the ocean by as much as 60 meters in the case of complete melting. This would require a rather large temperature increase since the temperatures are now well below freezing. The speed with which a melting of this ice might occur corresponds to a sea level increase of roughly 1 millimeter per year.

Is it possible to reach these high CO_2 levels? As indicated above, our knowledge of the systems regulating atmospheric carbon dioxide levels is less than perfect. It is now possible to estimate the effects, although with some uncertainty. The amount of coal required to double the level of CO_2 in the atmosphere is not very large compared to world coal deposits. Thus, we are not limited by the availability of coal, as in the case of oil or gas, but rather by the effects of atmospheric heating. We have an environmental limit which we should not exceed. The estimates of existing coal reserves do not take this into account; they deal with the physical presence of coal only. In planning for the future, man would be wise to consider these risks associated with fossil fuel burning.

Energy Policy for the Future

The primary cause of sulphur, carbon dioxide, and particulate matter emissions is the burning of fossil fuels. This fact forms the basis for concern about accelerated use of, and less stringent controls on, fossil fuel use. An effective solution to this problem would be to reduce the level of energy consumption and/or develop other energy sources, e.g., solar, geothermal, and nuclear. These sources are abundant although they still pose several environmental problems in the case of solar energy, and technological and social dilemmas in the case of large-scale use of nuclear energy.

Man is presently using energy at a rate much lower than the energy flux in natural systems driven by the sun. This suggests that man, by rerouting the great natural flux, can enjoy the consumption of even greater amounts of energy without worrying about its detrimental effects on the environment. This possibility is supported by the observation that much of the energy used is of the same order of energy intensity (that is, measured as watts per square meter) as natural energy fluxes.

The time delays involved in these questions are critical for an understanding of the problems and for design of action programs. A change from man's current dependency on fossil fuel burning to alternative energy sources will take many years. A slow build-up of pollutant levels in the soils, atmosphere, or the oceans may impose no big problem initially, but in the long run, basic changes in the ability of these systems to function may result. As this is an important problem of international dimensions, it calls for cooperative government efforts to establish effective mechanisms for handling it.

References

Bolin, Bert, ed. 1971. *Air Pollution across National Boundaries: The Impact on the Environment of Sulfur in Air and Precipitation.* Report of the Swedish Preparatory Committee for the United Nations Conference on the Human Environment, 1972. Stockholm: P. A. Norstedt and Söners Förlag.

Bolin, Bert. 1975. *Energy and Climate: A Summary of Our Knowledge about*

Those Mechanisms that Determine the Climate of the Earth and the Possibility that Man Directly or Indirectly May Influence the Climate. Stockholm: Secretariat for Future Studies. (Swedish edition: *Energi och Klimat.* Stockholm: LiberFörlag.)

Butcher, Samuel S., and Robert J. Charlson. 1972. *An Introduction to Air Chemistry.* New York: Academic Press.

Esmen, Nurtan A., and Morton Corn. 1971. Residence time of particles in urban air. *Atmospheric Environment* 5(8):571-578.

Friedlander, Sheldon K. 1973. Chemical element balances and identification of air pollution sources. *Environmental Science and Technology* 7(3): 235-240.

Lindén, Kurt, Sören Mattsson, and Bertil Persson. 1974. *Strålande Miljö.* Lund, Sweden: LiberLäromedel/Gleerup.

Rodhe, Henning, Christer Persson, and Ove Åkesson. 1972. An investigation into regional transport of soot and sulfate aerosols. *Atmospheric Environment* 6(9):675-693.

Whitby, K.T. 1973. Eighth International Conference on Nucleation (Leningrad, 28 September 1973).

17. The Impact of Urban Air Pollution on Remote Regions

John W. Winchester
Florida State University, U.S.A.

Introduction

Air pollution due largely to the profligate consumption of fossil fuels causes great damage to health and property in cities. It is not generally realized, however, that the pollution probably is doing irreparable damage to wilderness and remote marine areas as well. For, while the atmosphere can dilute and thereby lessen the immediate impact of pollution, it can also transport pollution to remote regions. The chronic exposure of ecosystems in remote regions to low levels of pollution affects the health of those ecosystems, but it is often difficult or impossible to obtain direct proof of this fast enough to provide for their protection. Therefore, we should regard the detection of pollution in remote regions as a warning of potential hazard.

Since it is difficult to distinguish pollutants from the atmosphere's natural components because they occur in low concentrations and because their presence is masked by the interference of other compounds, it is not feasible to investigate all potential pollutants. Therefore, we should endeavor to predict the occurrence of pollutants in remote regions on the basis of what we know about the behavior of pollutants in the areas where they originate. Moreover, we should take advantage of inadvertent or large-scale geophysical tracer experiments, first to determine pathways by which the atmosphere transports pollutants to remote regions, and then to reduce the emission of the pollutants at their sources.

Dilution and Long-Range Transport

The suggestion that "the solution to pollution is dilution" applies only to cities, not to wilderness areas threatened by the long-range transport of gaseous and particulate pollutants. For, although the atmosphere dilutes pollution near the sites at which fuels are consumed, this capacity is intimately related to the atmosphere's ability to transport pollutants fast and very far and to affect thereby the quality of air in distant areas.

We can now detect sulfur, lead, pesticides, industrial compounds, and radioactivity at remote continental and marine sites, in the polar regions, and in the stratosphere. In most known cases, no deliberate attempt had been made to forecast the speed with which, nor the distance to which the atmosphere would transport the pollution. Hence, the pollutants' effects on natural ecosystems and on people's health can be evaluated only after the fact.

Let us look closely at what is involved in evaluating the consequences of polluted air.

Evaluating the Consequences of Polluted Air

Pollutants released into the atmosphere are diluted considerably, and they ordinarily are transported horizontally quite rapidly. By contrast, pollutants deposited on land tend to remain in and contaminate only the immediate vicinity. Although the degree of contamination may be severe in the latter case, surrounding areas are not affected because the soil has the capacity to contain pollutants. Pollutants released into fresh water, on the other hand, are dispersed and diluted much more rapidly than they would be in soil, so that their concentrations drop rapidly in the immediate vicinity of their source but may increase rapidly in surrounding areas. This is so because water, unlike soil, has no natural capacity to contain pollutants. Hence, radioisotopes and petroleum spilled into the ocean eventually are dispersed worldwide; this kind of thing already is occurring. For example, radioisotopes from nuclear blasts set off in the Pacific Ocean have been found in the Atlantic Ocean, and oil-spills that occurred far at sea have contaminated beaches hundreds of kilometers away in all parts of the world.

Pollutants released into the atmosphere disperse even faster than those released into water, and may travel entirely around the Earth in a matter of days if they are emitted much above ground level.

However, under certain conditions, the atmosphere does retain pollutants for a long time. Because there tends to be a much slower exchange of material (mixing) between major air masses than within single air masses, pollutants are not thoroughly or evenly dispersed throughout the atmosphere. If we fail to take into account the atmosphere's natural capacity to retain pollutants in this way and to prevent thereby their complete dispersal, we may overestimate the atmosphere's capacity to dilute pollution. I now will describe three cases of the natural containment of pollution by the atmosphere.

Three Cases of Natural Containment

Three kinds of natural containment concern us here. The first type, which occurs during a reversal of the usual situation in which air nearest the ground is warmer than that above (a situation known as a "temperature inversion") is a within-air-mass phenomenon. The second type, due to the slow mixing of Northern and Southern Hemisphere air, is obviously a between-air-masses phenomenon. The third type of natural containment of pollution by the atmosphere involves the vertical mixing of air between the lower atmosphere (the troposphere) and the stratosphere.

Case 1: Temperature Inversions

Under stable weather conditions, colder air near the ground does not mix readily with warmer overlying air; hence, pollutants introduced into the air at ground level tend to stay there and to remain at high concentrations. The result is smog. Also, during stable conditions, smoke plumes do not dissipate into the surrounding air, but trail out in discrete, thin and concentrated streams.

Case 2: Inter-hemispheric Transport

Owing to the major, global-scale patterns of atmospheric circulation, air in the lower atmosphere moves very slowly between the Northern and Southern Hemispheres. Thus, for example, radioactivity released by the detonation of nuclear weapons, and lead emitted by automobiles and other vehicles, remain in one hemisphere for months, and are removed by rain before they diffuse into the other hemisphere to any significant degree.

Case 3: Stratosphere-to-Troposphere Transport

Generally, material moves between the lower atmosphere (the troposphere) and the stratosphere very slowly. Many pollutants remain in the troposphere for only a few weeks before they are removed by rain, but may remain in the stratosphere for several years before they slowly descend into the troposphere, where they are washed out by rain. Radioactivity from nuclear weapons set off years ago is still leaking from the stratosphere into the troposphere and causing new pollution whenever and wherever it rains.

The Effects of Pollutants on Remote Areas

In the long run, the key criterion for evaluating the impact of air pollution on the quality of air is the extent to which pollutants significantly affect people's health or the biosphere's ecological balance. In remote areas, pollutants may pose a direct hazard to man if, for example, people should eat tainted food or breathe impure air. Pollutants may also prove toxic to plants and animals. Acidic precipitation, for example, can reduce the fertility of soils and acidify streams beyond the point that fish can tolerate. Furthermore, components of the physical environment may be altered as well—such as temperature and the character (i.e., both the quantity and quality) of sunlight—which in turn could induce changes in organisms and ecosystems.

For example, a higher concentration of carbon dioxide in the atmosphere due to the accelerating use of fossil fuels, or contamination of the stratosphere with other compounds, may alter the atmosphere's composition and cause increases in its temperature or the amount of ultraviolet light that reaches the lower parts of the atmosphere and the surface of the Earth itself. The greater quantities of ultraviolet light would burn unprotected skin and could also drastically affect wilderness ecosystems. Unfortunately, these kinds of effects are very difficult to verify or assess, and it would be an impossibly large undertaking to investigate all of the potentially damaging effects of low levels of pollutants.

Therefore, except for a few impacts that now are believed to be critical—those of mercury and pesticides, for example—it is sufficient merely to have detected a pollutant in a remote region to justify being alert to signs of damage caused by it. Generally, it costs less to restrict the rate of pollution than it does to conduct research to determine whether it is potentially hazardous pollution. Hence, a key short-term objective ought to be the ability to predict the occurrence and to detect the presence of potentially harmful pollutants in remote areas before full damage has been inflicted.

Distinguishing Pollutants from Natural Substances

It often is difficult to detect pollutants in the atmosphere far from their origins in cities or industrialized areas, or to distinguish them from the natural components of the atmosphere. There are three principal types of difficult to detect or distinguish pollutants, namely, (1) pollutants that are also naturally occurring compounds, (2) pollutants whose chemical characteristics resemble those of naturally occurring compounds, and (3) pollutants that occur in such small amounts that they cannot be detected with current techniques.

Pollutants That Are Naturally Occurring Substances

A pollutant may be an otherwise naturally occurring substance such as a heavy metal or a trace gas, and its "natural-source strength" (or "background level") may not be known. Therefore, detection of this type of substance in a remote area is not proof that it is a pollutant transported there from a city or industrialized area. Carbon monoxide is an example of this type of substance. Until recently, the combustion of automotive fuel was believed to be the world's largest source of carbon monoxide, and the carbon monoxide detected in remote areas was assumed to have come from cities. However, several natural sources of carbon monoxide now are recognized, among them the soil, the surface of the sea, and reactions among the other gaseous components of the atmosphere. We now believe that only a very small proportion of the carbon monoxide in remote areas is pollution, and that it is of little ecologic consequence.

Pollutants That Resemble Naturally Occurring Substances: Organic Industrial Compounds

Some pollutants have chemical properties similar to those of naturally occurring substances; hence, they may be difficult to identify through chemical analysis. This problem is especially acute in the case of organic industrial compounds that have been transported by the atmosphere to the oceans and wilderness areas.

For example, until the late 1960s, we measured DDT and its degradation products in the oceans and the atmosphere without suspecting that the polychlorinated biphenyls (PCBs) were interfering with our analytical techniques; because of this, we overlooked PCBs. Now, however, we can identify PCBs with improved techniques, and we have discovered that they are more abundant in many remote areas than DDT is. Thus, our attention previously had been directed toward only one of two polluting industries.

Petroleum Hydrocarbons. The detection of petroleum hydrocarbons in the atmosphere and in marine organisms is another example of how hard it is to distinguish pollutants that resemble naturally occurring substances from the naturally occurring substances. Because hydrocarbons are a mixture of many different compounds, we have to determine the pattern of the compounds' relative abundances in the petroleum-hydrocarbon mixtures in order to measure them. Hydrocarbon compounds also occur naturally, but in patterns of relative abundance different from those of pollutants.

Pollution hydrocarbons can be distinguished from naturally occurring hydrocarbons only if they make up more than a small percentage of the mixture. When they are present in smaller amounts, they may still be significant as pollutants, but their presence is masked because the composition of the natural "background" material fluctuates somewhat.

It is a major undertaking to conduct a thorough investigation of organic pollutants with sensitive and sophisticated techniques, and since scientific manpower is limited, most environmental research on organic pollutants so far has been directed toward the critical pesticides and petroleum hydrocarbons. This means that, at present, many other industrial chemicals, some of which may prove hazardous in the future, are being overlooked. We did not realize the importance of chlorofluoromethanes in the atmosphere until 1974, for instance, because we were preoccupied with other work. To avoid such oversights in the future, we should require anyone with special knowledge about the consequences of introducing a particular commercial compound into the atmosphere to conduct a preliminary investigation of their impacts on the quality of air—not only of air in urbanized areas but of air in remote and wilderness areas as well. We should assign to polluters—and not to the scientific community—responsibility for environmental disturbances caused by new pollutants released into the atmosphere.

Pollutants That Occur in Very Low Concentrations

In many cases, a pollutant may not occur naturally, and its detection therefore is free of the interferences that complicate chemical analysis, but it may be present in concentrations so low that existing techniques cannot detect it. Radioactivity pollution, although very low in mass concentration, may be the easiest pollutant to detect because nuclear-particle detectors are very sensitive. Considerable research on the dispersal of radioactivity by the atmosphere has led to a fair understanding of the pathways by which it is transported across the globe. This is not true, though, for some nerve gases, which remain in the environment for long periods of time and may be toxic at concentrations well below those at which even the best analytical techniques can detect them.

The number of nonnatural compounds manufactured by our technology is extremely large and growing; many of them are becoming widely disseminated throughout the global environment and may be causing obvious damage even though they themselves may not have been directly detected. Environmental scientists are continuously on guard for new pollutants as advances are made in detection methods. The record of the last ten years shows that each year we become aware of pollutants that we did not suspect the year before. And next year's pollutants may be different from anything we know now.

Tracer Experiments

The effects of pollutants on wilderness areas depends upon the efficiency with which the atmosphere disperses them outward from cities and industrialized areas. The efficiency generally is high, but is variable; the variableness must be thoroughly understood before we can predict the pollutants' most probable trajectories through the atmosphere.

Fortunately, we have been able to study the dispersion of pollutants on continental and global scales by taking advantage of the inadvertent addition of selected pollutants. By watching their concentrations build up in remote areas over a period of time, we find out how the pollutants are transported by the atmosphere. Three major types of pollutant have been particularly useful for studying the long-distance dispersal of pollutants, namely, radioactivity released by nuclear explosions, fossil-fuel carbon, and trace elements released in the combustion of fossil fuels.

Nuclear explosions release radioactivity into the upper portion of the troposphere and the stratosphere; they permit us to trace the transport and removal of particles from the atmosphere by rain and to determine the rate of mixing between stratosphere and troposphere.

Fossil-fuel carbon, which, unlike that of the air, does not

contain carbon-14, is now being added to the atmosphere in large quantities and can be followed by radioactivity-counting techniques. The measurements obtained by those techniques can be used to determine the rate at which the ocean and the atmosphere exchange gases. The rate of mixing within the ocean also can be determined.

Trace elements from the burning of fossil fuels and from fuel additives, including sulfur and lead, have been found to contaminate the polar regions; hence, recently fallen snow in Greenland contains more of these substances than snow that fell a century ago. This clearly demonstrates that the atmosphere transports pollutants to remote areas.

We now have new types of contaminants that will be of value in determining how urban pollution affects remote regions, such as Freons (chlorofluoromethanes $CFCl_3$ and CF_2Cl_2—Freon-11 and Freon-12, respectively), which are used as propellants in aerosol-spray cans, as refrigerants, and so on. While Freons have been used for several decades now and therefore really are not new to industry, the serious hazards they pose have been recognized only very recently. Freons have been found to migrate upward into the stratosphere at a rate that will cause their steady buildup at high altitudes for decades, even if we stop releasing them immediately. While they are chemically inert in the lower atmosphere, in the stratosphere they enter into complex chemical reactions induced by ultraviolet light and reduce thereby the concentration of ozone in the so-called "ozone layer." It is unfortunate that this case of inadvertent pollution could not have been predicted long ago, for attenuating the ozone layer may permit so much additional ultraviolet light from the sun to reach the Earth's surface that people's skin will be burned, the incidence of skin cancer will increase, and other serious biological damage will occur. The problem appears to be a grave one and will almost certainly lead to a better appreciation of the degree to which remote areas are vulnerable to pollutants that have been transported by the atmosphere far from their sources.

Perhaps the largest deliberate atmospheric tracer experiment—exceeded in magnitude only by our inadvertent modification of the atmosphere—is the Isotopic Lead Experiment (ILE) conducted by the Joint Research Center of the Commission of European Communities in Ispara, Italy, which was scheduled to have begun in mid-1975. The ILE calls for replacing all lead used in the manufacture of tetraethyl lead with lead of special isotopic composition that can be traced over long distances because it is easily distinguished from background lead emanating from other sources. Thus, Italy will become the source of a tracer emitted in automobile exhaust. The ILE will contribute much to the world's knowledge—not only about the effects of fine-particle lead on people's health, but also about the transport of other atmospheric pollutants, and it should reduce substantially our ignorance about the impacts of urban air pollution on remote areas as well.

Selected Bibliography*

Committee on Impacts of Stratospheric Change. 1976. *Halocarbons: Environmental Effects of Chlorofluoromethane Release.* Committee on Impacts of Stratospheric Change, Assembly of Mathematical and Physical Sciences, National Research Council. Washington, D.C.: National Academy of Sciences.

Panel on Atmospheric Chemistry. 1976. *Halocarbons: Effects on Stratospheric Ozone.* Panel on Atmospheric Chemistry, Assembly of Mathematical and Physical Sciences, National Research Council. Washington, D.C.: National Academy of Sciences.

---

*Prepared by the editor.

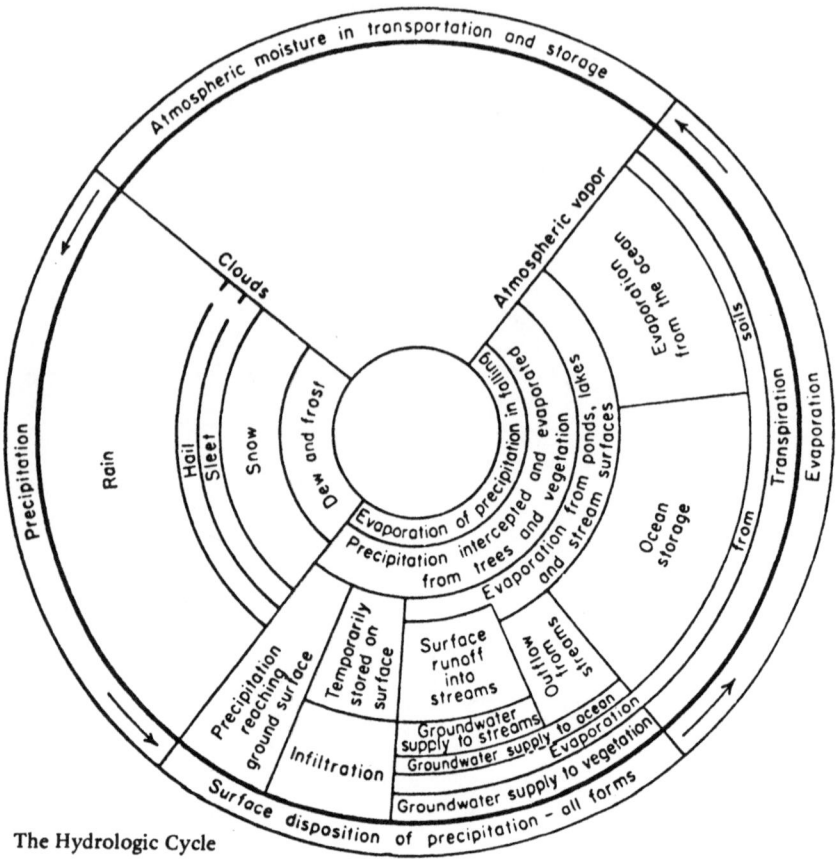

The Hydrologic Cycle

Tu estatua está extendida más allá de las olas.

—Pablo Neruda
"El Gran Océano"

The waters flow eastwards from their sources, resting neither by day nor by night. Down they come inexhaustibly, yet the deeps are never full. The small streams become large and the heavy waters in the sea become light and mount to the clouds. This is part of the Rotation of the Tao.

—Lü-shih ch'un-ch'iu

The Hydrologic Cycle: The Oceans

18. The Third United Nations Conference on the Law of the Sea and the Protection of the Marine Environment

Frank X. Njenga
Permanent Mission of Kenya
to the United Nations

Introduction

The entire international community has come to the realization that it is no longer a question of whether we should protect the marine environment—which occupies more than three-fourths of the Earth's surface—but one of whether we can any longer afford to treat the oceans as a cesspool and hope to sustain life in the world as we know it.

I believe it is the general realization, not only of environmentalists like those represented here, but also of all the participants at the Law of the Sea Conference, that protection of the marine environment is an urgent issue and that adequate solutions must be developed as soon as possible, before we reach the point of no return.

Nevertheless, the United Nations Conference on the Law of the Sea has been criticized as having so far relegated the issues of the

---

The views expressed in this paper are the responsibility of the speaker and do not necessarily represent the position of the Kenya government.

preservation of the marine environment to a secondary role. This seems to have been the prevailing view during the recent session of the Governing Council of the United Nations Environment Programme (UNEP) in Nairobi in April 1974; indeed, a resolution was adopted urging the United Nations Conference on the Law of the Sea to attach the highest priority to its efforts to incorporate in draft treaties effective provisions for the protection of the marine environment.

Problems Encountered at the Conference

The concern that it is necessary to take urgent measures to protect the environment by the Law of the Sea Conference is understandable, but I believe it would be unfair to accuse the conference of not having taken its work as seriously as it should have in this regard. The difficulties encountered in adopting effective and universally acceptable articles on protection of the marine environment stem from the very nature of the conflicting interests involved.

First, the fundamental issue concerns not merely the uses of the sea itself, which is the primary concern of the conference. It is the general understanding that approximately 80 percent of all pollution in the marine environment emanates from natural or human activities on the land environment. Consequently, no solution based purely on regulating the seas can be effective in protecting the environment. The conference has agreed that each state should take all necessary measures to prevent, reduce, and control pollution of the marine environment from any source, whether from the release of toxic, harmful, and noxious substances from land, the atmosphere, or vessels, or from installations and other devices used to explore and exploit the natural resources of the seabed and subsoil within and beyond national jurisdictions.

Second, another problem in the protection of the marine environment is the conflict between environmental and developmental considerations, since the latter, of necessity, have deleterious effects on the environment. A delicate balance has to be drawn in evolving regulations to protect the marine environment to ensure that they do not stifle development, particularly that of developing countries.

Still a third conflict exists, that between the maritime powers and navigational interests—whose interests lie in minimizing controls—

and the other coastal states, which stand to lose most from pollution generated by ships in the areas within their national jurisdiction. The former have emphasized that the primary role in the protection of the environment should be played by the flag states and the latter have been equally adamant that the coastal states should play the primary role in making and enforcing the regulations within their respective national jurisdictions.

Progress Made During the Third Session

Because of these problems, the work of the conference has proved difficult during both the second session in Caracas and the third session in Geneva. Essentially, the Geneva session was a continuation of the efforts begun in Caracas, and the working methods adopted there were the same. The methods involved a combination of informal consultations under the able chairmanship of Mr. José Vallarta of Mexico and formal sessions of the Third Committee under the chairmanship of the highly experienced Ambassador Yankov of Bulgaria. In both informal and formal sessions, participation was open-ended. In addition to these consultations at the official level, there were numerous group and intergroup meetings, which tended to complicate the process of consultation even further, but which were necessary to arrive at broadly acceptable formulations.

While no articles can be considered adopted until the work is completed, the Third Committee in Geneva arrived at a consensus on three basic articles in addition to the generally agreed obligation of states to protect and preserve the marine environment. The three articles deal with monitoring, environmental assessment, and standards for land-based sources of marine pollution.

Draft Article on Monitoring

It has been generally agreed that all states need to keep under surveillance pollution risks and the deleterious effects of all activities that they undertake, authorize, or control. UNEP has already launched the "Global Environmental Monitoring System Project" (GEMS) following a decision of the Governing Council at its first session in June 1973, and this has been welcomed by the conference.

The following article on monitoring was adopted by the Working Group of the Third Committee:

1. States shall, consistent with the rights of other States, endeavour, as much as is practicable, individually or collectively through the competent international organizations, to observe, measure, evaluate and analyse, by recognized methods, the risks or effects of pollution of the marine environment.
    In particular, States shall keep under surveillance the effect of any activities which they permit or in which they engage, to determine whether these activities are likely to pollute the marine environment.
2. States shall provide, at appropriate intervals, reports of the results obtained relating to risks or effects of pollution of the marine environment to UNEP or any other competent international or regional organizations, which should make them available to all States. (Document A/Conf.62/C.3/L.15/Add.1)

Draft Article on Environmental Assessment

The Working Group also considered that, in addition to monitoring pollution risks and effects, it is also necessary to assess pollution risks and effects before the activities are undertaken or authorized. The following provision is the result of the consideration of proposals on the subject of environmental assessment. It was originally proposed in connection with monitoring:

1. When States have reasonable grounds for expecting that planned activities under their jurisdiction or control may cause substantial pollution of the marine environment, they shall, as far as practicable assess the potential effects of such activities on the marine environment and shall communicate reports of such assessment in the manner provided in paragraph 2 or article VIII (monitoring).
2. States shall directly or through competent international or regional organizations, on request, provide appropriate assistance in particular to developing countries concerning the preparation of such environmental assessments.

Standards for Land-Based Sources of Marine Pollution

Approximately 80 percent of marine pollution originates from natural or human activities taking place on the land environment. It was previously agreed in Caracas that nothing in the convention shall derogate from the sovereign rights of a state to exploit its own natural resources pursuant to its environmental programs and policies for economic development and in accordance with its duty to protect and preserve the marine environment. During the Geneva session, a draft article reflecting this duty has been more or less agreed to as follows:

1. States shall establish laws and regulations to prevent, reduce, and control pollution of the marine environment from land-based sources including rivers, estuaries, pipelines, and outfall structures, taking into account internationally agreed rules, standards, and recommended practices and procedures.
   States shall also take such other measures as may be necessary to prevent, reduce, and control pollution of the marine environment from land-based sources.
2. States shall endeavour to harmonize their national policies at the appropriate regional level.
3. States, acting in particular through the appropriate intergovernmental organizations or by diplomatic conference, shall endeavour to establish global and regional rules, standards, and recommended practices and procedures to prevent, reduce, and control pollution of the marine environment from land-based sources, *taking into account characteristic regional features, the economic capacity of developing countries and their need for economic development.*
4. Laws, regulations, and measures, and rules, standards, and recommended practices and procedures referred to in paragraphs 1 and 3 respectively shall include those designed to minimize to the fullest possible extent the release of toxic and harmful substances, especially persistent substances, into the marine environment.

Developing countries consider the italicized words to be important, since global rules, standards, and recommended procedures for pollution control should not be used to unnecessarily hinder or interfere with a state's rights or needs for economic development. Nevertheless, each state should be expected to ensure that environmental factors are given adequate attention in making decisions concerning various activities. What is necessary therefore is that steps be taken on an international level, particularly to assist the developing countries to undertake economic activities, using environmentally sound methods or technology, but at the same time doing so efficiently and profitably.

Pollution from the Dumping of Waste at Sea

One of the most direct and common sources of pollution of the marine environment is pollutants originating from land or from man-made structures, through the dumping of waste and other materials at sea. Dumping was the subject of the 1972 London Convention on Dumping, which the Working Group of the Third Committee took into consideration. However, there is still disagreement on the definition of dumping. In the London Convention, dumping is defined as "any deliberate disposal of wastes and other matter at sea from vessels, aircraft, platforms or other man-made structures at sea, and any deliberate disposal at sea of vessels, aircraft, platforms or other man-made structures at sea. . . ." The definition does not incorporate the disposal of waste or other material incidental to or derived from the normal operation of vessels, aircraft, platforms, or other man-made structures. Many delegations consider this to be a serious omission because the second type of activity can be as harmful to the environment as the first. Consequently, the question of defining "dumping" remains. Nevertheless, the following article has been provisionally accepted:

> States shall establish national laws and regulations to prevent, reduce, and control pollution of the marine environment from dumping of wastes and other matter.
>
> States shall also take such other measures as may be necessary to prevent, reduce, and control such pollution.
>
> Such laws, regulations, and measures shall ensure that dumping is not carried out without the permission of the competent authorities of States.
>
> States acting in particular through the competent intergovernmental organizations or by diplomatic conference, shall endeavour to establish as

soon as possible and to the extent that they are not already in existence, global and regional rules, standards, and recommended practices and procedures to prevent, reduce, and control pollution of the marine environment by dumping of wastes and other matter.

Dumping of wastes and other matter... shall not be carried out without the express approval of the coastal State, which has the exclusive right to permit, regulate, and control such dumping....

Areas of Disagreement

Before I refer to the areas of disagreement, I should like to draw your attention to document A/CONF.62/WP.8/Part III of 6 May 1975, entitled "Informal Single Negotiating Text." This document contains forty-four articles on the protection and the preservation of the marine environment, incorporating the agreed articles referred to above. It was prepared at the responsibility of the chairman of the committee. But, drawing upon the results of the work done by the working group and the committee itself, as well as other proposals from various regional groups and other formal proposals submitted to the conference by delegations, it should not be read either as a negotiated text or as a consensus document. Its status as a document that was presented at the end of the conference, and consequently as one not yet discussed by the conference at all, should not be overemphasized. Where it reflects a broad agreement among the conference participants it can be considered a useful basis for the final agreement at the conference, but where it has taken a decidedly partisan approach to the issues involved it should be considered as only one of the numerous proposals before the conference.

This is said without any reflection on the good faith of the chairman of the conference, who had to work under very difficult conditions to meet a deadline imposed upon the chairmen of the three committees by the Plenary to provide a procedural device as a basis for negotiation before the end of the conference.

Pollution Due to Exploration and Exploitation of the Seabed

It is generally agreed that the coastal state should establish national laws and regulations to prevent, reduce, and control pollution of the marine environment arising from the exploration and exploitation of the natural resources, such as hydrocarbons, within the areas of national jurisdiction. It is also agreed that in enacting such laws and regulations each state shall take into account generally accepted rules, standards, and recommended practices and procedures and shall endeavour to harmonize their regulations at the regional level. The difficulties continue to persist as to the scope of the application of such rules, however, with some major powers insisting that the regulations should apply only to devices for the exploration and exploitation of the resources and not to non-resource-oriented activities in or on the seabed, such as placing research or security devices or equipment. The majority of the countries, however, believe that all activities that potentially could pollute the marine environment should be encompassed.

The controversy about the scope of the regulations extends to the area of the seabed beyond national jurisdiction. However, there is general agreement that in the international area of the seabed the international authority should enforce, in cooperation with the flag state, the rules and standards for the protection and preservation of the marine environment from pollution due to exploration and exploitation. In this disagreement about the scope of the application of the rules and regulations, the Single Negotiating Text has tended to side with the position of the "superpowers."

Vessel-Source Pollution

The controversy about vessel-source pollution concerns the capacity both to devise regulations to control vessel-source pollution within the territorial sea, the exclusive economic zone, and the high seas, and to enforce them. The majority of states have insisted that regulations to control this type of pollution should be agreed to internationally and that the coastal state should not have the power to make additional regulations except in certain clearly defined, special areas that should also be agreed to, and rules should be made by the competent international organization. This is in the recognition that pollution cannot be contained within national boundaries, and to be effective rules must be generally applicable without, however, prejudicing the need for regional regulations where applicable, as in the Baltic or the Mediterranean.

Certain countries have, however, insisted on the right to make national regulations affecting areas within their national jurisdiction where particularly severe climatic conditions create obstructions or exceptional hazards to navigation. This is the case, for instance, in the Canadian Arctic. Almost all coastal states, including developing coastal states, were in agreement with the need to have uniform regulations to control marine pollution and hence with the need to have internationally agreed standards. This cannot be taken to prevent a coastal state from augmenting the internationally agreed regulations within its territorial sea, where this right already exists. As for areas with severe climatic conditions, many countries sympathize with the rights of the coastal state to make regulations to combat marine pollution for the particular region, but believe that this has to be carefully circumscribed to prevent the proliferation of such special areas which discriminate against certain types of vessels. One safeguard might be the possibility of reviewing such rules and regulations, including the designation of such special areas by a competent international organization.

The great difficulties in controlling vessel-source pollution have been in connection with the question of enforcement. The major maritime powers and shipping interests insist that the present system of flag-state enforcement should continue to be the primary method of enforcing the generally accepted international regulations in the territorial sea, the exclusive economic zone, and the high seas. The only concession they seemed to have agreed to is the possibility of port-state enforcement, whereby the coastal state is to be given certain restricted rights to enforce voluntarily pollution regulations on ships within their ports or at offshore terminals within their jurisdiction. Even here, enforcement is restricted primarily to investigating the violation and notifying the results of the flag state, with certain limited rights or court proceedings where the flag state fails to enforce its primary jurisdiction.

The majority of the coastal states, including all the developing countries, have rejected this approach because it tends to negate one of the cardinal aspects of the exclusive economic zone: the right of the coastal state to protect the resources within its national jurisdiction. Such an approach also tends to ignore the nature of the exclusive economic zone, which is not part of the high seas.

At present, no one would challenge the fact that the coastal state has, under the existing law, the right to enforce pollution-control measures within its territorial waters. This seems to have been put into doubt, however, by some of the proposals before the conference, and in particular by the draft articles proposed by nine developed European powers (Document A/CONF.62/C.3/L.24, cosponsored by Belgium, Bulgaria, Denmark, the German Democratic Republic, the Federal Republic of Germany, Greece, the Netherlands, Poland and the United Kingdom). This proposal—which seems to have been used to the exclusion of proposals on vessel-source pollution that were already made in the provisions of the Informal Single Negotiating Text—would deny the coastal state any right of enforcement in the exclusive economic zone also. Such an approach offers no hope of resolving this issue at the forthcoming session of the conference.

To begin with, the flag-state jurisdiction favored by the maritime powers has proven totally ineffective in combating pollution, a fact to which anyone who has bothered to observe the oceans and the beaches can attest. There is no reason to believe that it will be more effective in the future. With the prevalence of the flag-of-convenience practice, most of the flag states are in no position to enforce any control over their ships (some of which have never even called at their ports of registration) nor, in many cases, do they have the will to do so. In any case, the direct interest that generates the will to enforce pollution-control measures does not exist in the case of a flag state that is many thousands of miles away from the areas likely to be affected by pollution. Even with the best of intentions, the state would be unduly hampered by the difficulties of obtaining acceptable evidence for conviction of the violators.

The coastal state, on the other hand, stands to be dramatically and irreparably harmed by major incidences of pollution off its coasts, and would therefore be motivated to ensure compliance with internationally agreed rules. The often expressed fear that the coastal state's enforcement would lead to unnecessary interference with navigation is real, but it has tended to be overstated. All the coastal states, and particularly all developing countries, depend heavily on navigation for their trade, and they stand to lose most from interference with navigation that would lead to a rise in freight rates. Therefore, they should be expected not to act irrationally.

If we are serious about preserving the marine environment, we cannot afford to allow irresponsible activities by some of the ship owners, many of whom, to maximize their profits, treat international regulations with utter contempt. The very possibility of enforcement by the coastal state would provide a deterrence and an incentive for compliance.

Responsibility and Liability

While there is general agreement that states have the responsibility to ensure that activities conducted under their jurisdiction do not damage the marine environment under the jurisdiction of other states or within the international area, there is disagreement as to the nature and scope of such responsibility. To be effective, however, many states believe that the states should bear the primary responsibility for activities conducted under their jurisdiction, including responsibility for ships under their registration, and should be responsible for the payment of compensation to clean the environment once it is damaged. The best safeguard for the marine environment, however, must remain in prevention rather than cure.

Conclusions

In conclusion, I hope I have demonstrated that whereas the Law of the Sea Conference has not as yet successfully negotiated the issue of protecting the marine environment, it is already engaged in very serious negotiations for this purpose. It should be borne in mind that the conference cannot be expected to develop comprehensive and detailed rules for protecting the marine environment, but only to provide the framework—a rational framework—within which effective means for the protection of the marine environment can be worked out by the competent international organizations. The best services that the states and interested organizations, such as the cosponsors of this conference, can render to the Law of the Sea Conference is to give it all possible encouragement to continue and conclude its efforts as soon as possible.

It is easy to be discouraged at the apparently slow rate of progress and to recommend that states take unilateral measures to protect the environment pending successful outcome of the conference. However, so long as the conference has reasonable chances of success (and I believe that at present the chances are very good), unilateral measures can have only a negative effect. Unilateral measures would encourage each state to take measures of its own to protect its own environment. The result would be a proliferation of dissimilar rules and regulations, leading to chaos in the protection of the marine environment. As I have tried to show, the marine environment is an integrated whole and requires uniform regulations for its effective protection. It is easy to take a unilateral action, but once one is taken, governments adopt defensive attitudes and fight against any contrary solutions that might imply the regulations they have adopted are irresponsible.

We do not have much time. We must begin thinking now about the period after the Law of the Sea Conference, the period during which the provisions adopted by the conference for protection of the marine environment will be implemented. We must start thinking of the best method of avoiding the spectre of "double standards," because the developing countries will find it difficult if not impossible to adopt any provision that might block their goal of economic development.

Yet the answer, in my opinion, is not to allow "double standards" whereby developing countries will be able to adopt standards lower than those accepted internationally. That would negate the whole effort. The answer lies in the international community's accepting the joint responsibility to assist the developing countries meet international standards by the meaningful transfer of technology, and in its providing technical assistance on both multilateral and bilateral levels. Serious consideration should be given to the possibility of establishing an "International Maritime Environmental Fund" through which such assistance could be channelled.

19. The Great Ocean Debate: A Clash of Realities

Arthur G. Bourne
Flitwick, England

The importance of the oceans to the welfare and even the survival of mankind in other than economic terms is all too often overlooked in the day-to-day approach to ocean affairs. Little or no consideration is given to the overwhelming threat to man's survival should anything go wrong with the great ocean system; indeed, the very likelihood of such an event is thought by many to be so remote as not to warrant serious, or still less, urgent discussion. And yet, large areas of the world's oceans are changing through the gross interference of mankind.

The negligence of a matter so vital to our interests is suicidal and shows the gulf that separates the politico-legal reality from the stark reality that is the planetary environment. This gulf is, I believe, responsible for the inability of political institutions to grasp the immediacy of the peril, its enormity and pervading presence. It is the reason, too, for our believing in the remoteness of the danger and the persuasion that we are free to deal with what, to us, are seemingly the more pressing social and economic needs of society.

The underlying reason for this state of affairs is that until very recently man-induced changes in the environment went unrecognized, and still go unrecognized, by the majority of mankind. The realization that there is a mounting threat to our survival has been a slow process and is only now being seen for what it is. Unfortunately, the response has been insufficient and many people remain unconvinced of the stature and immediacy of the problem. The politico-legal regime, cushioned as it is in fossilizing institutions, is little affected by the environmental reality. It is only on those occasions when environmental factors force themselves upon the attention of the political world view and threaten immediately and directly the political, economic, and military stability of a nation or nations that action is taken—and then usually in an ambience of panic, with the consequence that the situation is aggravated or the process accelerated.

An example of such a situation, though only indirectly environmentally generated, was that caused by the recent behavior of the Middle Eastern oil-producing states. Their action in turning off the oil faucets seriously threatened our oil-dependent civilization. Though it was perfectly well known that one day in the not too distant future the world's oil resources would run out, no one had made any serious moves towards developing alternative energy sources. As a result, when the oil supplies were suddenly curtailed, which was analogous to *a trial run for a future reality,* the oil-hungry world was caught unprepared. Even following such a warning, the immunological response, though violent, was short-lived. But this initial response demonstrated the sensitivity of the situation, and how governments overreact, with a consequent worsening of the environmental problems. A dramatic case of overreaction has been the almost obscene rush to extract the oil from beneath the sediments of the North Sea, with its concomitant interference with and potential pollution of the living resources of that valuable piece of ocean. This is not to say that governments never react in time or with forethought, but rather that, when they do, it is the rare occasion. Very often, in recent years, it has been the collective pressure of the citizenry who, having become aware of the pressing environmental problems, have made the politicians move.

The United Nations Conference on the Human Environment was a response of world governments to mounting public pressure and a worsening crisis, many of the symptoms of which had manifested themselves in the oceans. In several respects the conference seemed to offer some governments a means to defuse a troublesome situation. Subsequent history has not allayed that suspicion, but the conference had a graver flaw. The worm was in the Stockholm apple, for, because of either an inability to recognize it or through fear of admitting it, the most fundamental problem facing mankind was not discussed at that hopeful conference.

The greatest single threat to mankind is mankind. This danger is in two parts; first, man's numbers, which are outstripping the planet's resources, and second, the rate at which technological civilization is affecting his environment. Man is no danger to the environment, for it will adjust as it has done through the millennia; but the question we must ask is: can man himself survive the changes he has wrought? In principle, man's relationship with the environment is similar to that of other organisms; it is only the speed of change that is different. But this, plus his huge demand on the planet's resources, makes man an extremely vulnerable species. However well we recognize these facts, it is proving difficult if not impossible to persuade the mass of mankind in general and those in authority in particular to face the reality of our predicament.

The consequences of this inability or fear has led to the generation of a number of myths, among them that science and technology can provide indefinitely for the needs of the growing and consuming population. One hope born of this myth was the so-called "Green Revolution," which for a little while lived up to its promise but which was doomed in the long run to failure. In our situation, short-term palliatives are long-term catastrophes.

More pertinent to our interest in the oceans is the newer myth that the seas contain an almost infinite wealth of food, energy, and minerals—the premise upon which much of the present United Nations dialogue on the sea rests. This is a most dangerous myth, because it engenders false hopes in the minds of millions, and is beginning to figure larger-than-life in the thinking and planning of governments and industry. How can it be a myth when there are undoubtedly large stocks of minerals, fossil fuels, and living resources? It *is* a myth because even the wealth of the oceans cannot provide indefinitely for the human good and because the most important of those resources, the living resources, are already overtaxed. The advent of undersea mining and all the other activities now being considered will adversely affect them further. More important, the myth is dangerous because it further detracts from the pressing problems of population, wasteful use of resources, pollution, inadequate planning, and unfair distribution of wealth.

Let us for a moment consider the all-important food supply that the oceans provide. The history of fishery exploitation is a story of destruction, of the elimination of fishery after fishery. It is a tale of inadequate regulations inadequately enforced. The demise of the great whales is a sorrowful saga of human greed, though it is instructive—providing a lesson in what happens when a valuable resource is left in the hands of those most likely to benefit from its exploitation. Estimates of the potential food productivity of the oceans are multitudinous; optimists have placed it as high as 200 million tons annually, but the reality is that nowhere near that figure has been or will be achieved, even with the most up-to-date fishing methods. In fact, world catches are going down, not up.

Now that the whales are going, that myth has spawned a new one which says that we may now harvest the krill, the abundant crustaceans of the polar seas that feed the great whales. The argument goes something like this: if the whales have gone, then there must be a surplus of krill, and all that is necessary is for man to take the whale's place and directly harvest the crustaceans himself. Supposing that man takes the place of the whales—and let us imagine that it is feasible that he is successful; it leaves the no mean matter of the energy required to crop, process, and transport the krill. This pushes the idea over the edge of feasibility for many of the hungry, poorer nations. Even if we overcame that problem, which is unlikely, how long would it be before the krill followed the whales into oblivion?

The oceans are no cornucopia; they are limited by the constraints of the planetary environment, and the need for controls on the exploitation of the living resources is long overdue. However, as alluded to earlier, the threat to the oceans' biological resources does not come only from direct exploitation, but also through the indirect interference by man in the form of pollution, through the disruption of nutrient flow in estuarine and coastal zones, and through the activities of offshore oil and mineral extraction. The conflict between these interests is becoming acute in some regions, and the attitude of governments and industry towards the fisheries is far from healthy.

We should never lose sight of the fact that the oceans are the reservoirs for the Earth's water cycles, without which there would be no life. The most important interface on this planet is the boundary between ocean and atmosphere: this is the planetary "lung." It is there that the water enters the hydrologic cycle through evaporation, and it is there that the sun's energy is stored and released as and when the environmental machine requires it: the ocean is our thermal regulator.

The pollution of the Earth's atmosphere is a fact. No place is free of pollutant fallout, and as 70 percent of the planet's surface is ocean, the larger proportion of this fallout must be finding its way onto that vital surface layer, lowering its transparency to solar energy, reducing evaporation rates, interfering with the oxygen and carbon dioxide balance, and entering the oceanic food chains where it can do inestimable harm, with the consequence that life in the sea is poorer and an important food supply of man is impoverished. How callous are those countries which deliberately pollute the oceans near communities which are largely or solely dependent on their own fisheries!

We now have hurried and often ill-conceived plans to exploit the mineral and energy resources of the sea bed and, especially, of the continental shelves, where most of these resources are to be found and which, incidentally, are areas of intense biological activity and the nursery grounds of many commercially important fish. The pollution and the disturbance due to these activities could destroy what is left of the fishery potential in some regions and could have repercussions over a wide area of ocean, especially where migrant species are involved.

The tragedy is that few people question the need to exploit the resources of the oceans: in a world hungry for food, energy, and minerals, any new resource is inviting, but no one asks whether it is really necessary for us to be so ravenous, or whether by exploiting the oceans our hunger will be assuaged. There is sufficient evidence that the answer to both of these questions is no, but it is so much easier to exploit virgin resources than to conserve energy, distribute food more fairly, or recycle metals and other materials through our industrial civilization. We push aside the question of where we will go when the oceans have been exhausted—provided, of course, that by that time the changes brought about by pollution and other man-made disturbances have not already put an end to our existence.

One of the major areas of discussion within the United Nations at present is how we might equitably share the wealth, real or imaginary, of the oceans among the nations. The premise is taken that the oceans and the ocean bed can be thought of in terms of an extension of the land masses—in other words, in political rather than in oceanographic or environmental terms, betraying an ignorance of these realities. While efforts were being made to define continental shelves, continental rises, and continental slopes, the question was raised: Who owns materials of continental origin covering the real ocean bed? How absurd can we get? The saltiness of the sea also has its origins on the continental land masses; who, therefore, owns the water column? Taking this to its ultimate absurdity, who can claim the fish, for their very bones belong to the continents? If we assume for a moment some sort of sharing based on the political arguments, how can we hope to get an agreement when there is not, and never has been, an equitable sharing of the land resources? Therefore, the application and extension of political and legal precedents to the oceans is practically doomed from the start, and the indications are that the Law of the Sea negotiations will prove this point.

We are celebrating, this June day in 1975, a day we have put aside as the one on which we reconsider our attitudes toward and our actions upon this beautiful planet we call home. I suppose, too, that we are carrying on the hope that was generated at, and was the essence of, the Stockholm Conference: a hope that we would take greater care of our planet and that in the future we would be wise enough to discern those trends that were deleterious to its welfare; a hope that would enable us to take collective action to ensure a planet where the safety, health, and happiness of mankind, now and in the future, could be guaranteed, as far as man is able. Despite the flaws in the conference and the subsequent action by certain governments and groups to erode its spirit, we should do all in our power to maintain that hope. We should take stock of the human situation. We should not be satisfied with pernicious political palliatives and should be suspicious of cosmetic measures. Above all, we should not be lulled into complacency or apathy.

Let me try, then, to summarize the situation. First, the disparity between the needs of an increasing global population and decreasing resources is our main concern and constitutes the most intractable problem that mankind has ever faced. Second, although the worst-hit by this disparity are the Third World countries, the shortages are now having an increasing impact on the wealthier, industrial communities. Third, the proposals for solving the problems, which have been legion, have either been failures or fallen short of their predicted successes. The reason for these shortfalls and failures is that governments and the international community at large have not recognized the fundamental crises and have ignored the constraints placed on man by the environment.

Faced with such an unprecedented situation, governments have turned increasingly to the oceans as a potential provider of food, minerals, and energy. Unfortunately, the successes so far obtained in this direction have been achieved at the expense of the ecological integrity of the oceans and good relations between nations.

As we have seen, although the oceans do contain valuable resources, the most important of these, food, is highly sensitive to exploitation, and, extensive as they are, the oceans cannot satisfy the world demand for food. The uncontrolled exploitation of their mineral and fishery wealth can only deplete this potential. However, without the present contribution that the oceans make to the food and mineral requirements of the world's population, man's predicament would be exacerbated.

What, then, are the realities of the situation? In the first place, we have not taken, and are still not taking, sufficient account into our calculations of the natural constraints that the environment imposes upon us. Second, we are allowing the situation to drift because of cowardice, born of the fear of waking up to the reality.

Our problems are the direct result of our mismanagement of ourselves and of the Earth's resources. Unfortunately, a threat has to be tangible before we take action. Yet thousands of seabirds die every year, millions of fish perish, and a hundred thousand men, women, and children starve to death. Are these not tangible enough? Are these the abstractions of the statisticians' imaginations? Must we continue the sleepwalk to total disaster? Must we continue like badly programmed robots? The lawmakers and the politicians must be made to recognize the reality beyond their rarified regime.

If the present attitude prevails that the oceans represent a vast storehouse of resources that can and must be exploited at any cost, without any controls, then the position becomes hopeless. In such an ambience, if the economic and political requirements can be satisfied, then the technological means will be found and the ecological consequences will be hidden in the clamor to win the wealth. The growth of nationalistic ambitions—that malignant tumor—and the concomitant conflict will be a certainty. Nationalistic ambitions threaten the natural evolution of man and his institutions. The retrograde steps are already manifest at the Law of the Sea Conference, where they are threatening even the traditional free passage of ships going about their business. Politics frustrates the endeavors of those who study the seas in the interests of all mankind. But more dangerous than these are the irrational claims that can only lead to further problems and, ultimately, to despair.

However, if we can turn the tables of history and make a constructive beginning at the Law of the Sea Conference, recognize the environmental limitations, and work in a spirit of *real* cooperation, then there will be hope that the other problems of mankind might be treated in a like manner, and we can once again discern that hope that was the intent of the Stockholm conference.

Therefore, it is essential that the outcome of the Law of the Sea negotiations be the safeguarding of the health of the oceans. There must be an international regime of the sea by which the oceans and their resources are conserved and their exploitation—if they are to be exploited—controlled and the benefits of such exploitation distributed fairly among peoples. The alternative is catastrophe. If the lawmakers and political negotiators fail in their task, then we the people must somehow, by our collective efforts, make the reality of the oceans wash through the halls of the politico-legal establishment. Otherwise, our failure will be even greater than theirs.

We have to suppose that we wish man to inhabit Earth for thousands of years to come, and we can presume that if he does so his technological ability will increase so that in some far distant time he will be able to perform the miracles we can now only dream about. We must, therefore—for our own sake and for that of future generations—find a way out of our present difficulties. As a first step, we must ensure the integrity of man and the environment, and we must gear our population to the planet's resources, not chase futile and mythical cornucopias.

If we have the will, we can achieve the seemingly impossible. We have delved deep into the oceans; we have sent our instruments to the far corners of the solar system; we have looked down from space upon our beautiful planet, which gave us life and continues to sustain us. With care, our Earth will continue to serve us through the coming millennia.

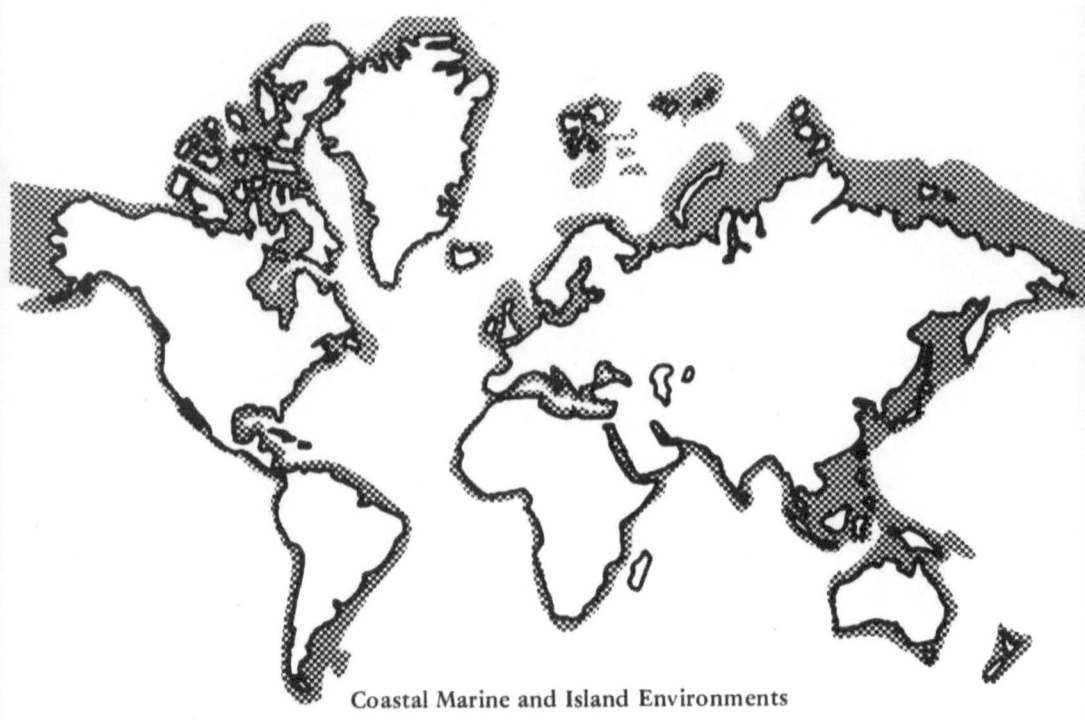

Coastal Marine and Island Environments

The creeks overflow: a thousand rivulets run
'Twixt the roots of the sod; the blades of the marsh-grass stir;
Passeth a hurrying sound of wings that westward whirr;
Passeth, and all is still; and the currents cease to run;
And the sea and the marsh are one.

—Sidney Lanier
"The Marshes of Glynn"

Junto a la sierra florida
Bulle el ancho mar.
El panal de mis abejas
Tiene granitos de sal.

—Antonio Machado

The Hydrologic Cycle:
Coastal Marine and Island Environments

20. Coastal Marine Environments, 1975

Rimmon C. Fay
California Coastal Zone
Conservation Commission, U.S.A.

Introduction

Primitive peoples found abundant food in coastal areas—the interface between land, sea, and air. Later, shoreline sites became centers of population, commerce, and manufacturing. Coastal areas have, in succession, been exploited for their fishery resources, developed for harbors and coastal works, used for the disposal of wastes, employed as sources of cooling water for powerplants, filled, dredged, diked, and degraded to the point that many are scarcely functional ecosystems. Coastal marine environments have suffered more at the hands of man than any other part of the Earth. In the context of resource management, this is truly puzzling—and it could also be considered prophetic of the ultimate outcome of man's capacity to affect nature and, therefore, his own well-being. Only when these delicate coastal areas are appreciated for their unique, irreplaceable qualities, will they receive the stewardship they require—the planning and protection needed for their continued existence.

There are many reasons why coastlines must be viewed in this light and why we must pay more attention to maritime coastlines now than to any other type of habitat on Earth. Sometimes the reasons given are fictitious—invented for political purposes—but usually they are quite legitimate. Often in the political realm, scientific reality and subjective desires cannot be accommodated equally nor their demands properly resolved in proposals and plans framed under economic pressure. When this happens, the consequence is further adverse impact on the coastal zone.

If we categorize the problems encountered in restoring, enhancing, and protecting the coastal zone according to the kind of human activity connected with them, we will provide ourselves with a guide for solving those very problems.

Fisheries

Marine fishing involves the extraction of high-quality animal protein—an essential component of the human diet—and of other natural sea products from the coastal zone. Many subsistence-level economies have relied upon this source of protein and do so to this day.

At the subsistence level, humans live in balance with the economy of the sea. The development of commercial fisheries, on the other hand, has resulted in the overexploitation of most fishery resources, bringing about a decline in the catches of individual species and, now, in the total catch of fish from the world ocean. In a few instances, fishery resources may be sustained at a "quasi-subsistence" or "put-and-take" level through such manipulations as the release of salmon from hatcheries and the mariculture of oysters and mussels. Aside from these, however, there are only a few other examples of success in the management of fishery stocks (including programs for the management of the northern fur seal, the yellow-fin tuna, and some shrimp stocks, and the harvesting of kelp off the coast of California), and these are not supported by long records of success. Most other currently exploited stocks of "fish" appear to be on the decline as a consequence of their excessive harvesting by man. Among the exceptions may be the antarctic krill *(Euphausia superba)* and squid, which may represent underutilized resources of major proportions. If past practice is a guide to the future, however, they too will be overutilized before appropriate management strategies have been developed.

Sedimentation

Human activities have long influenced the rate at which sediments accumulate in the inshore areas of the world ocean, a rate that has accelerated since the introduction of nuclear weapons in the 1940s and the temporary cessation of worldwide hostilities over the past three decades.

In most instances, the clearing of land for forestry or agriculture causes increased runoff and erosion of soil, often resulting in extraordinary flood damage. The initial impact of clearing is followed by development of the land for industrial or residential use, together with the construction of roads and other works; this increases runoff and reduces the amount of water that penetrates and is absorbed by the soil. Thus, excessive clearing, covering over, and paving have accelerated the loss of soils to the sea.

Unnaturally high increases in sediments in coastal areas harm the marine environment in a variety of ways. For example, turbidity caused by increased sedimentation may bury coastal marine organisms and cut off the light necessary for the growth of benthic "grasses" and algae, thereby eliminating critically important stands of marine plants. This is an extremely serious event where the plants are the basic components of an entire ecosystem, as are the kelp "forests" of offshore California, Chile, and Australia. Seagrass beds are important habitat for turtles and waterfowl. Benthic plants serve many essential ecological functions: for example, they provide food, shelter, and habitat, and they stabilize shorelines as well by halting or slowing erosion caused by waves and currents.

An excess of suspended sediments can diminish the fecundity of benthic organisms such as mollusks, impair the health of the animals that filter plankton from seawater by the fouling of their feeding apparatus, and reduce the abundance of light-dependent phytoplankton. Subtle variations in the chemical constituents of inshore waters unnaturally enriched by suspended sediments may be expected, but the impact of these changes has not yet been evaluated.

The converse—a reduction in the amount of sedimentation—has been observed in the eastern Mediterranean: a reduction in the usual seasonal inflow of nitrogen and phosphorus from the Nile River after the Aswan High Dam was built has impoverished the inshore waters and caused the decline of clupeid fisheries dependent on phytoplankton formerly sustained by the annual supply of nitrate and phosphate brought by the Nile.

Navigation

Maritime commerce conducted during favorable weather in well-charted waters is convenient, and less energy is used to transport goods by ship than by rail, truck, or air. While shipping by water generally is slower than the other modes of transport, for manufactured goods and many commodity items this usually is not a major consideration.

Today, shipping is done by increasingly larger and faster vessels that are subject to ever-decreasing control. In fact, some of the larger ships now navigating the ocean are automated to the point where the crew is required to act only when the automatic systems fail, or when the controls need to be altered for docking or sailing. This situation is dramatically reviewed by Noël Mostert in his book, *Supership* (1974), which calls attention to the many problems created by increasingly larger oil tankers.

The modern world's economy is based primarily upon the consumption of petroleum. As the supply of petroleum is depleted from terrestrial reservoirs, larger amounts of oil will be produced from deposits beneath the continental shelves of the world's coastal zones. And in addition, because economies of scale favor transporting more and more petroleum in ever-larger ships, the inevitable collisions or groundings of enormous tankers will have ever more serious and longer-range consequences.

It is estimated that some 0.1 percent of all the oil produced is willfully or accidentally spilled at sea. Much of the spilled oil finds its way to shorelines, where it is stranded by winds and tides. Some of it evaporates into the atmosphere, the rest settles to the ocean floor or is slowly degraded as the oil residues float upon the surface. Some of the oil stranded on beaches, as well as that which dissolves in seawater or is incidentally ingested by marine organisms, in one way or another enters the foodweb of the sea. Since some components of oil may be carcinogenic, the addition of petroleum to the marine foodweb cannot be beneficial to most marine organisms. The only exceptions occur among microorganisms (bacteria and fungi), some of which can break petroleum residues down.

We face the prospect of increased worldwide reliance on petroleum, which means that more petroleum will move by sea, both as crude oil and as refined products. More environmental damage will be caused by more frequent and larger oil spills. In many instances, the damage to organisms will not last long, yet there may be unknown long-term impacts upon the foodweb in general. Certain activities, such as commercial fishing and recreation, will be subjected to extraordinary inconveniences. In some places along the seashore, species that are restricted in distribution or abundance may become seriously endangered or lost. Flightless seabirds such as penguins and seabirds that normally do not rest upon the land and cannot escape from or avoid an oil slick are especially vulnerable. Similarly, such rare or endangered organisms as sea iguanas and sea turtles may be harmed by oil spills.

When oil spills occur in isolated, sparsely populated areas, the entire biota may be affected before emergency measures can be mobilized to mitigate the impact. Protective or cleanup efforts may be difficult in polar or subpolar regions during the winter.

Ever-increasing amounts of fossil-fuel energy also will move by ship in the form of liquified natural gas (LNG). In this method of transport, natural gas—largely methane—is compressed about 600-fold and thereby liquified. Then it is stored in special tanks on ships designed specifically for transporting natural gas.

Spilled LNG evaporates, and, mixed with the air, may explode. Leaks may occur as a result of a collision, the grounding of a vessel, or an accident during the loading or discharging of cargo. An accident with an LNG ship is very serious, for once one has been disabled at sea or has accidentally run aground, it cannot be conveniently or safely unloaded. A major accident in port during the transfer of LNG can have only devastating consequences.

Increased maritime commerce in ever-larger vessels not only presents the spectre of larger and larger accidents at sea or along the shore, but it requires as well the modification of harbors so that they can accommodate larger vessels with deeper drafts. Most of the world's harbors cannot accommodate vessels that draw more than 15 meters of water, and only two can accept very large crude carriers (VLCCs), which require 30 meters of water for navigation. Most

VLCCs must moor at "monobuoys" situated some 3 to 20 kilometers offshore. Huge pipes joined to the monobuoys transfer the cargo to or from shore.

Bringing large vessels into extant ports presents considerable navigational problems because the vessels are not easily maneuvered. But even before these ships could be brought into port, the harbors would have to undergo considerable dredging and other kinds of alteration. In most cases, increasing the volume of an enclosed body of water reduces its capacity to sustain populations of marine organisms. Also, dredging sediments and material from the bottom of harbors causes problems at the disposal sites, eliminates the dredged sites' biotas, and through the stirring up of sediments, degrades the quality of the water. The severity of these impacts must be assessed for each site where additional modifications are proposed.

Most artificial harbors were designed exclusively for navigation. Consequently, their relatively deep channels, steep pier walls, and dead-end slips and basins impede the circulation and exchange of water. Shallow areas with sloping banks, which are vital habitats for small fish, birds, other wildlife, and microorganisms that interact with the atmosphere at low tide, are missing. Interstitial microbiotas oxidize the organic and mineral compounds that usually are abundant on the bottoms of enclosed bodies of water, an action essential for maintaining high-quality water in these areas.

Pollution

Throughout human history wastes have been disposed of in the ocean. Even contemporary subsistence-level economies discharge their wastes directly into the ocean, but apparently without harming the biotas of the waters, people who come into contact with the waters, or local fisheries. Harm becomes apparent when the assimilation capacity of the waters is exceeded—that is, when materials discharged into them accumulate faster than the materials are transformed by resident organisms or dispersed. This reduces the amount of oxygen dissolved in the water, causes potentially septic conditions, and alters the biota. Shellfish contaminated with pathogenic organisms may become reservoirs of infection. Coastal waters can be protected by treatment, through oxidation, of all discharged wastewater; this reduces the "oxygen demand" of the materials as well as the number of pathogenic organisms introduced into the environment.

Some compounds are not transformed or dispersed, however, but instead concentrate in the foodweb up to levels that are toxic for the contaminated organisms and in concentrations potentially or actually toxic to people. Among such substances are radioisotopes, heavy metals, halogenated hydrocarbons (DDT, polychlorinated biphenyls [PCBs], and so on), and refractory carcinogenic compounds from petroleum or synthetic sources.

The quantity of lead in the biosphere has been increasing since the days of the first Roman smelters, but it has been accumulating more rapidly in the past thirty years through the combustion of leaded automotive fuels. Similarly, the addition of mercury to the biosphere increased dramatically in the same period of time—largely as a result of industrial processes that utilize this element—as well as DDT and its degradation products, and PCBs. Locally, they have had severe impacts on such animals as pelicans, Peregrine Falcons, cormorants, sea lions, and, it is suspected, a myriad of others. Substances such as these can only have more severe adverse impacts if they are permitted to continue contaminating the biosphere.

The chemical and biological pollution of shoreline waters has been joined by a third form of pollution, excessive heat, which is a physical, as opposed to a chemical or biological, pollutant. The construction of more and larger electricity-generating plants on the shorelines of the world is increasing the impact of their wastes on the biotas of nearby ecosystems. Seawater used for single-pass cooling carries along organisms that are then abraded, bruised, crushed, or impaled upon the surfaces of the tubes, pipes, pumps, and screens in the circulatory systems of the generating plants. In addition, periodic hot-phase cleaning or the use of biocides such as chlorine kills organisms resident in the circulatory systems, as well as those in the circulating water.

Seawater is used for cooling in some refineries and LNG facilities, when it is used as a heat sink, and unnaturally low water temperatures occur when LNG is regassified for storage until it is used in homes and industry. In each such case, the aquatic biota is affected.

Recreation

Every maritime society has used its oceanic shoreline for food, transport, and habitation. It is only natural, therefore, that man shows a strong urge for recreation there as well. There are shoreline resorts all over the world, except in Antarctica. Many of them exist primarily to provide people with a chance to enjoy sea air, sunlight, open space, and vistas with unlimited horizons. It is important that there be locations where this kind of valuable experience can be enjoyed.

Sometimes, however, especially near large centers of population, use for recreation exceeds the resiliency of shoreline environments, such as marshlands and tidepools, damaging both habitats and biotas. Other areas, such as sandy beaches, suffer little or no obvious damage, even when people use them intensively—although in some places selective hunting by recreational divers does reduce the abundance of the sought-after fish and shellfish. Also, some recreational facilities irreversibly and irretrievably modify critically important habitat areas. For example, the construction in wetlands of harbors for recreational watercraft has diminished the capacity of these ecosystems to sustain life. This is a type of modification that is well-known along the shorelines of Florida and California.

Management

Priority must always be given to protecting the ocean's capacity to function as a system according to its natural requirements by prohibiting uses that might interfere with its functioning. Doing so would not preclude any legitimate use; it would require, however, that all activities be designed and managed in such a way that the ocean's integrity is maintained. The attainment of this objective would allow the ocean to produce marine life sufficient for human nutrition as well as abundant natural products of value to the economy.

The ocean would still be available for commerce; beach and sea would still be open to both physical and aesthetic kinds of recreation. At the same time, however, certain other practices would have to be brought under control. The conservation of energy, reduction of erosion, careful treatment of wastewater, and proper designing of shoreline alterations are among the numerous measures that would assure the continued integrity of the oceans.

The Conservation of Energy

Nearly all contemporary uses of fossil and nuclear fuels damage natural systems, including the oceans: pollutants from the combustion of fossil fuels disperse unnaturally large amounts of sulfur, lead, and mercury into the air and thence to the land and the sea; heat from power plants kills zooplankton, fish eggs, and larval and adult fishes; and nuclear reactors leak and experience other unresolved problems of safe operation and the disposal of spent radioactive fuel.

The offshore production of oil and the ever-increasing transport of petroleum by sea bode ill for shorelines and waters because blowouts, spills, explosions, and other accidents may occur. Ever-increasing reliance upon vehicles propelled by internal-combustion engines requires that more land be paved for roads and parking spaces and diminishes the land surface available to receive rainfall; this results in increased runoff and in greater chance of damage from flooding.

Clearly, reliance upon the expanded use of fossil and nuclear energy is incompatible with either the restoration or the long-term maintenance of marine ecosystems. Mankind depends upon these living resources for a major proportion of its protein and natural products, and the ocean's phytoplankton has generated a large fraction of the atmosphere's oxygen. Activities incompatible with the restoration of the shoreline's natural functioning should be phased out unless there is overriding human need for them, and even then not until their adverse impacts have been assessed and means devised to mitigate them so as to avoid critically disrupting the ecosystems.

As an alternative to the increased reliance on fossil and nuclear fuels, other, nonpolluting energy sources should be utilized to the maximum extent possible, among them solar, geothermal, and wind energy—with application of the last to transport by sail. Economic and other incentives should be adopted to encourage the efficient use of energy, and should be applied wherever possible to construction, industry, transport, commerce, and agriculture.

The Reduction of Erosion and the Treatment of Wastewater

Management practices must be implemented to reduce the erosion of terrestrial sediments and structures; these might include, where feasible, the construction of sediment traps to reduce the transport of sediments into the ocean. Wastewater intended for discharge into the ocean should first be treated so that the ocean's assimilative capacity will not be exceeded. In general, this can be accomplished with an oxidative treatment that yields wastewater equal in quality to that of unpolluted seawater, which is naturally well-oxidized and contains low concentrations of dissolved organic substances, suspended organic particles, and bacteria.

Toxic and other environmentally dangerous compounds should not be discharged into the ocean in concentrations greater than those in seawater. This requires containing the compounds at their source (generally an industrial facility), before they are released to a waste-collection system that discharges into the ocean.

The Alteration of Shorelines

Modifications of natural shoreline features should be designed so as to have minimum adverse impacts upon ecosystems. When irreversible damage cannot be avoided, areas of equivalent functional capacity should be created nearby to provide suitable habitat for the species of organisms affected. Additions to or alterations of shoreline features should be designed and constructed so as to enhance the biological carrying capacity of the shoreline, and, wherever possible, the biological carrying capacity of extant shoreline features should be enhanced as well.

Fisheries

The animal components of coastal ecosystems (which are among the Earth's most highly productive) are varied and abundant, and constitute the greatest source of protein on Earth. Yet, as a consequence of overexploitation, pollution, and the loss of essential habitat, fisheries have failed to maintain themselves; however, implementation of the measures recommended above should restore both the quality of marine waters and habitat areas suitable for marine organisms. By the same token, lowering the demand for oil and reliance upon seawater for cooling or as a source of heat should further protect shorelines from oilspills and permit fisheries to regain their former richness.

Improved technology and equipment have increased the worldwide catch of fish, but only at the expense of areas that used to be too far away and too unsafe for the older, smaller, and slower vessels of former eras. Overexploited fishing grounds have failed to sustain their original levels of production. Inshore fisheries the world over—many already overharvested with relatively primitive techniques by an earlier, smaller world population—have suffered more from overexploitation than have high-seas fisheries, but this is a temporary situation because the pressure on high-seas fisheries is intensifying.

Even if the proposed 200-mile (approximately 320-kilometer) territorial sea is instituted, the continued unlimited access to limited fishery resources would make impossible the restoration of once-abundant stocks of fish and shellfish. Therefore, quotas must be set that are based upon conservative estimates of how large a catch can be without its impairing the capacity of a species to regain its former abundance. To implement this management strategy effectively, the catch and stock of each species must be accurately determined. If the stock of a species fluctuates too much or fails to increase, the catch quota should be lowered until the stock has increased. This is an extremely difficult but nonetheless vitally important approach.

It is difficult, however, to assess the catch of such users as recreational fishermen. Also, permitted catch, season, and method of capture differ from jurisdiction to jurisdiction. These kinds of differences must be resolved if stocks are going to be able to expand to the point where they achieve optimum sustained yields. Agreements on how to apportion catches or given species will have to be developed among the various parties (nations or interests) involved.

Summary

There must be stronger and better control of human activities in coastal zones. In order to achieve again the natural quality of water, restore the natural functional capacities of marine ecosystems, and use the living resources of the seas wisely, human activities in coastal zones will have to be diverted so they work with, not against, coastal ecosystems. To achieve these goals, people in many cases are going to have to change their patterns of living. In comparison with the alternatives available to most other organisms, to coastal ecosystems, and even to the human species as a whole, such changes are desirable. They will involve the internalizing of those costs of development, such as adverse environmental impacts, that formerly were considered externalities and that therefore were not included in the overall economic assessment of projects. Doing so will lead to enhanced and sustainable fisheries, safer maritime commerce and navigation, fewer oil spills at sea, the preservation of valuable agricultural soils, less pollution by toxic substances, greater efficiency in the use of water, and healthier recreation.

Reference

Mostert, Noël. 1974. *Supership*. New York: Alfred A. Knopf.

21. Conservation of the Coastal Marine Environment: A North European Viewpoint

Lars Emmelin
University of Lund, Sweden

Introduction

The processes of life seem to take on a special richness and diversity in ecological transition zones and at interfaces. This is true for whatever dimension is used in the investigation—whether one examines the global system that man inhabits at the interface between soil and atmosphere, a rocky shore, or a piece of growth medium for bacteria.

Slightly over two-thirds of the surface of the globe is covered by the ocean, but most of this vast area is a biological desert. Its productivity is low, as is the density of species and individuals. The often-mentioned "richness of the oceans" is a description that accurately applies to only two types of oceanic areas—the coastal zone and the upwelling areas.

The shallow coastal zone is that part of the ocean which lies within the 180-meter contour line. It occupies only about 7.5 percent of the total area of the ocean, but approximately half the total fish production of the oceans occurs here; according to present rough estimates, the rest comes from upwelling areas, which are areas where nutrient-rich waters rise to the sunlight-illuminated surface layers of the ocean (as off the western coast of South America), and in which photosynthesis can therefore occur. Thus, production in 90 percent of the ocean is so small that, for practical purposes, it can be neglected in computing the production of fish in the seas. The productivity of the coastal zone and upwelling areas is equal to or sometimes greater than that of well-managed agricultural land.

The importance of the ocean for the production of food for man and as one of the great stabilizing mechanisms that protect all life on Earth is undisputed. It is also well established that man has polluted significant parts of the oceans. This is true whether we examine pollution of the ocean horizontally or vertically. No coastal area adjacent to an industrialized nation is wholly unaffected by pollution. In certain areas, the problems are quite obvious without the aid of sophisticated scientific analysis. It has been shown that organisms from depths of 800 to 1,000 meters in the Atlantic contain concentrations of DDT and polychlorinated biphenyls in the same order of magnitude as do species from the upper layers of water.

The Scope of Marine Conservation

In its modern sense, conservation is usually equated with the rational management of natural resources (normally evading the questions "For whom?" and "Over what period of time?"). Marine conservation thus has to provide for a number of uses of the oceans, some of which are often seen by conservationists as threats to their goals rather than as part of their programs. Examples are the exploitation of renewable resources such as fish and of nonrenewable resources such as gravel, oil, and minerals.

Any population of animals produces a certain amount of waste, and it is arguable that using the marine environment to handle organic matter is both rational and legitimate. To disregard man's uses of the oceans in a marine-conservation program would seem self-defeating, since the problems created by this exploitation would make other protective measures meaningless.

A conservation program for the entire ocean must have, as its basic goals, pollution control and rational and equitable exploitation of resources. At present, the United Nations is working on this formidable task in a variety of ways. This is not the place to discuss the progress of that work. The dynamic nature of the ocean systems, the complicated legal and international problems involved in marine conservation, and the fact that the marine environment is basically alien to us should not, however, blind us to the needs of and possibilities for diversified local conservation action.

Since the coastal environment is influenced by a large number of human activities—fishing, the release of effluents, construction, and the extraction of oil, gravel, and minerals, etc.—the scope for natural-resource management must be greatest in this zone.

Marine Conservation in Northern Europe

The coasts of northern Europe are among those where severe regional pollution problems occur. Several commercially valuable fish species in the North Sea are definitely overexploited. Exploration for oil in the North and Norwegian seas is operating at the limits of present technology and under very difficult conditions. Several species of animals in the Baltic are threatened by environmental poisons, e.g., the Grey Seal (*Halichoerus grypus*) and the White-tailed Eagle (*Haliaëtus albicilla*). Fish from certain areas are considered unfit for human consumption because they contain high levels of contaminants such as mercury. Coastal erosion is a problem in parts of Britain and Scandinavia. The Dutch are conquering shallow coastal territory from the sea.

However, there are also hopeful signs of emerging action on the most fundamental conservation problems. A number of conventions exist to regulate dumping (the Oslo and London Conventions), pollution from land-based sources (the Paris Convention), pollution of the Baltic (the Helsinki Convention), and protection of the natural resources of the Baltic (the Gdansk Convention). National and international fishing regulations exist and are continually reviewed. Pollution from Swedish municipalities and industries to the Baltic has been cut drastically. The pulp and paper industry, for example, which is by far the most important polluter, has halved its total releases while more than doubling production in the last decade. Species of fish absent for a century are reappearing in the Thames estuary. In short, a situation is emerging in which other types of conservation action may be meaningful.

It is quite clear, however, that the concept of coastal-zone management is having difficulties in northern Europe. For example, it is obvious that in its present state of development, national planning in Sweden does not go beyond pollution control and a degree of control over exploitation of natural resources.

Protected Areas

Another approach to managing the natural resources of the coastal environment is the creation of marine parks and reserves. In a global survey made by the International Union for Conservation of Nature and Natural Resources (IUCN), only two countries in northern Europe—Denmark and the United Kingdom—are cited for taking this approach, and only the Danish reserves have received clear legal status. In several other countries the problems of marine reserves and wildlife presentation are receiving official attention. The inherent problems seem to be great—as shown, for instance, by a study made by the Natural Environment Research Council in the United Kingdom, the recommendations of which can at best be considered vague. Other organizations in the United Kingdom have promoted the concept of marine reserves, and a number of such areas do have a measure of protection and management plans—notably, Lundy Island at the entrance to Bristol Channel.

In Norway, a working group of the Norwegian Society for the Conservation of Nature has proposed a number of areas for conservation purposes. In Sweden, movements to get a similar working group organized under the Royal Academy of Sciences are under way. One area on the Swedish west coast will probably receive comprehensive protection against pollution, and a group of students at the University of Lund have made proposals for a marine reserve which meet with no legal problems.

Reasons for Protecting Marine Areas or Species

There are at least four reasons for protecting marine areas or species:

1. The protection of breeding areas or nursing grounds;
2. The preservation of areas of particular scientific interest;
3. The preservation of species or populations; and
4. The preservation and management of areas with recreational, aesthetic, and similar values.

The establishment of marine parks and reserves has been urged at a number of international conferences, notably the first and second World Conferences on National Parks (Seattle, 1962, and Yellowstone and Grand Teton National Parks, 1972). It was stressed by the United Nations Conference on the Human Environment (Stockholm, 1972) as one necessary component of marine conservation. As in so many other international conservation areas, the IUCN is doing an important job of collecting information and putting pressure on governments to get positive conservation action. As recently as May 1975, the topic was discussed at a conference in Tokyo.

The protection of areas as parks or reserves can provide concrete symbols of a comprehensive natural-resource-management policy; conversely, the establishment of only a few parks or reserves may be an easy or cheap way of seeming to have an adequate conservation policy, when in fact the policy may be inadequate.

Selected Bibliography

Björklund, Mona I. 1974. Achievements in marine conservation, I. Marine parks. *Environmental Conservation* 1(3):205-223.

Dybern, Bernt I. 1974. Water pollution—A problem with global dimensions. *Ambio* 3(3-4):139-145.

Hiscock, K., ed. 1971. *Report on the Proposal to Establish a Marine Nature Reserve around Lundy: Marine Biological Investigations.* Lundy: Lundy Field Society.

———, ed. 1973. A Seminar on Marine Conservation, April 13-14. Marine Science Laboratories Department of Marine Biology, University College of North Wales, Menai Bridge, Wales (mimeographed).

22. The Coastal-Zone Development Dilemma of Island Systems

Edward L. Towle
Island Resources Foundation, Inc.,
United States Virgin Islands

Introduction

The world's remote and varied islands are currently experiencing strong development based chiefly on tourism, on extractive enterprises, on their use as nodes for global air and steamship transportation, and on the increased demands for local resources generated by their rapidly growing populations. An unwanted by-product of this current development phenomenon has been a dramatic deterioration of island coastal environments, accompanied by a decline in the quality of life on islands, as measured by other means than such traditional economic indicators as gross national product (GNP) and per-capita income. On islands, as elsewhere, progress has its price.

The stresses and pressures of rapid population growth, unrestrained development, and modern technology are partly responsible for the decline of the environmental quality of island systems and their littoral zones, and for coastal-zone conflicts; another increasingly apparent cause of the decline is a serious methodological failure in the area of resource planning, allocation, and management.

One deficiency or failure has been the omission of environmental values from planning and development strategies. The principal question of the moment is how to incorporate environmental criteria to a far greater degree in the modernization of island communities, as they are shaped by local forces and by development agencies involved in or responsible for administration and growth on islands. A second deficiency has been the procedural and technical failure of the development-planning process to embrace, comprehend, or deal with coastal-zone resource-use conflicts, trade-offs, and externalities.

International and Regional Initiatives

Within the past decade, islands have been singled out at the international level—by the International Biological Programme (IBP), the International Union for Conservation of Nature and Natural Resources (IUCN), UNESCO, and, most recently, the 1972 Biosurvival Symposium—as areas that merit special attention owing to their extreme vulnerability to rapid change.

As early as 1966, scientists working with IBP, IUCN, and the Pacific Science Congress articulated a concern for the conservation of certain Pacific islands through a series of conferences, symposia, and resolutions that launched the Islands for Science program. Subsequent IUCN initiatives took the form of joint resolutions that were promulgated at the close of the Regional Symposium on Conservation of Nature—Reefs and Lagoons, held in Noumea, New Caledonia, in August 1971. These resolutions stressed the need for ecological guidelines in development projects (Resolution 1), and noted the vulnerability of island ecosystems to disturbances due to human activities (Resolution 5). The specific environmental impacts of tourist development on the resources of coastal zones of certain Pacific islands were also given attention (Resolution 12). (See the Annex to this paper.)

In the same year, one of the priority study projects listed by UNESCO's International Coordinating Council of the Man and the Biosphere Programme was the "ecology and rational use of island ecosystems." More recently, the initiatives of IUCN and UNESCO were reinforced at the United Nations Biosurvival Symposium, held in Stockholm in June 1972, where it was urged that certain islands be protected from development, for the exclusive use of scientific study. Other resolutions called for a consideration of aesthetic values in resource development and landscape "manipulation" and comprehensive land- and water-resource planning.

The most important regional statement to date has been that of the Caribbean Conservation Association, which approved fifteen resolutions on Caribbean island environments at its sixth annual meeting held in Saint Kitts, West Indies, in September 1972. Eight of the fifteen resolutions concerned the island coastal zone and its uses and management.

The Island Dilemma

Attracted by the undeveloped, pristine nature of islands and their recreational qualities, and encouraged by the reduced costs and ease of air transport, a virtual tidal wave of people from larger, urbanized, continental areas are travelling to previously isolated islands, threatening, inadvertently, to alter the very qualities that attracted them to the islands.

The anomaly of the island situation is twofold: on the one hand, increasing numbers of continentals regard islands as both idyllic sites for vacation experience and as superior sites for specialized industries and free ports; and on the other hand, they expect islanders to develop their resources to service these demands indefinitely, yet with no loss of the islands' unique character.

Islands are discrete points of conflict where developers directly confront fragile biological systems and circumscribed resources. The most dramatic evidence of this confrontation is found in the coastal zone. Evidence is mounting that the impact of this confrontation has upset the precarious biological balance conferred upon islands by their previous isolation and lack of exposure to modern technology. In a geographical sense, islands are becoming "an endangered species," destined to be rapidly incorporated into the larger, technologically dependent, culturally homogeneous, urbanized, and polluted systems of continents.

This threat to insular environments carries implications beyond irreversible changes in physical and biological assets. The environment is an integral part of the total island matrix that defines the quality of life for island peoples and visitors alike. The concept of "quality of life" is gaining favor as the focal point for converging economic, social, and environmental development objectives. It expands the narrow measure of progress by the standard economic indices of GNP and per-capita income by incorporating other social-welfare indicators and environmental indices and amenities that are a vital part of any island's life-support system.

By evaluating the current and projected stresses on the various environmental zones or features of islands, and by combining the results with traditional economic data, we can upgrade our planning and management of island resources. Unfortunately, the after-the-fact application of guidelines to already developed insular economies is more difficult and less rewarding for two reasons: (1) future land-use options are often largely foreclosed; and (2) present patterns of use are essentially irreversible because large amounts of capital have already been committed and because owners with vested interests require compensation. Nevertheless, guidelines can still be useful in planning future projects on developed islands if one resists the temptation to view past neglect as justification for future carelessness: there is no place within the complex interdependencies of the island system for the concept of a "degradation threshold," a point beyond which restoration of the environment becomes impossible or infeasible.

One cannot escape the question posed by well-meaning conservationists, namely, "Should islands be developed at all?" One view is that they ought to remain frozen in a state of nature, free from the contamination of twentieth-century development technology. Adherents of this homeostatic view assume that islands encumbered by an overdeveloped infrastructure and other displays of material "progress" are "spoiled."

This view may be based on genuine professional concern for the preservation of rare or specialized biotic communities that are well suited to research on evolution, genetics, population dynamics, and ecosystematics. In other cases, the preservationist's view may be prompted by a wish to return to a simpler, slower life with reduced responsibilities. Islands have long been cast as paradisiacal oases where, through a dialogue with nature, man can enjoy a continual rebirth of his spirit. This vision pervades the tourist-development phenomenon, supporting promotional literature and marketing materials. Some residents of islands also subscribe to this idyllic "keep things as they are," antidevelopment position.

Although certain isolated, sparsely inhabited islands are best used as preserves for science, most islands are not. To categorize them so would be to accept the simplistic premise that a majority of island peoples share the outsider's enthusiasm for a primitive economic state. Studies refute this premise.

It is assumed that the stewardship of islands is best vested in the hands of island peoples who gain their livelihoods over the long term from their lands. They alone have most to gain by the careful allocation and use of insular resources; they have the most to lose if their resource base is destroyed or diminished. It is their prerogative to influence development patterns. The question is, "What means should be chosen to effect that end, especially in the coastal zone, where the pressures for change are greatest?"

Ecological principles must be accepted if the environmental integrity of islands is to be maintained in the face of development. The principles articulated below respond to known adverse development impacts that have already occurred in island systems and which threaten to befall other islands in the future.

Oceanic Islands

The scope of this paper is limited by the obvious constraint of space and by the less obvious lack of precise data from detailed, island-by-island case studies of the impact of development on the coastal zone. This discussion will consider the development and conservation problems of the generic "island system," based on a review of the literature and on firsthand observation of only forty or so islands of Micronesia, the Arctic, and the Caribbean, out of the tens of thousands of small islands in the world ocean. The focus of attention will be directed more towards the smaller volcanic, coralline oceanic islands of the world, isolated or in groups, than towards the large island land masses referred to at the Noumea conference as "minicontinental" islands. Excluded are islands situated close to continents, whose politics, cultures, and economies are linked with those of the nearby continents, and which are best viewed as a special category of islands that requires different approaches to management.

The "Island System"

Development and its associated impacts are considered within the framework of the "island system," a concept that applies to individual islands as well as to groups of islands and their associated social, economic, and political systems. With regard to the encompassing environment, the concept of the "island system" is relevant in three respects:

1. It affirms the fact that an island is not a homogenous, discrete entity, but rather an assemblage of diverse sub-

aerial and subaqueous ecosystems in upland, littoral, sublittoral, and outer-shelf zones; most of these, in the case of small islands, are included in the coastal zone.
2. It stresses the importance of the interdependent linkages among island ecosystems: impacts in one ecosystem will have repercussions in another. Further, their areal extent will not conform to man's convenient geographical or political boundaries, such as "land" or "sea." The concept of island system also requires some perception of each individual island's relationship to and within associated island groups.
3. It allows for a "biocybernetic" view of island growth and development, wherein multiple-feedback, effect-to-cause phenomena can be added to the cause-to-effect phenomena usually considered.

The "Biocybernetic" Model

"Biocybernetics" is the science of the feedback relations between living and nonliving components of ecosystems. Biocybernetics makes it easier to see precisely how the output from one part of the system ultimately affects the input to the same part. Feedback can reinforce the original process that yielded the results; this is called positive feedback. Complex growth phenomena and associated environmental impacts can be more accurately understood by reference to a biocybernetic model. As a conceptual tool, this model stresses the importance of interacting links between various multiple causes and their several effects. It also indicates how initial actions may spawn consequences that accelerate through the system or, conversely, retard system processes.

The attraction of tourists to an island offers an example of feedback systems: with negative feedback, the consequence of an action acts upon its cause in a manner that diminishes the cause. One of the reasons for a declining tourist market might be explained in terms of the following interactions:

1. Increasing numbers of tourists come to an "unspoiled" island.

2. An infrastructure develops to service them.
3. The assimilative capacity of the environment is exceeded by "waste loading."
4. The natural environment deteriorates.
5. The number of tourists decreases.
6. The net quality of life is reduced.

However, it is more than likely that a complex variation of the above feedback loop will occur in conjunction with other positive and negative loops or processes: herein lies the threat to the conservation of islands. As the character of the island changes from unspoiled to "spoiled" (for example, becomes urbanized), the influx of tourists does not stop: it continues and may even accelerate, according to a positive-feedback loop that accelerates urbanization and leads to the influx of other tourists with different values. The implications of this process for island development are serious: it means that the first wave of tourists, who were attracted to an island for its unique environment, initiates a process that destroys the very assets that originally had lured them. This is not the end of the influx nor of local growth, however. The first group, acting in accord with the negative-feedback loop, will not return, but a second, quite different group of tourists will come for reasons other than the original assets, now lost. Acting in accord with the positive-feedback loop, the wave of urbanite tourists will return again and again, in gradually increasing numbers, altering the face of the entire island. Our concern is for the impact of this phenomenon on island environments and on the quality of life for island residents.

The biocybernetic framework can also be usefully employed as a methodological tool in the planning and management of the coastal zones of islands. Its use would represent a renunciation of narrow, sectoral planning for island coastal areas in favor of a more comprehensive, integrated, development approach. The feedback-loop framework induces a planning and decision-making process that is far more responsive to island values and far more likely to include aesthetic, cultural, and qualitative features that tend to fall by the wayside of traditionally planned development activity.

Characteristics of Oceanic Islands

Limitations of Resources

A basic characteristic of oceanic islands is their limited, fixed endowment of people, land, sources of energy, beaches, natural harbors, reefs, fresh water, and biotas. This limitation applies equally to the resources of the uplands, the coastal plain, the littoral and sublittoral zones, and the outer insular shelf. Because of their limited magnitude, island resources are particularly subject to overexploitation which degrades or completely destroys them. Coupled with their limited natural-resource bases, the circumscribed, fixed areas of oceanic islands come under increasing pressure as their populations grow. Islands have no hinterland except the sea, a situation that raises unprecedented, specialized questions in resource planning and management.

Vulnerability to Off-Island Influences

A second characteristic unique to oceanic islands is their extreme vulnerability to the destructive effects of modern, continental-scale development technology. This vulnerability is more important than the limitation of resources because, although islands have always been characterized by limited resources, they have not always been exposed to the formidable growth pressures of the modern world, to its machines, its mass-transport systems, and its mass media.

One reason oceanic islands are vulnerable is their remoteness and consequent isolation. In the past, this characteristic inhibited development; now, however, remoteness is largely immaterial because of the revolution in transport and communications. There is no longer an island in the world that cannot be reached in less than 6 hours by airplane from any city of a half million or more people. In addition to jets, hundreds of formerly isolated islands are now serviced regularly by seaplanes, short take-off and landing (STOL) aircraft, hydrofoils, cruiseships, yachts, and that ubiquitous means of transmitting culture, television.

The barrier of remoteness has been further overcome by factors far removed from the shores of the islands themselves—factors that set up wholly unanticipated technological encounters. Remoteness no longer guarantees freedom from pollution, offshore oil spills, dumped chemical and solid wastes, or the destruction of habitats by species that migrate to islands; all have affected oceanic islands in recent years.

An insidious aspect of the exogenous off-island influence is the global-development phenomenon itself: decisions that can affect an island and its coastal zone may be made thousands of miles away, in the boardrooms of multinational corporations. The tourist industry provides a good example: corporate hotel interests, international airlines, cruiseship lines, travel agents, the publishing industry, and banks have immense power to create tourism "booms." In many cases, islands that experience the impact of this development may not themselves be the chosen target islands; rather, they may be swept along in a development thrust aimed at a sister island, a neighboring archipelago, or even a broad oceanic region. Thus, the stimulus of development is often unpredictable and exogenous. Without planning, the vulnerability of an island is exacerbated. In many cases, the pace of change is incomprehensible and overwhelming, and it destroys an island's entire way of life.

Finally, isolation, circumscribed space, and other environmental factors also result in specialized biotic and even human communities. The specialized life forms of an oceanic island tend to be less tolerant of changes in environmental conditions than those on the margins of continents. Man's essentially short-term intrusion on an oceanic island can be permanently devastating to fragile, long-term, evolutionary interdependencies. Obviously, intensive and sophisticated scientific research is indispensible for understanding the dimensions of the problem.

Historical and Modern Development and the Coastal-Zone Systems of Islands

Historically, the development of islands has involved discovery, settlement, and subsequent wider human occupancy based on hunting, fishing, agriculture, and, less frequently, mining and trade. Modern development is founded on technological circumstances that allow island peoples to pursue livelihoods that are less dependent on the insular resource base. The balance of this paper describes the differences between the traditional and modern perspectives and presents reasons why the adverse impacts of modern development are fundamentally more threatening to the environmental quality of coastal zones.

The Nature of Historical Development

Land use is determined by environmental factors: an increase or a decrease in the quality or availability of the land resource causes a corresponding change in the fortunes of the islander who is using the land. This "environmental dependency" holds implications for both the use and abuse of natural resources.

When the natural, technology-free environment is the *sine qua non* of livelihood and survival, there is a built-in equilibrium or correlation between the human population and natural resources. Even when there are nominal introductions of technology from the outside, adverse impacts are regulated to a degree over the long term by a negative-feedback regulatory mechanism that checks serious man-environment dislocations. This is not to say that resources were not or cannot be abused, such as by the short-term overexploitation of coastal fishery resources; it is simply to acknowledge that, given a relatively stable population and limited technology, abuses and their consequences do have limits. In the case of islanders who rely on fish for a living, overfishing soon redounds to their disadvantage: as the fish supply is depleted, the islanders are driven to seeking other sources of food, other islands, and other modes of productive employment to avoid starvation. The dearth of fish acts to balance man with the resource.

In general, the adverse environmental impacts common to fishing, hunting, agricultural-subsistence, and export economies have been the depletion of resources and the despoliation of land. Soil infertility, erosion, and the degradation of land through overgrazing have been the major problems on land. Introduced exotic plants and animals have also caused ecological dislocations among indigenous species. While such problems persist on many tropical islands, only the more dramatic and destructive impacts associated with modern patterns of development in the coastal zone and the marine environment will be considered in this paper.

The "Technological Encounter"

In these few cases of islands that have been catapulted, figuratively speaking, into the technological world of the twentieth century within a mere decade or two, the environmental, cultural, and economic impacts have been severe. The experiences of these islands serve as harbingers of what other, developing islands may expect in the future.

The seriousness of the "technological encounter" is reflected in widespread changes in uses of coastal land and in concomitant environmental stresses and incompatibilities that the new activities create.

Exploring and drilling for oil and natural gas, tourism, and dredging are especially noteworthy in this regard because of their potentially pervasive impacts on coastal areas, their presumed economic benefits to traditional island economies, and the intensity with which they are being pursued at present.

The Implications of Introduced Continental Technology

The precipitous introduction of continental technology has fundamentally altered the patterns of land use on islands: no longer are island economies tied to the availability of intrinsic resources. For many developed islands, agriculture and fishing are less important than they once were. Some islands that once produced all of their own food now rely heavily on imported food, since the most productive lands have slipped into nonagricultural uses. The implications of using land without regard to an island's indigenous natural resources are twofold: first, it allows the island to support high, dense populations and to be developed beyond the level that its resource base could support; second, it creates a host of environmental, cultural, and economic incompatibilities that are most visible and pernicious in the coastal-zone area.

St. Thomas, the largest of the United States Virgin Islands, is an oceanic island that, through a reliance on technology, apparently has overcome the traditional barrier to development due to resource scarcity. Because of its poor soil and the unreliable rainfall, virtually all food for its burgeoning resident population must be imported.

Furthermore, fresh water, also scarce, must either be imported by tanker or manufactured in costly desalinization plants. Owing to the scarcity of construction aggregate, sand must be imported and dredged from inshore areas, often with adverse effects. Finally, because land itself is in short supply, land must be created by dredge-and-fill operations along the coast, disrupting fisheries, damaging reefs, and degrading the quality of inshore water.

While the island appears to flourish, with a high level of tourism and a high per-capita income, it is only at the expense of a nearly total dependence upon external capital, alien labor, tourism, imported food, and continental technology, and it occurs in the context of a deteriorating physical, ecological, social, and cultural environment. Progress of this type exacts its price, especially in the absence of sound planning and advance assessment of the environmental implications of development objectives, strategies, and tactics.

The Incompatibility Factors of Transplanted Technology

Some problems of incompatibility are due to a growing reliance on artificial environments, contrived or manipulated by technology; other incompatibilities are due to the irreversibility of previous land-use activities. The establishment of a petroleum-heavy industry complex in the Krause Lagoon on the south coast of St. Croix in the U.S. Virgin Islands is one example of environmental manipulation that has led to the obliteration of vast areas of coastline—coastline that was at one time one of the most biologically productive spots in the Virgin Islands. The industrial complex threatens to cause further decay along the coast through continued dredging, oil spills, and complementary industry. The Ponce-Guyama arc on the south coast of Puerto Rico is experiencing a similar metamorphosis, but on an even larger scale. The coasts of Bonair, Turks and Caicos, and Saint Kitts all are scheduled for development of this type.

Capital-intensive uses of land in island coastal zones are economically irreversible, and the magnitude of alteration and devastation of landforms makes them environmentally irreversible as well. Both the economic and environmental irreversibility of capital-intensive development on islands are important characteristics of modern development schemes that planners ought to confront.

Incompatibilities exist in the social and political spheres as well.

The influx of alien construction workers, industry executives, technicians, bankers, consultants to government, and permanent tourists, once set in motion, can quickly reduce local islanders to a minority on their own homeland, with obvious and unfortunate implications. The newcomers tend to espouse technically glamorous, environmentally damaging, and culturally inappropriate development projects and methods. Their choice of approaches, models, methods, and projects may be influenced by their predilection for continental modes of operations, or it may be due to their ignorance of the special cultural, technical, economic, and social requirements of small islands. Many continental developers, planners, and administrative officials who become involved with islands are often unaware of the immense differences of scale between insular and continental systems. Many of them assume that what is appropriate in a continental context is also appropriate on islands. Seldom is this assumption questioned, to the detriment of most island peoples and their environments.

The Impact of Tourism on Islands

We have singled out tourism for amplification in our discussion of contemporary development activities because of its current popularity and intensifying impact on island coastal zones. This does not mean that other activities, alone or in combination, may not prove more damaging to specific insular environments (oil pollution or dredging, for example); rather, our discussion of tourism simply points up the environmental hazards inherent in development and the internal conflicts and tradeoffs implicit in any comprehensive development strategy.

An island in need of foreign exchange, opportunities for employment, and general modernization regards tourism as a reasonable answer to its economic needs. Few people would argue that tourism cannot contribute to higher incomes and the long-term development goals of an island. What is open to question is the wholesale acceptance of mass tourism as the economic mainstay of an island's economy, with no strategy for meeting the adverse impacts of tourism on the island's society, environment, and economy. What is lacking is a proper concern for the aspects of tourism that degrade the quality of an island's environment and reduce the capacity of the island to absorb development. More obvious is the conflict between the development of its coastal zone for tourism and as a service area for industry (for example, as ports and airports, and for shipping, dredging of sand, and disposing of waste).

Tourist islands suffer adverse impacts because it is assumed that economic gain justifies any means to that end, even if the means entail the reckless exploitation of the islands' social and natural environments. In addition, there are important structural weaknesses in a tourist economy. For example, the industry is potentially unstable: because of the tax-incentive practices of competing islands, tourist expenditures have a low "multiplier effect"; there are inequities in the distribution of income; and indigenous cultural values are eroded by development that outsiders manage for outsiders. Lastly, both tourism-related construction and infrastructure services (power, ports, water supply, etc.) affect the coastal zone.

Studies in the Caribbean have shown that the "income multiplier" of tourist economies is often very small, on the order of 1 or 2, or even less. This is due to the large amount of "leakage" that occurs when foreign exchange enters the island economy. The benefits to an island's economy are less than is commonly assumed. Total commitment to tourism at any environmental cost cannot be justified as a course of action. Furthermore, a "low multiplier" raises questions about the economic desirability of certain uses of coastal land that have adverse consequences for the environment. Maintaining the quality of the coastal environment carries a price. The degree of impact that an island is willing to tolerate from a specific land use depends partly on the benefits the society derives from the activity.

A realistic view of environmental quality requires that it *not* be divorced from an economic framework. To be credible, ecological guidelines formulated for a specific development project should include a consideration of the economic benefits to an island of specific land-use activities. The economic benefits should then be compared with the adverse impacts, such as the intrusion into the coastal zone of overbearing, usually alien, architecture. Since condominiums have an affinity for beaches, their construction usually means the curtailment or complete loss of access by islanders to beaches, and of other traditional, nonconsumptive uses. Planners should enhance the contribution of tourism to the welfare of islands by judiciously favoring land uses that minimize the flow of money abroad. Planners should guide the development of tourism through the use of ecological principles that foster environmental harmony.

ANNEX*

Selected Resolutions of the Regional Symposium on Conservation of Nature—Reefs and Lagoons Held at Noumea, New Caledonia, 5-13 August 1971

Resolution No. 1:
Ecological Principles in Development Planning

The Symposium:
Believing that a major reason for the lack of success in many development projects has been the neglect of ecological principles during the conception and planning phases;
Being aware that IUCN in collaboration with development planners and other specialists has prepared a compendium setting out ecological principles with special reference to the problems of development planners;
Being aware that many agencies concerned with development intend to use this compendium to formulate ecological guidelines for the evaluation and elaboration of development projects;
Recommends that the South Pacific Commission and all governments and administrations concerned formulate and adopt such ecological guidelines as the basis for their own development projects.

Resolution No. 5:
Conservation of Islands for Science

The Symposium:
Realising that islands, because of isolation, limited size and other environmental characters, tend to develop specialised, and sometimes simple and fragile communities;

Source: IUCN Bulletin (New Series) 2(21) (Special Supplement).

Realising the special value to science of islands as locations for the continuing studies of evolution, genetics, population dynamics, interaction between species and many related topics;

Realising that island ecosystems, particularly of small and remote islands, having evolved in isolation, are vulnerable and easily disrupted by disturbances arising from human activities;

Being convinced that it is in the interests of mankind to conserve selected islands as sites for scientific studies, particularly remote and uninhabited islands;

Recommends to all governments concerned that they adhere to the proposed Convention on Conservation of Certain Islands for Science proposed by IUCN and scheduled for discussion at the UN Conference on the Human Environment (Stockholm, June 1972);

And recommends further that early attention be given to the selection of appropriate islands for conservation for this purpose.

Resolution No. 12:
Utilization of Coastal Marine Resources

The Symposium:

Being concerned at the extensive ecological damage that has been caused in certain Pacific islands by utilization of coastal marine resources, including reef materials, and by the development of tourism;

Being concerned particularly at the building of hotels and other developmental activities on beach fronts, at the gathering of sand and coral in large quantities for building and industrial use without thought to the disturbance of beaches, reefs and lagoons, extensive dredging operations, and overharvesting of marine shells and other forms of marine life;

Recommends to the South Pacific Commission and all governments and administrations concerned that relevant ecological studies should be a necessary prelude to approval of any development projects in coastal areas;

And further recommends that appropriate control measures linked to the findings of such surveys should be applied to regulate the building of hotels, the disposal of wastes, the gathering of sand and coral, the collecting and harvesting of marine shells and other forms of marine life as well as other types of exploitation so as to reduce or remove environmental disturbances resulting from such activities.

Delta of the Nile River and Surrounding Regions as Photographed from *Gemini VII*.

The river is within us, the sea is all about us . . .

—T. S. Eliot
"The Dry Salvages"

*All rivers run into the sea
yet the sea is not full: Unto the place from which
the rivers come thither they return again.*

—Ecclesiastes 1:7

The Hydrologic Cycle: Rivers

23. Water and Life: River Resources and Human Concerns

David H. Stansbery
The Ohio State University, U.S.A.

Rivers, Water, and Life

Since long before recorded history, rivers have been a valued part of Earth's landscapes. Evidence from prehistoric sites indicates that a source of fresh water was not only a prime factor in man's selection of the sites, but in many, perhaps in most, cases, it was the water supply that determined the success or failure of the venture.

We have been so generously supplied with this precious resource that it seems almost trite to say that life cannot exist without water. But how often do we pause to consider that each of us had our very origin and embryological development in an aqueous medium, or that we terrestrial beings are modified aquatics who have never lost our dependence upon water?

Each of us lives today in a sort of leathern sac of dead cells— our skin—which holds the living part of our being perpetually bathed in tissue fluid. We cannot taste or smell unless the substance of our interest becomes dissolved in the layer of moisture that covers those living cells capable of sensing its presence. Breathing, also, depends upon the inhaled gases' first dissolving in the moisture-lined alveoli of our lungs. We spend the waking hours of our lives looking at the world—but always through a film of water. Drying is indeed dying. But to look only at our individual need for water puts us in the position of not being able to see the forest for the trees. Our physiological dependence upon water is only the beginning.

Rivers and the Grand Cycles of Nature

Rivers play a dual role in two of the most fundamental processes on Earth: since their very beginning, these ribbons of flowing water have been the agents of transport in both the hydrologic and the geologic cycles. Thanks to the evaporative power of the sun's rays and the relentless pull of Earth's gravity, both the water and the land across which it courses to the sea are kept in constant motion: the water that falls on the crust of the Earth carries with it in its downward journey those soluble substances it encounters, as well as the more obvious sediments of clay, silt, and sand. These are eventually transported to the slack water of the continental shelf and beyond. Were it not for the net uplift of the continental masses, the Earth long ago would have become completely covered by the seas.

To what extent does man understand the many ramifications of these intertwined phenomena—or his relationship to or dependency upon them? The amounts of matter and energy involved in these cycles are almost beyond comprehension. If we should significantly alter the cycles in any way, surely it should be done with deep understanding, care, and caution.

As important and far-reaching as the values mentioned above are, modern man typically does not put them on his list of the benefits of rivers. They are, quite naturally—though somewhat dangerously—taken for granted as permanent features of our existence.

Rivers and Modern Man

The present-day values of rivers include the use of their water for agriculture, industry, recreation, and domestic supply. The larger rivers are increasingly used as major arteries of transportation, while the burden of other services on almost all streams grows as the human population expands. Efforts to increase the utility of rivers to man have included channelization, dredging, diversion, and impoundment. These modifications have led to the loss of many of the natural values of rivers, without any appropriate compensation.

Despite our advanced technology, the use of rivers as recipients of refuse has continued over the years. The raw sewage of yesteryear has been largely replaced by sewage-plant effluents with nitrates and phosphates in eutrophic concentrations. Legislators have been modestly successful in eliminating some toxic chemicals from industrial outfalls—but increasing amounts of herbicides, fungicides, insecticides, and fertilizers from our farms now wash into streams. The river biota has *not* been improved.

Man is more aware today than ever before of the damage being done to rivers, and there is abroad the general belief that substantial progress is being made toward correcting the situation. This belief may be well-founded, but as yet the biota of our rivers reflects almost no improvement.

Points of No Return

Currently there is great concern in many areas of environmental science that we take care not to pass any points of no return. This approach is commendable because experience has shown that solutions, once effected, may be greater problems than the problems they were designed to correct.

It is a decided advantage to have "a second time at bat." The physical and chemical alteration of our rivers and their floodplains has been destructive and costly. All of these changes, however, are correctable in large part if we but remove the cause and give nature time.

But this is not always true of the life in or associated with flowing streams. A species, once extinct, can never be brought back, regardless of our need. This fact is more significant when we realize that many, perhaps most, of our medicines and pharmaceuticals, and *all* of our foods come from plants or animals that either are wild species or were domesticated from wild species. While the rate of extinction is increasing, systematic examination of wild species for potential human benefits has really only begun. Recently, for example, we discovered that a mold produces an enzyme that catalyzes the transformation of cellulose into glucose. Perhaps we could change our waste paper into sugar, a major step forward in recycling. This advance would not be possible without this heretofore "worthless" species of mold.

We should be sincerely grateful that the ancestor of the common cow did not become extinct with the mastodon. We should also be thankful that we did not use one of our fungicides to rid ourselves of the mold *Penicillium* before we discovered the value of penicillin as an antibiotic.

The Need for Action

The need for action to preserve river species becomes especially clear when we realize that only about 2 percent of the Earth's surface is fresh water and that most of this is lakes in which most river species are absent. Few people are aware that the oceans of the world, vast as they are, contain only 7 percent of the animal species of the world, while 12 percent of our animal species are found in fresh water. Expressed per unit of habitat area, the species diversity of our rivers is sixty-five times greater than that of the seas.

Whatever the reasons for this striking difference, we are left with the realization that the smallest major habitat area on Earth—our rivers—has the greatest number of aquatic species. The additional observation that the river habitat has undoubtedly suffered the greatest damage points up the need for prompt, effective measures to prevent our passing additional points of no return. Preserving a river in its natural or near-natural state will result in the preservation of most of the river biota, as well as of other natural values.

Today we have the knowledge and the technology to save our rivers. The question is, "Do we have the necessary concern—both for ourselves and for those who will come after us—to save them?"

24. Conservation Problems in the Amazon Basin

Ghillean T. Prance
The New York Botanical Garden, U.S.A., and
Instituto Nacional de Pesquisas da Amazônia, Brazil

Introduction

The Amazon River, said to be the longest in the world, extends 6,740 kilometers to the sea, draining a watershed of 7 million square kilometers. The Amazon also transports the greatest volume of water of the world's rivers. Seventeen of its tributaries are over 1,500 kilometers long. In all, there are 30,000 kilometers of readily navigable waterways below the first falls of the rivers. Most of the Amazon's watershed is low-lying and covered predominantly by tall rain forest. I will concentrate on the lowland area since my experience is confined to that part of the watershed region.

Despite a long history of biological exploration, the Amazon region remains one of the world's least-known ecosystems. The tragedy of Amazonia is that its natural vegetation is being destroyed before we gain the understanding that would enable us to utilize the area in an ecologically sound manner. In Brazil, the Amazon region is rapidly being cut up by a network of new highways. The most ambitious projects are the trans-Amazonia highway, which extends 6,368 kilometers from east to west, right across the region of the Amazon itself, and the Northern Perimetral Highway of 2,465 kilometers, which runs east to west, just south of Brazil's northern frontier. Peruvian Amazonia is also being traversed by new highways.

By themselves, these highways do comparatively little damage to the region, and there are strong justifications for them as aids to the integration and development of the countries involved. However, the highway programs are accompanied by large-scale colonization programs, which involve the cutting and burning of vast areas of forest by primitive methods unsuited to the local conditions. It is this slash-and-burn agriculture which really threatens the Amazonian ecosystem.

The largest fire in the world during the year 1974 occurred in the Amazon forest, an area that does not catch fire naturally. The Volkswagen Company of Brazil, trying to establish a ranch, sprayed an area of 10,000 square kilometers with a defoliant, waited for the leaves to dry, and then set fire to the entire area—in spite of the fact that they had signed an agreement, in conformity with Brazilian law, to cut only half the area and to preserve the other half as forest. They were attempting to diversify their interests by setting up an enormous cattle ranch because they felt the motor industry was threatened by the worldwide petroleum crisis. It is the consensus of agronomists and soil experts that the area will be able to sustain cattle economically for at most fifteen years. If this kind of destruction is allowed to continue, and if United States chemical companies continue to supply defoliants in the quantities used for the Volkswagen project, the Amazon forest will quickly disappear.

The Amazon Ecosystem

Before discussing in more detail the threats to the Amazon ecosystem and making a few suggestions for their correction, I would like to say a little about the natural ecosystem and its equilibrium, since there have been many exaggerated and erroneous reports in the popular press and in some semiscientific articles.

Vegetation

Approximately 90.5 percent of the region is covered with dense forest. One of the gravest misconceptions about the area is that it is a vast, uniform mass of vegetation. Much of the area covered with forests does have a uniform physiognomy, but there is rich botanical diversity varying with local conditions, such as rainfall, degree of flooding, and type of soil. Apart from the tall rain forest, about 9.5 percent of the area is covered with other types of vegetation, such as savannahs and *campinas,* and by open-water surfaces. (A summary of the vegetation types is given in Pires [1973].) Nearly 90 percent of Amazonia is covered with forest on nonflooded ground. This does not mean that the same few plant species occur throughout the region. In fact, while a small group of common plant species of wide ecological amplitude is distributed throughout the forest areas, the vast majority are of quite local occurrence, which means that an entire population of a particular species could be eliminated by the deforestation of a fairly small area. This must be kept in mind when one is planning reserves for the area. If only a few widely separated tracts of Amazonian forest are preserved, then species that do not naturally occur in those areas could be eliminated.

The factors influencing the present distribution of tree species in Amazonia are complex and still far from obvious. First, the sheer number of species is important. There are usually over 120 tree species in a single hectare of forest, but in another hectare only a few kilometers away, half of the species may be different. The number of species at a site depends on the soil's fertility and depth, drainage, water table, rainfall, etc. In addition to these present-day variables, the history of the region has played an important role in the diversity and distribution of species. Worldwide climatic changes during the Pleistocene Epoch and in more recent times also affected the Amazon region profoundly. During periods of drier climate, the amount of savannah increased, isolating the forest into patches. This in turn isolated species into two or more disjunct populations, increasing the chances for evolutionary change before the continuous forest cover returned with a moister climate. (This aspect of the forest was first described by Haffer [1969] and is summed up by Prance [1974]).

Although only about 2 percent of the total area of forest is on ground subject to flooding, this area is important both ecologically and in the future development of the region. The flooded forest is divided into two main types: periodically flooded forest (called *várzea*) and permanently flooded swamp forest *(igapó)* (Figure 1). The vegetation in the flooded areas tends to be less diverse than that of the upland forest, as shown in Table 1.

Many species in the flooded forest are widely distributed throughout the region, and the species variation between sample plots from widely separated flooded areas is much less than that of samples from forest on high ground. In addition, there is the influence of the type of water flooding the forest. There are very muddy white waters, clear waters stained black by plant compounds, and less frequent clear water. Certain riverine plants grow only in one water type, mainly because of the difference in acidity. For example, the famous *Victoria amazonica* water lily grows only in white-water

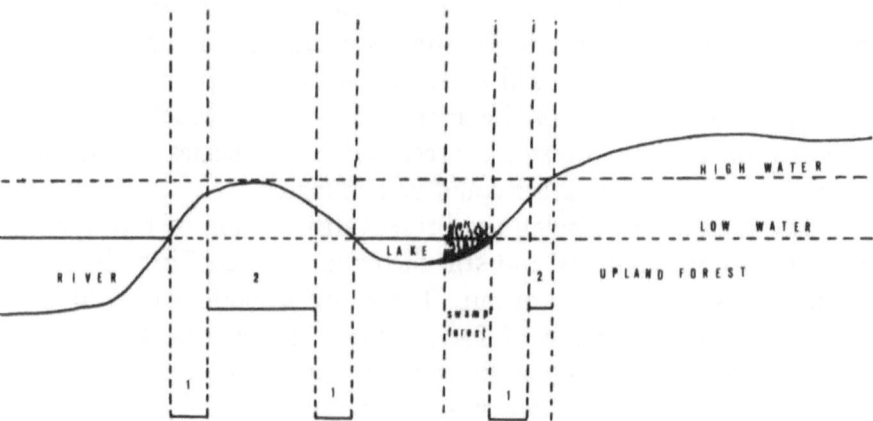

Figure 1: Diagram of the Relationship Between Forest Types and the River. There are two principal types of forest: upland (nonflooded) forest, and lowland forest—either permanently flooded swamp forest or periodically flooded *várzea* forest. (1 = often flooded; 2 = flooded for only a few months each year.)

areas. The white water of the Rio Solimões has a pH of 6.9 to 7.4 (Schmidt 1972a). In addition, there is considerable variation in nutrients and humic matter. [Schmidt (1972a) gives the humic content of the three main types of Amazon water as follows: Rio Negro (black water)—26.6 mg/liter; Rio Solimões (white water)—14.1 mg/liter; Rio Tapajós (clear water)—2.26 mg/liter. Further details about Amazonian water contents are given in various studies of the Max Planck Institute for Limnology in Plön, for example, in Anonymous (1972a); Schmidt (1972a, 1972b); Sioli (1965, 1967); and Williams et al. (1972).]

Table 1. Comparison of Three Forest Types in Eastern Amazonia (sample plot of 5 hectares)

Type of Forest	Number of Plant Families	Number of Genera	Number of Species	Number of Individuals
Upland forest	52	136	224	2,607
Periodically flooded forest	45	124	196	2,912
Swamp forest	47	118	180	2,792

Source: João M. Pires, unpublished thesis.

Figure 2 shows the annual fluctuation in the level of the Rio Negro at Manaus Harbor between 1965 and mid-1974; it reveals that the yearly range in level became increasingly less during the last three years.

Since there is less variation in the number of species in the riverine areas, fewer conservation areas would be required to maintain this vegetation than in the upland forest. However, there is also a notable amount of endemism in the flooded areas. To give an example, *Polygonanthus amazonicus* (Rhizophoraceae) is known only from the swamp forest around Maués in Amazônas, Brazil. This species, which has been placed in five different plant families, apparently has a very local distribution: the only other species of the genus has a similar local distribution in the swampy areas of the upper Rio Negro.

Whereas it is foolish to introduce agriculture on much of the upland soil, the alluvial soil beside the white-water rivers has high

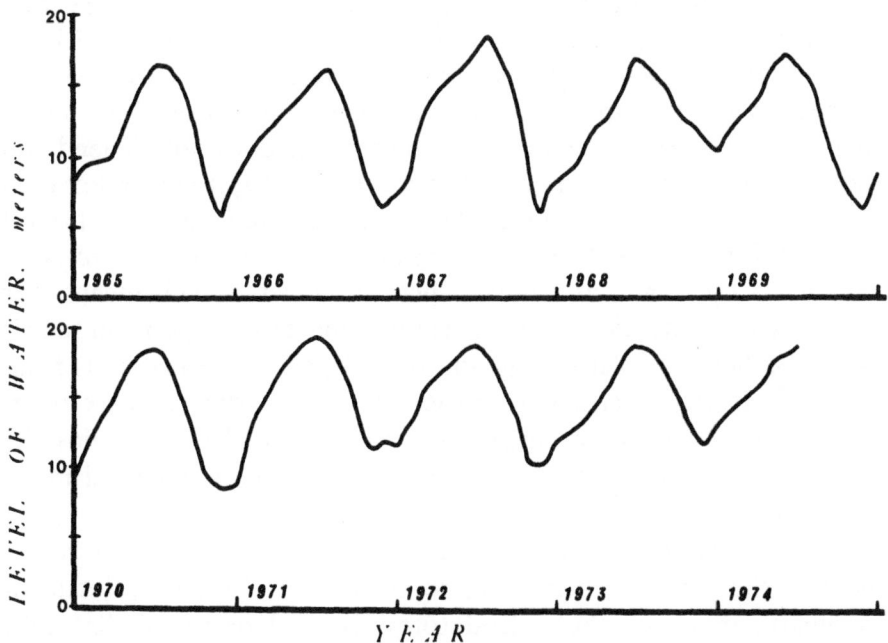

Figure 2: Graph of the Annual Rise and Fall of the Rio Negro at Manaus Harbor, 1965 through Mid-1974. Notice that during the last three years the low-water level was progressively higher.

potential; that of the black-water areas has very little agricultural potential. Efficient utilization of the alluvial areas could help conserve more of the forest on higher ground. These areas obviously have special utilization problems, but they are highly suited to flood-tolerant crops, such as rice and jute. The potential of rice has already been demonstrated in the Jarí project in eastern Amazonia, where exceptionally high yields of rice are being obtained. Conversely, upland rice grown by colonists along the trans-Amazonia highway in the region of Altamira has proved a complete failure. The value of floodplains has long been pointed out by Professor Felisberto C. de Camargo (1951). In the floodplain areas the need is for fast-maturing crops such as beans or corn. Riverine agriculture has the added advantage of the river transport system.

Soil and Biomass

Conservationists have placed much emphasis on the poverty of the Amazonian soil. But there is considerable variation in soil type within Amazonia, and recent surveys have emphasized this point and have, in fact, found some large areas of quite fertile soil. However, it is correct to say that the majority of the Amazonian forest is on a very poor soil. But even the forest on better soil does not depend too heavily on the soil: the rich vegetation exists in spite of, rather than because of, the soil. Analysis shows that the biomass of a typical tropical forest consists of 44 percent carbon, 45 percent oxygen, 6 percent hydrogen, and only 5 percent minerals. This means that 95 percent of the biomass is made up of materials fixed from the air with the help of solar energy and only 5 percent of materials from the soil. It is this efficient fixation of energy, the complex interactions of the different organisms, and so on, that maintains the forest. Bearing this in mind, we can see how such a poor soil maintains such a rich forest.

The soils of Amazonia are described by Falesi (1972), but it is beyond the scope of this paper to treat the soil types in detail. In summary, much of the upland forest is on laterite soil (latosol), which is poor acid soil (pH 3.5 to 4.2); it should not be used for agriculture based on present techniques. Similarly, it is wasteful to cut down forest on white-sand areas (regosol), although this has happened in many such areas around Manaus. A much more fertile soil is that called *terra roxa* by Brazilians (alfisol). Recent surveys, e.g., that of Silva and Carvalho Filho (1973), have shown that perhaps 10 percent of the forest is on this much better soil.

So far, the areas devastated by colonization along highways have not properly taken into account the distribution of the better soils. Much of the forest already felled—for example on the Manaus-Pôrto Velho highway—is on latosol. In areas subject to slash-and-burn agriculture, the amount of soil leaching is enormous. Brinkmann and Nascimento (1973) studied the effect of slash-and-burn agriculture on plant nutrients in a latosol area near Manaus. They showed that although burning initially releases a lot of nutrients to the soil, these are rapidly lost by leaching. The soil will yield a reasonable crop only in the first two years after burning; even then it is too acid for many crops.

Sioli (1970) points out that the absence of colloids in the soil to fix fertilizer salts makes much fertilization uneconomical because the salts are quickly washed out by the heavy rainfall.

Another important reason why rich forest can exist on poor soil is the existence of an efficient natural recycling process. This has been demonstrated by Stark (1971a, 1971b, 1972), who showed that the trees in the forest obtain nutrients from dead organic litter on the forest floor through the action of mycorrhizal fungi, which pass the nutrients directly to the living tree roots. This means that nutrients are recycled without ever entering the soil. Obviously, cutting and burning the forest destroys the organisms involved in the nutrient-cycling process. Studies of Amazonian rainwater from the Manaus region (such as those of Schmidt 1972a and Anonymous 1972b) reveal the extremely low mineral-ion concentration of the Amazonian rivers. For example, the litter falling to the forest floor contains 18.4 kilograms of calcium per hectare, yet the streams nearby have a calcium content too small to be detected. This demonstrates the efficiency of the recycling process in the Amazon ecosystem.

When an area is cut and burned, most of the original biomass is lost by the burning and by subsequent leaching, which alters the texture of the soil and destroys essential organisms. Thus, it is not surprising that in agricultural projects there is a progressive decline in yield, or that when the ground is left fallow it takes more than a century for the original forest cover to return. For proper regeneration, the area must be left untouched for about 100 years. Further, since regeneration requires a source of seeds of the original forest trees, it occurs with difficulty in large areas with little forest nearby. In many felled areas, the secondary forest is cut during regeneration and after only a partial recuperation of the soil. This impoverishes the soil further. The area then can no longer sustain even secondary forest and becomes bamboo thicket or bracken-covered and completely useless wasteland.

Another interesting aspect of the forest is that it supports comparatively little animal life in proportion to plant life. One hectare of upland forest contains about 900 metric tons of plant biomass and about 0.2 ton of animal biomass (Fittkau and Klinge 1973). There is a shortage of edible plants in the rain forest, partially because many plants have chemical defenses against herbivores. Man has always existed in the forest in low numbers, supported by slash-and-burn agriculture. We should not regard this area as one that could support dense populations; rather, we should look for the many forest products which come from an area maintained in its natural state. We must also remember the vital role of animals in maintaining the ecological equilibrium; without them, the high plant biomass would not exist.

Atmosphere

Many fallacies have been spoken about Amazonia as the "lungs of the world." This is a myth based on a misinterpretation of a statement made by Dr. Harald Sioli to the effect that the world depends on the Amazon forest for oxygen. In a climax forest, where there is a constant biomass, the amount of oxygen produced by photosynthesis is approximately equal to that consumed by plant respiration. On the other hand, nonclimax areas of secondary forest or agricultural fields, because they are not in equilibrium, produce more atmospheric oxygen. The forest of the Amazon plays the much more important role of absorbing carbon dioxide from the air. [*Editor's Note:* Amazonia's rain forests apparently do play an important role in regulating local-, regional-, and even global-scale weather patterns, however. For discussions of this interesting possibility, the reader is referred to the following papers: (1) Irving Friedman. 1977. The Amazon basin, another Sahel? *Science* 197(4298):7; (2) Gerald L. Potter, Hugh W. Ellsaesser, Michael C. MacCracken, and Frederick M.

Luther. 1975. Possible climatic impact of tropical deforestation. *Nature* 258(5537):697–698; and (3) Reginald E. Newell. 1971. The Amazon forest and atmospheric general circulation. In *Man's Impact on the Climate*, eds. William H. Matthews, William W. Kellogg, and G. D. Robinson, pp. 457–459. Cambridge, Mass.: The MIT Press.]

Threats to The Amazon Ecosystem

Some of the threats to the Amazon ecosystem have already been touched upon in the general discussion, but in a symposium of this nature it is important to treat them in more detail. Basically, areas of Amazonia are threatened by (1) colonization based on the new highways; (2) industries unsuited to the region; and (3) poorly planned hydroelectric projects.

Since 1970, the Brazilian government has had an ambitious project to construct a network of highways in Amazonia (Figure 3). The original justification for these roads was colonization and farm-

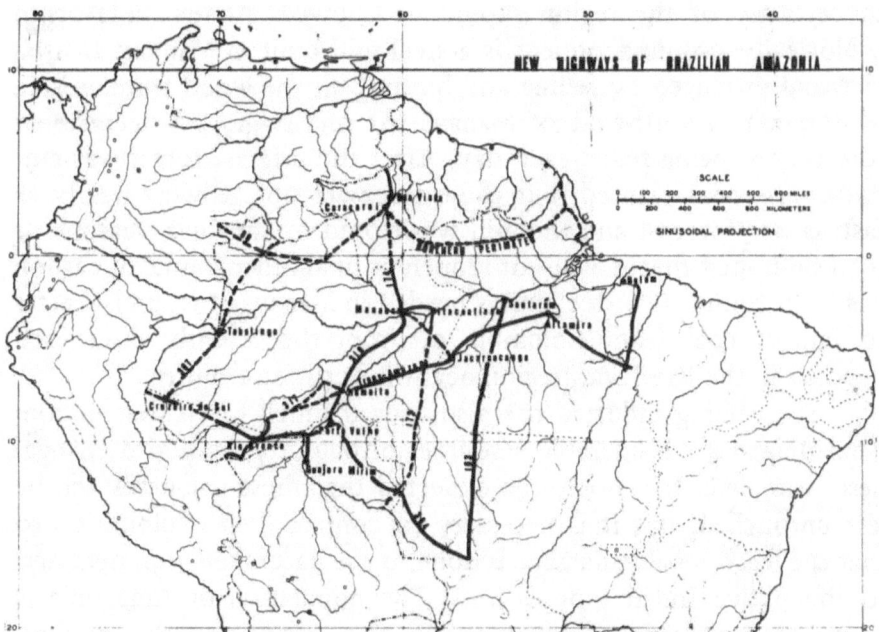

Figure 3: Map of the Principal Highways in Amazonia. Solid lines, passable in 1974; broken lines, under construction.

ing to relieve the dry, famine-stricken northeast region of Brazil. Organized colonization was implemented along parts of the trans-Amazonia highway. While the roads may really be necessary for the integration and security of Brazil, the earlier colonization programs have not relieved the northeast and have not been too successful in other ways, either. A considerable amount of forest has been cut and burned along some of the highways, and already some of the colonists have abandoned the cleared areas. Because of the methods used and the poor soil, virgin forest has become abandoned secondary forest or low-yield agricultural fields. Since these early efforts have not worked as planned, a new emphasis has been given to creating large ranches instead of small farms. The large ranches being formed are also using slash-and-burn methods. Much of the area around the Manaus–Pôrto Velho road has been divided up into lots for these ranches.

To date, industry has done comparatively little damage to the region, but we should be watchful. Only industries compatible with the ecology of the region should be allowed. An example of an ecologically unsound project is a steel mill built in Manaus. It used charcoal produced by felling and burning all the wood from an area of forest just northwest of Manaus. At one stage, 2.5 hectares of forest were being felled each day. After the original forest was cut, *Eucalyptus* was planted; but these trees are not growing nearly as fast as was forecast and the mill was closed for financial reasons. It is to be hoped that it will not reopen until another source of carbon has been found. If a steel mill is needed in Manaus, it is surely better to import coal from Colombia or from the recently discovered deposits in the Rio Javari region near the Peruvian frontier.

The mining industry has also caused much forest destruction. Amazonia has considerable resources of bauxite, cassiterite, manganese, iron, etc. It is not to be expected that these resources will be left untouched, but it is necessary to control their exploitation so that the least possible damage is done to the natural environment and to the native Indian populations. The population of Amazonia is bound to increase, and industries will be needed to provide employment. It is to be hoped that more suitable projects can be worked out.

Seven hydroelectric projects are being considered for Amazonia. None of them is actually as destructive or as drastic as one dreamed up in North America for Amazonia by the Hudson River Project.

Ecologically speaking, there are two sides to hydroelectric development: we have to balance the destruction of areas flooded behind the dam against the energy produced by falling water rather than the burning of fossil fuels. It is probable that some hydroelectricity would be ecologically better for the region than the present system of fossil-fuel-burning diesel generators and wood-burning thermoelectric plants. Wood-burning plants can destroy a nearby forest; for example, the forest around the small town of Lábrea has been devastated since one of these wood-burning plants was introduced. In contrast, hydroelectric dams create lakes in which fish farming is possible. The chief danger in planning these hydroelectric plants is politics. If dam sites are planned which minimize damage to the forest, avoid flooding areas of high plant or animal endemism, and include animal-rescue work (as was done in the Brokopondo dam in Surinam), only then will these hydroelectric projects benefit Amazonia.

Related to these threats to the ecological balance in Amazonia is the problem of disease—a subject I cannot treat in detail here. The medical aspects of man's spread into Amazonia have not been explored. Perhaps the most important diseases to study are malaria, schistosomiasis, Chagas' disease, and onchocerciasis (African river blindness). Both schistosomiasis and onchocerciasis are closely linked to the problems of river settlements and changes in the river's ecological equilibrium. Amazonia is naturally protected from schistosomiasis by the acidity of the water. But in some city streams, for example in Manaus, where lime has been used in agriculture, the water is no longer acidic, giving the opportunity for the vector snails to flourish. These and other medical problems must be considered before Amazonia is opened for extensive development.

Summary of the Dilemma

The tropical forest ecosystem is the most productive in the world in spite of poor soils, heavy rainfall, and a series of pathogens. The interactions among the plants, mycorrhizal fungi, insects, other animals, climate, etc., are extremely complicated. We are only beginning to understand the ecosystem. Present methods of utilizing the forest, often based on techniques developed for the temperate regions, are not efficient. Low agricultural production is the result of a process which first destroys the forest, then erodes the soil and renders large areas useless for many years. In a healthy forest the soil loses about 1 kilogram of organic matter per hectare; in a cleared area it loses 34 metric tons! We cannot expect an area as large as Amazonia to remain untouched as an "unproductive" biological reserve while population and famine increase worldwide. But it is unreasonable to cut down the forest for short-term gains when it has so much potential for solving man's long-term problems. The paucity of information about the ecosystem means that we do not know how to utilize the area without destroying the biomass and the gene pool of plants and animals of potential use for man.

Suggestions

Since it is not known how to exploit the Amazon rain forest in an economically and ecologically sound way, the most important need is for conservation. The ideal is to conserve the forest as it is or, failing that, to conserve large areas that will maintain the equilibrium of the forest. In selecting these reserve areas, we must bear in mind the regional variations of the forest, the minimum area needed to maintain various organisms, the maintenance of an adequate gene pool for regeneration of disturbed areas, and the selective use of the more productive areas for development. The conservation measures for the trans-Amazonia highway oblige each colonist to conserve 50 hectares of forest for every 50 hectares he cuts. The chess-board effect this produces does not take into account the minimum area various organisms need to survive, the resultant weed problems, or the change in the balance of phytophagous insects.

To organize an effective conservation program, a unified legal regime is needed. In Brazil (as in other countries) there are so many different government organs interested in preserving or utilizing the forest that areas reserved by one section are planned for development by another. Brazil does have an environment secretariat that is directly answerable to the president. This is commendable and would be even more effective if that one secretariat, or any other single institution (such as the Forest Service), had full power to act on conservation matters.

Another necessity is to intensify basic and applied research in Amazonia. Fortunately, this has begun with Brazil's *Programa do Trópico Úmido* (Humid Tropics Program) and the Organization of

American States' *Programa Interamericano de Desenvolvimento dos Trópicos* (the latter based too much on "green revolution" principles). Recently, Brazil began awarding a large amount of money to research projects in Amazonia. This research must emphasize interdisciplinary approaches involving ecology, pedology, agronomy, forestry, botany, zoology, chemistry, and many other sciences. With the results of these programs man will be in a better position to use the Amazon region wisely, through techniques devised in the tropics and not in the temperate zone.

Together with the need for conservation is the obvious need to divert the immediate colonization pressure from Amazonia. First, the present colonization program is not keeping up with the annual population growth rate in northeast Brazil and so is not relieving that problem. Second, it does far less harm to the environment to concentrate the populations of Amazonia in cities rather than in rural areas. Cities can be planned with industries which do not destroy the surrounding forest. Many new industries in Amazonia draw very little on local resources. In the free port of Manaus, for example, 28,000 new jobs have been created since 1968 in watch factories, television factories, etc.

It is to be hoped that the number of industries which use local resources in a sound manner can gradually be increased; when this occurs, it would not be necessary, for example, to ship all the Amazonian rubber to south Brazil for processing, or ship Brazil nuts to Britain to be tinned and then exported to the shops of the Manaus free port. Local industries in Manaus are using some local resources—for example, the jute industry, or the jewelry factories using local gold from the Santarém area.

The single most destructive industry in Amazonia is farming. To conserve the forest, farming must be concentrated in other areas. The *campo cerrado* to the south of Amazonia is much more suitable for agriculture, as are the savannahs of the northern frontier around Boa Vista and northern Pará. This development would have to be accompanied by properly planned transport facilities, including freezer trucks, etc. There is already considerable research being done in agricultural techniques for savannah and *campo cerrado* areas at the Centro Internacional de Agricultura Tropical (CIAT) in Cali, Colombia.

Pressure to produce meat could be reduced by a well-planned

fishing industry, turtle farming, the use of local animals, etc. In Brazil, research aimed at this is now under way in Amazonia, at the urging of the present minister of agriculture.

I have already mentioned that some parts of Amazonia have more agricultural possibility than others. Farming that does take place should be rigidly controlled; the areas chosen should be, in order of preference: floodplain areas *(várzea)*, the patches of forest on better soil *(terra roxa)*, and patches of savannah or *campo cerrado* within the forest. The rain forest on latosol and the white sand areas should be left untouched except for the actual roadways themselves. In advocating the use of two Amazon habitats (floodplain and rich soil), we must remember to reserve some of these areas to maintain the natural gene pool of biota confined to them.

As part of a program to select priority areas for economic development, the Superintendency for the Development of Amazonia (SUDAM) inaugurated in March 1975 a plan for utilizing 100,000 square kilometers of *várzea* in the Amazônas and Pará regions. The project was stimulated by successful use of floodplain land in the Rio Jarí project. While this represents progressive planning, development must be accompanied by rigid protection of both nonpriority areas and sufficient reserves of national floodplain areas to maintain the original forest habitat.

Since well-managed areas represent such a small part of Amazonia, research efforts must concentrate on how to use the much larger portion of forest on clay latosol. Such utilization will obviously include rational use of timber, including extraction of cellulose, and many other forest products. There are many economic resources in the forest: nuts (e.g., the Brazil nut, *Bertholletia excelsa*), fruits, seeds (e.g., tonka bean, *Coumarouna odorata*), tanbark (bark containing tannin—tannic acid), latex, resins, oils, etc. These products can be collected without killing the trees. Rosewood (*Aniba* sp.) and sorva (*Couma* sp.) trees are seriously threatened with extinction because they are felled to extract rosewood oil and latex, respectively. In the Old World tropics, some noneconomic species in the forest are killed to allow commercial species room to grow. This selective agriculture does much less damage than slash and burn. Refining the natural forest is preferable to monocultures. The scattered distribution of trees in the forest protects them from insects and disease, as was dramatically shown in the 1920s, when

Amazonian rubber plantations were attacked by the leaf fungus *Microcyclus (Dothidella) ulei*. It is too early to predict the results of some of the monocultures recently introduced in Amazonia, such as the *Gmelina arborea* in the Rio Jarí project and the *Eucalyptus* in the Manaus steel mill project. In looking to the future perhaps the Amazonian forest will produce methane and alcohol to aid world fuel problems and leaf proteins to alleviate world hunger.

So far, I have concentrated on plant products, but there are many other aspects to the conservation and utilization of the forest. The fishing industry, for example, faces a similar dilemma between conservation and utilization and is in desperate need of further research. Undoubtedly, the well-planned farming of the 2,000 species of fish in Amazonia will become increasingly important. Some native species of turtles face extinction because of exploitation by local people. Recent legislation prohibits the use of wild turtles, so turtle farms have been started in the city of Belém. Are domestic cattle the best source of meat for Amazonia? Perhaps in the future we should farm deer or tapir, both overhunted animals at present. Another marine animal, the manatee, faces extinction due to overhunting. The manatee is an important animal in controlling aquatic weeds in the river ecosystem. Perhaps manatee farming in lakes behind hydroelectric dams is possible in the future.

In concluding, I want to stress again the importance of conserving the whole range of organisms to maintain the equilibrium of the Amazonian ecosystem. Many plants depend on pollinators (e.g., insects, birds, and bats), dispersal agents, or protectors (e.g., ants protect some plants from other insects). This fascinating and complex balance can only be maintained by rigid conservation programs.

I have not treated in depth many of the problems in Amazonia which lie outside of my field of botany, but which must be studied in any ecological plan for Amazonia. Others are more qualified to discuss fisheries, health, zoology, climate, etc., but we scientists are linked together in our desire for the wise use of Amazonia.

In considering the Amazon region we must understand the desire for development in the countries involved. We need to work towards a balance between developing what is currently needed and conserving what is valuable and a future source of information to man.

References

Anonymous. 1972a. Die Ionenfracht der Rio Negro, Staat Amazonas, Brasilien, nach untersuchungen von Dr. Harold Ungemach. *Amazoniana* 3(2): 175-185.

Anonymous. 1972b. Regenwasseranalysen aus Zentralamazonien, ausgeführt in Manaus, Amazonas, Brasilien, von Dr. Harold Ungemach. *Amazoniana* 3(1):186-198.

Brinkmann, Wilhelm L. F., and J. C. Nascimento. 1973. The effect of slash and burn agriculture on plant nutrients in the Tertiary Region of central Amazonia. *Acta Amazonica* 3(1):55-61.

Camargo, Felisberto C. de. 1951. Reclamation of the Amazonian flood-lands near Belém. In *Proceedings of the United Nations Scientific Conference on the Conservation and Utilization of Resources* (Lake Success, New York, 17 August to 6 September 1949), vol. VI, Land Resources, pp. 598-602. United Nations, New York: Department of Economic Affairs.

Falesi, I. 1972. O estudo atual dos conhecimentos sobre os solos da Amazônia brasileira. In *Zoneamento Agriícola da Amazônia I.* Boletim Técnico No. 54, pp. 17-67. Belém: Instituto de Pesquisas e Experimentação Agropecuarias do Norte.

Fittkau, E. J., and H. Klinge. 1973. On biomass and trophic structure of the central Amazonian rain forest ecosystem. *Biotropica* 5(1):2-15.

Haffer, Jürgen. 1969. Speciation in Amazonian forest birds. *Science* 165(3889): 131-137.

Pires, João M. 1973. Tipos de vegetação da Amazônia. In *O Museu Goeldi no Ano Sesquicentenário,* Mário F. Simões, ed., pp. 179-202. Belém: Publicações Avulsas Museu Paraense Emílio Goeldi, No. 20.

Prance, Ghillean T. 1974. Phytogeographic support for the theory of Pleistocene forest refuges in the Amazon basin based on evidence from distribution patterns in Caryocaraceae, Chrysobalanaceae, Dichopetalaceae, and Lecythidaceae. *Acta Amazonica* 3(3):5-28.

Schmidt, G. M. 1972a. Chemical properties of some waters in the tropical rain-forest region of Central-Amazonia along the new road Manaus-Caracarai. *Amazoniana* 3(2):199-207.

———. 1972b. Amounts of suspended solids and dissolved substances in the middle reaches of the Amazon over the course of one year. *Amazoniana* 3(2):208-223.

Silva, Luiz Ferreira da, and R. Carvalho Filho. 1973. *Solos do projeto "Ouro Preto."* Ilhéus, Bahia: Centro de Pesquisas do Cacau.

Sioli, Harald. 1965. Bemerkungen zur Typologie amazonischer Flüsse. *Amazoniana* 1(1):74-83.

——. 1967. Studies in Amazonian waters. In *Atas do Simpósio sôbre a Biotica Amazônica* (Belém, 1966), vol. 3, Limnologia, H. Lent, ed., pp. 9-50. Rio de Janeiro: Conselho Nacional de Pesquisas.

——. 1970. Ecologia de paisagem e agricultura racional na Amazônia brasileira. In *II Simposio y Foro de Biología Tropical Amazónica* (Florencia and Leticia, Colombia, 21 to 30 January 1969), J. M. Idrobo, ed., pp. 268-279. Asociación pro-Biología Tropical. Bogotá: Editorial PAX.

Stark, N. 1971a. Nutrient cycling I. Soils. *Tropical Ecology* 12(1):24-50.

——. 1971b. Nutrient cycling II. Vegetation. *Tropical Ecology* 12(2):177-201.

——. 1972. Nutrient cycling pathways and litter. *BioScience* 22(6):355-360.

Williams, W. A., R. S. Loomis, and P. T. Alvim. 1972. Environments of evergreen rain forests on the lower Rio Negro, Brasil. *Tropical Ecology* 13(1):65-78.

25. Development Projects in the Nile Basin, with Special Reference to Disease Repercussions

Emile A. Malek
Tulane University, U.S.A.

Introduction

A river is a natural stream—usually a large one—that flows in a well-defined channel, drains a large area of land, and (usually) discharges into an ocean, a lake, or another river. In some cases, when a river reaches porous soil it soaks into the ground or, in certain arid regions, evaporates. The rivers important to us in the context of this conference are of the majority type, in that they discharge into another body of water.

The area drained by a river is its basin. The Amazon River,

which is 6,270 kilometers in length, has the largest basin in the world (7.2 million square kilometers); the Missouri-Mississippi is 6,260 kilometers long and drains a basin of 3.25 million square kilometers. The longest river in the world, the Nile, is 6,700 kilometers long and drains a basin of 2.85 million square kilometers. Some of the principal rivers in the world, their lengths, and the sizes of their basins, are listed in Table 1.

Many nations have developed along rivers, and their histories have been linked with the rivers. Man, recognizing his dependence on rivers and their significance to his welfare, has tried since the beginning of history to make full use of them and to guard himself against their dangers by controlling them. River works, varied in character and purpose, have included the prevention of flooding and the reduction of its effects, the improvement of nagivation, and the use of the rivers' water in cultivating land. Since time immemorial, the works have been constructed—first of earth, then of masonry—to form reservoirs for the storage of water for domestic purposes and for irrigation. For example, the Akkadians or the Babylonians built a huge earthen dam on the Tigris, the ancient Egyptians constructed a masonry dam on the Nile, and the Romans built several large masonry dams in Italy and northern Africa.

Table 1. Selected Data on Some of the Major Rivers of the World

River	Approximate Length (kilometers)	Size of Drainage Basin (square kilometers)
Amazon	6,270	7,175,000
Congo	4,660	3,695,000
Danube	2,780	830,000
Indus	2,740	965,000
Missouri-Mississippi	6,260	3,250,000
Niger	4,190	1,765,000
Nile	6,700	2,850,000
Ob'	5,150	2,980,000
Río de la Plata	3,750	3,100,000
Saint Lawrence	3,025	1,465,000
Volga	3,700	1,459,000
Yangtze	5,800	1,980,000
Zambezi	3,540	1,330,000

In recent times, engineering skills have been employed to construct various types of dams to harness river waters for multipurpose projects. The dams have provided hydroelectric power for industry and water for agriculture and transport. They have created large lakes for recreation and fisheries. Several more dams are projected in various developed countries, as well as in developing countries. (In the developing countries, such projects are greatly needed, especially to produce food for their ever increasing populations.)

Unfortunately, while there have been beneficial returns from the harnessing of rivers, there have also been unforeseen and unfavorable consequences. The balance among river beds, their banks, and their waters can be easily disturbed by man's projects. It does not follow, however, that dams and the harnessing of rivers should be condemned; ecological conditions that control biotopes (the smallest geographic units of habitats) vary from one area to the other, and it is thus unfair to assume that dams in general, or even most of them, cause irreparable malfunctions that hurt man and his environment. Rather, by thorough, multidisciplinary, and comprehensive preconstruction studies of each area, man can guard development projects against unforeseen and unfavorable effects—or at least minimize their dangers.

Turner (1971), himself an engineer, has shown that development projects and ecology can be made compatible. In my work over the years, I have collaborated with engineers for the control of parasitic diseases, schistosomiasis in particular. They have provided me with new ideas, and I have stimulated their interest in our work. I shall get back to dealing with these problems later in this paper, but first, I want to review case histories of the development projects that have been undertaken in some of the world's major river basins. I will emphasize projects on the Nile River, with which I am familiar.

The Zambezi River and Lake Kariba

The Zambezi River is the fourth largest of Africa's rivers and the largest of those that flow eastwards to the Indian Ocean. It is 3,540 kilometers long and drains 1.3 million square kilometers. Its main channel is clearly marked from beginning to end. A well-marked belt of highland separates the Zambezi's basin from that of the Congo, the latter river flowing west. Important tributaries of the

Zambezi are the Kabompo, the Lungwebungu, the Kwando, and the Shire, the last representing the drainage of Lake Nyasa.

Lake Kariba on the Zambezi, formed behind the Kariba Dam, began to fill towards the end of 1958 and reached its maximum extent in 1963. It is situated on the Zambia-Rhodesia border and has a surface area of about 5,200 square kilometers. It is 269 kilometers long, has a maximum depth of 120 meters, an average depth of 45 meters, and a gross capacity of over 158.6 billion cubic meters (billion in the American and French usages—i.e., a thousand millions [10^9]). The lake is oriented approximately northeast-southwest and is situated in the middle section of the Zambezi. The Kariba Dam was the first of the major African impoundments built for hydroelectric power and irrigation.

Scudder (1972) discussed the consequences of the lake, in particular the unavoidable relocation of about 57,000 people. Fish production in Lake Kariba has been disappointing, probably for ecological reasons that apply only to this lake. First, the fish are protected in the areas that are still heavily vegetated and in large areas where there are submerged trees. A rapid buildup of the aquatic fern *Salvinia auriculata* created, together with other vegetation, a swamp or *sudd*. By 1962, the sudd had covered about one-tenth of the lake's surface; thereafter, its extent diminished but is still significant. Second, the fish fauna was adapted to living only in the free-flowing Zambezi, a river that has no natural lakes.

In spite of several control measures taken against Bovine trypanosomiasis and tsetse flies, before and during the resettlement of the human population and its cattle herds, the situation was aggravated by the early 1960s. Hira (1970a, 1970b) reported on the increased prevalence of human schistosomiasis in the Lake Kariba area on the Zambian side, at Siavonga. He gave prevalence figures of 31 percent for the urinary form of the disease caused by *Schistosoma haematobium,* and 1.8 percent for the intestinal form of the disease caused by *S. mansoni.* The snail hosts, which are breeding in the lake, are *Bulinus (Physopsis) africanus* and *Biomphalaria pfeifferi,* respectively.

Volta River and Lake Volta, Ghana

The Volta River is the largest river on the coast of upper Guinea, West Africa. It extends between the Gambia and the Niger Rivers and is 1,400 kilometers long. The name Volta was given on account of its winding course. It has two main upper branches, the White Volta and the Black Volta.

The Akosombo or Volta Dam, constructed on the Volta River at Akosombo, Ghana, is 123 meters high; the gross capacity of its reservoir is 146 billion cubic meters. The length of its crest is 700 meters. It has been reported that 1 percent of the people of Ghana have been relocated in connection with the Volta Dam.

As to disease repercussions, the change in the ecology of the area and the relocation of the people have resulted in a rise in schistosomiasis. The construction of the dam ended in 1964, and snails of the species *Bulinus (Bulinus) truncatus rohlfsi* were found in the inundated regions in 1966. The explosive growth of aquatic weeds, particularly *Pistia stratiotes*, the later *Ceratophyllum demersum*, created favorable conditions for the survival and growth of this snail in crowded and heavily polluted shores near towns. This was followed by an outbreak of schistosomiasis in several townships alongside the newly formed lake (Paperna 1970).

Infection was sporadic in communities on the eastern bank of the Volta, where the mean infection rate of urinary schistosomiasis was below 5 percent. Within two years after the onset of the infection in 1967, almost all local children had become infected. In other sites, however, stabilization of the environment in the lake rendered conditions favorable for the snails. At present, a World Health Organization/United Nations Development Program team is investigating the situation and collecting excellent baseline data in order to recommend the most effective measures for controlling schistosomiasis. The snail *Biomphalaria pfeifferi*, which is the host of *Schistosoma mansoni*, has not been found in the lake; it is believed, however, that it may be present in the northern part of the lake.

The creation of the lake displaced about 80,000 people, most of whom were resettled in fifty-two villages. The lake's shore, about 6,400 kilometers, has attracted between 70,000 and 90,000 fishermen of the Ewe tribe, who have settled in about 1,000 new villages along the shore.

Onchocerciasis, or river-blindness, has been endemic in Ghana for some time. The situation with regard to this disease has changed: in some foci its prevalence has decreased, while in others it has increased as a consequence of the new lake. The transmitters of the parasite are black flies, *Simulium damnosum*, whose larvae are aquatic and survive and metamorphose in flowing waters, primarily among rapids. Some of the breeding places have been inundated, whereas new breeding places have been created elsewhere.

Detailed reports published before the Akosombo Dam was constructed predicted the repercussions with reference to disease (MacDonald 1955). Moreover, Hughes (1964) studied the damsite and the surrounding area. He saw the vector of onchocerciasis, *Simulium damnosum*, around the damsite and especially along the Kpong and the Senchi rapids, which are some 8 to 11 kilometers downstream. Larviciding of the Volta River with DDT above the dam caused a considerable decline in the number of flies rather quickly; however, small spillways not far north of the lake created new breeding places for the black fly vectors.

Burton and McRae (1965) found that a small dam at Soe in Upper Region, Ghana, with a vegetation-free and obstruction-free concrete spillway, was a manmade breeding place for the black fly, *Simulium damnosum*. The area is endemic for onchocerciasis, but had previously been free of vector breeding. Those authors referred to several other similar sites in Ghana and elsewhere.

Niger River, Nigeria

The Niger River, one of the great rivers in West Africa, is 4,200 kilometers long and drains a large basin. There is a vast curve below its headwaters, not far from the ocean. It flows northeast, east, and then southeast, until it enters the Gulf of Guinea through an immense delta. The Kainji Dam in Nigeria, the first dam constructed on the Niger River, was completed in 1968. It has a concrete gravity section and an earth section. The concrete section is 71 meters, and that of the earth section is 79 meters high. Its crest is 8.75 kilometers long. The dam created Lake Kainji, which covers about 1,475 square kilometers and has flooded an area formerly occupied by about 50,000 people. Several United Nations organizations have been studying the ecological and disease consequences of Lake Kainji.

Dez River, Iran

The Dez River arises in the Zagros Mountains in northern Iran and flows south into the Persian Gulf at Khorramshahr. The Pahlevi

(Dez) Dam, built on the Dez River in the Zagros Mountains, was completed in 1964. The dam is of the arch type and is 223 meters high. Because it is situated in a narrow area between the mountains, its crest is only 262 meters long. The gross capacity of its reservoir, which extends for about 12 kilometers above the dam, is 3.3 billion cubic meters.

About 12,100 hectares are irrigated with water from the reservoir, and 50,500 hectares more are undergoing irrigation. The irrigated area is situated between Dezful and Ahwaz, in Khuzestan Province, an eastern extension of Mesopotamia. Because of effective control measures against several diseases in the area—in particular schistosomiasis and malaria—there have been no apparent unfavorable repercussions of this development project. On the contrary, the incidence of schistosomiasis, for example, is now far less in Khuzestan than it was before the dam was built. The World Health Organization supported a schistosomiasis research and control team in the area, which has been operated entirely by very capable Iranians since 1967. The Plan Organization in Iran, the Khuzestan Water and Power Authority, and the Near-East Foundation, have a great interest in Khuzestan, and have been actively supporting the control of parasitic diseases in the region.

Mekong River

The Mekong River is 4,160 kilometers long and is the world's eleventh longest river. It arises in the Himalayan plateau of China, near Tibet, and divides and waters the Indochinese Peninsula before it flows into the China Sea.

The Mekong River Development Project plans several dams along the Mekong basin for power, irrigation, and flood control, and to attract foreign investment to the participating countries—namely, Thailand, Cambodia, Laos, and South Vietnam. Some of the dams projected for the Mekong basin have been completed (two in Thailand, one of them the Ubol-Ratana or Nam Pong; the Nam Ngum in Laos; and one in Cambodia near Phnom Penh). The biggest and most ambitious of those planned is the $1 billion Pa Mong Dam between Thailand and Laos. Billions of kilowatts are to be produced from these dams, and it is expected that agricultural production will be raised to feed the region's rapidly increasing population.

Although one of the goals of the Pa Mong and other dams on

the Mekong is to improve fisheries for the people of the area, there might be some losses in fisheries in the Mekong itself through interruption of spawning and other needs. There will also be losses in marine fisheries off the Mekong Delta, because of the loss of sediments and nutrients essential for some marine organisms. Fortunately, the Mekong river project has been under the supervision of and under extensive study by the United Nations and the Southeast Asia Bureau of the United States Agency for International Development, and adequate consideration has been given to unfavorable effects.

Ubol-Ratana (Nam Pong) and Nong Wai Dams, Thailand

The Ubol-Ratana (Nam Pong) and Nong Wai dams were built on the Mekong River in northeast Thailand, in Nam Pong District of Khan Kaen Province—the former for hydroelectric power and irrigation, and the latter for irrigation. They were constructed in 1963 and have been in operation since 1966-1967. The Ubol-Ratana dam was built at a cost of about $60 million. It is yielding electricity at the rate of $1 million per annum, and the fish yield from the reservoir is about 2 million kilograms per annum (a yield of 8 million kilograms had been predicted). The reservoir is a manmade lake of about 405 square kilometers. The topography of the area—forests, fields, villages, swamps—was changed, as were the natural fauna, insects, water animals, poultry, and cattle. Development of irrigation schemes, which are supporting about 10,000 families downstream, and resettlement of the villagers from flooded areas resulted in a movement of the population in and out of the area.

A study conducted by Harinasuta et al. (1970) to determine the impact of the dams on the incidence of disease indicated that there was a high prevalence of intestinal-parasitic and liver-fluke infections (opisthorchiasis) in all the areas studied in the vicinity of the dams, and also downstream. The snail *Parafossarulus striatulus*, first intermediate host of *Opisthorchis viverrini*, is breeding very well in the lake. There was no evidence of schistosomiasis (it would have been due to *Schistosoma japonicum*), or of paragonimiasis (it would have been due to *Paragonimus westermani* and other species in Thailand). *Angiostrongylus cantonensis* infections were high among rats (the nematode worm occurs in the lung), and in the aquatic-snail hosts, *Pila* sp., and thus it is a potential threat because the natives frequently eat raw snails. Malaria is unlikely to be a risk, because its incidence among the inhabitants at the time of study was very low in all the areas.

The São Francisco River, Brazil

In the American tropics and subtropics several development projects are proposed for areas where certain important parasitic diseases are endemic. This region is discussed in the preceding paper by Dr. Ghillean Prance. However, I shall only briefly mention the northeast of Brazil and the São Francisco River, also in Brazil, because I have made some personal observations in that region, and while in Brazil again last year (1974) I had the opportunity of visiting some parts of the São Francisco Basin where development projects are planned.

Of all the regions of the vast country of Brazil, only the northeast is arid: it receives very low rainfall and is subject to long drought periods, in some places of three months' duration. The aquatic snail hosts of *Schistosoma mansoni* are adapted to such conditions, and some can survive the drought by aestivation. A government agency, Departamento Nacional de Obras contra Seca (DNOCS) has built in the last two decades about 150 public water-storage reservoirs, some as large as 20 million cubic meters, in addition to about 350 private reservoirs, with a minimum size of 300,000 cubic meters. The stored water is used mainly for water supply. Most of the dams are in the state of Ceara. Many of them harbor the snail hosts of schistosomiasis mansoni.

In recent years, Brazil's Ministry of the Interior has planned several irrigation schemes in the São Francisco River Basin. The Superintendencia do Vale São Francisco (SUVALE)—which is somewhat similar to the Tennessee Valley Authority—coordinates all the planned projects and their possible repercussions with regard to disease, resettlement, and socioeconomic problems. Some projects involving 25,200 hectares are under construction or in operation in four states. Most of these have been implemented as experimental projects to be followed by other projects involving a much larger acreage.

In the Barreiras region of the valley, in the state of Bahia, I visited an experimental area at São Desideiro, where the reservoir behind the dam fortunately harbors the poorly susceptible snail host, *Biomphalaria straminea*. However, above the reservoir a narrow and shallow natural stream supports a large population of the highly susceptible snail host, *Biomphalaria glabrata*, some of which were infected. At another village, Catolandia, canals in cultivated areas harbor *B. glabrata*, of which I found 25 percent to be infected with *Schistosoma mansoni*. This is considered to be a high infection rate. The incidence of *S. mansoni* among the inhabitants of Catolandia and vicinity is as much as 40 percent.

The Nile Basin

The name "Nile" comes from the Latin *nilus* and the Greek *neixos*. In *The Odyssey* (ca. 600 B.C.), the river's name, as well as that of the country (Egypt) through which it flows, is "Aiyurros," a word that survives in both the name "Egypt" and in the name "Copt" (*Gupti*, in the Arabic of Upper Egypt). The history of Egypt and its ancient civilization is intimately linked with the Nile River—"Father Nile." The communities that developed around the Nile, even during the Neolithic period, were not mere gatherers of food or hunters: they practiced agriculture.

The Nile is the longest river in the world, 6,693 kilometers from the source at the head of the Luvironza, about 64 kilometers east of Lake Tanganyika, to the entry of the Rosetta branch into the Mediterranean. (The Missouri-Mississippi is 6,260 kilometers long according to the United States Army Corps of Engineers, and the Amazon is 6,275 kilometers long.) The Nile is interesting to geographers because it is so long and because, unlike other great tropical rivers, it flows into the Mediterranean, originating south of the equator and flowing north across thirty-five degrees of latitude. It thus passes through regions with a variety of climates, biomes, and peoples.

The Nile River Basin (Figure 1) covers approximately 2.6 million square kilometers, or about one-tenth the area of Africa. The basin encompasses the countries of Burundi; Uganda; parts of Tanzania, Kenya, and Zaire; most of the Sudan; part of Ethiopia; and the cultivated portion of Egypt. The Luvironza River (about 2,200 meters above sea level) in Burundi joins the Kagera River, the principal river flowing into Lake Victoria; the Victoria Nile (in Uganda) flows northward out of Lake Victoria,

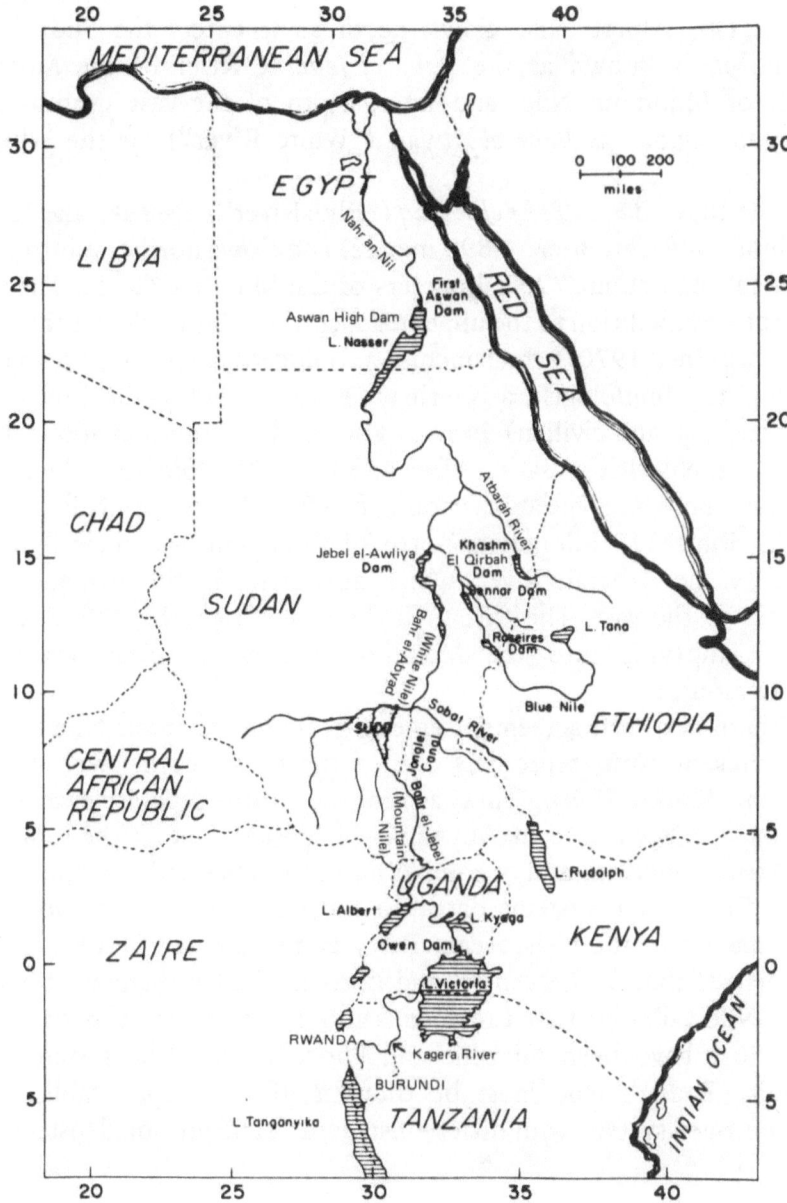

Figure 1: The Nile River Basin. Map shows the political boundaries, the Nile's headwaters in the Central East African Plateau, and the reservoirs and dams constructed since 1902.

proceeds through the swampy Lake Kioga, and then into Lake Albert. The Albert Nile flows north as it enters the Sudan at Nimule; it is known as the *Bahr el-Jebel* ("River of the Mountain"), or Mountain Nile, and later, north of the vast swamps of the Sudd region, as *Bahr el-Abyad* ("White River"), or the White Nile.

The Blue Nile, or *Bahr el-Azraq* ("Blue River"), rises at Lake Tana in Ethiopia (elevation ca. 1,890 meters) and flows northeast into the Sudan through about 770 kilometers of the Blue Nile Gorge. The recent Abbai Expedition to the upper reaches of the Blue Nile in Ethiopia (Blashford-Snell 1970), which included zoologists, archeologists, an ornithologist, a limnologist, a veterinary surgeon, and medical officers (both military and civilian), provided detailed information about this part of the Nile River Basin. After another 770 kilometers, the Blue Nile joins the White Nile at Khartoum, forming the "main" Nile *(Nahr an-Nil)*. Some 310 kilometers north of Khartoum, the main Nile is joined by the Atbarah River, which also arises in the highlands of Ethiopia to the east. The Nile then flows north another 2,500 kilometers, emptying into the Mediterranean at the Damietta and Rosetta mouths.

There has been agreement among geologists that the Nile River in its present form represents one of the most recent rivers (Ball 1939 and Shukri 1950). They assume that three separate river systems once existed—one in Abyssinia (Ethiopia), one in the Central East African Plateau, and one in Nubia and Egypt—and that the linking of these three systems dates no farther back than the second major pluvial of the Pleistocene. Berry and Whiteman (1968), however, believe that in the central and northern Sudan there must have been a Nile valley in Late Cretaceous and Early Tertiary times. Also, there must have been Miocene and Pliocene Blue Nile systems in Ethiopia. Today's Nile must be thought of as a long-established, complex river system with different structural and erosional histories.

Hydrology of the Nile

If we are to deal with the effects of development projects on the Nile's ecosystems and with the conservation of the Nile, we must understand the river's hydrology. As Worthington (1972) has emphasized, one should treat freshwater environments from the perspective of entire catchment areas, rather than from that of individual bodies of water such as particular lakes, rivers, or reservoirs.

The hydrological studies that have been made on the Nile are of theoretical and practical interest, and are the basis of the great scheme of projects that has recently been drawn up for the full utilization of the Nile's waters for irrigation.

The Nile catchment includes the Nile River—with its barrages, dams, and reservoirs, all its tributaries, the great and small lakes and swamps, and the mountain streams in whose waters it originates. In the aquatic ecosystems of this great complex may be found examples of development in which the possible malpractices as well as benefits of modern technology were not considered, as well as examples of development in which ecological considerations have been successfully applied.

Dr. H. E. Hurst, then Director of the Egyptian Department of Irrigation (now the Nile Control Department), and his Egyptian colleagues have contributed considerably to our knowledge about the hydrology and conservation of the Nile (Hurst et al. 1946; Hurst 1952). Information on the hydrology of the Nile has also been contributed by the present writer (Malek 1958a), by Hammerton (1972), and by Worthington (1972).

The hydrological characteristics of the Nile River are better known than are those of any other big river, at least those of its lower reaches. Four or five millennia ago, the ancient Egyptians recorded river levels on "nilometers," some of which can still be seen today. There is a nearly complete series of records of the river's maximum and minimum annual levels between 641 A.D. and 1450 A.D., and later, with interruptions, on the Roda Island gauge in Cairo. On Elephantine Island near Aswan, levels have been recorded since 1870.

The Nile's waters are derived mainly from the Ethiopian Plateau and the Central East African Lakes Plateau, minimal contributions being derived from the Nile-Congo divide. According to Hurst (1952) the annual discharge in the main Nile reaching Egypt is 84 billion cubic meters, out of which 48 billion cubic meters are contributed by the Blue Nile, 24 billion cubic meters by the White Nile, and 12 billion by the Atbarah. Thus, 84 percent of the Nile's discharge is provided by the Ethiopian Plateau (also taking into account the Sobat), and 16 percent by the Central East African Lakes Plateau. The Blue Nile and the Atbarah and their tributaries flow from Ethiopia and are the source of the annual Nile flood in Egypt, where the Blue Nile rises to a peak flow from July through September. The Atbarah is dry for six months, yet it contributes one-seventh of the annual flow.

The torrential Blue Nile, because of the high rate of erosion in the Ethiopian Plateau, carries about 100 to 130 million tons of silt during the flood season; the Atbarah carries a smaller amount. This silt is now retained behind the Roseires (ar-Rusayris) Dam in the Sudan and the Aswan High Dam.

The White Nile provides a very steady contribution that, for six months of the year, is about three times that of the Blue Nile. About half the flow of the White Nile is lost in the swampy Sudd region.

Control of Nile Waters

The lives of millions of people in Egypt and the northern Sudan depend entirely on the Nile, for without the Nile their countries would be desert. There is a long history of attempts to harness the waters of the Nile for flood control and for irrigation. Tradition has it that Menes, the first king of Egypt (ca. 3400 B.C.) was the first to build banks to control the Nile and that he dammed the Nile to build the city of Memphis. The pharaoh Omhotep proposed the use of Lake Karoun in Fayoum Oasis, west of the Nile Valley, to store the floodwaters for irrigation and for regulation of the flood. His statue now stands at Aswan not far from the new High Dam.

Irrigation engineering was a developed science in ancient Egypt, and devices used at that time are still used in the lower Nile in Egypt. Among them are devices used for lifting the water, of which the *shadoof* was the earliest (it is pictured on the walls of a tomb in Thebes, ca. 1250 B.C.). The "Archimedean screw" appeared much later than the *shadoof* and was introduced by the Greeks.

Early in the nineteenth century, a series of barrages was started, not to store water but to lift the level of the river in order to command a system of canals at all times of the year. The system of canals made it possible to change over from the old basin irrigation (annual flooding) to perennial irrigation, so that, instead of one crop, two, three, or even more crops a year could be grown on the same land. The first of the barrages, the Delta Barrage, was built a little north of Cairo, beginning in 1843, to control the Rosetta and Damietta branches of the Nile; later, with the aid of the Zifta Barrage on the

Damietta branch, it commanded the whole delta. The next barrages were constructed in Upper Egypt to provide perennial irrigation—at Asiut in 1902 (enlarged in 1938, with a canal flowing north to supply Fayoum with Nile water) and at Nagah-Hamadi in 1930.

The first Aswan Dam was completed in 1902. Its height was increased in 1912 and again in 1934 to store Nile waters during the flood of the Blue Nile and the Atbarah, so that the water would be available at low stage. Even this "Small Aswan Dam" was composed of a little more than half as much masonry as the great pyramid of Giza.

The next dam to be built in the Nile River Basin was the Sennar Dam on the Blue Nile, which was completed in 1925 to provide irrigation waters to the Gezira (el-Jazirah) scheme, the basis for the Sudan's economy. The scheme has now been expanded to include the Managil area, with an addition of another 400,000 hectares. The Jebel el-Awliya Dam (White Nile Dam) was completed on the White Nile in 1937, about 40 kilometers south of Khartoum. It was built and operated by Egypt to control the White Nile for Egypt's needs, but also allowed the Sudan to establish a large number of small irrigation schemes operated by pumps. The Owen Falls Dam, built at the outlet of the Nile from the 67,000-square-kilometer Lake Victoria for hydroelectric power and to control the outflow of Lake Victoria was opened by Queen Elizabeth II in April 1954. The dam has raised the level of water in the lake, altering its natural ecosystems. For this reason, Lake Victoria is now considered a man-made lake. On the Atbarah close to the Ethiopian border, the Khashm el-Qirba Dam was the next to be constructed in the Nile Basin. It was completed in 1964, its purpose being to control the Atbarah during its flood period and to irrigate a gravity scheme. The Roseires Dam followed and was completed in 1966 on the Blue Nile close to the Ethiopian border, nearly 290 kilometers upstream from the earlier Sennar Dam. Its goals are to greatly increase the Blue Nile storage, previously possible only at Sennar, and also to produce hydroelectric power.

The Aswan High Dam, the latest and mightiest in the Nile Basin, was opened by President Sadat of Egypt on January 15, 1971. The dam, about 7.2 kilometers south of the first small Aswan dam constructed in 1902, has impounded the water in the form of the 480-kilometer-long Lake Nasser, of a maximum capacity of 164 billion cubic meters. The lake will never be drained. It will enable Egypt and the Sudan to utilize fully the waters of the Nile. Table 2 presents hydrological and other information about the Nile dams, which are compared to major dams which have been constructed for development projects elsewhere.

Consequences of the Nile Projects

The one principal difference between the Nile River and rivers in industrialized European and North American countries is that sewage and chemical wastes do not contaminate the Nile's water as they do those of rivers in Europe and North America. Industrialization has barely begun along the Nile and its tributaries, and the density of the population is low except in the Nile Delta. Such consequences of industrialization are inevitable, and are bound to occur a few decades from now. Until that happens, we should realize that all changes in the hydrology are caused by development projects.

Man's impact on the ecology of the river cannot be determined in the cases of the first Aswan (1902), the Sennar (1925), or the Jebel el-Awliya (1937) dams because no studies were conducted before they were built. Postconstruction studies were undertaken by the University of Cairo on the first Aswan reservoir and by the University of Khartoum Hydrobiological Unit on the Jebel el-Awliya and Sennar reservoirs. These investigations included surveys of the water chemistry and the phytoplankton and zooplankton. Their results are summarized in Hammerton (1972), as are those of preconstruction and postconstruction investigations on the Roseires Dam.

I have studied the hydrology, ecology, and disease repercussions of construction of the Jebel el-Awliya and Sennar reservoirs in the Sudan. Some of my results are as yet unpublished, while others have already been reported (Malek 1958a, 1958b, 1959a, 1959b, 1960a, 1969, and 1971).

In my investigations, I surveyed the sites of the Khashm el-Qirba and Roseires reservoirs and their vicinities, on the Atbarah and the Blue Nile, respectively, before the dams were constructed (Malek 1958a). Moreover, I described the influences on the incidence of schistosomiasis of pump schemes irrigated from the Jebel el-Awliya reservoir and from the main Nile north of Khartoum, in some of my published papers (Malek 1958a, 1958b, 1960b, and 1962). I made

Table 2. Characteristics of the Dams on the Nile River and of Selected Other Major Dams

Name	Type	River	Country	First Year of Operation	Height (meters)	Length of Crest (meters)	Maximum Depth (meters)	Approximate Length (kilometers)	Reservoir Capacity (cubic meters or square kilometers)
First Aswan	Masonry, granite	Nile	Egypt	1902	--	--	5.3	80	5.0 billion[a]
Sennar	Masonry, rockfill	Nile (Blue Nile)	Sudan	1925	--	--	17	77	0.9 billion[a]
Jebel el-Awlia	Masonry, rockfill	Nile (White Nile)	Sudan	1937	--	--	15	480	3.5 billion[a]
Khashm el-Girba	Earthfill buttress	Nile (Atbarah)	Sudan	1964	77	535 + 3,955 (wings)	20	80	2.0 billion[a]
Roseires	Concrete buttress	Nile (Blue Nile)	Sudan	1966	68	18,371	50	80	3.0 billion[a]
Aswan High (Sadd el-Aali)	Earthfill, rockfill	Nile	Egypt	1971	127	4,133	90	480	164 billion[a] 5,850 (area)
Kariba	Earthfill, rockfill	Zambezi	Zambia, Rhodesia	1960	--	--	120	270	160 billion[a] 5,180 (area)
Akosombo	Rockfill	Volta	Ghana	1964	130	735	--	320	148 billion[a] 8,800 (area)
Kainji	Earthfill, rockfill	Niger	Nigeria	1968	75 (rock) 83 (earth)	9,187	--	--	1,475 (area)
Pahlevi	Concrete arch	Dez	Iran	1963	234	275	--	58	3.5 billion[a]
Ubol-Ratana (Pa Mong)	Concrete	Mekong	Thailand	1966	--	--	--	--	400 (area)
Bhumiphol (Yanhee)	Concrete arch	Ping	Thailand	1964	177	536	--	--	12.3 billion[a]

[a] "Billion" attributing to American and French usage (i.e., 1×10^9).

observations at the first Aswan dam, and upstream from it to the northern Sudan. I also investigated cultivated areas under perennial irrigation in the Nile Delta in relation to the epidemiology and transmission of schistosomiasis. (I presumed that the areas were irrigated by waters stored in the first Aswan reservoir.) In Uganda and Tanzania, I conducted only ecological observations on Lake Victoria and on the Nile, and, in Ethiopia, on the Blue Nile and its tributaries and on some development projects. Recently, twice during 1974, I undertook preliminary studies on the effect of the new Aswan High Dam on the ecology of the Nile in the delta, and on cultivated areas in the delta with reference to the transmission of schistosomiasis.

My own work, and the work of other investigators on the Nile reservoirs and irrigation systems, are described in the following pages.

First Aswan Dam. Studies of algae and other plankton, conducted long after the reservoir of the first Aswan dam was established, showed that the reservoir was a moderately eutrophic one. There was a succession of algal species, their abundance or scarcity were closely related to the annual flood—that is, to the abundance or scarcity of silt in the reservoir. No quantitative measurements were made, and no comparison was made with the algae and other plankton upstream of the dam basin.

A survey conducted by the Bilharzia Snail Destruction Section of the Egyptian Ministry of Public Health (1947) revealed the presence of large populations of *Bulinus truncatus*, the snail host of *Schistosoma haematobium*, in areas under perennial irrigation and in other areas irrigated by pumps upstream from the dam. I found *Bulinus truncatus* snails in *khors* and ponds near various villages and at steamer stops in the Nubian section of Egypt, south to the Sudan border (this section now forms part of Lake Nasser upstream (south) of the new Aswan High Dam). At the time of my studies, there were only a few schemes under perennial irrigation in this section of the river, mainly at El-Dakka and Ballana; otherwise, cultivated land exists only along the narrow banks of the river and by floating pumps. Near this section of the river (now Lake Nasser) there are a few hills on both sides of the river. During periods of high water, some low areas with many palm trees along the banks become inundated.

The water that used to be stored in the first Aswan reservoir helped to expand perennial irrigation in a few districts in Upper Egypt, but mainly in the delta. In one of these few schemes in Qina and Aswan provinces in Upper Egypt—that is, below the dam—it was demonstrated that the introduction of perennial irrigation led to a rise in the incidence of urinary schistosomiasis (the only form present in the area) within three years. For example, in the village of Mansouria, the proportion of the population infected with urinary schistosomiasis rose from 11 percent in 1934 to 64 percent in 1937; in the village of Binban, it went up from 2 percent in 1934 to 75 percent in 1937.

In the delta region of Egypt, both forms of schistosomiasis exist—the intestinal form, due to *Schistosoma mansoni* (6 percent in the southern section and 60 percent in the central and northern sections), and the urinary form, due to *S. haematobium*. In a representative area in the Qalyub district near Cairo, we investigated the epidemiology, transmission, and control of the disease, in 1952 and 1953. Our work included surveys to determine the incidence of the two forms of the disease among the populace; a survey of the distribution and ecology of the intermediate hosts and their infection with the schistosomes; treatment of some 900 patients with a chemotherapeutic drug, Fouadin; and a certain degree of sanitary and snail-control measures. Typical of cultivated areas in the Nile Delta, the relatively small project area (about 2,125 hectares) in the Qalyub district (Project Egypt 10, supported by the World Health Organization), contains more than 430 kilometers of canals and drains. In these watercourses, the incidence of schistosome infection among the snails was, as expected, very low, yet the few infected snails were, as in other endemic areas—especially with the urinary form— enough to maintain high endemicity among the human population (about 50 percent in most villages and 70 percent in one of them). Of the 14,289 *Bulinus truncatus* snails collected over a year throughout the project area, 71 (0.4 percent) were infected with *S. haematobium*, and of the 3,263 *Biomphalaria alexandrina* collected, 19 (0.6 percent) were infected with *S. mansoni*. The cold winter weather and forty days of drought during the winter affected the seasonal densities of the snail populations, and their infection with the schistosomes and other trematodes.

In another project supported by the World Health Organization (Project Egypt 49), further data were obtained on the epidemiology and control of schistosomiasis in another part of the delta, near Alexandria, especially on the prevalence, incidence, and correlation of the disease with personal and cultural habits and the effectiveness of certain molluscicides (Farooq et al. 1966a, 1966b; Farooq and Mallah, 1966).

Sennar Reservoir. Figure 2 is a map of the dams and reservoirs on the Nile River in the Sudan; it also shows locations of agricultural schemes irrigated by gravity or by pumps: the Gezira (el-Jazirah) and its extension (about 800,000 hectares) and the Kenana (Abu Naama, Millet; 485,000 hectares), and those irrigated by pumps (648,000 hectares). The total area of the gravity and pump schemes is thus about 2 million hectares. Recently, a small dam was built on Rahad River, a tributary of the Blue Nile, and a 8,000-hectare gravity scheme is projected there.

The Sennar Reservoir is about 77 kilometers long and is up to 4 kilometers wide. I described this reservoir, as well as the Jebel el-Awliya Reservoir, as large "aquaria" for snails, aquatic weeds, and plankton, and the snails, weeds, and plankton as stock from which they are propagated in the downstream stretch of the Blue Nile, in the Gezira gravity scheme, and in pump schemes irrigated by waters stored in the reservoir (Malek 1958a).

Some of the constituents of the plankton, especially several species of algae, are the main source of food necessary for the survival, growth, and production of large populations of the snail hosts of schistosomiasis, fascioliasis, and other animal diseases. The aquatic weeds, in addition to supplying oxygen in the water, provide surfaces on which the snails crawl and deposit their egg-clutches.

I demonstrated the presence of several species of blue-green algae, diatoms, and green algae in the water of the reservoir and the Gezira and in the digestive tracts of snails, and Talling and Rzòska (1967) showed that the Sennar Dam enhances development of plankton. They demonstrated that the dense plankton produced in the reservoir not only maintained itself through several generations in the reservoir, but also for 335 kilometers downstream to the junction with the White Nile at Khartoum.

During my studies, I found the following aquatic weeds in both the reservoir and in the Gezira canals: *Potamogeton perfoliatus, P. crispus, P. pectinatus, Polygonum glabrum, Ceratophyllum demersum,*

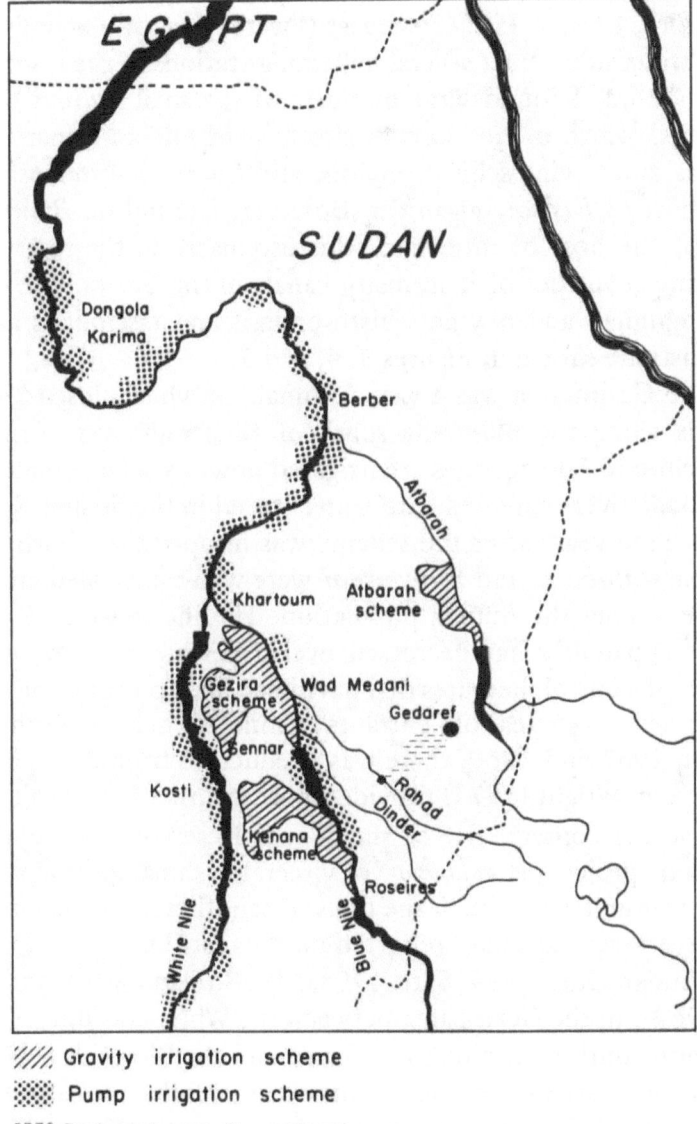

%%%% Gravity irrigation scheme
▓▓▓▓ Pump irrigation scheme
≡≡≡ Projected gravity scheme

Figure 2: Dams and Reservoirs on the Nile River in the Sudan. The areas in the central and northern Sudan that are irrigable from the Nile River. The total area that can be irrigated by either gravity or pump irrigation is over 2 million hectares. The projected Rabad scheme will involve some 8,000 hectares. (Updated from Malek 1958a.)

Najas pectinata, and *Vallisneria aethiopica.* In the reservoir, but not the Gezira, I found *Pistia stratiotes* ("water lettuce") and *Nymphaea* spp. ("water lilies"). At several collecting stations in the reservoir and in the Gezira, I found large numbers of the snail *Bulinus (Bulinus) truncatus,* which is the intermediate host of human urinary schistosomiasis and bovine schistosomiasis, and *Lymnaea natalensis,* which is the host of *Fasciola gigantica.* However, I found no *Biomphalaria ruppelii,* the host of intestinal schistosomiasis, in the reservoir, but abundant quantities of it in many canals in the Gezira. The distribution of human and bovine schistosomiasis and fascioliasis and their snail hosts are shown in Figures 3, 4, and 5.

The Gezira area was a vast savannah, of which close to 400,000 hectares along the Blue Nile south of Khartoum were irrigated in 1925 (close to 800 hectares are irrigated now) by a large main canal— a manmade river—supplied with water stored in the Sennar Reservoir. Within three years after the scheme was in operation, both *Schistosoma haematobium* and *S. mansoni* were well-established and on the increase among the human population. The incidence of *S. Haematobium* apparently has decreased over a period of years, probably because of control measures that have been in effect for some decades. Among new seven-year-old entrants to nineteen schools in the Gezira between 1957 and 1960, there was a reduction from 28.3 percent to 3.3 percent. Wright (1973) provides more information on this subject.

Another consequence of the Sennar Reservoir and adjacent development projects is *kala azar,* or visceral leishmaniasis, a protozoan disease transmitted through the bites of sandflies, *Phlebotomus orientalis.* Several workers have reported on the distribution of this disease in the Sudan (Kirk 1939; Satti 1958a, 1958b; Hoogstraal and Heyneman 1969). In the Gezira area between the White and Blue Niles, and the region south of it along the White and the Blue Niles—both situated in the heart of the Sudan—*kala azar* is endemic. *Kala azar* also occurs beyond the inter-river region, from Kassala (Kassala Province) in the north, south to Malakal and Nasir in Upper Nile Province, and occurs as well along the Ethiopian frontier of Blue Nile and Kassala provinces in eastern Sudan.

Watercourses and irrigation developments are in themselves not directly influencing the spread of *kala azar,* because it is not a waterborne disease. However, a danger lies in the development of new villages which brings in laborers from nonendemic areas in conjunction with irrigation and crop-growing schemes. Moreover, such

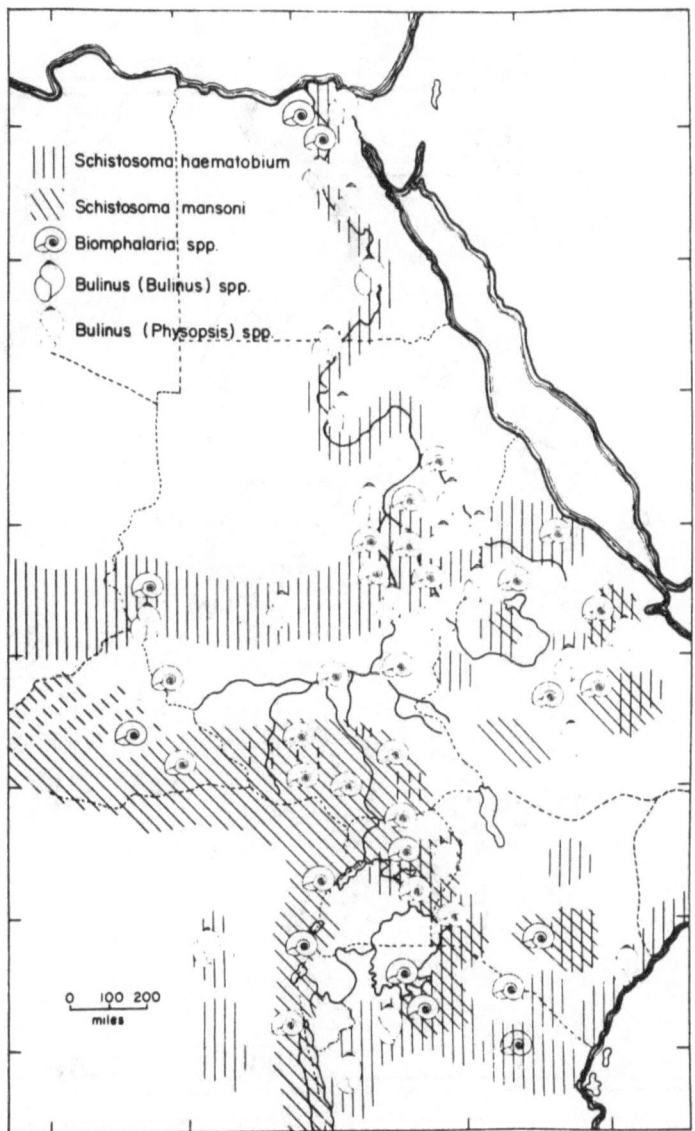

Figure 3: Distribution of Schistosomiasis and of the Snail Hosts of Schistosoma *spp. in the Nile Basin and Adjacent Territories.* The distribution of human schistosomiasis (the urinary form, due to *Schistosoma haematobium,* and the intestinal form, due to *S. mansoni);* of *Bulinus (Bulinus)* spp. and *Bulinus (Physopsis)* spp., the snail hosts of *S. haematobium;* and of *Biomphalaria* spp., the snail hosts of *S. mansoni.*

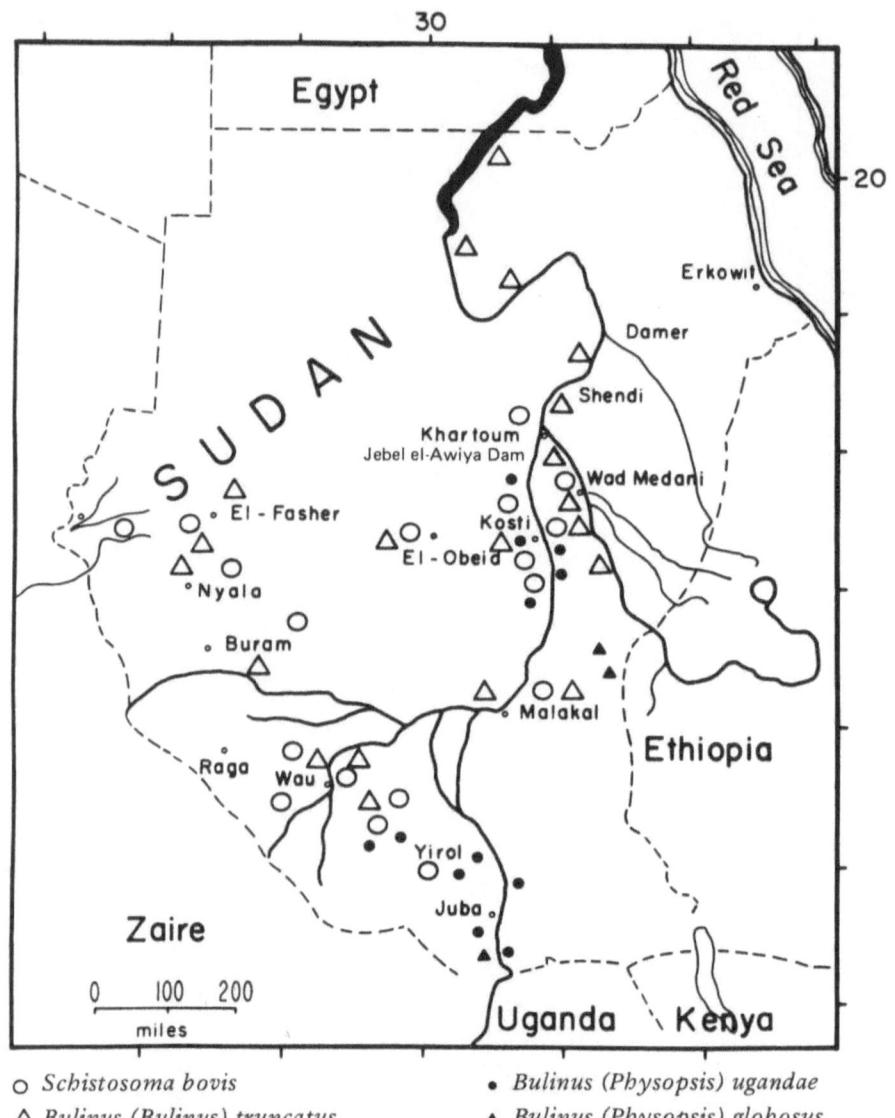

○ *Schistosoma bovis*
△ *Bulinus (Bulinus) truncatus*
● *Bulinus (Physopsis) ugandae*
▲ *Bulinus (Physopsis) globosus*

Figure 4: Distribution of Manmade and Animal Schistosomiasis in the Sudan, Caused by Schistosoma bovis *and Its Snail Hosts.*

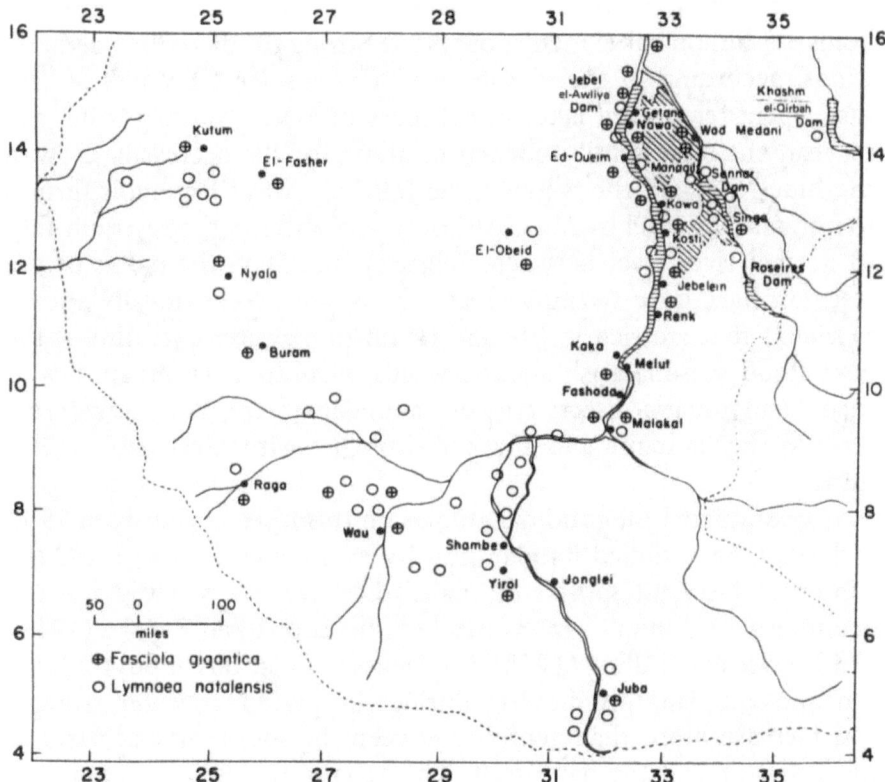

Figure 5: Central and Southern Sudan—The Distribution of Natural and Manmade Fascioliasis and Its Snail Host.

schemes and their canals influence the immediate environment and attract wandering animals, which might be reservoirs of *kala azar*.

The Blue Nile Basin in the Sudan—mainly south of Wad Medani, for about 480 kilometers upstream—is a center for another disease, onchocerciasis. This filarial disease is caused by *Onchocerca volvulus*, whose larvae are transmitted through the bites of black flies, certain species of *Simulium*, *S. damnosum* in particular.

Jebel el-Awliya Reservoir. The Jebel el-Awliya Dam, constructed in Egypt in 1937, has converted the White Nile into a slow-flowing lake, 480 kilometers long. The lake's width varies, depending on the height of the banks and the extent of the floodplain; in places, it is up to 4.8 kilometers wide. Usually, the dam is closed each July, and the reservoir begins to store water; when full, usually by September, it holds 3.5 billion cubic meters of water. Strong southerly or northerly winds, according to the season, produce considerable waves on this vast, open, temporary lake. In February or March, varying from year to year, the dam is fully opened to allow for the decreasing flow in the Blue Nile, and the water in the Jebel el-Awliya Reservoir shrinks to normal river level by May. Although the water in the reservoir falls to normal river level between February and May, the banks of the reservoir have large swampy areas, ponds, and creeks (locally known as *khors*) that are rich in life and retain some water until almost the next flood season. Lush aquatic weeds, plankton, several species of snails, and aquatic insects are very abundant there. Such large areas provide the biologist and the parasitologist with ideal habitats for study.

I conducted longitudinal studies on this reservoir between 1953 and 1959, and studied its plankton by examining the water and the contents of several species of snails' digestive tracts. I did not use quantitative techniques in my studies, but Brook and Rzòska (1954) and Prowse and Talling (1958) did. Dense populations of phytoplankton and zooplankton develop during the period of water storage, and increase more than tenfold between the southernmost parts of the reservoir and the dam itself. The aquatic weeds in it are similar to those in the Sennar Reservoir. However, the water hyacinth (*Eichhornia crassipes*) was abundant in the Jebel el-Awliya Reservoir. *E. crassipes* was introduced into the Sudanese Nile and was present in the Jebel el-Awliya Reservoir as early as 1955—notwithstanding the statements of some botanists, who have reported that it was introduced two years later (Gay 1958). Moreover, there are small islands of papyrus (*Cyperus papyrus*) in the reservoir as far north as Kosti. Some of these are anchored near the banks, while others are shifting.

The Jebel el-Awliya Reservoir supports a rich fish fauna, and the fisheries industry has proven to be highly productive and profitable. Monakov (1969) has reported on animals other than fish, such as decapods, copepods, cladocerans, conchostracans, chironomid larvae, oligochaetes, parasitic copepods, and hemipterans-heteropterans. Williams (1969) studied the distribution of river-crabs (Potamidae) in the Nile and reported *Potamonautes nilotica* to be the most widely distributed, occurring from the delta to the streams flowing into Lake Victoria.

In some parts of west Africa, namely, Zaire, Nigeria, the Cameroons, and Liberia, lung-fluke infections due to the trematode *Paragonimus africanus,* caused by eating raw crabs infected with the metacercarial cysts of the fluke, are prevalent. The first intermediate host of this fluke, melaniid snails, is present in the Nile River in Uganda, the Sudan, Ethiopia, and Egypt. Although the Sudanese, as a rule, do not eat crabs, thousands of west African immigrants settle in the Sudan to work, temporarily or permanently. So far as I know, this consequence of the reservoir has not been looked into in the Sudan. I did not come across any cases of the disease when I conducted my studies some years ago in that country.

The following members of the rich snail fauna of the Jebel el-Awliya Reservoir are hosts of these disease-producing trematodes of humans and animals: *Biomphalaria alexandrina, B. sudanica (Schistosoma mansoni); Bulinus (Bulinus) truncatus, Bulinus (Physopsis) ugandae (S. haematobium); Bulinus (Bulinus) ugandae, Bulinus (Bulinus) truncatus, Bulinus (Bulinus) forskalii (S. bovis); Bulinus (Bulinus) truncatus, Bulinus (Physopsis) ugandae (Paramphistomum microbothrium); Bulinus (Bulinus) forskalii (Gastrodiscus aegyptiacus);* and *Lymnaea natalensis (Fasciola gigantica).* I have reported elsewhere on the distribution of these snails in the Jebel el-Awliya Reservoir, in other reservoirs, and in other parts of the Sudan, and on their effectiveness in transmitting human and bovine schistosomiasis, paramphistomiasis, gastrodisciasis, and fascioliasis (Malek 1958a, 1958b, 1959a, 1959b, 1959c, 1966, 1969, and 1971).

Human Schistosomiasis. I gained my information on human schistosomiasis in the Jebel el-Awliya Reservoir by examining urine and stool samples of many residents of the area and by going through the records of hospitals and clinics. Both urinary and intestinal forms of the disease occur. The intestinal form is much more prevalent (60 to 90 percent) in this part of the Sudan than elsewhere. Moreover, this is one area of the Sudan where clinical manifestations of schistosomiasis mansoni—including hepatosplenomegaly, ascites, and cor pulmonale—are very evident.

It is difficult to determine accurately the prevalence of human schistosomiasis in this and in other parts of the Sudan because the annual reports of the medical services list in one figure the combined prevalence of both forms of the disease. Moreover, the figure is based only on the examination of hospital patients and school children. In the annual report for 1952 and 1953, the prevalence of the urinary disease for the White Nile is stated to have been 5.4 percent. I have determined the capacity of *Bulinus (Bulinus) truncatus, Biomphalaria alexandrina,* and *Biomphalaria sudanica* in the reservoir to transmit human schistosomiasis by infecting white mice in the laboratory with cercariae that emerged from naturally infected snails and recovering the adult worms later.

The Jebel el-Awliya Reservoir is of interest to schistosomiasis investigators because it is situated along the route of human populations moving in both eastern and western directions. From several countries in western and central Africa, many pilgrims bound for Mecca settle here temporarily before they complete their journey. Others settle permanently in the area as laborers. These migratory people carry different strains of *Schistosoma* spp. with them. Thus, a good number of strains of both *Schistosoma mansoni* and *S. haematobium* infect the snails, and later humans, in the reservoir, increasing the complexity in the transmission of the disease and, even more so, in its clinical manifestations and in the efficacy of the chemotherapeutic drugs used to treat it.

Figure 3 shows the distribution of human schistosomiasis in this and in other parts of the Nile River Basin.

Bovine Schistosomiasis. I determined that bovine schistosomiasis in the Jebel el-Awliya Reservoir, as well as elsewhere in the Sudan, is caused by one species, *Schistosoma bovis.* The distributions of *S. bovis* and of its snail hosts in the Sudan are shown in Figure 4.

S. bovis infects cattle, sheep, goats, camels, and equines throughout the entire area of the Jebel el-Awliya Reservoir (as well as in the Sennar Reservoir and the Gezira irrigation scheme), causing considerable pathology in various organs of the animals, which become emaciated. The animals usually become infected when they are brought to the reservoir's swampy banks to drink. Bovine schistosomiasis has become more prevalent since the Jebel el-Awliya Reservoir was established, a significant fact because the Sudan depends greatly on its livestock industry.

Bovine schistosomiasis occasionally infects humans. I reported on one case of eggs in the urine of a boy in the White Nile area (Malek 1961 and 1969), and probably there are more cases which go unnoticed. Cattle are also susceptible to *Schistosoma mansoni*, the species that infects humans (Barbosa et al. 1962; Saeed and Nelson 1969); therefore, it is of interest to determine the extent of such infections in cattle in this part of the Sudan. The implications are obvious, in that cattle would act as reservoirs for the human-infecting species and help in its propagation.

I found that a representative sample of domestic animals in the vicinity of the Jebel el-Awliya Reservoir was infected with *S. bovis* in the following proportions: cattle, 12/26 (46.1 percent); sheep, 7/12 (58.3 percent); horses, 3/5 (60 percent); and camels, 1/3 (33.3 percent). The snail *Bulinus (Physopsis) ugandae*, collected near the Kosti railway station and north of the Kosti bridge for two-hour periods during certain months of 1958 and 1959, showed the following infection rates with *S. bovis*: July, 0 of 0 snails; September, 0 of 0 snails; November, 0 of 50 snails (0.0 percent); December, 0 of 274 snails (0.0 percent); January, 2 of 692 snails (0.29 percent); March, 12 of 1,304 (0.92 percent); April, 1 of 194 snails (0.5 percent); and May, 0 of 33 snails (0.0 percent). The last two collections, in April and May, were obtained from residual pools, but aestivating snails were also collected in areas from which the water had receded a month earlier. Of 486 snails of all sizes, 106 revived in the laboratory and two were found to be infected with *S. bovis*. In addition to *B. (P.) ugandae*, which is the main transmitter of *S. bovis* in the reservoir, *B. (B.) truncatus*, which is much less common, is also a host, as two snails out of 216 were found infected on Aba Island, also in the reservoir.

Fascioliasis. The species of liver fluke which infects cattle, sheep, and goats in most of the Nile Basin, as well as most of Africa, is *Fasciola gigantica*. This species utilizes aquatic lymnaeid snails as an intermediate host; in Africa, the most prevalent lymnaeid is *Lymnaea natalensis*. *Fasciola hepatica*, which occurs in Europe and North America, on the other hand, utilizes mainly amphibious lymnaeid species—for example, *Lymnaea truncatula (Fossaria truncatula)*, *Fossaria cubensis*, and others. (Some aquatic lymnaeids in North America are also susceptible.) Thus, in Africa, fascioliasis is a water-borne disease. It follows, then, that water-development projects would aggravate a preexisting low prevalence of the disease. This is well demonstrated in the area of the Jebel el-Awliya Reservoir, because cattle in this area are fed fodder gathered on its banks. During periods of low water, the animals are taken for grazing and drinking to the swampy banks, and, after the water has receded, to the dried-up areas.

Fascioliasis is an important disease of great economic impact in the Sudan, where livestock is the largest export next to that of cotton. The disease is sometimes fatal, but the loss is mainly due to the hundreds of thousands of livers that are condemned daily, in addition to emaciation, which results in carcasses of reduced value. As in other areas where fascioliasis affects sheep and cattle (Europe and South America), humans become infected by eating watercress or other plants cultivated near or in water. Usually, the disease is manifested in the form of epidemics.

I did not encounter any human cases of fascioliasis in the Jebel el-Awliya Reservoir or in other parts of the Sudan, but it is very likely that they do exist but have not been recognized or reported. The cercariae of *Fasciola gigantica* (or *F. hepatica*), in addition to encysting on vegetation to form metacercarial cysts, also encyst freely in the water and settle to the bottom. Thus, unfiltered and untreated drinking water might also be the source of human fascioliasis in many parts of Africa and Asia. The adult worms occur in the bile ducts of the liver of cattle and humans, but in humans they are also found in ectopic sites, causing lesions in many organs but without releasing eggs in the stools; thus, their presence goes unrecognized except during surgery. In the cattle I examined in the area of the Jebel el-Awliya Reservoir, immature or mature worms were not uncommon in ectopic lesions, especially the lungs, and such lesions contained calcified concretions. The prevalence of infection among cattle which I examined at abattoirs and slaughtering places in the reservoir area was as high as 58 percent. The distribution of man-made and of naturally transmitted fascioliasis in the Sudan, as well as of its snail hosts, are shown in Figure 5.

Since *Lymnaea natalensis*, like other aquatic lymnaeids, requires high concentrations of dissolved oxygen, it follows that the planorbid snail hosts of human and animal schistosomiasis have a wider geographical distribution in the Sudan than does *L. natalensis*, since they adapt to a wider range of habitats. *L. natalensis* occurs in large or small reservoirs in the Sudan, where large populations of the snail exist among thick growths of weeds. The overflow on both banks of rivers and creeks harbors the snails, as do seepages along the banks below dams; they frequent rocks and gravel on which thick layers of algae accumulate. Such is the case in development projects in the central Sudan. However, irrigation canals do not seem to favor the establishment of large colonies—as on Aba Island in the White Nile Reservoir, or in the Gezira scheme. In other parts of the Sudan, the habitats of lymnaeids in naturally transmitted foci of fascioliasis are the rain pools in Kordofan and Darfur provinces in the west; creeks with vegetation and moderately moving waters in the Jebel Marra area in the west; and the floodplains of several Nile tributaries and large swamps (so-called *toich*) in Bahr el-Ghazal and Upper Nile provinces.

Throughout the year in Jebel el-Awliya Reservoir, the size of colonies of *Lymnaea natalensis*, as well as of the snail hosts of schistosomiasis, and their infection with *Fasciola gigantica* and the schistosomes vary according to the opening and closing of the dam. I encountered large populations along the banks and in the overflow from November through February, during high water. From February or March through April, there were still large populations in large ponds, swampy areas, and *khors* in the floodplain after the water had receded with the opening of the dam. In May and June, the populations were greatly affected by drought, and although many perished, some remained alive and active in small bodies of water, while others had aestivated under vegetation, in the shade and on moist mud. Some of them recovered when placed in water in the laboratory. The few survivors in the puddles of remaining water, in addition to the aestivating snails, apparently constitute the "seeds" for the snail colonies when the dam is closed in July and the reservoir fills again. The snail populations build up from August through October, as does aquatic vegetation, and apparently start being exposed to the infection from human and animal feces and human urine at this time of the year. Table 3 gives the proportions of *Lymnaea natalensis* in the reservoir that I found infected with *Fasciola gigantica*.

Paramphistomiasis and Gastrodisciasis. Paramphistomiasis and gastrodisciasis are diseases of domestic animals. They also are snail-borne and have become very prevalent as a result of the establishment of the reservoir. Paramphistomiasis is due to parasitism in the rumen of cattle, sheep, and goats with the fluke *Paramphistomum*

Table 3. Fascioliasis in the Snail *Lymnaea natalensis* in the Jebel el-Awliya Reservoir, White Nile, Sudan

Date	Number of Snails Examined	Number of Snails Positive for Fascioliasis	Proportion of Snails Positive for Fascioliasis (in percentages)
April 1954	223	9	3.86
January 1955	228	11	4.82
April 1955	104	12	11.53
October 1958	18	0	0
November 1958	63	0	0
December 1958	126	2	1.58
January 1959	292	9	3.08
March 1959	334	17	5.08
May 1959	56	1	1.78

microbothrium (Malek 1959). Gastrodisciasis is caused by the fluke *Gastrodiscus aegyptiacus*, which occurs in the colon and cecum of horses, donkeys, and mules. I worked out its life cycle from material obtained in the reservoir (Malek 1960 and 1971).

Leishmaniasis. I discussed the significance of visceral leishmaniasis *(kala azar)* as an endemic disease in the Sudan in the section on the Sennar Reservoir. The endemic area also includes the White Nile, especially the southern end of the Jebel el-Awliya Reservoir. This is an *Acacia-Balanites* woodland typical of the habitat of the sandfly vector, *Phlebotomus orientalis*. In the 1940s, several outbreaks occurred in the Melut and Kaka areas, not far from Malakal (Figures 4 and 5).

Khashm el-Qirbah Reservoir. The dam constructed on the Atbarah River close to the Ethiopian border was opened in 1964, and the water stored in the 80-kilometer-long Kashm el-Qirbah Reservoir now is irrigating a 200,000-hectare agricultural scheme. I had a chance to survey the damsite and its vicinity five years before construction began, and found the snail *Bulinus (Bulinus) truncatus* at low water in the Atbarah River, at the damsite, and at several foci upstream, and in some small tributaries and *khors* in the area (Malek 1958a). This species of snail, which is the intermediate host of urinary schistosomiasis, had not been encountered previously in that drainage system; hence, there was prediction that it will be introduced and propagated in the new irrigation scheme. I found no *Biomphalaria* snails, the hosts of *Schistosoma mansoni*.

Since 1964, about 50,000 residents of the former city of Wadi Halfa, now flooded by Lake Nasser, have been resettled in the new scheme irrigated by the Khashm el-Qirbah Reservoir. The prevalence of urinary schistosomiasis was high where the Wadi Halfans came from, and thus there was the possibility that the disease would spread in the Atbarah scheme. Of more significance, however, is the fact that they went from a nonendemic to an endemic area of *kala azar*. Thus, because they were not immune to it, several outbreaks of *kala azar* occurred in 1967. (It is possible, however, that the outbreaks were due to increased contact in the area with sandflies, the vectors of *Leishmania*.)

There were many casualties among the resettled Wadi Halfans, and it is because of this that many of the survivors returned to their homeland. They may have returned also because they could not adjust to the new way of life. In any event, they returned to live in a small town at the edge of Lake Nasser to await proposed agricultural and fishing schemes.

Roseires Reservoir. The Roseires Reservoir, completed in 1966, is about 80 kilometers long and 50 meters deep and is situated on the Blue Nile near the Ethiopian frontier. It has been estimated that, of a total of 130 million tons of silt carried by the Blue Nile each year, some 20 million tons are deposited in the Roseires Reservoir. This is another area which I surveyed for its snail fauna in 1955, before the dam was constructed and the reservoir created.

Bulinus (Bulinus) truncatus, but not *Biomphalaria* spp., occurred in ponds and *khors* in the area, but was absent from the Blue Nile itself. To my knowledge, no information has been gathered on the relationship between the snails and the prevalence of schistosomiasis on the one hand, and the reservoir on the other. However, the University of Khartoum's Hydrobiological Research Unit has gathered detailed information on the impact of the reservoir on the biology of the

Blue Nile, and Hammerton's studies (1972)—conducted before and after the construction of the dam—revealed a seasonal 100- to 200-fold increase in both phytoplankton and zooplankton.

The water of the Roseires Reservoir stratified soon after filling, and the lower layers became completely deoxygenated. In 1967, when deoxygenation affected all the water temporarily (Worthington 1972), there was high fish mortality. Large beds of the Nile oyster (*Etheria elliptica*) were eliminated because large amounts of silt were deposited in the upper reaches of the reservoir. *E. elliptica* is not of economic importance, however.

Chironomidae have increased in the Roseires area, while the black fly *Simulium damnosum,* vector of human onchocerciasis caused by *Onchocerca volvulus,* breeds in the turbulent waters near the sluice gates of the dam. Breeding sites for *Simulium* and mosquitoes also developed in low-lying areas flooded by waters of the reservoir.

The Aswan High Dam and Lake Nasser. The Aswan High Dam is the latest of the major development projects in the Nile River Basin. It is situated in Egypt about 8 kilometers south (upstream) of the first Aswan Dam, which was built in 1902 and heightened in 1912 and again in 1934. Although the Aswan High Dam was opened officially in January 1971, the reservoir (that is, Lake Nasser) has been filling slowly since 1964, and for several years almost no water has flowed into the Mediterranean from the Damietta (Dumyat) or the Rosetta (Rashid) mouths of the Nile.

The dam was constructed of earthfill and rockfill. It is 109.2 meters high, and its crest is 3542.4 meters long. The maximum depth of Lake Nasser is 90 meters. When the lake, which has been forming since 1964, attains its full capacity, it will cover some 5,000 square kilometers and will extend for about 480 kilometers upstream (south) of the dam. It should store 164 billion cubic meters of water, almost twice the flow of the Nile into Egypt (84 billion cubic meters).

On the basis of extensive hydrological studies of the Nile over many years, including detailed mathematical work, Dr. H. E. Hurst and his Egyptian colleagues at the Physical Service, Irrigation Department (now Nile Control), developed the concept of "century storage" for the full utilization of the Nile waters (Hurst et al. 1946; Hurst 1952). In their proposals, Lake Victoria and Lake Tana should be used as major reservoirs, while Lakes Kioga and Albert would be used as regulating reservoirs. A Jonglei Canal (shown in Figure 1) should be dug to bypass the Sudd region of the Sudan, saving the huge quantities of water (about half the flow of the White Nile) wasted in this area. A flood-protection dam was also proposed at the fourth cataract.

Of all the proposed projects, only one was constructed—the Owen Falls Dam at the outlet of the Nile from Lake Victoria, completed in 1954. Later, a decision was made to proceed with the construction of the Aswan High Dam, in spite of an alternative project—that is, the Jonglei Canal and the Lake Albert Dam. The latter project is based on thorough investigations contained in several large volumes. Hurst and his colleagues had favored the proposed Aswan High Dam, but also recommended that it could be combined with the equatorial projects. A diagram showing all projects proposed for the Nile Basin and the concept of "century storage" is illustrated with clarity in Hurst (1952).

Another alternative to the Aswan High Dam was made by the late Dr. Abdel Aziz Ahmed (1960), former Chairman of the Egyptian Hydroelectric Power Commission. Dr. Ahmed suggested that the nearby Wadi Rayan depression could be used as a natural reservoir for over-year storage, and that a canal could be constructed to connect it with the Nile and the depression filled with free-flowing water. Water would be withdrawn from the depression by electrically driven pumps. However, the planning of the Aswan High Dam project was too advanced for Egypt to have adopted Dr. Ahmed's proposal.

Purposes of the Aswan High Dam. The Aswan High Dam was constructed with the following goals:

1. To store the Nile's floodwater in order to replace annual flooding (basin irrigation, yielding only one crop annually) with perennial irrigation in some parts of Upper Egypt, making possible triple cropping—a needed boon for the Egyptian economy and Egypt's increasing population
2. With the impounded water, to irrigate all available arid areas in Egypt
3. To provide water to the Sudan under the terms of the Nile Waters Agreement of 1959 between Egypt and the Sudan.

4. To protect Egypt, and to some extent the northern Sudan, from unusually high floods
5. To obtain, from twelve giant turbines, about 10 billion watts of hydroelectric power for several projected industrial plants
6. To initiate a fishing industry, and probably also tourism, in Lake Nasser

Perennial Irrigation in Egypt. As stated above, one of the main objectives of the Aswan High Dam is to enlarge the area in Egypt that is under perennial irrigation. A historical note on this type of irrigation is in order.

Before the Aswan High Dam was built, about four-fifths of Egypt was under this system of irrigation. An area in Upper Egypt, between Aswan and Asiyt, was under irrigation by flooding (basin irrigation), a practice dating from the time of the ancient Egyptians until last century. Perennial irrigation differs from basin irrigation in that much smaller quantities of water are run onto the land at regular intervals of two or three weeks throughout the year. It must have been practiced, on a very small scale, in very early times in addition to basin irrigation. Lifting gadgets working from pools, wells, or the river, must have been used, as they are still being used in many places in Egypt. However, the present perennial-irrigation system depends on an intricate network of canals (primary, secondary, tertiary, etc.), which started early last century. Some of these latter canals, leading from the Nile to the Mediterranean and the Red Sea, followed the course of older canals, which had existed in pharaonic times.

It was not until the middle of last century, and again early this century, that perennial irrigation was intensified and expanded, especially after a system of barrages had been constructed. These barrages were built to raise the level of the water upstream, so that water can still flow into the canals when the river is low. Thus, barrages differ from dams in that their function is not to serve as storage reservoirs.

Beneficial Returns of the High Dam. There have been obvious benefits to Egypt, and to some extent to the Sudan, as a result of the Aswan High Dam. Water stored in Lake Nasser has allowed Egypt to make use of any irrigable land on the outskirts of the Nile Valley, and to increase arable land from 2.8 to 3.5 million hectares. Also, the waters retained behind the dam enable Egypt to convert about 280,000 hectares that had been under basin irrigation to perennial irrigation. This has been done gradually since the lake started filling. Egypt is now self-sufficient in wheat and is an exporter of rice, which means millions of dollars in hard currency. The country has been obtaining billions of kilowatts every year, and the estimate for future years is 10 billion kilowatts annually.

The Aswan High Dam has, like other previous dams on the Nile, increased the biological productivity, both within the lake, and probably in the whole river, downstream to the delta. This change in the aquatic ecosystem of the lake has its bearing on the fisheries industry. Beneficial returns in the form of a fisheries industry have already been helpful to both Egypt and the northern Sudan. There are no problems in the lake of submerged trees or other vegetation, such as there are in most tropical manmade lakes, such as Lake Kariba. Moreover, because Nile fishes are already adapted to Nile lakes and reservoirs (Lake Albert in Uganda, Lake No in the Sudan, and the Jebel el-Awliya Reservoir), they find it easy to adapt to lake conditions behind the Aswan High Dam. Therefore, although it might be true that the annual yield of fish has not yet met expectations (Sterling 1972), the yield is still high and is satisfactory to both Egypt and the Sudan, and it is expected to increase in the next decade.

The northern Sudan is also to make use of the waters stored in Lake Nasser to irrigate present and projected schemes. This is according to the 1959 Nile Waters Agreement between Egypt and the Sudan, which enables both countries to utilize all the Nile waters flowing into them: Egypt, 65.5 billion cubic meters and the Sudan, 18.5 cubic meters—a total of 84 billion cubic meters.

One of the greatest benefits of the Aswan High Dam has been its protection against floods, an advantage that is not often included among the beneficial returns of the dam, in spite of its significance. It has been realized since the time of the ancient Egyptians that Egypt is a flat country which lies at a level below that of the Nile. Dangerous floods have occurred throughout history (during last century they occurred in 1863 and 1878). During this century, the most dangerous flood occurred in Egypt in 1946, when the Blue Nile and Atbarah had their second highest flood on record.

Predicted and Unforeseen Adverse Ecological Effects of the High Dam. In spite of the obvious gains which have been and which will be obtained as a result of the construction of the high dam at Aswan, certain unfavorable consequences have already been reported; some of them had been expected, while others were unforeseen. These repercussions have been in the news, and several popular articles have been written about them. Unfortunately, few of them were documented by strong evidence or thorough studies.

Several factors—namely seepage, evaporation, and utilization of waters by Egypt and the Sudan—are preventing Lake Nasser from attaining its full capacity, and it is believed that it is now only half full. Lake Nasser's entire western bank is composed of porous Nubian sandstone through which water seeps underground. Evaporation is considerable in this arid zone of the Nile Valley, and with the lake now menaced by large masses of aquatic weeds, mainly water hyacinth, more water is believed to be lost through evapotranspiration. The total water loss is estimated at 10 billion cubic meters—not 30 billion, as stated in some publications (Sterling 1972).

The silt brought in from Ethiopia by the Blue Nile's and the Atbarah's floodwaters is now retained behind the dam and is settling to the bottom of the lake. This development was predicted, and it was hoped that it would prevent or minimize the seepage. At any rate, the silt settling behind the dam, an estimated 50 to 100 million tons annually, has had the following three effects: deprivation of cultivated land in Egypt of a rich and a free fertilizer; cutting off the delta shoreline at the Mediterranean; and failure of the fisheries industry in the eastern Mediterranean.

1. *Deprivation of Cultivated Land.* The delta itself was formed over the years by silt deposited with the Nile flood. Throughout Upper Egypt and the delta, the silt was always deposited on the land, which made the land among the most fertile in the world. Now, chemical fertilizers have to be used to compensate for the loss of silt, and it has been proposed that such fertilizers be produced locally.

The age-old industry of brickmaking with the Nile silt has been known and practiced since pharaonic times. The silt had always been obtained from the Nile, and from canals and streams. Now, the industry has to depend on other raw material, unless it is feasible and economical to transport the silt from Lake Nasser.

2. *Cutting off the Delta Shoreline.* A detailed study by Kassas

(1972) showed that the coast of the Nile Delta is actively retreating. The study involved the examination of maps, the review of historical documents concerning the sites of old coastal fortresses and lighthouses, and Kassas' own observations. The brackish-water lakes, Burullus and Manzala, are at present separated from the Mediterranean by narrow and fragile bars of sand, which are all that remains of much broader belts of land that once separated the lakes from the sea. If the bars collapse the lakes will be transformed into marine bays; if the embankments now protecting the cultivated land near the coast erode and break, the land may be inundated by seawater.

3. The Decline of Fisheries in the Eastern Mediterranean Sea. George (1972) reported a decline in the fisheries industry in the eastern Mediterranean and in the coastal brackish-water lakes of Egypt. Before 1964, some 18,000 tons of sardines used to be taken from the eastern Mediterranean every year; in 1969, however, only 500 tons were taken. The decline occurred because the eastern Mediterranean was no longer being freshened or enriched by the great quantities of fresh water and plant nutrients flowing into it in the floodwaters of the Nile.

Displacement of People. About 50,000 people were displaced from Wadi Halfa and nearby areas of the Sudan by the rising waters of Lake Nasser. This consequence of the dam was considered very early in the planning stage.

As early as 1954, I attended several meetings that were held in Khartoum at the request of the Wadi Halfans, who wanted their case to be heard. Like all northern Sudanese, they were proud people, highly determined, and closely attached to their ancestral land, and worried about the traditions and culture to be lost. In those meetings in the early 1950s, they defended their case vigorously. In the end, however, before the lake started filling, they were resettled in a new agricultural scheme which is irrigated by waters stored behind the Khashm el-Qirbah Dam on the Atbarah, close to the border of Ethiopia. Some of them refused to move and stayed behind in the same area after their town and villages were under water. Many of those who were resettled at Khasm el-Qirbah returned to their ancestral land because they could not adjust to the new place, far away from their homeland, and because they had suffered many casualties as a result of *kala azar,* which is prevalent in the new area.

Diseases. Another predicted impact of the Aswan High Dam was that certain parasitic diseases would become more prevalent because the ecological conditions created by the dam provide for far more suitable modes of transmitting such diseases than did the previous conditions. This consequence has not been thoroughly assessed, but on the basis of preliminary studies, the prevalence of malaria is expected to increase because new, suitable breeding habitats have been created for the vector mosquitoes, *Anopheles gambiae*, which had menaced Upper Egypt for many years.

Onchocerciasis has long been reported in the northern Sudan, close to the southern section of Lake Nasser (Haseeb et al. 1962). It is very likely—but again it has not been determined—that new breeding habitats have been provided for the larvae of the vector black fly, *Simulium damnosum*.

Another important disease, schistosomiasis, is bound to increase because it is directly related to water-development programs. New favorable habitats for its snail intermediate hosts have been created behind the dam and in all irrigated areas upstream and downstream. Yet all that has been published on the subject are several newspaper articles and a few articles by scientists (van der Schalie 1960, 1972; Heyneman 1971) of general coverage. All of the articles contained only speculations based on the fact that more suitable habitats for the snail hosts will be created by the expected reclamation of land and by the conversion of the basin to perennial irrigation in Egypt.

What should be noted, however, is that the area to be reclaimed from arid and desert areas is comparatively small. The change from basin to perennial irrigation is certainly something worth worrying about; but again, this is nothing new, because it happened before in Upper Egypt, as I have said.

However, three aspects not considered in the articles are more significant than others, with respect to the expected increase in the transmission of the disease; therefore, these became the subjects of a preliminary study I conducted during two visits to Egypt in 1974. They are: (1) the changes (relevant to the disease) that have taken place in the Nile River itself, (2) the expected stability of the snail populations, and (3) the changes in the transmission season(s) in irrigation canals in areas that had been under perennial irrigation. The objective of my study was to determine whether, as a result of the Aswan High Dam, schistosomiasis was more common in Egypt because of increased transmission of the disease in the Nile and in irrigation canals that used to be under perennial irrigation. I had hypothesized that the increase in transmission was due to elimination of the so-called "winter closure" and to the stability of the snail populations throughout the year.

The results of my survey and ecological observations follow.

The Nile in the Delta Region. I examined the Nile in the delta region at the Delta Barrage, Warraq el-Arab, Imbabah, and Cairo; El-Mansurah, Talkha, and vicinity on the eastern, Damietta Branch; and at widely separated sites on the western, Rosetta Branch (Figure 6).

At Cairo and immediately north of Cairo, at Imbabah, Warraq el-Arab, and the Delta Barrage, I found moderate quantities of aquatic vegetation, but few or no snails. The water was only minimally turbid, and slowly moving. This is especially noteworthy because the first study was conducted during August, which is the flood season of the Nile.

The Eastern (Damietta) Branch of the Nile at El-Mansurah, Talkha, and vicinity, about 70 kilometers south of the mouth of the river at the Mediterranean, is representative of present conditions in several sections of the lower Nile in the delta. Long stretches of the lower Nile have been changed to what look like drains. In the center of the river are large islands with earth and sand bottoms that support cattails (*Typha* sp.) and reed grass (*Phragmites* spp.). In addition, large masses of water hyacinth *(Eichhornia crassipes)*, as well as *Potamogeton crispus* and *Ceratophyllum demersum,* are very common. I collected many specimens of the snail *Bulinus (Bulinus) truncatus* and its egg clutches, and a few specimens of *Biomphalaria alexandrina,* the intermediate hosts of *Schistosoma haematobium* and *S. mansoni,* respectively.

The level of the water was noticeably low, and the current had decreased in velocity. The absence of silt in the Nile below the dam

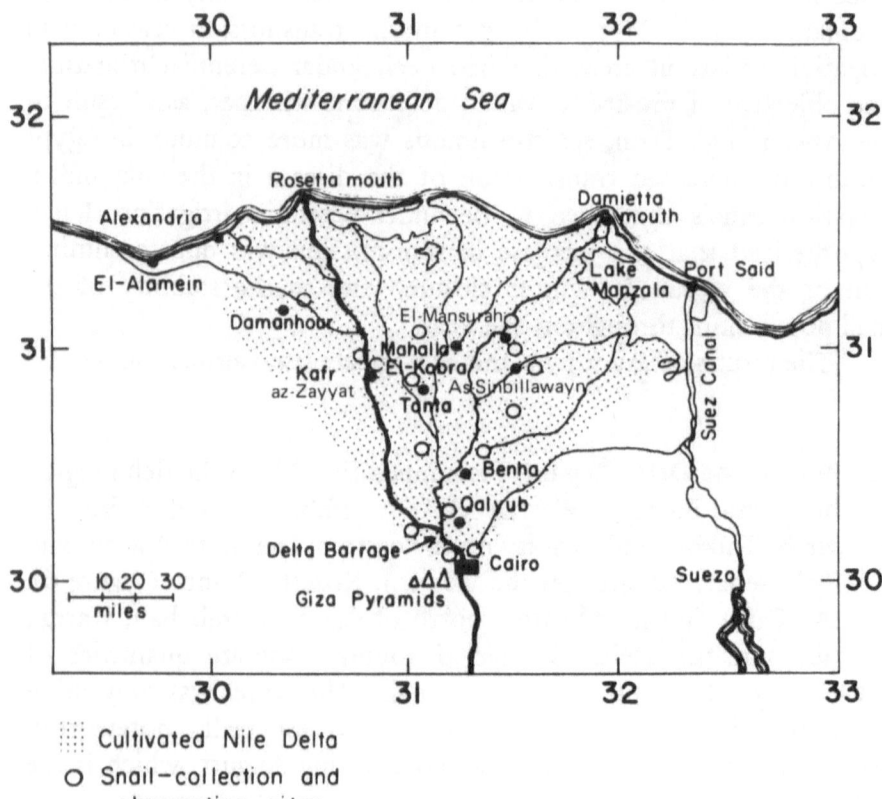

Figure 6: Schistosomiasis in the Nile River Delta. Sites (open circles) at which collections and observations of schistosomiasis-transmitting snails were made during 1974.

evidently enhances the growth and propagation of aquatic weeds and microflora through the increased penetration of sunlight. I saw a rich, scummy periphyton layer on the weeds, an essential diet for the snails. The absence of silt and other causes of turbidity makes the water blue.

On account of the ecological changes in these areas of the delta, the Nile there has become a year-round center of human activity, with the result that pollution and human-water contacts have increased. Before the construction of the dam, no one dared to get close to the Nile in those same areas during the three or four months of high water, except to cross by large boats or ferries.

I examined the western branch of the Nile at widely separated localities, among them the Delta Barrage and vicinity, and Kafr az-Zayyat. I saw a moderate amount of vegetation, mainly water hyacinth, and although I did not see large islands of cattails and reed grass at the points I examined, I believe that they exist in other areas of the western branch, as they do in the eastern branch of the Nile.

I examined irrigation canals and drains in the Delta during the study at the following localities: Qalyub, Benha, Mit Ghamr, As-Sinbillawayn, and El-Mansurah, east of the Damietta Branch of the Nile; Mahalla el-Kobra, Tanta, and Kafr az-Zayyat in the central part of the delta; and near Damanhour and Alexandria in the western part of the delta. In the canals and drains of the delta, as in the Nile, there are abundant aquatic weeds and microflora, the latter being essential as food for the snail hosts. As a consequence, there are denser populations of the snail hosts of both urinary and intestinal schistosomiasis, both among the weeds and throughout the watercourses.

I collected *Bulinus (Bulinus) truncatus* and *Biomphalaria alexandrina* from irrigation canals, both large and small, near El-Mansurah, east of the Damietta Branch. I was familiar with this area and with its snail fauna before the Aswan High Dam was constructed and detected a noticeable increase in the densities of the snail populations of the two intermediate hosts.

The high densities of the snails evidently are due to the fact that the numbers of snails are more constant throughout the year than they were some years ago. This stability in numbers is due to the elimination—after the construction of the dam—of the so-called "winter closure," whereby the canals used to be closed and drained for about forty days in December, January, and February. During that time, the silt deposited there by the Nile flood would be dredged out, along with the snails and vegetation—usually by hand sickles, hoes, and rakes.

Before the Aswan High Dam was constructed, there was little or no transmission of schistosomiasis in the Egyptian Nile. It was transmitted, instead, mainly in the irrigation canals and drains—especially in the delta, which had been under perennial irrigation since the end of last century. My study indicated that, at least in some sections of the lower Nile, ecological changes caused by the dam enhanced the transmission of the disease. Furthermore, the snail hosts now are abundant in the Nile. Also, with the absence of silt (which is now retained behind the dam) and the slower water curent in the lower Nile, there now is a higher probability than there was before the dam was constructed that the miracidia of the schistosomes will come into contact with the snails and that the cercariae of the schistosomes will infect human beings. Moreover, since I was familiar with the areas studied in the Egyptian Delta, I realized that year-round human activity in and near the Nile has increased considerably because of the low, clear, and slowly moving waters. There is now more fishing, seining, swimming, and washing of domestic utensils and clothes; such activities used to occur only in irrigation canals.

The 480-kilometer-long manmade Lake Nasser, south of the dam, is also menaced by water hyacinth and other weeds. This is not surprising, because aquatic weeds, mainly water hyacinth, *Ceratophyllum* and *Potamogeton*, are abundant in the White Nile, especially south of the Gebel el-Awliya Dam, and in the Blue Nile in the Sennar Dam Reservoir.

One brief survey was conducted in some villages in Upper Egypt, close to the Aswan High Dam, to determine its effects on the prevalence of schistosomiasis in irrigation schemes. An increase in the prevalence rates was detected by the examination of the urine of some inhabitants of these villages; unfortunately, their study was of short duration and the results were not published.

In the irrigation canals that I studied in the cultivated areas of the Nile Delta, which had been under perennial irrigation since the end of the last century, the habitats of the planorbid snail hosts of the schistosomes (Malek 1958b) evidently have become more favorable for the establishment of stable (and thus large) populations of these snails. The absence of silt has permitted more sunlight to penetrate the water, enhancing the growth and propagation of aquatic weeds and microflora, the latter of which is essential as food for the snails. The same has always been true of the Gezira and other irrigation schemes on the Blue Nile, and of others on the main Nile in the Sudan, where aquatic weeds are fewer when the Nile is in flood and more abundant thereafter.

At present, a significant factor in the ecology of the snail hosts in the delta irrigation canals has been the elimination of the winter closure, since it is no longer necessary to clean the canals of silt deposit during the Nile's flood period. Barlow (1937) suggested, on the basis of results he obtained from an experimental area in the Nile Delta, that while continual clearance of the canals over a period of years would not completely eradicate the snail hosts, it would constitute the major factor in the control of the disease.

Another change in the epidemiology and transmission of schistosomiasis in the Nile Delta due to the elimination of the winter closure is alteration of the transmission season(s) of the disease. This possibility has not been investigated so far. In 1952 and 1953, we demonstrated in a representative area of the delta near Qalyub, north of Cairo, that the transmission period of both urinary and intestinal schistosomiasis was from May through November, but that during this period it became minimal during the flood season. Naturally, there was no transmission during the forty-day winter closure, starting about the end of December. After the closure period was over, the snail populations increased during February and March from the few snails that had survived drought by aestivating in cracks in the canal bed and from snails introduced with the incoming water after the closure period was over. Snails with patent infections, from which cercariae emerged, began to appear in May. With the present elimination of the winter closure, environmental factors—mainly seasonal and diurnal variation in temperature, food, the density of aquatic weeds, and an absence of silt—are probably the principal influences on the undisturbed snail populations and their infection with the schistosome miracidia. These aspects are greatly in need of investigation by those interested in schistosomiasis research; they will supplement the projected studies that will provide a computerized "data bank" on the environmental effects of the Aswan High Dam.

Discussion and Conclusions

I have tried to present an account of development projects in various river basins, in particular the Nile River Basin. These projects have had certain benefits as well as unfavorable consequences. Common to most of these projects are hydroelectric power, especially for industrial projects, and more cultivated land, which means food for increasing populations and a boon to economies. In the Sudan alone, over 2 million hectares have been placed under irrigation by gravity and by pumps. Moreover, throughout the Nile there has been an increase in the biological productivity of the river as a result of the seven dams constructed on it in Uganda, the Sudan, and Egypt.

The boon to food production and the economy has been in the form of the fisheries industry. To increase fish production even further, fisheries research is being conducted by (1) The Inland Fisheries Research Institute, Khartoum, Sudan, operating since 1956; (2) The Institute of Freshwater Biology, Cairo, Egypt, operating since 1955; and (3) the United Nations Development Programme (UNDP), Lake Nasser Development Center, Aswan, Egypt, operating since 1969.

Only a few studies have been conducted to assess the disease repercussions of development projects in river basins throughout the world; nevertheless, those that have been conducted are convincing in the message they carry. Some of the projects, however, had few repercussions on disease because of (1) certain ecological features of the localities in which they are situated; (2) adequate preconstruction studies; or (3) the institution of concentrated control efforts before and after the projects were established. Moreover, Hughes and Hunter (1970 and 1972), after reviewing the situation in Africa and determining that there has been an impact on the prevalence of disease, very correctly cautioned that systematic epidemiological data are rare, and that the data that do exist for certain regions or for Africa as a whole are not comprehensive in their coverage. Macdonald (1964) made more or less similar recommendations.

It is well known that development projects in Africa, the Middle East, and South America—in particular reservoirs and irrigation schemes—have aggravated preexisting low prevalences and sporadic foci of schistosomiasis. This is because the network of canals create ideal habitats for the snail hosts, which are essential for the development and propagation of the parasites, the schistosome worms. I have treated the relationship of snail ecology to manmade habitats in more detail elsewhere (Malek 1972).

Development projects also foster dense human, usually immigrant, populations in the form of labor forces and permanent settlers. This increases human-water contact and, consequently, the transmission of the disease. McMullen et al. (1962), on the basis of their extensive travels and surveys in several countries for the World Health Organization (WHO), very admirably reviewed the subject of water-resources development. They forecast an increase in prevalence of schistosomiasis in almost all the countries visited. It should be noted that schistosomiasis is contracted more in irrigation canals and reservoirs than in rivers. Not all parts of the main rivers are dangerous. There are safe portions with sandy beaches where no vegetation accumulates and no snails exist, and where there are swift currents at least during some months of the year. The Blue Nile is torrential, and except for reservoirs and protected bays and pools along the banks, the river is safe. Long stretches along the main Nile north of Khartoum are safe. In these areas, swimming and bathing by many expatriate Europeans and Americans and their families during the period 1953-1959 and under my supervision did not produce infections in this group, in which all ages were represented. The creation of hundreds of villages, settled mainly by fishermen along Lake Volta, Ghana, has increased the prevalence of the disease. Many of the fishermen on Lake Nasser are infected with the urinary form of the disease, and a study by Eyakuze et al. (1974) showed that the fisheries industry, and to some extent the steamers, have increased the prevalence of the disease in the Lake Victoria area. The economic loss due to schistosomiasis in the various countries of the world where the disease is endemic was elaborately treated by Wright (1972). Dr. Wright also very ably provided figures and analyses of partial disability, complete disability, and estimates of the number of people throughout the world who are exposed to the risk of infection.

Water-resources developments also enhance the transmission of malaria by providing more habitats for the larvae of the mosquito vectors; moreover, the dwellings in the new villages (especially the thatched-roofed ones) provide suitable shelters for the adult insects. Also, the ending of a pastoral, nomadic living pattern, as well as the migration of laborers, facilitates the spread of malaria. Hughes and Hunter (1972) show how development programs have aggravated not only malaria, but malnutrition as well, in both rural and urban situations in Africa.

Any ecological changes in Africa south of the Sahara that bring man, tsetse fly, and parasite *(Trypanosoma)* together are bound to favor the transmission cycle of trypanosomiasis, or sleeping sickness—a protozoan disease. Both bovine and human trypanosomiasis are major obstacles to economic and social development in Africa. Huge areas on the continent are infested by the tsetse flies *(Glossina palpalis, G. morsitans,* and other species). In the most highly infested belts it is impossible to raise cattle. An outbreak of sleeping sickness, both bovine and human, occurred shortly after Lake Kariba was established on the Rhodesia-Zambia border and the people transferred (Apted et al. 1963).

Another trypanosome, *Trypanosoma cruzi*, prevalent in South and Central America, causes Chaga's disease. It is transmitted by a rudivid bug, triatoma, which inhabits the dwellings of villagers; population concentrations favor its breeding. The larvae of the black fly *Simulium* breed in flowing waters, and bites of the adult flies cause onchocerciasis ("river blindness") in west and central Africa. It also occurs in some Central and South American countries. *Kala azar* is indirectly related to water-resources developments and is transmitted through the bites of sandflies, *Phlebotomus orientalis*, and other species. Throughout the history of the Sudan, for example, tribes have been essentially destroyed by *kala azar* in the Fung District of Blue Nile Province, where an outbreak occurred as recently as 1956 (Satti 1958a and 1958b).

In view of the diseases and other consequences of water-resources developments, one might ask whether the projects have been worth all the problems they have caused, whether they really were necessary, and whether the nations in which they were installed could have done without them. In most cases, the answer is that the projects were necessary because of the benefits they provided.

Nonetheless, planners should take precautions before projects are constructed, and multidisciplinary studies should be undertaken to prevent or minimize unfavorable consequences. If unintended effects still occur, then measures should be instituted to control them.

One of the principal causes of such problems is the fact that developed countries, which usually suggest and often support projects in developing countries, have not set a good example in their own territories. They have committed ecological mistakes and ignored the effects on the environment of their own activities: the use of insecticides, synthetic detergents, and inorganic nitrogen fertilizers pollute their waters and endanger many organisms, besides being important health hazards. Cancer contracted from contaminated water over a long period may be just as bad as some of the slow-killing diseases of the tropics.

The Earth's rivers have to be conserved and kept clean. For

example, before a dam is constructed or a scheme established, all disciplines should take part and contribute their expert advice. The engineer, the economist, and the ecologist should work together and show an awareness of and a concern for each other's views. This collaboration should continue after the project is in operation. There has been a tendency among biologists in the last few years to blame harmful repercussions on the ecology, including disease, on engineers, and to recommend at the same time, in speeches and articles, collaboration with them. They also call for international collaboration, and they note a lack of concern on the part of international organizations.

How can such biologists recommend something that has been in operation and effective for many years? Collaboration among several engineers and biologists to combat disease has been taking place—to mention only a few, there are the engineers Buzo, Araoz, Lanoix, Francot, Rainey, Jobin, and McJunkin. On the basis of what they learned from biologists, they have put forward excellent recommendations for preconstruction and postconstruction measures to eliminate or minimize disease and other ecological consequences.

As to international organizations, several United Nations agencies have shown concern and have played an active part—not only in financing studies related to development projects—but, as well, in coordinating efforts to obtain the best out of projects with minimal adverse effects on the environment. When a government asks the help of the United Nations Development Programme in starting a new project, UNDP asks other U.N. agencies (for example, WHO, FAO, and UNEP) to assess the ecological and health hazards of the project before it takes any action. When I worked for two years for WHO in Geneva, one of my duties was to report to UNDP on all the predicted health hazards of the projects under its consideration and on what was known in the way of precautions and control measures.

At present we have the tools to control certain diseases. The engineers have provided us with control measures for schistosomiasis and malaria, measures designed to be implemented before development projects are under way. In cases where these are not implemented before dams, reservoirs, and irrigation schemes are constructed, other engineering measures are available to correct the mistakes and lessen the dangers. In the case of schistosomiasis, engineering measures can modify, reduce, or eliminate snail habitats;

prevent or reduce human contact with potential transmission sites; and facilitate the treatment of water bodies with molluscicides (World Health Organization 1965, Buzo 1972). Gravity-irrigation schemes, with their canals and drains and flow of water, can be designed by experienced engineers in such a way as to reduce or discourage snails. For example, the induced fluctuation of water levels in reservoirs and canals effectively controls schistosomiasis and malaria. The planning of an irrigation scheme should result in the most efficient use of land and water resources. The lining of canals with hard materials, such as concrete and asphaltic concrete, contributes to the control of the snail hosts, and the use of underground drainage to reduce the extent of exposed watercourses has recently been introduced in some irrigation schemes—for example, in Egypt. But the effect of underground drainage on the transmission of schistosomiasis and malaria has not yet been fully evaluated.

Changes in the method of growing rice, such as those in Japan and the Philippines, would eliminate or reduce the incidence of waterborne diseases. For example, rice, like other crops, can be grown under intermittent irrigation: the ponding of water in the paddies is not necessary. In addition, the direct delivery of irrigation water to paddies has been proposed as an alternative to relying on the flow of runoff from one paddy to another.

In addition to engineering measures used to ensure a healthy environment, chemical measures can also be taken—for example, the use of a number of chemical molluscicides that are effective against the snails at very low concentrations, with minimal effect on the biota (World Health Organization 1965 and 1973, Malek and Cheng 1974). Their efficacy has been demonstrated in certain irrigation schemes and reservoirs in several parts of the world, where they had a marked impact on the prevalence of schistosomiasis. However, this method was used only in situations where fish are not an essential part of the diet of the population, because most of the available molluscicides have, with varying degrees, some piscicidal properties.

Integrated control measures are often recommended for most of the snail habitats. After the breeding places of snails in big reservoirs and some watercourses have been located through thorough surveys, and it is feasible and economical to focus the control measures—i.e., apply "focal control." The same is also feasible for the breeding sites of mosquito larvae (malaria and filariasis) and of black-fly larvae (onchocerciasis).

There is, however, a need for the training of technicians who will be able to properly use the tools for control that are presently available. Therefore, it is the job of the medical biologists to convince funding agencies and international organizations of the need to establish training centers in one or more centrally located developing countries. It might be advisable to recommend support for development projects only if the countries receiving the funds commit themselves to implementing disease-control measures, before and after establishment of the projects. It might also be advisable for funding agencies to insist on the introduction of human population-control measures (birth control) by the countries involved, at the time the projects are being planned. Naturally, funds for disease control should be included in the benefit-cost estimates.

There are projects for which efficient control measures, applied before and after establishment of the projects, have eliminated or minimized disease repercussions. One of them is the Pahlevi Dam and irrigation project in Khuzestan, Iran. Of course this project has had the advantage of being situated in an arid, mountainous, and uninhabited area from which the snails are entirely absent. Nevertheless, the success of the Pahlevi project shows that harmful repercussions are not an inevitable result of all development projects, but that, rather, each project should be considered on its own merits, on the basis of the local topography, biota, and existing diseases.

To draw parallels among the environments of the Pahlevi Reservoir, the reservoirs in the Nile River Basin, and the projected reservoirs in the Mekong River Basin is, no doubt, unjustified. The planorbid snail hosts *Biomphalaria* spp. and *Bulinus* spp. of the Nile River valley adapt very well to irrigation canals and reservoirs. But the aquatic hydrobiid snail *Lithoglyphopsis aperta*, which is the snail host of a *Schistosoma japonicum*-like blood fluke in Southeast Asia apparently is very limited in distribution and might not adapt to conditions in reservoirs and irrigation canals. In the small focus of oriental schistosomiasis discovered on Khong Island in the Mekong River in Laos, only the snail occurs, and it is abundant in very small areas, and not along the whole shore of the island (Kitikoon et al. 1973). Moreover, neither the snail nor the disease was found in the entire area around the Ubol-Ratana Dam in Thailand, and the snail has not been located in any of the small foci where the disease is present in Thailand. I do not mean to imply it is impossible that the existing low prevalence of schistosomiasis in Southeast Asia could be increased, but only that I believe the situation in the Mekong River Basin is not as alarming as it is in other places.

There are proposals for projects which will harness the Nile River's waters. Such proposals must consider the above recommendations for multidisciplinary preconstruction studies, correct planning and designing of reservoirs and irrigation schemes, and the use of existing tools for control of several waterborne diseases. Already, a small dam was recently constructed on the Rahad River, a tributary of the Blue Nile in Sudan and a small gravity-irrigation scheme of about 8,000 hectares is planned (Figure 2), to be irrigated from waters stored behind the dam.

Other dams have been proposed for the Nile, the possible sites being Lake Tana, Lake Kioga, and Lake Albert. These dams were proposed in the plans of Hurst et al. (1946) and Hurst (1952) for storage, flood control, and river regulation. A dam for hydroelectric power—one of several—has been suggested at the sixth cataract in the Sudan. With the construction of the Aswan High Dam, there is no need to control the Blue Nile flood. A dam at Lake Tana would only regulate the flow of the river in the case of a very low flood.

The project planned for the Sudd region in the Sudan is, in my opinion, a significant one because of the possible repercussions of disease and other ecological effects. The project would have a considerable impact on the ecology of the region and the Nilotic tribes who live there. A Jonglei Canal (Figure 1) would increase the flow of the White Nile, by bypassing the extensive and famous Sudd swamps. Accordingly, these swamps will be drained, and the close to 1 million people of Nilotic tribes—the Dinka and the Shilluk in particular—would have to be displaced. They would have to change their way of life, which is intimately connected to the swamps and the river.

With the continuously growing human population of the Nile River Basin, urbanization and industrialization are bound to take place. The consequences of such changes, which are very well known in economically developed countries, would be the pollution of the Nile's waters by sewage and industrial wastes. So far, fortunately, the Nile remains free of those pollutants.

References

Abdin, G. 1948. The conditions of growth and periodicity of the algal flora of the Aswan Reservoir (Upper Egypt). *Bulletin of the Faculty of Science, Faoud I University* 27:157-175.

Ahmed, Abdel Aziz. 1960. Paper No. 6120. In *Proceedings of the Institution of Civil Engineering*, volume 17.

Apted, F.I.C., W. E. Ormerod, D. P. Smyly, B. W. Stronach, and E. L. Szlamp. 1963. A comparative study of the epidemiology of endemic Rhodesian sleeping sickness in different parts of Africa. *Journal of Tropical Medicine Hygiene* 66:1-16.

Ball, J. 1939. *Contributions to the Geography of Egypt*. Survey and Mines Department, Ministry of Finance, Egypt. Cairo: Government Press, Bulag.

Barbosa, F. S., I. Barbosa, and F. Arruda. 1962. *Schistosoma mansoni:* Natural infection of cattle in Brazil. *Science* 138(3542):831.

Barlow, Claude H. 1937. The value of canal clearance in the control of schistosomiasis in Egypt. *American Journal of Hygiene* 25(2):327-348.

Berry, L., and A. J. Whiteman. 1968. The Nile in the Sudan. *Geographical Journal* 134(1):1-37.

Blashford-Snell, J. N. 1970. The conquest of the Blue Nile. *Geographical Journal* 136(1):42-51.

Brook, A. J., and J. Rzòska. 1954. The influence of the Gebel Aulyia dam on the development of Nile plankton. *Journal of Animal Ecology* 23(1):101-114.

Burton, George J., and Thomas M. McRae. 1965. Dam-spillway breeding of *Simulium damnosum* Theobald in northern Ghana. *Annals of Tropical Medicine and Parasitology* 59(4):405-412.

Buzo, Z. J. 1972. Engineering measures for control. In *Symposium on Future of Schistosomiasis Control* (New Orleans, 1-6 February 1972), ed. Max J. Miller, pp. 63-68. New Orleans: Tulane University.

Egyptian Ministry of Public Health. 1947. Bilharzia Snail Destruction Section. Third Annual Report, 1944-1945. Cairo: Government Press.

Eyakuze, V. M., A.B.C. Dallas, S. S. Baalawy, and R. S. Mtoi. 1974. The role of steamers and itinerant fishermen in the dissemination of *Schistosoma mansoni* to the western shore of Lake Victoria. *East African Journal of Medical Research* 1:47-57.

Farooq, M., J. Nielsen, S. A. Samaan, M. B. Mallah, and A. A. Allam. 1966a. The epidemiology of *Schistosoma haematobium* and *S. mansoni* infections in the Egypt-49 project area. II. Prevalance of Bilharziasis in relation to personal attributes and habits. *Bulletin of the World Health Organization* 35(3):293-318.

———. 1966b. The epidemiology of *Schistosoma haematobium* and *S. mansoni* infections in the Egypt-49 project area. III. Prevalence of Bilharziasis in relation to certain environmental factors. *Bulletin of the World Health Organization* 35(3):319-330.

Farooq, M., and M. B. Mallah. 1966. The behavioural pattern of social and religious water-contact activities in the Egypt-49 Bilharziasis project area. *Bulletin of the World Health Organization* 35(3):377-387.

Gay, P. A. 1958. *Eichornia crassipes* in the Nile of the Sudan. *Nature* 182(4634): 538-539.

George, C. J. 1972. The role of the Aswan High Dam in changing the fisheries of the southeastern Mediterranean. In *The Careless Technology: Ecology and International Development*, eds. M. Taghi Farvar and John P. Milton, pp. 160-178. (Record of the Conference on the Ecological Aspects of International Development, Washington, D.C., 8-11 December 1968.) Garden City, New York: The Natural History Press.

Greany, W. H. 1952. Schistosomiasis in the Gezira irrigated area of the Anglo-Egyptian Sudan. I. Public Health and field aspects. *Annals of Tropical Medicine and Parasitology* 46(3):250-267.

Hammerton, D. 1972. The Nile River—A Case History. In *River Ecology and Man. Proceedings, International Symposium on River Ecology and the Impact of Man*, 1971, at the University of Massachusetts, eds. Ray T. Oglesby, Clarence A. Carlson, and James A. McCann, pp. 171-214. New York: Academic Press.

Harinasuta, C., S. Jetanasen, P. Impand, and B. G. Maegraith. 1970. Health problems and socio-economic development investigation on the patterns of endemicity of the diseases occurring following the construction of dams in northeast Thailand. *Southeast Asian Journal of Tropical Medicine and Public Health* 1:530.

Haseeb, M. A., M. H. Satti, and M. Sherif. 1962. Onchocerciasis in the Sudan. *Bulletin of the World Health Organization* 27:609-615.

Heyneman, D. 1971. Mis-aid to the Third World: Disease repercussions caused by ecological ignorance. *Canadian Journal of Public Health* 62(4):303-313.

Hira, P. R. 1970a. Schistosomiasis at Lake Kariba, Zambia. I. Prevalence and potential intermediate snail hosts at Siavonga. *Tropical and Geographical Medicine* 22(3):323-334.

———. 1970b. Schistosomiasis at Lake Kariba, Zambia. II. Transmission of *Schistosoma haematobium* and *S. mansoni* at Siavonga. *Tropical and Geographical Medicine* 22(3):335-344.

Hoogstraal, H., and D. Heyneman. 1969. Leishmaniasis in the Sudan Republic. 30. Final epidemiologic report. *American Journal of Tropical Medicine and Hygiene* 18(6):1087-1210.

Hughes, C. C., and J. M. Hunter. 1970. Disease and "development" in Africa. *Social Science and Medicine* 3:443-493.

———. 1972. The role of technological development in promoting disease in

Africa. In *The Careless Technology: Ecology and International Development*, eds. M. Taghi Farvar and John P. Milton, pp. 69-101. (Record of the Conference on the Ecological Aspects of International Development, Washington, D.C., 8-11 December 1968.) Garden City, New York: The Natural History Press.

Hughes, James P. 1964. Health aspects of the Volta River Project in Ghana. In *Industry and Tropical Health V*. Proceedings of the Fifth Conference of the Industrial Council for Tropical Health, sponsored by the Harvard School of Public Health, Boston, 29-31 October 1963, pp. 43-52. Boston: Harvard School of Public Health for the Industrial Council for Tropical Health.

Hurst, H. E. 1952. *The Nile. A General Account of the River and the Utilization of its Waters*. London: Constable and Co., Ltd.

Hurst, H. E., R. P. Black, and Y. M. Simaika. 1946. The Future Conservation of the Nile. *The Nile Basin, vol. 7. Physical Department Paper no. 51*. Ministry of Public Works, Egypt. Cairo: S.O.P. Press.

Kassas, M. 1972. Impact of river control schemes on the shoreline of the Nile Delta. In *The Careless Technology: Ecology and International Development*, eds. M. Taghi Farvar and John P. Milton, pp. 179-188. (Record of the Conference on the Ecological Aspects of International Development, Washington, D.C., 8-11 December 1968.) Garden City, New York: The Natural History Press.

Kirk, R. 1939. Studies in leishmaniasis in the Anglo-Egyptian Sudan. I. Epidemiology and general considerations. *Transactions of the Royal Society of Tropical Medicine and Hygiene* 32:533-544.

Kitikoon, V., C. R. Schneider, S. Sornmani, C. Harinasuta, and G. R. Lanza. 1973. Mekong schistosomiasis: 2. Evidence of the natural transmission of *Schistosoma japonicum*, Mekong strain, at Khong Island, Laos. *Southeast Asian Journal of Tropical Medicine and Public Health* 4:350-358.

Macdonald, G. 1955. Medical implications of the Volta River project. *Transactions of the Royal Society of Tropical Medicine and Hygiene* 49:13-27.

Macdonald, George. 1964. Health problems in developing Africa. In *Industry and Tropical Health V*. Proceedings of the Fifth Conference of the Industrial Council for Tropical Health, sponsored by the Harvard School of Public Health, Boston, 29-31 October 1963, pp. 28-34. Boston: Harvard School of Public Health for the Industrial Council for Tropical Health.

McMullen, D. B., Z. J. Buzo, M. B. Rainey, and J. Francotte. 1962. Bilharziasis control in relation to water resources development in Africa and the Middle East. *Bulletin of the World Health Organization* 27:25-40.

Malek, Emile A. 1958a. Distribution of the intermediate hosts of bilharziasis in relation to hydrography, with special reference to the Nile Basin and the Sudan. *Bulletin of the World Health Organization* 18:691-734.

———. 1958b. Factors conditioning the habitat of bilharziasis intermediate hosts of the family Planorbidae. *Bulletin of the World Health Organization* 18:785-818.

———. 1959a. Natural and experimental infection of some bulinid snails in the Sudan with *Schistosoma haematobium*. *Proceedings of the Sixth International Congress of Tropical Medicine and Malaria* 2:5-13.

———. 1959b. Trematode infections in some domesticated animals in Sudan. *Journal of Parasitology* 45 (Suppl.):21.

———. 1959c. Check-list of helminth-parasites of domesticated animals in Sudan. *Indian Veterinary Journal* 36:281-288.

———. 1960a. *Bulinus (Bulinus) forskalii* Ehrenberg, 1831: Intermediate host of *Gastrodiscus aegyptiacus* (Cobbold, 1876) Looss, 1896. *Journal of Parasitology* 46 (5, Section 2):16.

———. 1960b. Human and animal schistosomiasis in Khartoum Province, Sudan, N. Central Africa. *Journal of Parasitology* 46(1):111.

———. 1961a. The ecology of schistosomiasis. In *Studies in Disease Ecology*, vol. 2, ed. Jacques M. May, pp. 553-568. New York: Hafner Publishing Co.

———. 1961b. The biology of mammalian and bird schistosomes. *Bulletin of the Tulane University Medical Faculty* 20:181-207.

———. 1962. Bilharziasis control in pump schemes near Khartoum, Sudan, and an evaluation of the efficacy of chemical and mechanical barriers. *Bulletin of the World Health Organization* 27:41-58.

———. 1966. The snail in the epidemiology of schistosomiasis in the Sudan. *Hakeim* [Students' Medical Society, Faculty of Medicine, University of Khartoum, Khartoum], no. 20:26-30.

———. 1969. Studies on bovine schistosomiasis in the Sudan. *Annals of Tropical Medicine and Parasitology* 63:501-513.

———. 1971. The life cycle of *Gastrodiscus aegyptiacus* (Cobbold, 1876) Looss, 1896 (Trematoda: Paramphistomatidae: Gastrodiscinae). *Journal of Parasitology* 57:975-979.

———. 1973. Snail ecology and man-made habitats. In *Epidemiology and Control of Schistosomiasis (Bilharziasis)*, ed. N. Ansari, pp. 57-60. Published on Behalf of the World Health Organization. Baltimore: University Park Press, and Basel: S. Karger, A.G.

Malek, Emile A., and Thomas C. Cheng. 1974. *Medical and Economic Malacology*. New York: Academic Press.

Monakov, A. B. 1969. The zooplankton and the zoobenthos of the White Nile and adjoining waters in the Republic of the Sudan. *Hydrobiologia* 33(2): 161–185.

Paperna, I. 1970. Study of an outbreak of schistosomiasis in the newly formed Volta Lake in Ghana. 2. *Tropenmedizin und Parasitologie* 21:411–425.

Prowse, G. A., and J. F. Talling. 1958. The seasonal growth and succession of plankton algae in the White Nile. *Limnology and Oceanography* 3:222–238.

Saeed, A. A., and G. S. Nelson. 1969. Experimental infection of calves with *Schistosoma mansoni*. *Transactions of the Royal Society of Tropical Medicine and Hygiene* 63:456–458.

Satti, M. H. 1958a. Early phases of an outbreak of kala-azar in the Southern Fung. *Sudan Medical Journal* 1:98–111.

———. 1958b. Kala-azar in the Sudan and tropical Africa. *Proceedings of the Sixth International Congress on Tropical Medicine and Malaria* 3:646–657.

Scudder, Thayer. 1972. Ecological bottlenecks and the development of the Kariba Lake Basin. In *The Careless Technology: Ecology and International Development*, eds. M. Taghi Farvar and John P. Milton, pp. 206–235. Garden City, New York: The Natural History Press.

Shukri, N. M. 1950. The minerology of some Nile sediments. *Quarterly Journal of the Geological Society of London* 105:511–534.

Sterling, Claire. 1972. Superdams: The perils of progress. *The Atlantic* 229(6): 35–41.

Talling, J. F., and J. Rzòska. 1967. The development of plankton in relation to hydrological regime in the Blue Nile. *Journal of Ecology* 55:637–662.

Turner, D. J. 1971. Dams and ecology: Can they be made compatible? *Civil Engineering* 41:76–80.

van der Schalie, H. 1960. Egypt's new High Dam—Asset or liability. *The Biologist* 42:63–70.

———. 1972. World Health Organization Project Egypt 10. A case history of a schistosomiasis control project. In *The Careless Technology: Ecology and International Development*, eds. M. Taghi Farvar and John P. Milton, pp. 116–136. (Record of the Conference on the Ecological Aspects of International Development, Washington, D.C., 8–11 December 1968.) Garden City, New York: The Natural History Press.

Williams, T. R. 1969. Freshwater crabs of the River Nile. *Sixteenth Annual Report of the Hydrobiological Research Unit, University of Khartoum*. Khartoum.

World Health Organization. 1965. The Snail in the Control of Bilharziasis. *World Health Organization Monograph Series*, no. 50.

———. 1973. Schistosomiasis control. *World Health Organization, Technical Report Series*, no. 515.

Worthington, E. B. 1972. The Nile catchment—Technological change and aquatic biology. In *The Careless Technology: Ecology and International Development*, eds. M. Taghi Farvar and John P. Milton, pp. 189-205. Record of the Conference on the Ecological Aspects of International Development, Washington, D.C., 8-11 December 1968. Garden City, New York: The Natural History Press.

Wright, W. H. 1972. A consideration of the economic impact of schistosomiasis. *Bulletin of the World Health Organization* 47(5):559-565.

———. 1973. Geographical distribution of schistosomes and their intermediate hosts. In *Epidemiology and Control of Schistosomiasis (Bilharziasis)*, ed. N. Ansari, pp. 32-249. Published on behalf of the World Health Organization. Baltimore: University Park Press, and Basel: S. Karger, A.G.

Freshwater Wetlands

The Hydrologic Cycle: Freshwater Wetlands

26. Freshwater Wetlands

Arthur R. Marshall
Interlachen, Florida, U.S.A.

Introduction

I now present myself as your final panelist. In contrast to Dr. Amini and to Professor Matthews, I am—from the world view of EARTH-CARE—a neophyte. My life involvement (twenty-five years as an environmentalist and ecologist) has to this moment been with "FLORIDACARE." Therefore, some may justifiably ask whether my parochial experiences can have meaning in the world view. I believe they can—for these reasons:

1. All plant, animal, and human life on Earth depends upon the maintenance of viable life-support systems;
2. Life-support systems are threatened now as never before in Earth's history;
3. All such systems are interconnected and all operate under the same natural laws; and
4. The widespread malfunctioning of life-support systems is due largely to the human population explosion and to the ever-increasing demands being made upon them.

Because these problems are occurring nearly everywhere, we can profit from lessons learned anywhere—however parochial.

Assault on the Wetlands of Florida

The wetlands of Florida—salt, brackish, and fresh—have been under assault by technology and the longest sustained migration any American state has experienced. It is a sad process to behold. An endless flow of newcomers creates a market demand for homesites on or near the shores of lakes, rivers, and bays—a demand which entrepreneurs are pleased to accommodate by dredging, filling, and draining the wetlands of the state, Florida's most valuable life-support systems. Ironically, many of these same people become Florida's most conscientious environmentalists—about five years too late.

High-Speed Destruction

Many observers of the Florida environment have noted the speed at which it is being degraded. In much less than a hundred years, many of its major environments—both wild and urban—have been degraded to levels experienced in other states, only after two or three years of human disturbance.

Florida is a grandiose, uncontrolled environmental experiment. Underlying the technology, the growth, and affluence which it engenders is the little known fact that the Florida peninsula is much less able to withstand the disturbances of man than most other states.

The peninsula is surrounded by sea water which is "eager" to rush in. It is underlain by porous limerock strata which are also filled with residual salt water left there by the events of its geologic history. As the shallow freshwater pools overlying the residual salt-water layers are consumed, those ancient chloride waters rise into the freshwater strata.

The peninsula is a freshwater "island." Freshwater supplies are replenished solely by local rainfall during a three- to four-month rainy season and must last the duration of an ensuing six- to eight-month dry period.

The bays, lakes, and rivers of the peninsula have an average depth of less than fifteen feet. About one-third of the peninsula—the coastal plain—lies less than twenty-five feet above mean sea level. Thus, the lowering of the surface-water level by as little as a foot or two can drain thousands of acres of wetlands in that nearly level land.

As a result of this array of severe environmental constraints, coupled with massive human disturbances, I have been able to observe over a twenty-five-year period, degradations of natural resources to a degree which would require several lifetimes in many other states. My experience is comparable to that of a manufacturer who subjects, in a few days or weeks, a proposed new product to stresses which it would sustain only after months or years of normal use.

Observers should pay close attention to the changes taking place in the delicate wetlands of Florida and to other sensitive wetland areas around the world (e.g., the Sahel region in Egypt), for those changes occur quickly—within the lifespan of one generation—and they tell us much we need to know about the wetland ecosystems in general.

Why Protect Wetlands?

The first question confronting this symposium is, Why protect wetlands? We can't really gain anything by protecting them; we can, at best, merely retain their values which have been there all along.

Wetlands are among the most ancient life-support systems. In eons past, they were the life systems through which countless species passed in their evolutionary migration from ocean to land. Since evolution is virtually endless, this migration through wetlands is ongoing.

Wetlands sustain an enormous diversity of aquatic and semiterrestrial plants and animals. This diversity of species, as we now know, provides resiliency against diseases and other disturbances which monotony cannot do.

Wetlands are enormous natural absorbers of solar energy. They are effective collectors and repositories of the nutrients which slide and percolate down into them from uplands. They convert those nutrients into living matter under the drive of the sun's energies.

Wetlands absorb and dissipate the energetic forces of storm and flood. They recharge groundwater levels; they hold out the seas; they are the primary generators of the organic soils of the world; they contribute moisture to the atmosphere through evaporation and transpiration, thus helping to maintain the hydrologic cycle from Earth to air and back to Earth. They assimilate many of the wastes of man; and they provide food as well as intrigue, great beauty, and diverse recreations for the soul of man.

What Does the Loss of Wetlands Mean to Man?

The assault on Florida's wetlands is teaching us just how much man has to lose when the wetlands perish.

Careless dredging has caused salt intrusions into freshwater aquifers. Drainage of wetlands has so completely reversed the processes by which wetlands have built up a store of peat and muck soils, that great agricultural enterprises based on those soils will end soon because the peat and muck deposited over a period of 5,000 years will be exhausted within the next twenty-five.

Man has destroyed much of the capacity of wetlands to assimilate wastes and, consequently, has helped convert several major freshwater lakes into eutrophic nutrient sinks. He has lowered water levels in wetlands so much, and has so altered their seasonal regimes, as to place great stress on a major national park, the Everglades National Park. The existence of dozens, perhaps a hundred, Florida-wetland plant and animal species is threatened; and floods which technology cannot prevent alternate with droughts and wildfires which are unavoidable.

By draining wetlands, man has also paved the way for rapid urbanization and, along with it, high energy demands which wetlands cannot long sustain. These energy demands are required in part to protect the urban developments against the "floods" they will experience because the wetlands were never meant to support such developments.

This brings me to the bigger question confronting this symposium: How do we protect the wetlands?

Protecting the Wetlands

I must tell you in all fairness to the environmentally concerned people of Florida, that they have played a major role in generating an awareness of the values of wetlands. They have set an example for developers and agriculturalists who, consequently, are following a similar course toward environmental awareness. The effects of this new awareness have spread to government agencies—local, state, and federal—which have genuinely absorbed injections of environmental hormones. Clearly, the struggle to save the wetlands continues. Nevertheless, I believe that the struggle in Florida can teach us some meaningful lessons—lessons of potentially wide application. That is my reason for coming to this conference.

A Program for Effective Environmental Action

The balance of my statement consists of principles, methods, and philosophies which I have assembled over a period of twenty-five years: an environmental "code of action," so to speak, which I believe is applicable to environmental struggles anywhere. These are my own formulations and I alone am responsible for them.

I put these thoughts before you, perhaps with more audacity than sagacity, for the reason that they deserve wider attention, analysis, and strengthening for the protection of wetlands and many other besieged life-support systems. The list is not complete; nor are its elements perfected. I doubt they ever will be, simply because no one—not all of us together—is a master of the intense and complex game we play. I also recognize that ecologic, social, and governmental differences will require local modifications of my basic concepts, but I doubt that fundamental revisions will be necessary. The items are in no especial order.

1. *All life-support systems—wild and urban—operate under the same laws of nature.* All life depends on the continued viability of those systems. I propose, therefore, that people and governments everywhere adopt the preservation of life-support systems as their primary goal and purpose for at least the next twenty-five years.

2. *Certain repetitive phenomena are inherent in any life-support system that is significantly disturbed by man.* I believe these phenomena are universal—applicable to all life systems. They can be summarized in the following maxims.

   a. We are dealing with integrated systems in every environmental problem.
   b. No pristine system is as efficient in maximizing the production of desirable forms of life as man can induce by rational intervention.
   c. Intervention by man can bring life systems to a higher level of efficiency—to an optimum in the production of desirable species of plants and animals. This is the "creative intervention" concept proposed by Dr. René Dubos.

d. Continued intervention by man along a given course of action will cause a decrease in the efficient production of desired species, because man doesn't know when to stop. That decrease in efficiency will inexorably follow the accelerating path of an exponential curve. Its origin will be announced to us by a series of symptoms of distress as the system undergoes increasing strain.
e. After a period of time during which its viability will become increasingly precarious, the system will reach a critical level of efficiency. Pressures on the old system will then be relieved—like steam pressure through a pop-off valve—by a precipitous decline and conversion of the old system into a new one.
f. The new system will be very productive, but the species and other values which were prized in the old system will be largely eliminated and replaced by new species and other values which man does not highly prize.
g. These concepts are applicable to the behavior of such assorted systems as lakes, bays, wetlands, rivers, farms, forests, and cities. They are also applicable to the behavior of that system we cherish above all—the human body—and are in fact the primary basis of modern medical diagnosis and prognosis. Ill human bodies produce signs and symptoms of distress for which the physician searches in developing diagnoses and prognoses.

These concepts are universal in their meaning; they indicate universal similarities of function in what have commonly been regarded as dissimilar systems. While simple accumulation of fact is a salient purpose of scientists, a higher purpose is to recognize and define similarities between functions, phenomena, and systems which were never known to have them.

These concepts provide a simple mode of comprehending changes due to human disturbances in life-support systems. They enable us to diagnose the condition of ecosystems by observing the symptoms of distress. They also tell us that in developing prognoses we are often forced to make an about-face. Finally, they announce that once a system has released its pressures—i.e., collapsed—"treatment" will be impossible, at worst; costly in dollars and energy; and traumatic in the sense of surgery, at best.

3. *In addition to the ecological side of ecosystem issues, there is a societal side.* This is clearly the tougher arena. The diagnoses and prognoses of scientists and of an array of other essential professionals are nothing more than informational if society fails to make effective, rational responses.

4. *There are only three ways in which society can respond to the realities of life-systems: through* voluntarism, coercion, *or by* force. By *voluntarism,* I mean the voluntary alteration of lifestyles by individuals, organizations, or institutions. *Coercion* is especially the province of governments in the exercise of their powers to dictate what people can or cannot do. By *force,* I refer to the unavoidable penalties generated by the systems themselves when voluntarism and coercion both fail.

There is no more appropriate example than the forces at work at this moment in the city around us. Voluntarism and coercion have been suddenly displaced by the stringent forces of the city system which neither the city government nor its people can escape. Another force will constrain us beyond choice when the gasoline pumps again run dry because the natural resource—fossil fuel—has diminished. In Florida, the force of intermittent drought compels curtailment of water uses; the force of eutrophication of lakes compels us to seek other water sources; the forces generated when the peat and muck are gone from the Everglades will sharply narrow the options of the agriculturalists. The force brought into play by the extinction of one of the plant or animal species in an ecosystem will cause the ecosystem to collapse and prevent us from fully restoring it.

5. *It is impossible to find effective responses to our current predicaments solely in science and technology.* Scientists are constrained by the ethics of their profession from probing the very questions society must ultimately ask and answer. Science and technology, in fact, have contributed materially to the predicaments we find ourselves in—pollution, population growth, drainage of wetlands, destruction of ecosystems, gross impoverishment of the poor, grosser enrichment of the rich, nuclear weapons, and the intolerable demands of cities. Mankind survived more than a million years without science and technology and without our panoply of bewildering problems. Some few primitive societies still do.

When governments, in their indifference or innocence, devote their liberal resources solely to science and technology—as I have often seen them do—in an effort to resolve environmental problems, they are ignoring and often evading the central issues. Science and technology can help ease us back from the precipice we are on. Science can enlighten us if we can make meaningful diagnoses and prognoses of our problems, but it cannot make the subjective choices that mankind must now make.

There are, to be sure, scientists who do participate in making those choices, both personally and by encouraging others to do so. But in this sense they are no longer performing as scientists—they are performing as concerned human beings who know a little more of the realities than others do. As such, they relinquish, however momentarily, the constraining ethical mantle of their chosen field.

The meaningful answers—answers on which our futures and those of ecosystems turn—do not reside in science and technology; they lie in the cultural evolution of mankind. They reside in matters which are not objective, not quantifiable, and thus far-too-little observable; matters which are not repeatable in controlled experiments; matters which are totally beyond the purview of science as proscribed by its own ethics. They are matters of the human spirit and the human heart. Just as we have learned to recognize the reality of biological evolution, and more recently of scientific and technological evolution, we must now concern ourselves with cultural evolution. Therefore, should not government, for instance, support this conference and countless individuals and organizations of like bent who struggle with meager personal resources to effect changes that *are* relevant to the issues, that can serve to alleviate the array of problems in the world?

6. *One of our greatest problems is our enormous drive to specialize, to try to do only one thing.* That is impossible in a systematized world. Such single-mindedness, tragically, has produced many of our present environmental failures and dilemmas.

We see specialist tendencies in education and the professions; in individual drives for wealth and power; in corporate and municipal growth patterns. Government itself is composed of agencies whose legislated goals are specialized: to build roads, to construct flood control projects, to go to the moon. This is not evil necessarily, except that each project ignores the complexities of life-support systems and its effects on them.

We often speak of interdisciplinary efforts wherein concern with effects on whole systems is the central goal. Our actions, however, bear scant witness to such good intentions.

Specialization over time has proceeded in an ever-narrowing pattern. This process, coupled with overdraughts on Earth's resources, has produced situations in which interdisciplinary examinations of issues are sorely needed. We must bring essential specialists together in holistic groups in order to reverse our philosophic direction and create a broader, integrating base.

7. *There are no renaissance persons—and perhaps there no longer can be.* But there can be renaissance "groups" composed of specialists, each of whom contributes knowledge of his speciality to the interacting brain mass of the group. Such a group calls for a nontraditional structure in order to creatively and effectively diagnose environmental issues.

We tend in all our institutions to employ a standardized pyramidal hierarchy to accomplish designated goals. This is a singularly effective method of accomplishing specialized and routinized goals. It is a singularly destructive method with regard to reaching integrated ecosystem goals—creative goals.

In dealing with processes, in seeking to maintain the viability of life-support systems, a totally new organizational approach is needed. A discussion leader is needed to present the particulars of a problem to a cohesive body of concerned and qualified specialists for their inputs, their digestion, and diagnosis. There can be no chiefs and no Indians in this renaissance group. Communications must move fully and freely in all directions; the product must be that of the group. I have worked in a dozen such efforts. They are at first exhausting. Through determination and persistence, members of the group ultimately come together as one mind. All share exhilaration from knowing that the product is soundly conceived from the systems view.

8. *American specialists on environmental issues torment themselves unnecessarily by going directly with their diagnoses and prognoses to elected legislative groups.* Their solutions to environmental ills often involve a complete reversal of traditional social goals. No elected official can on his own sponsor changes if he wishes to survive. He is caught in a web of ambitions for growth, higher energy consumption, faster production, etc., from which he cannot escape.

Changes of such magnitude have always come about in the United States through cultural evolution of the society as a whole. Thus, the product of any holistic renaissance group must be used primarily to inform the electorate of the realities of our predicaments.

In this way, we can liberate the elected official from the philosophical tyranny which binds him and move toward an enlightened legislative resolution of our problems.

9. *Environmentalists should contact and support certain precedent-setting individuals in society, and their techniques and successes should be carefully analyzed for future use.* These individuals are concerned professionals, citizens, public employees, and especially concerned women, who have accomplished environmental goals. We should be especially aware of concerned young people and support and train them as necessary, for they have greater energy and flexibility than their elders to modify intolerable lifestyles.

10. *It is not necessary to beat the drums.* The truth is enough, and you can live with it firmly through any siege.

11. *Even though you need not beat the drums, you must remember that as an environmentalist you are engaged in a struggle.* You are not just a guest at a tea party: the struggle is for the survival of Earth and man, which can be won only through a profound reordering and redirection of dominant social goals. You will, therefore, encounter some people who will vehemently disagree with you. Resistance to change is the source of all pain (from Buddha, via Timothy Keyser). Purring will get you nowhere.

12. *In reaching environmental goals, we cannot expect to go from A to Z in one step.* We can go from A to B to C, and occasionally we can skip a step or two. But due to the hard realities of environmental planning, we should always keep goal C in mind while reaching for B, and the moment goal B comes into view, reset our goal to C or D or E.

13. *On compromise: do not accept compromise as a settlement of an issue because it is the most popular accommodation of a series of fallacious propositions.* The product of such a compromise will itself be fallacious. Compromise is valid only in the step-by-step process along a new and viable path—from A to B and so on.

14. *It is regrettable but true that the more precipitous our environmental predicament becomes, the more likely we are to change things.* Hang on a little longer. We all need the threat of increasing threat to goad us.

15. *Aim to set precedent, no matter how small the case.* It is far easier to expand a precedent than it is to set it. But you must, in the first instance, carefully pick a case with winning ingredients,

for if you lose you shall have to live with a miserable precedent for a long time.

16. *Keep in mind that ecosystems have maintained man for more than a million years.* It's reassuring.

17. *Environmental programs must become more people-oriented than before.* People who, with good reason, are fearful today are not likely to be ardent supporters of environmentalists who present only negative positions. We must promote employment, self-dependence, craftsmanship, the joy of work, abandonment of the rat-race in favor of pleasurable low-energy lifestyles, concern for others, and a sense of community and of heritage—all as valid both for the environment and for people.

18. *Environmentalists should meet often with their colleagues—*not only to learn, but also to reinforce their spirits and their convictions through contact with kindred souls.

19. *Avoid with all your worth the foreclosure of options on life-support systems.* Once one is destroyed, it is pure hades to restore it.

20. *Promote the idea that we cannot develop all of our lands and waters.* Nature provides some valuable free services to us—services which we will increasingly require as energy supplies dwindle.

21. *Remember a lesson hunters and fishermen learned, after some antipathy, some years ago—the concept of "bag limits."* Before the coercive invocation of bag limits by government, hunters and fishermen were sure they could not live happily with such a law. Now they brag about getting their "limit." Society needs "bag limits" with regard to the lifestyles of individuals, corporations, and governments if we are to avoid the forcing functions of strained life systems. Further, we might brag about it!

22. *Think first of how to achieve environmental goals through government.* Of course, we need all the support we can get from government. We should note, however, that many admirable goals have been accomplished by individuals working in their own communities and regions. We should honor them and regard them as the effective people they really are. Government often tends to follow the paths set by these people, if only through "reverse" coercion! For this reason, I recommend the establishment, the support, and the analysis of the effectiveness of the methods of at least a small core of local leaders, which would constitute a Local Environmental

Action Program (LEAP). They should be maintained, for instance, as a citizen-effort counterpart to the federal Department of the Interior or to the United Nations Environment Programme.

These are some thoughts on saving freshwater wetlands and other vital life-support systems.

I offer these thoughts freely to you and to this conference—and especially to young environmentalists, in the hope that my concepts will shorten their time of maturation to much less than my own twenty-five years.

27. Wetlands: Assessment and Access

G. V. T. Matthews
International Waterfowl Research Bureau,
England

Introduction

Our ideal should be to set aside wetland wildernesses sufficiently large and comprehensive so that they may remain natural, evolving ecosystems. This is seldom feasible in crowded Europe. It is rapidly becoming difficult in other parts of the globe as the human population proliferates and modern technology destroys. We must save what we can, and when that is inadequate, manipulate what remains to make the best and most varied use of it. We must consider the ecological requirements of *all* the wetland animals and plants we seek to conserve.

The Convention on Wetlands of International Importance especially as Waterfowl Habitat, the Ramsar Convention [see Appendix B—Editor], is ably described by His Excellency Mr. Eskandar Firouz in this same session. I will not, therefore, go into its history, provisions, and effectiveness. Suffice to say here that the Ramsar

Convention has immense possibilities for good. I would like to develop two considerations. First, how to decide what *is* a wetland of international importance, and second, how to persuade the governments concerned—the politicians, planners, and the people behind them—that wetlands *are* worth conserving.

The Assessment of a Wetland's Importance

The successor conference to that at Ramsar in 1971 was held at Heiligenhafen, Federal Republic of Germany, in December 1974. One of its main activities was directed towards agreeing a better set of criteria of a wetland's importance, in international terms. The deliberations of a committee of experts were accepted by the conference, and its report presents a summary of them, which appears as the "Annex" to this paper.

It will be readily appreciated that many of these criteria, sensible and acceptable as they may be, are defined very broadly and often must rest on the opinion of an expert rather than on the presentation of precise, scientifically gathered data. To some extent, this is a realistic acceptance of the fact that measuring an area's biological productivity and its importance to a wide range of organisms is an immensely laborious and time-consuming task. There simply are not enough trained scientists, nor the money to employ them, to make the detailed assessments required. Even if there were, so much time would be required for the investigation to be made and the results analysed that the wetland concerned could well be destroyed in the interim. Wetlands are one of the easiest habitats for modern technology to destroy, corrupt, or modify beyond recognition.

The Use of Indicator Species

The International Waterfowl Research Bureau (IWRB) has therefore concentrated on obtaining information on those "indicator species"— waterfowl and shorebirds—that are relatively easily assessed and whose presence in large numbers shows that a wetland must have a high biological productivity. (The converse does not necessarily hold true, for some biologically suitable areas do not support waterfowl because of over-shooting or other forms of incidental disturbance.) Another reason for concentrating on waterfowl is that in many countries there are numbers of highly competent amateur ornithologists who can be persuaded to carry out regular censuses without charge. In areas where such observers are scarce, a great deal of information has been amassed by small, mobile teams of visiting ornithologists. Professional surveys have, of course, also been made, especially where the nature of the terrain makes aerial survey the only feasible way of obtaining adequate cover.

Since these international waterfowl censuses were begun in 1967, some 40,000 records have been received from 13,000 wetland sites in fifty-five different countries in Europe, Asia, and Africa. The observations are usually made around the middle of January, but November counts have also been made, and the program will be extended to other months.

Papers submitted by Atkinson-Willes and Prater to the Heiligenhafen Conference (Smart, in press) summarize the present situation within the Palaearctic migratory system. Distribution maps for the different waterfowl species were presented and, using the numerical criteria 1(i) and 1(ii) set out in the Annex to this paper, a list of 166 internationally important wetland sites was drawn up, 80 in northwestern Europe, 56 around the Mediterranean, and 30 in southwestern Asia. Information from Africa is less precise as yet, but a more general paper by Roux (in Smart) gave a lot of information about western Africa in particular. This list is only a start, in that it refers to the value of wetlands in mid-winter. Many others are equally important during the autumn and spring migrations and during the breeding season.

With the information that IWRB has gathered in this way, governments can begin to set aside wetlands for conservation, especially as they sign and ratify the Ramsar Convention. This calls for the appending of an official list of wetlands in each country destined for specific conservation. Some countries do not have the specialists necessary to make these assessments and are glad to accept the guidance of the international bodies. Further assessments on botanical, hydrological, and other physical or biological grounds are still needed, but are beyond the scope of IWRB to provide.

Education and Access

We may refer to the third group of criteria, those concerned with the research, educational, or recreational values of wetlands. At first sight, these may seem wholly contradictory to the concept of a wetland wilderness. Perhaps research could be accommodated without disturbance, but a question arises in the case of education and, especially, recreation. In fact, a formula of compromise has been developed. First, however, we should examine the need for educational and recreational incursion into the wilderness. We should remember, too, that many wetlands, in Europe at least, owe their present structure to human use and controlled exploitation, and that if these were removed, a rapid, probably deleterious, change would follow.

One of the hardest battles that the conservationist fights is that against the popular image of wetlands as damp, flat, disease-ridden places. The view that wetlands are wastelands is very widespread. Most people consider them fit only to be drained for agricultural use, to be filled by the staggering rubbish of our profligate civilization, to serve as soakaways for the poisons and filth from our manufacturing processes, or to provide spaces for noisy airports away from centers of population. Until a majority of politicians, planners, administrators, and taxpayers are convinced that there is intrinsic value and beauty in a marsh or mudflat, there is little hope of keeping even a fraction of our wetlands intact. We may gain a momentary stay of execution by a happy coincidence of interests and circumstances, but even legislative protection is reversible when the equation of national need and national forbearance changes.

Undoubtedly the best long-term security is a well-educated public brought up from infancy to think of the environment as an integrated whole, tampered with at our own ultimate peril. But the destruction of wetlands is proceeding so rapidly that we cannot wait for the next generation to save them: they will not be there to be saved. We must appeal to and influence the present generation, which has the power to make the decisions now, to stay the hand of the ruthless planner and engineer by pressure on the politicians and legislators.

The ordinary person is not impressed when we talk of "biological productivity"—when we quote statistics showing that a coastal marsh produces far more biomass than a carefully cultivated field of corn. "Biomass for whom?" is likely to be one of the less objectional retorts to what laymen consider elitist prating. To some extent their interest and enthusiasm can be captured by finely illustrated books and really good movie films showing the beautiful variety of animals and plants that depend on wetlands. Undoubtedly television has done a very great deal to awaken public feeling for wildlife in general. But there really is no substitute for personal involvement: nothing makes so much impact as seeing and hearing what is actually taking place, rather than when recorded on paper, film, or tape. And of all the denizens of wetlands, the most easily viewed, heard, and appreciated are the waterfowl.

Access and Observation: The Concealed Approach

The difficulties are great and obvious. Waterfowl, especially the quarry species, are very wary and easily disturbed by the sound—let alone the sight—of humans. The intrusion of great crowds of people would disrupt one of the essential attractions of wetlands, their lonely vastness. The very passage of vehicles and feet would disrupt the particularly sensitive substrate of tundra, marsh, or bog. Indeed, it is often difficult to get people over such substrates in reasonable safety and comfort.

Nevertheless, it does seem that a successful method of reconciling the apparently irreconcilable has been evolved by the Wildfowl Trust in Britain, under the inspiration of Sir Peter Scott.

The formula is relatively simple: it is the attention to essential details that makes for success. In brief, the technique is to bring the public to the outskirts of the wetland and conduct them into it along a few screened corridors leading to well equipped observation posts and towers.

If the subsoil is of sufficient depth and consistency, by far the best way of screening the approach corridors is to throw up parallel earth banks, 2 to 3 meters high. The initial construction may be rather expensive, but once built and grassed over, such walls need very little maintenance. Moreover, they blend with the background, and such "seawalls" are to many eyes a familiar and unobtrusive feature of the wetland scenery. The earth walls not only hide the people but muffle the sounds that even the most carefully wardened party will make. An overhead screen of camouflage netting will prevent the birds from "flaring" at the sudden sight of people below.

Excavating the soil to build the banks will leave shallow lagoons, to which the waterfowl will come to drink and bathe, right close to the observation facilities. Shorebirds will find food in the mud and shallow margins, again presenting themselves to the inspection of concealed watchers.

On many wetlands the construction of earth walls will not be feasible. In such cases, screens of dried reeds most readily fit into the scene, and their material is generally readily to hand. However, the cutting of the reeds and the construction of screens is laborious and time-consuming. Moreover, they will last only between ten and twenty years, however carefully they are made. They must also be protected by other fencing if there are cattle grazing in the wetland. Recently there has been developed a fiberglass "thatching" for house roofs in England. Though the thought of such an artificiality in a natural area may well be anathema, it would have the advantage of being nonbiodegradable. Wooden or wattle screens are very expensive, and there are real problems due to their resistance to the wind. These problems can be reduced, but at further expense, by constructing a wall of overlapping slats.

Probably the best compromise is to make reed screens and to plant belts of wetland shrubs alongside. These can grow up and be thickened out to provide sufficient cover by the time the original reed screen has decayed. Evergreen shrubs are preferable because cover is particularly needed during the winter months. If deciduous trees, such as willow, are used, they must be pruned repeatedly to keep plenty of low cover, and the belts will have to be much wider.

Observation Posts and Towers

At intervals along the screened corridors there should be small observation posts and at their end a more elaborate observatory. Suitable designs for these structures have been evolved by the Wildfowl Trust through long trial and error. The dimensions of observation slits, benches, and elbow- and footrests have all been carefully worked out to suit the average observer; comfort is conducive to enjoyable observation, especially when one is dealing with the lay public, which has yet to be convinced of the wonders of the wetland. Little details, like hinging the shutters over the observation slits along their lower edge, are important. They can then be opened slightly if the birds are very near and lowered progressively the farther away they are. A length of plastic chain and a hook permits one to fix the shutter in any position. Binoculars can then be steadied on its top.

The walls and roof can be of wood or reeds, but such a structure should be raised on pillars to minimize contact with the wet ground. The space between the floor and the ground must be screened as well, so that the birds do not see the legs of people as they climb in—an obvious point, but one that is often forgotten. Similarly, the entrance must be screened from the side and, if opposite an observation slit, must be sufficiently low to prevent people from being silhouetted against the sky.

The problem of rotting—implicit in the wetland environment—can also be overcome by foundations of concrete, which will certainly be necessary for any major observatory structure. However, another compromise with modern technology can be reached over the smaller observation posts. Premoulded fiberglass rainshelters for road workers have been found to be easily adaptable for the purpose, and they have the advantage of being completely nonbiodegradable. They are light enough to be easily moved to another, more suitable, site if seasonal or longer-term changes make it necessary to do so. Again, there are aesthetic objections to the intrusion of completely artificial structures into the natural landscape, but one has to face the facts of economics, especially in countries where maintenance labor is prohibitively expensive.

A main observatory can be made more luxurious than the posts en route. Quite large windows can be incorporated. Vertical glass reflects sky to birds below, helping to conceal the people within. Venetian blinds can be gradually adjusted as the birds regularly present outside become more confident. Within the observatory, chairs can be provided instead of benches. Carpeting on the floor deadens sound. Space heating and window de-misting facilities can be added. This may sound overluxurious, but we are aiming to convert the average "man-from-a-car" who will not squelch a muddy mile to peer through a draughty slit. He may come to such pleasures eventually, but he must be led to them gently. Also we must cater for the elderly citizen—not forgetting that they are often better equipped to finance our conservation projects than are the hardy young.

The siting of observatory and observation posts must be done with an eye to the landscape and so that their main outlooks are on the wilderness, not on other observation slits. Also, they must face north so that for much of the day the sun is behind the observers, giving them the best light for viewing and photography.

In the wetland flatness observation towers have an obvious advantage. Moreover, the occasional vertical feature is visually attractive—witness church steeples. A tower near the beginning of the system of screened approaches will not only give the visitor an overview and foretaste of the delights available, it will also serve as a watchtower for the wardens. Another good place for a tower is at a point where the screened corridors must end because of substrate problems. Powerful binoculars on stands can then give good views of distant birds and reveal the full sweep of the wilderness.

Towers up to 20 meters high can be safely constructed of steel scaffold tubing, clad with wood or composition board and guyed against high winds. A more solid tower is essentially an upward multiplication of a concrete and wood observation post. In any case, we should think in terms of a structure capable of holding at least fifty people at a time—one bus-load. Again, it is necessary to emphasize that the visitors must be screened to reach the observation platforms without showing themselves to the birds. There are too many openwork towers with a cabin on top situated in the middle of an open plain.

Enticing the Birds

Having conducted the public down an extensive (and expensive) system of screened corridors to observation posts, towers, and, especially, the main observatory, it is necessary to ensure that they have waterfowl to see. Any disturbance in the area covered by the viewing system must be rigorously excluded. There must be no shooting within a kilometer of the edge of the central sanctuary. Shooting even farther away should be forbidden during the middle hours of the day, when the public uses the facilities more heavily. Low-flying aircraft can cause severe disturbance and must be discouraged by edict, with proper emphasis being laid on the dangers of striking birds. Any cattle or sheep that need tending or supplementary feeding should be removed before the waterfowl flocks arrive in season.

Positive measures should also be taken to ensure high-quality feeding opportunities for waterfowl in proximity to the viewing facilities. Thus, summer grazing by cattle and sheep will keep the grass sward in a suitable condition for the wintering flocks of geese or widgeon. Short-term rotational flooding of grassland can be engineered to make food concealed in the sward and soil available to the birds. Special crops (of grain or potatoes, for instance), left unharvested and shallowly flooded, can be especially grown for waterfowl.

As the spectacular concentrations of waterfowl needed to impress the public build up, they will eat out the natural or artificial crops near the observatories. It may then be necessary to put out regular supplies of feed to keep the birds coming. There has been quite a lot of controversy about this technique, largely on the grounds that it destroys the essential wildness of the birds. Yet it is no different, in essence, from putting food out on a garden bird-table. It has just the same fascination for the tyro birdwatcher. The same precautions must also be observed. Food must be restricted so that the birds do not become wholly dependent on it. Surpluses that might rot and engender disease must be avoided. The surrounding sanctuary area must be able to support the number of waterfowl that are attracted by the food provided and by the general lack of disturbance; otherwise, conflicts are bound to arise with agricultural interests on the periphery. Also, unscrupulous sportsmen there can exact an unduly high kill of birds lulled by security.

Another technique over which there has been argument, after its successful implementation in England, is that of floodlighting the assembled waterfowl on the dark winter evenings. There is no doubt that this provides a breathtaking spectacle which has enormous success in awakening the public to an interest in the birds and their habitat. It has the advantage that the workingman can come and see the waterfowl after his working hours; and for conservation to be a success it must be accepted and supported throughout all strata of the community. A possible risk in floodlighting is that birds might hit obstacles when flying onto the illuminated lagoon. These should, of course, be eliminated or, in their turn, illuminated. A more subtle question is whether the extra light modifies the delicate photoperiodic mechanisms, based on day length, that trigger migratory and reproductive cycles. This factor is being carefully watched, but so far no differences have been noted between the breeding success or migratory timing of birds that come to floodlit lagoons and those of birds that experience normal hours of darkness.

Displays of Captive Waterfowl

However much cunning we employ to ensure that wild birds are readily visible to the public, we cannot provide a spectacle if the birds are not there. In the temperate zones migratory waterfowl are present only for part of the year—in mid-winter or on autumn and spring passage, according to location. Yet the majority of the public we seek to influence is unwilling to emerge into the countryside except during the summer. Wetlands are then at their dullest, though a breeding colony of herons can be a spectacle, and even a gull colony is exciting for those unused to a swirling bird metropolis. Where no such natural summer display exists, the only answer is to provide one.

The concept of a tastefully presented display of captive waterfowl on the edge of a natural wetland is due, again, to Sir Peter Scott. A famous painter of waterfowl and landscapes, he considers it one of the most creative of his artistic endeavors to design and lay out natural-looking complexes of ponds and grazing areas so that the birds are seen to their best advantage. Again, there must be the attention to detail. Buildings should be grouped in a convex line so that the main-entrance viewpoint looks out only on birds and landscape, not on other buildings. Car parks and other ancillaries must lie behind the buildings or screening earth banks. Ponds must be edged with soil-colored, rough concrete to prevent erosion. Paradoxically, ponds should be on the firmer soil, not on the marshy areas, and the water flow through them should be as rapid as possible to prevent the build-up of disease. Internal fences must run away from the main line of vision, not across it. The perimeter fencing needed to keep out predators must be made as inconspicuous as possible (incidentally, this fence keeps people *in,* preventing them from spilling out over the wetland and ruining its wilderness quality).

Most waterfowl settle into captivity and appear perfectly content walking and swimming in the open paddocks and ponds. The necessary prevention of flight by the cutting of one wing tip does not have the traumatic effect that might be expected. Such pinioned birds remain healthy and breed freely. Their downy young are an especial attraction to the visitors, and a proportion can be left full-winged. Geese, in particular, raised by grounded parents do not acquire the migratory tradition. They fly around freely but seldom stray far, providing the incomparable thrill of geese against the backdrop of clouds and sky for those visitors who cannot come in winter to see the wild flocks. Both full-winged and grounded birds perform yet another function—decoying in the wild birds close to the main observation facilities. Indeed, many duck species will become quite indifferent to humans, mingle freely with the captive birds, and take food with them. The visitors can have birds at their feet that are wild or captive and be unable to distinguish between them. This is a powerful lesson of the relationship that man can establish with the animals if he has the forbearance and will so to do.

Some Final Considerations

Having attracted the public into such a "honey pot," no opportunity must be lost to educate them discreetly. This is done not only by showing them the wilderness and the birds, but by restrained and informative exhibitions, by illustrated lectures, by films, and by making available books and pamphlets. The advantages of such facilities for the instruction of young people of all ages cannot be overstressed.

The capital cost of setting up the system of screened approaches and observation facilities, together with the landscaped area containing the captive birds, is undoubtedly substantial. But it is money very well spent in educating a very large number of people to appreciate the need for the conservation of wetlands in particular and of the natural environment in general. Once established, there is no reason why the facilities should not pay for themselves through the medium of entrance fees. The public tends to value things for which it must pay and, more and more, people are coming to realize that the best things in life are not free.

ANNEX

Summary of Criteria for Establishing the International Importance of Wetlands

Adopted at the International Conference on the Conservation of Wetlands and Waterfowl, Heiligenhafen, Federal Republic of Germany, December 1974

1. *Criteria Pertaining to a Wetland's Importance to Populations and Species.* A wetland should be considered internationally important if it:

i. regularly supports 1 percent (being at least 100 individuals) of the flyway or biogeographical population of one species of waterfowl;
ii. regularly supports either 10,000 ducks, geese, and swans; or 10,000 coots; or 20,000 waders;
iii. supports an appreciable number of an endangered species of plant or animal;
iv. is of special value for maintaining genetic and ecological diversity because of the quality and peculiarities of its flora and fauna;
v. plays a major role in its region as the habitat of plants and of aquatic and other animals of scientific or economic importance.

2. *Criteria Concerned with the Selection of Representative or Unique Wetlands.* A wetland should be considered internationally important if it:

i. is a representative example of a wetland community characteristic of its biogeographical region;
ii. exemplifies a critical stage or extreme in biological or hydromorphological processes;
iii. is an integral part of a peculiar physical feature.

3. *Criteria Concerned with the Research, Educational, or Recreational Values of Wetlands.* A wetland should be considered internationally important if it:

i. is outstandingly important, well situated, and well equipped for scientific research and for education;
ii. is well studied and documented over many years and with a continuing program of research of high value, regularly published and contributed to by the scientific community;
iii. offers especial opportunities for promoting public understanding and appreciation of wetlands, open to people from several countries.

4. *Criteria Concerned with the Practicality of Conservation and Management.* Notwithstanding its fitness to be considered as inter-

nationally important on one of the criteria set out under 1, 2, and 3 above, a wetland should only be designated for inclusion in the list of the Ramsar Convention if it:

i. is physically and administratively capable of being effectively conserved and managed;
ii. is free from the threat of a major impact of external pollution, hydrological interferences, and land-use or industrial practices.

A wetland of national value only may nevertheless be considered of international importance if it forms a complex with another adjacent wetland of similar value across an international border.

Reference

Smart, M., ed. (In press). *Proceedings of the International Conference on the Conservation of Wetlands and Waterfowl* (Heiligenhafen, Federal Republic of Germany, December 1974). Slimbridge, Gloucestershire, England: International Waterfowl Research Bureau.

28. Wetland Conservation and Management Problems

Eskandar Firouz
Department of the Environment, Iran

Introduction

We are all aware that wetlands and wetland ecosystems are centers of very high biological productivity and of great faunistic and floristic diversity, and that they thus constitute a natural resource of inestimable value to mankind. At the same time, they are often areas of great scenic beauty, with considerable potential for both consumptive and nonconsumptive forms of recreation. To pollute, degrade, or destroy wetlands is indefensible on either ecological or economic grounds.

In arid zones, where water is generally in short supply for man and wildlife alike, wetlands assume an even greater importance. The ever-increasing demand for water for industry and agriculture, coupled with the sensitivity of the wetlands to the irregular hydrographic conditions characteristic of arid and semiarid zones, renders them particularly vulnerable and ever more prey to man's diverse schemes, which are so often implemented without thought to their environmental consequences. It is not surprising, therefore, that despite a rapidly growing awareness of the need to conserve our wetlands, their destruction and elimination continues.

Iran has long recognized the importance and, at the same time, the vulnerability of its wetlands, and during the past eight years or so has taken numerous steps to alleviate or circumvent the inherent threats to their existence. Initially, concern for the nation's wetlands was focussed on their importance for migratory waterfowl, a national resource of great economic importance in many parts of the country. Thus, from the beginning, one aspect of the international significance of wetlands became apparent—namely, that to safeguard

---

Mr. Firouz's paper was presented at the conference by Dr. Mohammed Reza Amini, Head of the Aquatic Ecology and Fisheries Group of the Department of the Environment, Tehran, Iran.

our population of wintering wildfowl we had to ensure that adequate conservation measures were taken at wetlands throughout their flyways. It was obvious that the destruction of wetlands in one country affected many countries, for a reduction of waterfowl populations in one would mean a reduction for all.

It was in this context that Iran gladly offered to host an International Conference on the Conservation of Wetlands and Waterfowl at Ramsar in 1971, a meeting that resulted in the passing, after almost a decade of meetings and discussions at various European venues, of a "Convention on Wetlands of International Importance especially as Waterfowl Habitat." In this convention, known as the Ramsar Convention, the contracting parties agree to designate suitable wetlands within their territories for inclusion in a list of "Wetlands of International Importance," and to observe their international responsibilities for the conservation, management, and wise use of migratory stocks of waterfowl. [See Appendix B.—Editor]

It is further provided that each contracting party shall promote the conservation of wetlands and waterfowl by establishing nature reserves within wetlands; encourage research; endeavor, through management, to increase waterfowl populations; and promote the training of personnel for the wardening, management, and study of wetlands (Carp 1972).

It was in recognition of the international nature of wetland resources that our sovereign, the Shahanshah, advanced in his message to the Ramsar Conference the magnificent proposal to place one of the country's wetland ecosystems of global significance in joint trust with an international agency "to conserve and administer for all mankind." The area subsequently selected for this reserve, namely Dasht-i-Arjan marsh, Lake Parishan, and intervening and adjacent mountain ranges and valleys (an area of some 70,000 hectares) in central Fars, was chosen on the basis of recommendations made by the delegates to the Ramsar Conference themselves. Not only is this an area of exceptional importance for a wide variety of wildlife species, particularly waterfowl, but it is also a region of outstanding scenic beauty, one that compares favorably with some of the world's finest national parks.

Problems of Wetland Conservation

Before I elaborate on the Ramsar Convention, I will return to the problems of wetland conservation and attempt to place the complex ramifications of these problems in their proper context. In so doing, I should state that the arid and semiarid zones pose no new problems for wetlands. Rather, the problems here are merely intensified or magnified by virtue of a harsher environment and are thus placed in a more discernible perspective.

Because wetlands are highly sensitive ecosystems, their survival everywhere, and in the dry regions in particular, indisputably hinges both on careful land-use planning and a multidisciplinary approach.

The problems enumerated below, with which we are constantly trying to come to grips in Iran, are no doubt universal, and as such they must be duly considered in the ecological equation for the conservation of all wetlands.

Water

Shortage of water is becoming a worldwide phenomenon, a problem that is, of course, proportionately magnified in the more arid regions. Burgeoning human populations, unprecedented rates of development —both in agriculture and industry—and increasing extravagance and wastage at many levels in the utilization of water supplies combine to create unrealistic demands that place many wetlands of arid regions in great jeopardy.

Land

Wetlands and adjacent lands are often very fine agricultural land, and thus they become prey to drainage and reclamation. Equally, they attract urbanization and industry.

Pollution and Eutrophication

Wetlands, particularly those situated in inland drainage basins, are highly susceptible to the concentration of urban, industrial, and agricultural pollutants from runoff. Where uncontrolled, the result is both eutrophication and a build-up of biocides in aquatic food chains, which end in waterfowl, predators, and man. Left to itself, eutrophication will eventually "suffocate" such wetlands with aquatic vegetation, transforming a fine wetland into a biological desert.

Conflicts of Use

The numerous important uses to which wetlands are put are often inherently antagonistic to one another. The following types of uses may be cited:

1. Traditional uses: grazing, reed cutting, primitive transport, and subsistence fishing and wildfowling;
2. Recreational uses: sailing, powerboating, water-skiing, sport fishing, sport hunting, birdwatching, and so on;
3. Industrial uses: transport, generation of electricity, cooling, and so cn; and
4. Conservation uses: the preservation of species, habitats, and ecosystems; wildlife management; and game and fish farming.

Considerable attention is now being given to the concept of multipurpose use, but all too often implementation of several diverse programs without adequate prior investigation has caused irreparable damage.

Pest Control

In many regions, wetlands have been seriously modified, if not completely drained, to alleviate a disease problem—notably in programs aimed at controlling malaria. Such programs have often involved extreme contamination with pesticides or the destruction of vegetation with herbicides, resulting in total destruction of the natural ecosystems.

Impermanence

In arid and semiarid regions, the natural situation is one of great fluctuations in the extent and condition (and thus in the productivity) of wetlands owing to irregularities in the rainfall regime. This is often aggravated by damming and irrigation projects, as well as by the diversion of water and man-induced lowering of the water table. As seasonal fluctuations in water levels are exaggerated by man's interference, natural vegetation—and hence wildlife populations—may be destroyed.

Management Philosophy

Assuming that adequate measures are taken to protect a wetland, there are many ways in which it might be managed, depending upon the aims of management. Even among conservationists, however, there is considerable disagreement as to the objectives of management. For example, there is disagreement about whether, and to what extent, to allow the exploitation of wildlife resources; to what extent to manage for certain economically important species (for example, wildfowl and sport fish) when this might significantly alter the natural state and result in unnatural associations and concentrations of wildlife; whether to introduce exotic flora and fauna; whether to conduct artificial-propagation programs with wildlife that will be released into the wild; and whether to carry out predator control. These are very real issues that have been given little attention by the international conservation bodies.

The Dependence of Wetlands on Their Watersheds

A wetland is rarely a discrete entity, but should be regarded in the context of its entire watershed. Watershed degradation, industrial development, and environmental pollution hundreds of kilometers and often several countries away from a wetland may have significant effects on its integrity and viability. Wetland conservation programs must therefore be viewed in the light of entire regional land-use and development plans, both at the national and international levels.

Migratory Waterfowl—A Shared Resource

As mentioned earlier, migratory waterfowl have long been recognized as an extremely important natural resource, one that is truly international. The waterfowl that visit Iranian wetlands originate in breeding areas as distant as northeastern Europe and eastern Siberia, and may move on to winter in areas as far afield as South Africa and the Bay of Bengal. If we in Iran wish to ensure that our waterfowl populations receive adequate protection throughout their migratory routes, we must negotiate with something on the order of thirty governments. You, in North America, are fortunate indeed that the great majority of your migratory waterfowl populations spend their entire lives within three nations, Canada, the United States, and Mexico.

Basic Information

Before we can begin to formulate wise policies for the conservation of wetlands, we must accumulate a vast amount of basic information—biological, hydrological, geographical, cultural, and so on. In many countries, notably those in North America and Europe, and in parts of the Soviet Union, this poses no outstanding problems. With many thousands of active professionals and amateurs providing a constant stream of data, and an informed general public often eager to help, your problems seem trivial as compared with those in vast portions of Asia, Africa, and South America, for which knowledge is almost entirely lacking and where there are very few trained workers. Recently, great emphasis has been placed on surveying the poorly known areas and on training local personnel, but, unfortunately, the basic information cannot be gathered overnight.

In Iran, after some eight years of relatively intensive wetland surveys, we are still making new discoveries with almost embarrassing frequency. While some of these discoveries are of little more than academic interest, others will doubtless have important repercussions on our management policies and programs. And herein lies a perennial problem: the threats facing freshwater wetlands in rapidly developing arid and semiarid areas are immense. Wetlands are virtually disappearing overnight. The conundrum is that we must have the basic information upon which to formulate our conservation policies, but it may be several years before we can acquire such data, and by that time the wetlands in question might well have disappeared. Therefore, our decision in Iran has been that if we are to save our wetlands, we must often act long before we have accumulated the basic information: we must use inspired guesswork and intuition, at the risk of being criticized by scientists and politicians. Yet, a management policy implemented now but later found to be inappropriate can usually be remedied, while a wetland that has been destroyed can rarely be recreated, and a species once extinct is gone forever.

Ecologically, geographically, and scenically, Iran's wetlands are of great interest. Their diversity is such as to include shallow lakes that freeze solid each winter, as well as mangrove swamps; freshwater marshes with alders and reeds, as well as ephemeral and brackish desert wetlands. As stated before, we have made substantial efforts in the reconnaissance and investigation of the wetlands of Iran as well as in the accumulation of numerous data on the status of our waterfowl. During the past seven years, over 1 million hectares of the most significant areas have been provided complete protection, and these wetlands are situated in all of the major geographical areas of Iran.

Eighteen of our wetlands (or wetland systems) have been designated wetlands of international importance and appended to the Ramsar Convention. Ten of these are included in the above system of protected areas. But this is somewhat misleading, since the ten protected wetlands together comprise more than four-fifths of the total area of the eighteen and are also unquestionably the most important. Another way of underlining the importance of these protected areas is to cite the proportion of the waterfowl counted during our last three nationwide, midwinter censuses of these ten wetlands. In 1972-73, 72 percent of all waterfowl counted in Iran were found within these protected areas; in 1973-74, about 40 percent (owing to very unusual flooding conditions); and in 1974-75, 70 percent (Scott 1971, 1972, 1973, and 1974).

The status of these protected wetlands has improved notably since protection was begun, but it would be idle to claim that the abovementioned threats have suddenly disappeared. They are present and real. But the relevant laws have been dramatically strengthened by the passage of the Environmental Protection Act in July 1974; and the understanding that is happily coming into being between the Department of the Environment and other agencies of the Government of Iran in relation to environmental considerations is proving more and more efficacious.

Another legal implication of great importance is the provision in the Environmental Protection Act that places all wetlands in the public domain under the complete jurisdiction of the Department of the Environment. But for this provision, many important wetlands would by now have been impaired or wiped out. As an illustration, one freshwater-saltwater wetland complex of great international importance in the Mian Kaleh Protected Area on the southeastern shores of the Caspian Sea would have been irrevocably damaged by a large power plant that was originally planned for construction within the reserve. Owing to the high-sulfur fuel projected for its use, it would have emitted over 170 tons of SO_2 per day. However, the Ministry of Energy responded to the request of the Department of the Environment and agreed to move the plant some 20 kilometers to the west of the protected area and, moreover, with the consent of other relevant agencies, to use natural gas rather than fuel oil.

Just like the Wadden Sea in Holland and Germany, Mian Kaleh owes something to the Ramsar Conference—namely, international recognition. Undoubtedly one of the most important achievements of the conference, and also of the follow-up Conference on the Conservation of Wetlands and Waterfowl, held at Heiligenhafen, West Germany, in December 1974, is that it helped draw the attention of the participating nations to the international importance of their wetlands and thereby to their international responsibilities in wetland conservation.

Since 1961, the International Union for Conservation of Nature and Natural Resources (IUCN) had been cosponsoring three major wetland projects aimed eventually at identifying and describing important wetlands throughout the world. They were:

1. Project MAR for the conservation and management of temperate marshes, bogs, and other wetlands, under the joint sponsorship of IUCN, the International Council for Bird Preservation, (ICBP), and the International Waterfowl Research Bureau (IWRB)
2. Project AQUA for the conservation of aquatic habitats (lakes and rivers), under the joint sponsorship of IUCN, Societas Internationalis Limnologiae (SIL), and the International Biological Program (IBP)
3. Project TELMA for the conservation of all actually or potentially peat-forming ecosystems, under the joint sponsorship of IUCN, UNESCO, and IBP

At Ramsar and at Heiligenhafen, these three projects were brought together under a single umbrella, and are now providing the basis for the preparation of a "Directory of Wetlands of International Importance." The directory has been designed specifically to monitor the conservation status of wetlands of international importance, to provide up-to-date reference works listing these wetlands, and to assist in the task of protecting as many as possible.

It is to be hoped that the directory, currently being compiled by a special task force in collaboration with the IUCN Secretariat, will constitute a valuable working tool for a wide range of groups concerned with conservation, biological sciences, education, recreation, and tourism. It should also serve as an important guide to governments, with respect to both the status of the wetlands in their countries and to the measures they should take to conserve them.

By focussing attention on the importance of wetlands and publicizing the recognition they deserve, the Ramsar Convention and the resulting "Directory of Wetlands of International Importance" undoubtedly constitute milestones in wetland conservation at a global level. Nonetheless, there were (and still are) those among the delegates at several United Nations meetings who were critical of the Ramsar Convention because it "lacked some teeth." This is most unrealistic, however, for such imperfect dentition is the price of multinational consensus and certainly applies to all similar international agreements. If their countries have yet to sign the convention, it is not inappropriate to remind them that, in many international areas of concern (one of which is surely conservation), it is better to have something to adhere to than nothing at all. Happily, the Ramsar Convention has a sufficient number of signatures and will enter into force in the near future. [The Convention entered into force on 21 December 1975.—Editor]

Experience has shown over and over again that nations cannot develop comprehensive and fully effective conservation policies simply at the national level. And yet, a considerable degree of isolationism still persists in many regions. Networks of reserves are established, wildlife legislation created, and programs of biological research and pollution control implemented by governments without thought for what might be taking place in neighboring countries. Only through the regular interchange of information and ideas and through the development of multinational research programs can we gain a proper understanding of the significance of our resources and develop a rational conservation policy.

At both the Ramsar and Heiligenhafen conferences, considerable attention was given to the formulation and standardization of criteria for the classification of wetlands and qualification in terms of international importance. It was largely owing to a reappraisal of our wetlands on the basis of these criteria that we began to appreciate the significance of some of our coastal sites. Three such areas were included in our list of wetlands of international importance, and one of these, a part of the Khouran Straits near Bandar Abbas in the Persian Gulf, has subsequently been given complete protection.

Discussions at Ramsar drew Iran's attention to the importance of these sites for wintering shorebirds. Later, and in a quite different context that was to become the central theme of the Regional Meeting on Marine Parks and Reserves, held in Tehran in March 1975, it became apparent that these same areas are of great importance as breeding grounds and "nurseries" for a variety of fish species of considerable significance to the Persian Gulf fisheries.

In a review of problems in the conservation of freshwater wetlands in arid regions, reference to the Persian Gulf fisheries may seem a digression. But surely this is not so, for the conservation of the world's natural resources is not a series of problems, each of which can be considered and solved separately. Rather, it is a single problem with many facets, each intricately interwoven with all the others.

References

Carp, E., ed. 1972. *Proceedings of the International Conference on the Conservation of Wetlands and Waterfowl* (Ramsar, Iran, 30 January–3 February 1971). Slimbridge, Gloucestershire, England: International Waterfowl Research Bureau.

Scott, D. A. 1971, 1972, 1973, and 1974. Mid-winter Wildfowl Censuses in Iran 1971, 1972, 1973 & 1974. *Iran Department of the Environment Project Completion Reports.*

Terrestrial Ecosystems: Temperate Forests

29. Uses and Problems of the Temperate Forest in the United States

Eleanor C. J. Horwitz
Massachusetts Audubon Society, U.S.A.

Introduction: The Charge Before Us

We have before us today a monumental charge. In the short time allotted to us we are asked to identify and address ourselves to the uses and the problems of the temperate forest. It is a major undertaking, for there are plenty of both. I will, of course, not solve any problems nor provide any solutions in these few minutes. Instead I shall raise questions; questions which you must, in time, answer for yourselves; questions which you might wish to keep in mind as we explore some specific problems of these vital regions.

It is, perhaps, presumptuous to tell you about the temperate forests, for these are the areas best known to all of you here. These are the forests of the land around New York, and throughout most of North America. They are the forests of the temperate zone, the most heavily settled and industrialized region in the world. Thus the pressures on these forests are extreme. We need these forests for their space and their resources. We must use them. But how? And once that has been decided, how heavily will we use them? Use, especially intense use, implies a need for management. How shall we manage these woodlands? And how much of the land shall we leave totally unmanaged—as wilderness?

How Shall We Use the Forests?

Let us begin by addressing the first issue. How shall we use the forests? Why do we need them?

We need them for wood products. In an ecologically conscious age, there is heavy emphasis on using resources that are renewable. The population grows, and with it the need for housing. The material of choice for that housing is wood. And wood is also the material of choice for hundreds of other items that we construct to embellish our lives. We also need the forests for paper. This is an era and a society in which we are wedded to the idea of inexpensive mass communication. That, after all, is the stuff of which a democracy is made. We write freely and seek to make all manner of materials public and available to all. But cheap pulp means heavy cutting of the forest.

We need forests for the mineral deposits under them. Traditionally we have solved many of our problems (yes, even environmental problems) with technological solutions. Most of these solutions require more and better machinery. Thus, in 1975, there is a need for an increased supply of metals, and there is a renewed interest in the mining of coal to power our homes and factories. Many of the still available deposits of minerals and coal lie under the forests. We cannot tap them without seriously disrupting the forest.

We need the many green plants of the forest—trees and others—to help maintain the many chemical balances on Earth. They play a vital role in the production of oxygen and in the recycling of carbon dioxide. They play an important part in the cycles of such nutrients as nitrates and phosphates, sulfates and silicates. And it is the green plants, which, through photosynthesis, fix the solar-energy fuel for the biological system.

We need forests for the protection of watersheds. The forest floor provides an important filtering mechanism, and water that travels through it to streams or groundwater reservoirs is thereby purified. Thus, in the last analysis, the quality of our water begins with the health of the watershed and the forest upon it. The forest also plays an important role in controlling the water regime of any given area: during the summer growing season, forest plants take up large amounts of water, which are transpired and evaporated from the leaves of the plants. This water is returned to the atmospheric water reserve and does not flow into streams. This means that, in areas receiving large amounts of precipitation during the growing season, the healthy forest controls floods.

We need forests for wildlife habitat. Forests for this purpose can be of many kinds, for there are animal species associated with every stage of forest development. To mention but two, deep-forest areas, such as might be found in wilderness regions, are needed as a possible home for wild turkeys and red squirrels. A more heavily used type of forest is new growth of the sort found in managed woodlands and on farms that are reverting to forests. Here there are the deer, raccoons, and gray squirrels that are so highly prized by outdoor-recreation enthusiasts.

We need forests as places for recreation—a need that becomes more important as urban pressures mount and people seek the forest for its peace and solitude.

Finally, we need the forest to maintain some degree of wilderness as a record of particular types of communities and as reserves. And the wilderness, too, serves as a recreation area.

Our Changing Needs

This multiplicity of needs and uses has long been recognized and even institutionalized in the United States Forest Service's multiple-use policy. However, some of these uses inevitably come into conflict. When this occurs, one use must dominate another. But which use will that be? Traditionally, the dominant use has been determined by need—and most often the need has been for timber and pulp.

At present, however, our perceived needs are changing. What uses do we wish to accord the highest priorities? And having decided what uses are to have priority, we must then decide how intensely we will use the forest for those purposes.

Determining Intensity of Use

How heavily shall we log? In 1975, there is a rapidly escalating demand for wood and wood products. It has been suggested that we may be cutting more than is regenerated each year. If this is so, we are indeed in trouble. But to date, little is known about the relative amounts of harvest and regrowth. More information is needed. And more information is needed on the question of whether our cutting methods are always, or even usually, appropriate.

Much has been said and written about clearcutting, which is one particularly controversial method of harvesting trees. Is it good? Is it harmful? I would submit that it is both and it is neither. A cutting technique is a tool rather like a hammer. Clearcutting is such a tool. It is meaningless to debate the merits of the method in a vacuum. It is as good or as bad as each particular operation: there have been disastrous cuts and there have been good ones. The more meaningful question is whether the amount and type of cutting is appropriate to the specific situation in which it is used.

How much mining shall we permit? The increased need for coal and metals still lying beneath the forest floor has already been noted. The least expensive way of recovering these deposits is through strip mining. Such mining, however, disrupts large sections of forest and leaves them devoid of topsoil, often almost impossible to restore.

How much shall we accede to the ever-increasing demand for roads? Roads are essential for logging, for mining, and for recreation. Bad road-building techniques cause more damage and more erosion to the forest than any other use. Poor road-building techniques cause soil to wash away unimpeded; they muddy streams and may cause entire hillsides to wash out; and they may become eyesores, leaving permanent scars on the landscape.

This raises another difficult question: Do we want roads that are expensive, well-engineered, and permanent, or roads that can be built quickly and inexpensively, which are used briefly, and which disappear rapidly once they are no longer needed?

The increase in roads through the forest has provided access to places hitherto beyond the reach of all but the hardiest visitors; the increased mobility couples with the surge of interest and concern for the environment that we have fostered. Thus, more people come to the woods and they, in turn, demand more and better roads and other facilities. The stress on the forest increases.

Do we want this? In time, education and intensified public interest in natural areas may prove to be the hardest blow we have dealt this resilient ecosystem. How much use is permissible before we are forced to say, for reasons of forest health, this much and no more?

What Kind of Forests Do We Want?

Next we must ask what kind of forests we want and how we can achieve our aims. Some people would have primarily even-aged forests, which loggers find easy to manage. Visitors find their park-like evenness soothing and pleasing. In some cases, these second-growth forests are actually preferred to untouched and unused forests.

As an alternative, we can have uneven-aged forests, which are more diverse and thus support a greater diversity of animal life. These are the forests of choice for most "wildlifers" and for many foresters, but they are often hard to move through and may, in places, be spindly or decadent. Uneven-aged forests are natural to many of the eastern forests regions, but not to many western forests, which have always been composed of clusters of even-aged stands due to fires and windstorms.

Today we can manage for either kind of forest, or we can leave them unmanaged. If we select the latter alternative, we must accept the fact that the forests may decay through human agency, either through actual use, or indirectly through changing the environmental conditions (air, water, and climate) that determine forest growth patterns. Once we know the sorts of forests we want, we must decide how to obtain them. Shall we do it through programs of cutting, burning, or both? Again, public controversy on the issue is intense.

Administering Forests

All of these questions and problems are compounded by the problem of administering forest lands. A large portion of temperate-forest lands in the United States are controlled by agencies of the federal and state governments. These agencies are designed to respond to expressed public desire, their charge to manage the forest in accord with the needs of the majority of their constituents. This they have done.

Unfortunately, public desire can change very quickly as we have seen over the last five years. Large organizations respond slowly, but they do respond; trees respond more slowly still. Thus, even when the management programs change to reflect new values, we must be patient until the changes appear in the forest, and that may take twenty or more years.

Lands owned by industries are used heavily: they are managed very effectively, especially those owned by companies that plan to be around in the future. But they are managed for a single purpose only.

A last group of forest lands are held in private ownership. Not heavily used today, they will be come increasingly important when the need for wood rises. They are not always well managed, and often yield only short-term economic gains.

How can individual owners be impelled to consider the health of the forest beyond their meager human tenure? It is a delicate problem in which we must always be keenly aware of the forest owner's rights—rights which must be protected at all costs despite our eagerness to "save the forests."

Balanced Policies

It is obvious that the temperate forest poses many problems and that there are many choices to be made. As we establish forest-management policies, we must consider the needs of as many people as possible. We must consider those needs on a national and a global basis—social, economic, and also environmental needs. No doubt everyone here today is concerned about social inequities, about standards of living, and prices. They also should be concerned about environmental health and about aesthetics—subjects that are not necessarily the same. To some extent, they pull us toward different solutions for dealing with the temperate forest. It is our task to balance these considerations as we, through our elected representatives, determine goals and methods for managing the forest.

30. EARTHCARE and Temperate Forests

F. Herbert Bormann
Yale University, U.S.A.

Introduction

Forests, grasslands, deserts, and oceans play a major role in maintaining the "balance of nature" upon which all life on this planet is utterly dependent. These natural ecosystems, although modified in varying degrees by man, still occupy most of the Earth's surface and exercise considerable control over patterns of climate, hydrology, and circulation of nutrients; over the cleansing functions of air, water, and soil; and over reliable patterns of food production on land and in the sea.

Yet all of these ecosystems are subjected to ever-increasing stress as human populations press into every usable corner of Earth space and as the use of technology—motivated primarily by short-term gain—tremendously increases man's capability to destroy nature. This conference should therefore emphasize not only protection for endangered species and natural areas, but the urgent need for overall policies that view natural ecosystems as vital and irreplaceable assets. Such policies constitute not elitism, but hard-headed realism: the welfare of mankind and natural ecosystems will rise or fall together.

Temperate forests provide an example of the conjugal relationship between man and nature where divorce is impossible. They are widespread, diverse, and among the most important components of the biosphere and of the "balance of nature." They occur on all continents with the exception of Antarctica and include the temperate rain forests of northwestern North America, eastern Australia, and southern Chile; the broadleaf and needle evergreen forests of Australia, New Zealand, and western and southeastern North America; and the great temperate deciduous forests of Europe, North America, and China (Walter 1973). Almost all of the fertile land within the temperate-forest region has been converted to permanent agriculture, yet the remaining forests contain about 29 percent of the mass of all living organisms on the land surface even though they occupy only 12 percent of the Earth's land area (Whittaker 1970). Most important of all, temperate forests capture about 21 percent of all sun energy fixed into plant tissues by land-based plants.

Temperate Forests and Stability of the Global Environment

Forests play a major role in the maintenance of worldwide environmental stability. Although they appear quiet and relatively static, forests are sites of intense activity: each year, millions of liters of precipitation, trillions of calories of energy, and vast tonnages of gasses flow into each square kilometer of humid forests, where they are used, altered, and dispersed. Thousands of species of plants, animals, and microorganisms use the water, energy, and nutrients (chemical elements necessary for life) to live and reproduce; by so doing, they contribute to the regulation of the flow of energy, nutrients, and water through the biosphere (Figure 1).

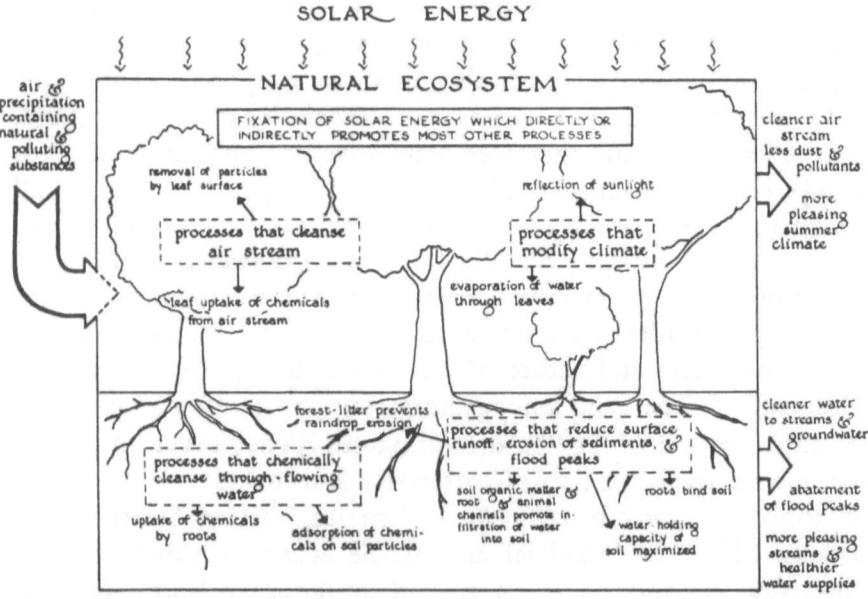

Figure 1. Diagrammatic Representation of the Effects of Various Processes that Occur in Forest Ecosystems on Climate and on the Streams of Air and Water that Pass through Forests.

The functions of forests in regulating environment have been intensely studied within the temperate region, and a number of general principles have emerged. By its normal biological activities, the forest makes important and highly predictable contributions to such worldwide cycles as the hydrologic, oxygen, carbon, and nitrogen cycles (Likens and Bormann 1972).

The forest ecosystem exerts considerable control over local aspects of the hydrologic cycle. During the growing season, forests use substantial quantities of water in transpiration—the evaporation of water through plant leaves. This conversion of liquid water to water vapor reduces the summertime flow in streams that drain the forest, thus reducing the height of summer storm flows and minimizing flooding and erosion (Bormann et al. 1969, 1974).

As a consequence of their evaporative activities and reflective qualities, forests modify local climates by reducing temperature extremes and producing a local climate more pleasing to man. Forests may also affect regional or world climates (Newell 1971).

Forest soils tend to absorb rainwater and direct it downward to groundwater reservoirs, while minimizing surface runoff (Bormann et al. 1969, 1974). Groundwater reservoirs represent an important water supply for man and are generally free from evaporative losses. This is not the case for water stored in surface reservoirs.

The forest ecosystem exercises very strong control over the processes of erosion; thus, the water in streams flowing from most forests is essentially free of sediments (Bormann et al. 1969, 1974). The control of erosion is very complex, but much of it is centered in the upper layer of the soil. This layer, often the product of hundreds of years of development, is the layer most often destroyed by human activities.

Forests act as extremely important chemical filters. The chemistry of rainwater is completely altered as it passes through the forest ecosystem. The concentrations of some elements in streamwater is higher than in rainwater, while those of other elements are much lower (Likens and Bormann 1972). Often those held within the ecosystem are pollutants, such as lead, radioisotopes, cadmium, and hydrogen ions (Likens and Bormann 1972; Smith 1974, 1975). Thus, the forest ecosystem purifies water and yields chemically "cleaner" water to both streams and groundwater. Although natural ecosystems have a fair capacity to filter out pollutants, there is a limit to this beneficial function, and too much pollution can disrupt the ecosystem and lead to the loss of its filtration capacity (Smith 1974).

As forest soil acts as a chemical filter for water flowing through it, so the living trees, shrubs, grasses, and soil act as a filter for the stream of air that moves through the forest. The forest lowers the content of dust and such gaseous pollutants as carbon monoxide, sulfur dioxide, and ozone, yielding a cleaner and more healthful air stream (Smith 1974).

In all stages of their development, forests provide habitats for a large variety of species of plants and animals, and thus contribute to the preservation of species and the maintenance of species diversity. This is an extremely important biological consideration.

Finally, forests and their successional stages provide a diversity of landscapes and add to the quality of man's life by offering a variety of recreational and aesthetic opportunities.

The remarkable thing is that forest ecosystems yield these immense human benefits—air and water filtration, erosion control, groundwater recharge, the maintenance of biogeochemical cycles, climate modification, and the preservation of species diversity—essentially free of charge, or at least with no expenditures of fossil fuel. Of course large amounts of energy are required to support these activities, but the sole source of the energy utilized is sunlight.

Stresses on Temperate Forests

The large proportion of solar energy fixed by temperate forests indicates that they play a major role in regulating regional and global environment. Nevertheless, they are subject to ever-increasing stresses generated by human society that diminish their extent and reduce the efficiency with which they perform natural functions. Levels of stress are particularly related to technology and are severe because much of the area of temperate forest is within the borders of highly developed, technological countries (Figures 2, 3, and 4). Coupled with the stresses due to technology is a philosophy, deeply ingrained in modern industrial society, that discounts or ignores the importance of the vital services rendered by nature and that believes natural functions can easily be replaced by technological substitutes.

The current emphasis on temperate forests is great, and closely

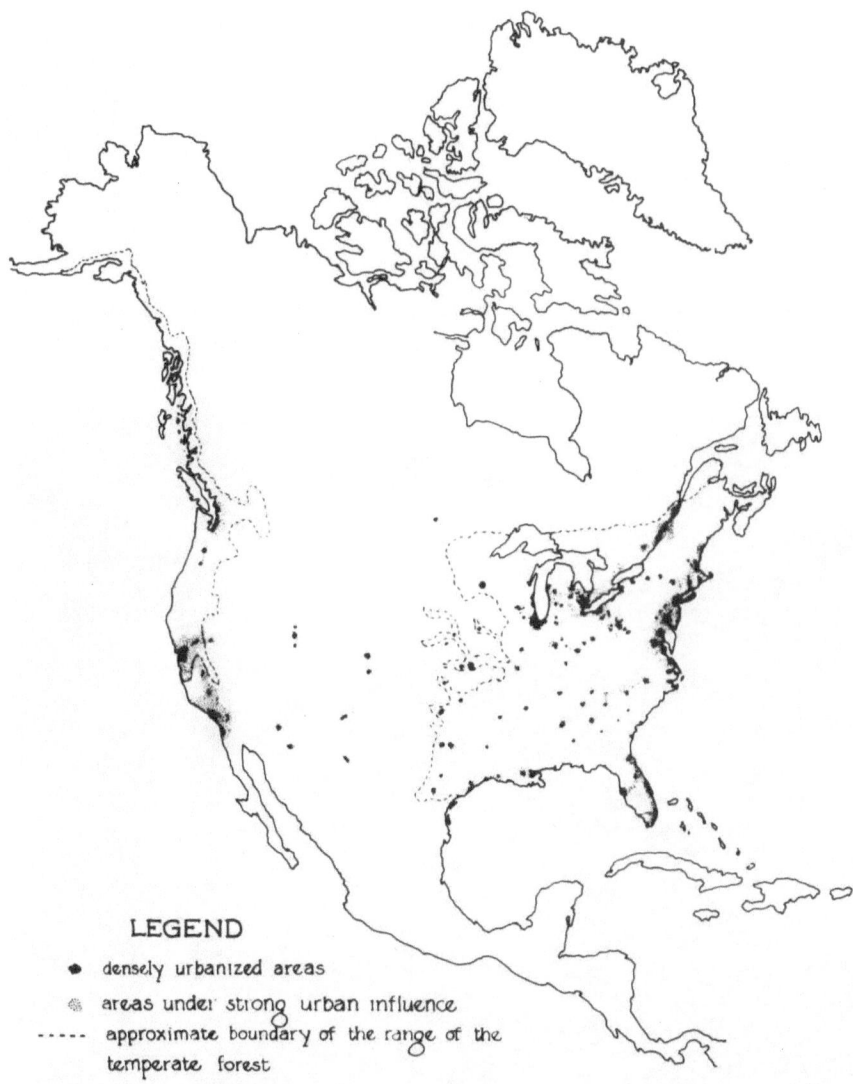

Figure 2. Map Showing the Relationship Between the Temperate Forests and Major Urban Areas of North America. Densely urbanized areas are in black; areas under their immediate influence are stippled; approximate distribution of temperate forests is indicated by the broken line. Note that most of the urbanized areas are situated within the temperate-forest regions. *Sources:* Urbanized areas after Tunnard and Pushkarev (1963); temperate forests after Oosting (1956).

related to the growth of regional populations and economies, and the development and use of new technology that far outstrips our understanding of the environmental impact involved. The magnitude of the problem can be understood only when we consider the entire range of stress to which much of the temperate forest is simultaneously subjected.

Categories of Stress

Using data from the United States, we will consider three major categories of stress: (1) competing land uses—urbanization and

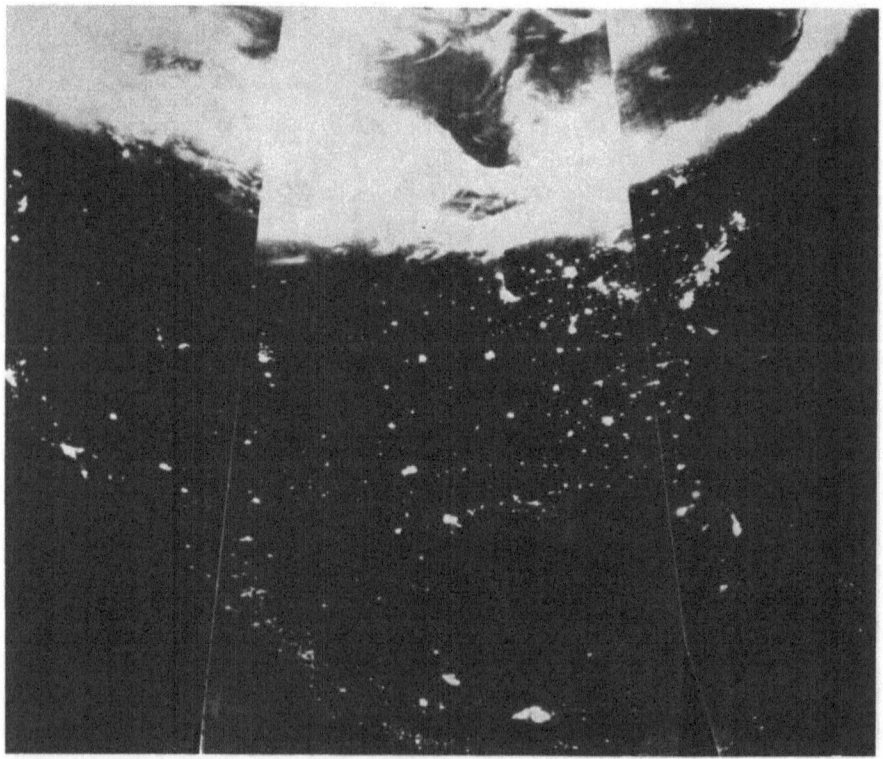

Figure 3. Satellite Photomosaic of North America at Night. Aurora Borealis is the bright area in northern latitudes (top of the photomosaic), and centers of fossil-fuel-energy expenditure are the bright spots in more southerly latitudes. Most of the fossil fuel is expended within the temperate-forest region (compare with Figure 2). (Courtesy U.S. Air Force.)

Temperate Forests

transportation; (2) mismanagement of the land surface in pursuit of renewable and nonrenewable resources; and (3) regional pollution.

Land Use. If it is assumed that almost all land suitable for agriculture has already been preempted, the major competing uses for the area still occupied by temperate forest are urbanization and transportation. For the United States as a whole, it is estimated that roughly 1,300 square miles of rural land are partially converted to urban and transportation uses each year (Council on Environmental

Figure 4. Satellite Photograph of the Eastern United States at Night. The bright areas are centers of concentrated use of fossil fuel situated within the eastern deciduous forest (compare with Figures 2 and 3). Photographed from an altitude of 1,000 kilometers.

Quality 1972). Most of this land is at the periphery of urban regions and is consumed primarily by "urban sprawl." However, most urban-industrial regions are located within and fairly well distributed throughout the temperate forest region (Tunnard and Pushkarev 1963; Figures 2, 3, and 4). This distribution is an extremely important consideration, since it indicates that much of the temperate forest is fairly close to the environmental pressures generated by urban regions.

Mismanagement of forests or land surface caused by the pursuit of renewable and nonrenewable resources can markedly reduce the efficiency with which the temperate forest carries out its natural functions. Before we explore this point, however, it is important to note that voluminous research in temperate regions indicates that, with appropriate harvesting and management plans, much of the temperate forest can be managed for the production of wood products and/or increased yields of water or wildlife with little loss of the natural functions of the forest ecosystem (Seaton et al. 1973). However, abuses occur and poor harvesting techniques can lead to relatively longterm degradation of forest sites (Seaton et al. 1973; Barney 1974). Few data exist on the extent to which forest lands in the United States have been degraded by poor harvesting practice, particularly of private forest lands, which constitute 72 percent of all commercial forest land in the United States.

We also need more information about the impact of temporary roads, constructed to harvest a forest and afterwards abandoned, on the subsequent productivity of a site. There are few data on the quantity of land disturbed by temporary roads in either federal (Cliff 1972) or commercial forest lands, but probably tens of thousands of kilometers of temporary roads are constructed each year. Finally, thousands of square kilometers of diverse temperate forests have been and are being converted to forest monocultures, which are increasingly composed of highly genetically selected trees. Although monoculture is an old and almost universally accepted forest practice, new questions are being raised about its environmental implications and the long-term stability of such forests (Dahlston 1972).

Renewable and Nonrenewable Resources. One of the severest effects on temperate forests results from strip mining for nonrenewable resources. About 13,000 square kilometers of the United States, much of it in the temperate-forest zone, had been stripped through 1965 (Council on Environmental Quality 1972). Only a third of this area has been revegetated. In 1970, about 400 square kilometers were stripped for coal alone. As we substitute coal for hard-to-get oil, even greater areas will be stripped. The revegetation of strip-mined areas is often difficult because of the total disruption, erosion, and instability of the surface; loss of topsoils; exposure of sulfur-bearing minerals that react with air to form sulfuric acid; and the harshness of the microclimate. Most so-called "reclaimed land" bears faint resemblance to its premined condition. It is improbable that within the lifetimes of people now living, the efficiency of natural processes in most strip-mined areas will be remotely equivalent to what it was before the areas were stripped.

George Woodwell (Woodwell 1970) has noted that forests subjected to stress from pollution or radiation respond in a typical pattern. First, sensitive species are diminished and lost; then, larger organisms, such as trees and shrubs, die out, followed by smaller shrubs and herbs. Only some species of lichens and mosses can tolerate intensely stressful situations. This progressive loss of species and ecosystem structure involves a shift away from the complex functions that typify the intact natural forest: there is a change from stable, predictable relationships toward instability; from "tight" (closed) nutrient cycles and substantial control of the ecosystem's hydrology and erosion by the forest, to open cycles and less control by the forest; from highly specialized species to rapidly reproducing, less specialized species, such as weeds, rodents, and starlings; and from an ecosystem that runs itself to one that requires a constant input of fossil and human energy to achieve standards of human acceptability.

Regional Pollution. Numerous cases of severe air-pollution stress have destroyed entire ecosystems in North America. Many such ecosystems are located within the temperate-forest zone. At the turn of the century, smelters at Copper Hill, Tennessee, destroyed 28.5 square kilometers of deciduous forest and replaced an additional 15.5 square kilometers of forest with grass species. Near Kennet, California, all vegetation was destroyed by smelter fumes in an area of more than 270 square kilometers. Recently, at Wawa in northern

Ontario, air pollution from an iron-sintering plant, transported by strongly prevailing southwest winds, caused extensive damage to the boreal forest situated northeast of the plant. Trees were very severely damaged up to 39 kilometers away. In Sudbury, Ontario, sulfur dioxide emitted from three large smelters is severely injuring trees over an area of 1,865 square kilometers (Smith 1974), while in Palmerton, Pennsylvania, zinc fumes from smelters have denuded 5.2 square kilometers (Jordan 1975).

The loss of sensitive species is somewhat harder to document, but several are well known. An aluminum-reduction plant near Spokane, Washington, has killed Ponderosa Pine *(Pinus ponderosa)* over much of a 130-square-kilometer area, while the mortality of Douglas Fir *(Pseudotsuga taxifolia)* has been documented on a 1.3-square-kilometer area next to a phosphate-reduction plant in Georgetown Canyon, Idaho, where fluoride was the stress agent (Smith 1974). Oxidants, primarily ozone, are seriously damaging trees in the San Bernardino, Angeles, and Cleveland National Forests in southern California. In the San Bernardino National Forest alone, oxidants from distant urban regions are presently damaging 1.3 million trees on an area of more than 160 square kilometers. Its most important species of tree, Ponderosa Pine, is also the most susceptible species; these trees are killed by direct air-pollution damage and by insects that breed in the bark of those that are weakened. If air pollution from the Los Angeles basin continues to insult this forest, Ponderosa Pine may be almost eliminated from the area, and well-stocked stands will be converted to poorly stocked stands vegetated by more resistant species such as Sugar Pine *(Pinus lambertiana)*, Incense Cedar *(Libocedrus decurrans)*, White Fir *(Abies concolor)*, and Douglas Fir. Conditions of exposed ridge crests will probably favor the regeneration of shrubs over other conifers.

The loss of trees to air pollution has also been reported around Tokyo and in the Rhine Valley (Wenger et al. 1971). Smith (1974) speculates that a list of species damaged by air pollution compiled for temperate forests of the world would be of very impressive length!

The more subtle effects of air pollution on forest ecosystems are exceedingly hard to quantify but are potentially of enormous importance. Research has shown that air pollutants can reduce growth, alter the rate of necessary decomposition processes, affect the reproductive processes of plants, and predispose plants to attack by insects or fungi. In some instances, pollutants act synergistically, reinforcing each other's effects (Smith 1974). Even though these subtle effects on temperate forests may presently occur over very widespread areas, they are hard to detect because of the "background noise" attributable to natural fluctuations in weather. The worry is that subtle degradations occur and may accumulate, leading up to a fairly sudden massive breakdown of the forest system.

Although we do not have adequate measures of the subtle effects of air pollutants or of pesticides, which may act like pollutants in diminishing the role of particular species within the ecosystem, we do know that potentially damaging concentrations of pollutants are widespread. For example, oxidants such as ozone are not limited to southern California, but occur in the hinterlands of many urban areas as well. Recently, it has been discovered that rainfall over thousands of square kilometers in the east-central region of North America (Likens and Bormann 1974) and of north-central Europe (Bolin 1971) has become a mixture of nitric, sulfuric, and hydrochloric acids. In the United States, the most recent increase in the acidity of rainfall is probably related to the increased introduction into the atmosphere of nitrogen compounds from automobile exhaust (Cogbill and Likens 1974). On a smaller scale, Smith (1975) has estimated that approximately 306,000 square kilometers of roadsides in the United States contain high concentrations of lead derived from automobile exhaust. How much of this is in forest regions is unknown.

All of these stresses—urbanization, poor forestry practice, strip mining, smelting, the regional application of pesticides, and regional air pollution—collectively represent an enormous and increasing stress on temperate-forest ecosystems. As the temperate forest is degraded under this human pressure, it loses its capacity to carry out its natural functions, to maintain stable landscapes and biological diversity, filter air and water, recharge groundwater, control erosion, yield sediment-free water to streams, and provide aesthetic and recreational enjoyment. These natural functions are powered by solar energy, and, to the degree that they are lost, they must be replaced by extensive and continuing investments of fossil-fuel energy and other natural resources in order to maintain the quality of life. Erosion-control works must be built, reservoirs enlarged, air-pollution-control technology upgraded, flood-control works installed, water-purification plants improved, air conditioning increased, and substitute recreational facilities provided. These substitutes represent an enormous tax burden and an unnecessary drain on the world's supply of natural resources.

What can a conference on EARTHCARE recommend for temperate forests under widespread, complex, and increasing stress? Piecemeal recommendations, such as the preservation of selected natural areas or the protection of endangered species through restrictive measures, fail to recognize the dimensions of the problem. How then do we deal with the true dimensions?

Energy Utilization

Two principles that govern the interaction between natural ecosystems and the utilization of fossil energy provide a guide for more comprehensive recommendations. The first principle is that an industrial society's increasing use of energy is directly related to increasing stress on natural ecosystems. The second is that a decrease in nature's "use" of solar energy necessitates the increased use of fossil energy if the quality of human life is to be maintained.

The causes of the increasing use of energy in industrial societies are a complex mix of population growth, rising standards of living, and its profligate use. The use of energy is an index of the use of natural resources, and growth in energy-use indicates an increased utilization of other resources—for example, more mining and smelting and more timber harvesting. Not only does this lead to greater stress on natural ecosystems through the increased generation of wastes and pollutants, but the costs per unit of production rise drastically as easily available natural resources are depleted and as increasingly expensive pollution-control technology must be applied merely to stay in the same place.

What are the options for limiting growth in the use of energy and hence for maintaining a less stressful environment? In the longer term, stabilization of human populations is essential to any planning, while right now the most obvious point of attack is to curtail our own profligate use of energy.

As to the level of individual lifestyle, every gallon of gasoline or fuel oil that we save, every light we turn out, and every energy-expensive commodity we reject at the marketplace represents a fraction less stress on our natural ecosystems. At the level of political action, support for such energy-conservation policies as better mass transportation, energy-conserving architecture, and less-energy-intensive agriculture and forestry will accomplish the same result. The key point is that conservation of energy and conservation of nature are directly linked.

The "energy crisis" of 1973-74 has promoted an intense new interest in the utilization of our most certain source of energy—the sun. The pages of scientific journals are filled with descriptions of technological schemes for capturing sunlight, yet the forest—finely tuned to its task by eons of evolution—remains one of the most efficient solar converters. The forest does not convert solar energy to electrical power; rather, it uses it directly to carry out natural functions of great benefit to man. When we reduce the quantity of solar energy captured by natural systems, we must to some degree substitute fossil energy and other natural resources to take over the functions formerly performed by nature. The unnecessary destruction or degradation of natural ecosystems wastes natural resources and increases the stress on the natural ecosystems that remain.

Conclusion

The relationship between the conservation of energy and the conservation of nature suggests that it would be a wise policy to maintain as much fixation of solar energy by natural ecosystems as possible. Obviously, some natural ecosystems inevitably will be destroyed or degraded as man converts or utilizes them, but the large amount of unnecessary disruption can be substantially reduced. The elimination of needless destruction should be a major objective when we evaluate, for example, the choice between strip mining and deep mining. Although less obvious, the same consideration is of great importance in the choice between development by low-density urban sprawl and high-density, planned communities. Urban sprawl not only consumes more land and requires more money and materials per unit of development, but, once completed, it requires more fossil energy and generates more air and water pollution. The same arguments hold for the choice between new highways and mass transportation (Council on Environmental Quality 1974).

Careful land-use planning can contribute greatly to the dual objectives of conserving natural ecosystems and conserving energy.

By conserving solar-powered natural systems, planning can significantly lower the demand for fossil energy and simultaneously reduce direct and indirect stress on natural systems.

The proposals I have suggested must be recognized for what they are: stop-gap measures—measures that will buy time by slowing man's overwhelming assault on our basic life-support systems. In the longer haul, a fundamental restructuring of society is required: man must learn to live in harmony with nature—not as master, but as one element in a complex and interdependent life-supporting system (Peterson 1975). Only in this way can man conserve natural ecosystems and enable his descendants to live a reasonable life on this Earth.

References

Barney, D. R. 1974. *The Last Stand: Ralph Nader's Study Group Report on the National Forests.* New York: Grossman Publishers.

Bolin, B., ed. 1971. *Air Pollution across National Boundaries: The Impact on the Environment of Sulfur in Air and Precipitation.* Report of the Swedish Preparatory Committee for the Conference on the Human Environment, 1972. Stockholm: Boktrykerier P. A. Norstedt & Söner.

Bormann, F. H., G. E. Likens, and J. S. Eaton. 1969. Biotic regulation of particulate and solution losses from a forest ecosystem. *BioScience* 19(7): 600–610.

Bormann, F. H., G. E. Likens, T. G. Siccama, R. S. Pierce, and J. S. Eaton. 1974. The export of nutrients and recovery of stable conditions following deforestation at Hubbard Brook. *Ecological Monographs* 44(3):255–277.

Cliff, Edward P. 1972. *Report of the Chief, Forest Service, 1970–71.* Washington, D.C.: U.S. Department of Agriculture.

Cogbill, C. V., and G. E. Likens. 1974. Acid precipitation in the northeastern United States. *Water Resources Research* 10(6):1133–1137.

Council on Environmental Quality. 1972. *Environmental Quality: The Third Annual Report of the Council on Environmental Quality.* Washington, D.C.: U.S. Government Printing Office.

———. 1974. *The Fifth Annual Report of the Council on Environmental Quality.* Washington, D.C.: U.S. Government Printing Office.

Dahlston, D. 1972. Forest management and the protection of forest resources—A critique. Paper read at the symposium "Temperate Climate Forestry and the Forest Ecosystem" at the Annual Meeting, American Association for the

Advancement of Science, Washington, D.C., 28 December 1972. Audiotape available from American Association for the Advancement of Science, 1515 Massachusetts Ave., N. W. 20005.

Jordan, M. J. 1975. Effects of zinc smelter emissions and fire on a chestnut oak woodland. *Ecology* 56(1):78–91.

Likens, G. E., and F. H. Bormann. 1972. Nutrient cycling in ecosystems. In *Ecosystem Structure and Function. Proceedings of the Thirty-first Annual Biology Colloquium*, ed. J. Wiens, pp. 25–67. Corvallis: Oregon State University Press.

———. 1974. Acid rain: A serious regional environmental problem. *Science* 184(4142):1176–1179.

Newell, R. E. 1971. The Amazon forest and atmospheric general circulation. In *Man's Impact on the Climate*, eds. William H. Matthews, William W. Kellogg, and G. D. Robinson, pp. 457–459. Cambridge, Mass.: MIT Press.

Oosting, H. J. 1956. *The Study of Plant Communities*, 2nd ed. San Francisco: W. H. Freeman and Co.

Peterson, R. W. 1975. The quest for quality of life. *BioScience* 25(3):166–171.

Seaton, F. A., M. Clawson, R. Hedges, Jr., S. Spurr, and D. Zinn. 1973. *Report of the President's Advisory Panel on Timber and Environment*. Washington, D.C.: U.S. Government Printing Office.

Smith, W. H. 1974. Air pollution: Effects on the structure and function of the temperate forest ecosystem. *Environmental Pollution* 6(2):111–129.

———. 1975. Lead contamination of the roadside ecosystem. Unpublished manuscript.

Tunnard, C., and B. Pushkarev. 1963. *Man-Made America: Chaos or Control. An Inquiry into Selected Problems of Design in the Urbanized Landscape.* New Haven: Yale University Press.

Walter, H. 1973. *Vegetation of the Earth in Relation to Climate and the Eco-Physiological Conditions.* Trans. J. Wieser. Heidelbert Science Library, vol. 15. London: The English Universities Press; New York: Springer-Verlag.

Wenger, K. F., C. E. Ostrom, P. R. Larson, and T. D. Rudolph. 1971. Potential effects of global atmospheric conditions on forest ecosystems. In *Man's Impact on Terrestrial and Oceanic Ecosystems*, eds. W. H. Matthews, F. E. Smith, and E. O. Goldberg, pp. 192–203. Cambridge, Mass.: MIT Press.

Whittaker, R. H. 1970. Communities and Ecosystems. London: Collier-Macmillan.

Woodwell, G. M. 1970. Effects of pollution on the structure and physiology of ecosystems. *Science* 168(3930):429–433.

31. Wilderness Fire Management in Yosemite National Park

Jan W. van Wagtendonk
Yosemite National Park, U.S.A.

Introduction

Yosemite National Park encompasses a broad range of vegetation types, from chaparral stands at 2,000 feet to alpine fell-fields at 13,000 feet. Not only the vegetation varies; also included within this range are variations in the nature of fuel and, consequently, the probability of fire as well. In these varieties of vegetation, natural fire plays different roles—from a dynamic force in the chaparral and mixed conifers to an insignificant factor in certain alpine types. The National Park Service recognizes these differences and is managing each type so that fire can play its natural role. This has not always been the case.

Evolution of a Fire Policy

Historical Conditions

Before national parks were established, natural fires were prevalent in many of the parks that later joined the park system. Lightning has been the primary source of natural fire ignitions, particularly in the mountainous West (Komarek 1967). These fires burn with varying intensities and periodicities depending on the nature of the vegetation and climate. Among the important roles played by fire in these vegetation types are the recycling of nutrients, the removal of undergrowth, and the perpetuation of fire-adapted species. Dominant trees such as ponderosa pine (*Pinus ponderosa* Laws.) evolved with a regime of periodic surface fires and were favored in many areas. Lodgepole pine (*Pinus contorta* Dougl.), through adaptations such as serotinous cones, was able to reproduce in areas subject to less frequent high intensity fires.

Cultural burning by Indians was also a factor in some park areas.

The reasons for this burning are not clearly understood, but it is thought that fires aided in food procurement and played a part in religious ceremonies (Reynolds 1959). These fires also varied in intensity and served to maintain certain vegetation types.

With the advent of European man, a period of indiscriminate burning began. Settlers used fires to clear the forest for farming and grazing purposes, and loggers set fire to timber slash and remaining trees. In California, it was a common practice for miners to ignite large areas of timber and brush to aid in prospecting and mining activities. Such fires not only destroyed timber and structures which were valuable to the new inhabitants, but also changed natural environments for which they had little or no value.

Suppression Policies

As a consequence of widespread fires and the concern for preserving park values, a policy of suppressing all fires on national park lands evolved. It began in 1866 in Yellowstone National Park and in 1890 in Yosemite National Park (Agee 1974). The purpose was to reduce the damage from human as well as natural sources. The thought was that fire was a destructive force in the forest, rather than a dynamic element of the forest system. The act which established the National Park Service in 1916 implicitly reiterated the suppression policy by stating that the "natural and historic objects and wildlife therein" would be left "unimpaired for the enjoyment of future generations." A total fire exclusion policy was justified by stressing the protection of objects rather than the perpetuation of natural processes.

Coupled with the suppression policy was a vigorous fire prevention campaign designed to reduce the number of man-caused fires. An outgrowth of this program was the idea that all fires were bad, the only good fire being a dead fire.

Leopold Report

The management of national park lands under the policy of fire exclusion caused many changes. Overprotection from natural surface fires permitted the forest floor to become a tangle of understory vegetation and accumulated debris. Thickets of shade-tolerant species increased and caused a shift in the flora and fauna away from the natural successional pattern. This increase in undergrowth and debris became so thick that the inevitable wildfire soon reached catastrophic proportions.

These changes were noted by the Special Advisory Board on Wildlife Management for the Secretary of the Interior (Leopold et al. 1963). The report, known as the Leopold Report, recommended that each park be restored as nearly as possible to the conditions that existed when European man first visited the area. It went on to state that the controlled use of fire is the most "natural" means of accomplishing such a restoration.

Current Management Philosophy

In 1963, the Leopold Report was incorporated as Park Service policy for natural areas of the national park system. The new policy included natural and prescribed fire management as well as the traditional suppression of wildfires. The presence or absence of natural fire in an ecosystem was recognized as one of the ecological factors perpetuating that system. Under this policy, natural fires were allowed to run their course as long as they could be contained within predetermined management units and when the burning was consistent with approved management objectives.

Prescribed fires may be used as a substitute for natural fires to achieve resource management objectives, if it is determined that natural fires cannot meet the objectives and if prescriptions can be developed for such fires. In addition, prescribed fires may be used for fuel hazard reduction and to reinforce fire management zone boundaries.

Any fire which is caused by man, or threatens human life, the destruction of historic or natural resources or park facilities, or resources or facilities outside of park boundaries will be extinguished. This would include the extinguishment of fires which do not further management objectives or which exceed the burning prescriptions.

Wilderness Fire Management

Within wilderness areas, management objectives are determined by the Wilderness Act. Specifically, the act directs that wilderness be "protected and managed so as to preserve its natural conditions."

This is interpreted to mean that natural processes such as fire are part of the wilderness ecosystem. Both the U.S. Forest Service and the National Park Service have established natural fire zones in areas designated or proposed as wilderness.

Yosemite's Experience

Fire histories indicate the importance of natural fire in the vegetation types of Yosemite's wilderness areas. Over the past nineteen years, there have been a total of 1,135 known lightning fires. The least number during any year was two, while the maximum number was 121. Although there is some variation in the areas of incidence from year to year, the distribution over time is uniform, with the mixed conifer and red fir types receiving the preponderance of lightning fires. This is to be expected since fuel and weather conditions in those types are more conducive to fire ignition and spread.

In 1972, the first natural fire management zone was established in the park, comprising a total of 187,007 acres. During that summer, two fires occurred in the zone while an additional six fires were allowed to burn outside the zone after it was determined that they would meet management objectives. The largest of the fires in 1972 burned 0.1 acre, and the total acreage was 0.31 acres. Five of the fires were in snags; the remainder burned in brush or duff.

The zone was expanded in 1973 to encompass 465,651 acres. Twenty-five natural fires occurred within the zone boundaries, and two fires were allowed to burn outside of the zone. A total of 35.09 acres were covered of which 30 acres were on one burn near Ostrander Lake. All of the twenty-two natural fires in 1974 burned within the zone. Nineteen of these fires were less than one acre in size and two less than 100 acres. The remaining fire, called the Starr King Fire, spread across nearly 4,000 acres before being extinguished by an early winter storm.

Perspective. Near Yosemite National Park, fire-scar histories dating from 1390 indicate the continued presence of fire in the region (Wagener 1961). These fires were either set by lightning or by Indians, and the interval between them averaged eight years (Wagener 1961, Biswell 1963, Kilgore 1972). Shorter periods are indicated by computer simulation models integrating weather, fuel, and fire variables (van Wagtendonk 1972). During such a relatively short period, fuel build-up was not sufficient to produce a high intensity fire. The resultant forest was open with patches of tree reproduction and brush interspersed with needles or various grasses for forbs. Ponderosa pine and giant sequoia (*Sequoia gigantea* [Lindl.] Decne.) are dependent on frequent low-intensity fires to reduce competition from the more shade-tolerant incense cedar (*Libocedrus decurrens* Endl.) and white fir (*Abies concolor* [Gord. and Glend.] Lindl.) in the mixed conifer zone of the park.

From the time Yosemite Valley was first seen by European man in 1851, to 1890, settlers replaced the Indian as a fire starter. At that time the park was established, and attempts were made to eliminate all fires. Active protection began in 1900, when army troops were detailed to the park to prevent trespassers and intruders from entering. The establishment of the National Park Service in 1916 transferred the protection duties to park rangers, and during the next fifty years, the policy of total fire exclusion was continued.

With the incorporation of the Leopold Report into the administrative policies, a new era of fire management began in Yosemite National Park. The first prescribed fire was set in the spring of 1970. Since then, twenty-four prescribed fires have burned over 2,500 acres. These fires were set to reduce fire hazards and to restore meadow and forest land to their natural condition. In 1972, a natural fire management zone was established, and fifty-seven fires have been allowed to run their course during the past four years.

Total Fire Management. Fire management in Yosemite National Park presently falls into six operations: routine wildland fire control, structural fire control, prescribed burning, natural fire management, conditional fire management, and loose-herding of wildland fire. Within the natural fire zone, all naturally caused fires will be allowed to burn. In the remaining area, all fires will be suppressed unless it is determined that a specific fire would meet management objectives. In that case, the fire would be loose-herded by minimal crews. In addition, in certain areas designated as conditional fire management zones, natural fires will be allowed to burn if specific weather and fuel conditions have been met. Air reconnaissance flights are made throughout the fire season to detect new fires and to monitor fires which are either being suppressed or allowed to burn.

The Starr King Fire

Chronology. The Starr King Fire was started by a lightning strike on August 4, 1974, in the vicinity of Mount Starr King, just south of Yosemite Valley. An air reconnaissance reported one snag burning and a small surface fire on the west side of a small drainage basin at an elevation of 8,000 feet. The basin was surrounded by open granite ridges with sparse fuels above, opening below to a continuous forest with sufficient fuel to allow the fire to back down the drainage indefinitely. The primary vegetation in the area included Jeffrey pine (*Pinus jeffreyi* Grev. and Balf.), red fir (*Abies magnifica* A. Murr.), and western juniper (*Juniperus occidentalis* Hook.), with some lodgepole pine, white fir, and whitebark pine (*Pinus albicaulis* Englem.). Brush patches of green leaf manzanita (*Arctostaphylos patula* Greene) and huckleberry oak (*Quercus vaccinifolia* Kellogg) extended from the creeks to the ridge tops.

For the first ten days, the fire burned with such a low intensity that it went virtually unnoticed from the ground. By August 14, columns of smoke were visible from Glacier Point and Yosemite Valley, and by August 27, the fire had spread to the north and south covering 130 acres.

From August 27 until September 9, the fire spread an additional 930 acres and doubled in size to 2,250 acres during the next ten days. The greatest expansion was to the east and west through brush and Jeffrey pine growing in conjunction with some red fir, white fir, and western juniper. The fire spread more slowly to the north, where the major fuels were lodgepole pine and red fir and to the south where dense stands of lodgepole pine occupied the creek bottoms.

On September 28, a small control action was made to keep the fire inside the natural fire zone boundary. A fireline connecting Illilouette Creek and the granite faces of Mount Starr King was constructed and 420 acres were backfired. By October 1, the fire had reached its maximum extent of 3,920 acres, and after several days of rain and snow was declared out on October 31.

From 1930 through 1974, there have been nine other fires within the area burned by the Starr King Fire. All were suppressed. Three of these fires had approximately the same point of origin as the Starr King Fire.

Fire Effect. Three trips were taken into the fire to measure fire, fuel, vegetation, and weather variables. The dates of these trips were August 27, September 11, and September 24. The weather during this period did not fluctuate greatly. Daytime relative humidities ranged from 41 percent to 16 percent, with the average being 31 percent. Temperatures were in the 60s and 70s during the day and dropped to 40 degrees Fahrenheit at night. Windspeeds never exceeded 12 miles per hour and the average fuel stick moisture content was 5.5 percent.

Ground fuel depth measurements were taken before and after the fire passed over sample plots, in order to determine fuel consumption. The average consumption was 7.045 tons per acre. The least amount of fuel was lost underneath western juniper trees where only sparse fuels existed. The greatest consumption was in a red fir stand where a relatively intense fire consumed 34.17 tons per acre.

Rates of spread were measured at each plot and varied from 1 inch per hour to 120 feet per hour. Area rate of spread for the fire as a whole was greatest on September 20, when 200 acres burned. During much of August, the average area rate of spread was 5.42 acres per day.

Intensities were calculated using caloric values from Agee (1973) and the fuel consumption and rates of spread from the plots. They varied from .29 BTU/second/foot for the smoldering fire in the western juniper stand to 679.82 BTU/second/foot for the red fir stand. Underneath the Jeffrey pines, intensities varied from 29.49 BTU/second/foot to 539.48 BTU/second/foot. The moist creek bottoms where some lodgepole pine burned did so with an intensity of 2.67 BTU/second/foot. The fire was characterized by patches of intensely burned areas separated by lightly burned or unburned areas.

The effects of the fire on vegetation were related to the variations in burning intensity. Pockets of huckleberry oak and manzanita were completely consumed, and downed Jeffrey pine logs and snags burned intensely. Some standing green Jeffrey pines burned in cat faces. In some cases, clusters of red fir from 70 to 100 feet high were completely killed. It was estimated that not more than 3 percent of the mature canopy was actually killed by the fire. Only 60 percent of the area inside the fire perimeter was burned, most of the unburned area being open granite with little or no fuel accumulation.

The differential effect of fire in the various vegetation types was well illustrated by this fire. Brush fields were almost entirely consumed as they would be periodically, since they are flammable within a few years after a fire. Jeffrey pine is much like ponderosa pine with low- to moderate-intensity fires burning the surface layers and only occasionally killing a live tree. Red fir, on the other hand, burned intensely when conditions were favorable, but not at all under more moist conditions. Since the proper conditions do not occur frequently, fuel accumulates and the eventual fire is intense. This is also true of the lodgepole pine, as evidenced by the low-intensity fire near the creek bottom. Another stand of lodgepole pine along a ridge top burned intensely, however, since conditions were drier there. Fire is not a significant factor for western juniper since fuel accumulations are small and individual trees are surrounded by granite.

Because of the low intensities and slow rates of spread, wildlife was little affected by the fire. Stacks of vegetation which are used by the Sierra mountain beaver *(Aplodontia rufu californica)*, were seen before and after the fire. California mule deer *(Odocoileus hemionus californicus)* were grazing within the burn during and after the fire. Three black bears *(Ursus americanus californiensis)* were also seen in the burned area.

Since the Starr King Fire burned within the watershed which is used to supply Yosemite Valley, water quality measurements were available for analysis. A comparison between post-fire measurements and similar periods in previous years showed no significant difference for water temperature, suspended sediments, major dissolved ions, or suspended organic elements.

From mid-August until the end of September, considerable amounts of smoke were produced by the fire. Each evening smoke would drift down into Yosemite Valley and would usually clear by 10 a.m. when up-slope winds would disperse it. One qualitative measure of the smoke came from a long-time resident who stated that there was much less smoke in the valley from the fire than there had been from campfires before the number of campsites had been reduced. Radiometer measurements of solar radiation at a research site ten miles northeast of the fire showed great reductions in langleys received when the wind was from the southwest.

The Starr King Fire is in an area which receives low recreational use. When the fire neared and crossed the only trail in the area, backcountry travel was restricted. Although no direct measures of visitor reaction to the Starr King Fire were made, letters and comments for or against the program give a clue. Only one letter of complaint was received concerning the fire. Public comments at meetings held for the master planning process have been twenty-to-one in favor of the natural fire management program.

Other Wilderness Areas

The 100-square-mile White Cap Fire Management Area of the Selway-Bitterroot Wilderness in northern Idaho was the first such area established by the Forest Service. Studies began in 1970 to determine operational strategies for a more natural occurrence of fire in the wilderness area (Mutch 1974). As a result of these studies, prescriptions and guidelines were developed, and in 1972, the fire management zone was approved. The management plan is specific to the many different vegetation types in the wilderness and calls for suppression actions to protect human life and property or to contain fires in the management unit. In 1972, there was only one fire in the unit, while six fires occurred during 1973. One of these, the Fritz Creek Fire, burned more than 1,200 acres within the unit and 400 acres outside it. Control action was taken on the portion outside of the fire management unit. In 1974, another unit of 135,000 acres was designated in the Selway-Bitterroot Wilderness as a fire management area. In the summer of 1974, each unit had a single fire of less than one-quarter of an acre. The only other Forest Service wilderness to have an approved fire management area is Gila Wilderness in Arizona where 33,000 acres were designated.

Portions of the natural fire zones in ten Park Service areas have been proposed as wilderness and await enactment. Fire management within these zones would not change upon enactment since present policies are in harmony with the Wilderness Act. The Park Service already recognizes fire as a natural process; the Wilderness Act reiterates the philosophy of perpetuating such processes.

Natural fire management in the Park Service started in 1968, when Sequoia-Kings Canyon National Parks set up the first natural fire zone (Kilgore and Briggs 1972). Since that time, the zone has been expanded to include nearly 600,000 acres. Ninety-seven fires covering 7,800 acres have burned in the zone during the last seven years. The largest of these was the Comanche Fire, which burned 3,060 acres in 1974.

Everglades National Park was the first park to use prescribed fire, beginning their program during the 1950s (Robertson 1962). Natural fires have been allowed to burn since 1968, with seventy-three fires burning about 9,425 acres during the past seven-year period. Some of these fires were man-caused and were determined to meet management objectives.

A modified natural fire program is in use in Saguaro National Monument in Arizona (Gunzel 1974). There, natural lightning-caused fires (or wildfires) are allowed to burn under prescribed conditions. By analyzing the various fire and weather factors concerned, it was determined that fires from mid-July to mid-September would be of the desired low to moderate intensity. A total of twenty-eight out of fifty fires have been allowed to burn without control over 900 acres under this program.

The most widely publicized natural fire to occur has been the 3,700-acre burn in Grand Teton National Park during the summer of 1974. This fire was ignited by lightning in July and burned until it was put out by late autumn snows (Kilgore 1975). Two other smaller fires burned in the park last summer, while in nearby Yellowstone National Park, six more fires burned some 300 acres.

Other national park areas which have established natural fire management zones include Rocky Mountain with five fires, Carlsbad Caverns with one fire, Guadalupe Mountains with two fires, and North Cascades National Park with no fires (Kilgore 1974).

Conclusion

The management of natural fires within wilderness areas of the National Park Service will continue to be an active program. The end objective is to restore fire as a natural process in those parks where it was historically present. In Yosemite, it is hoped that, as fuels are modified through prescribed burning at the lower elevations, the natural fire zone will increase in size. This is currently happening, as the zone has been enlarged to 481,500 acres for 1975, with 132,738 additional acres in the conditional fire management zone. Eventually, all but the developed areas of the park will be managed with natural fire as a part of the environment.

With increased experience and research, there will be a better understanding of the role fire has played in the diverse ecosystems of the National Park System. By restoring fire to its natural role, the

Park Service is guaranteeing the perpetuation of these unique natural resources.

References

Agee, J. K. 1973. Prescribed fire effects on physical and hydrologic properties of mixed-conifer forest floor and soil. Ph.D. dissertation, University of California, Berkeley. *Also:* Contribution Report 143. Water Resources Center, University of California.

———. 1974. Fire management in the national parks. *Western Wildlands* 1(3):27-33.

Biswell, H. H. 1963. Research in wildland fire ecology in California. *Proceedings of the Second Annual Tall Timbers Fire Ecology Conference* (Tallahassee, Florida, 14–15 March 1963) 2:62-97.

Gunzel, L. L. 1974. National policy change—natural prescribed fire. *Fire Management* 35(3):6-8.

Kilgore, B. M. 1972. The ecological role of fire in Sierran conifer forests. *Quaternary Research* 3(3):496-513.

Kilgore, Bruce M. 1974. Fire management in national parks: An overview. *Proceedings of the Montana [Fourteenth] Annual Tall Timbers Fire Ecology Conference* (Fire & Land Management Symposium, Missoula, Montana, 8 to 10 October 1974) 14:45-57.

———. 1975. Restoring fire to national park wilderness. *American Forests* 81(3): 16-19, 57-59.

Kilgore, B. M., and G. S. Briggs. 1972. Restoring fire to high elevation forests in California. *Journal of Forestry* 70(5):266-271.

Komarek, E. V. 1967. The nature of lightning fires. *Proceedings of the California [Seventh] Tall Timbers Fire Ecology Conference* 7:5-42.

Leopold, A. S., S. A. Cain, C. M. Cottam, I. N. Gabrielson, and T. L. Kimball. 1963. Wildlife management in the national parks. *Transactions of the North American Wildlife and Natural Resources Conference* 28.

Mutch, R. W. 1974. I thought forest fires were black! *Western Wildlands* 1(3): 16-22.

Reynolds, R. D. 1959. Effects upon the forest of natural fires and aboriginal burning in the Sierra Nevada. Master's thesis, University of California, Berkeley.

Robertson, W. B. 1962. Fire and vegetation in the Everglades. *Proceedings of the Annual Tall Timbers Fire Ecology Conference* 2:67-80.

van Wagtendonk, J. W. 1972. Fire and Fuel Relationships in Mixed Conifer

Ecosystems of Yosemite National Park. Ph.D. dissertation, University of California, Berkeley.

Wagener, W. W. 1961. Past fire incidence in Sierra Nevada forests. *Journal of Forestry* 59(10):739-748.

Tropical Forests

. . . the sun pours down gold
fountains pour out green water
colors touch us like fingers
of green quetzal wings . . .

—"A Song of Nezahualcoyotl"

Y en el fondo del agua magna,
como el círculo de la tierra,
está la gigante anaconda
cubierta de barros rituales,
devoradora y religiosa.

—Pablo Neruda
"Algunas Bestias"

Terrestrial Ecosystems: Tropical Forests

32. Emphasis on Tropical Rain Forests: An Introduction

Lawrence S. Hamilton
Cornell University, U.S.A.

Introduction

While the title of this session of the EARTHCARE Conference is "Tropical Forests," the two panelists of the session will emphasize rain forests. There are many kinds of tropical-forest formations, ranging from unusual thorn forests, through several kinds of seasonal forests, to many kinds of evergreen wet forests. These forests have been shrinking rapidly in recent years as tropical countries clear more land for expanding populations.

To meet the wood demands of developed temperate countries, tropical countries are viewing their forests as a means of achieving economic development. Under these pressures, the welfare of the forests as important renewable resources, with their wildlife and watersheds, wilderness and scientific values, often gets lost in the rush for short-term exploitation.

It is the tropical rain forest to which this panel will devote most of its attention, because the rain forests of the world are, in the first place, unusually vulnerable and, in the second place, under an unprecedented assault. In these introductory remarks, I will say something about both the vulnerability of tropical rain forests and the assault on them.

The Unusual Vulnerability of Rain Forests

While there are many kinds of rain-forest communities, some of which, because of high water or soil condition, consist of only a few species, the general situation in these forests is one of floristic richness. There are many species of plants and often a correspondingly great diversity of animal life. In tree species alone, there may be 50 to 120 species within an area of one acre (0.4047 hectare)—many more than in our temperate forests. There are many layers or canopies in the rain forest and many forms of plant life. The situation is one of great diversity, with a correspondingly very large gene pool. Some species are scarcely known; nor are their properties and utility to man. These forests, which have developed over millennia in relatively nonstressful environments from which cold and drought were absent, are in a sense, the genetic cradle of the world—the center of evolutionary diversity.

The very diversity of species means that on any given area there are usually only a few individuals of any one species. This is a definition of rarity. It means that species are easily lost if they are exploited with little restraint. In the following papers, Dr. Gómez-Pompa will explain the regeneration difficulties that lead to the loss of tree species, and Ing. Mondolfi will point out the rarity problem with tropical-forest wildlife. The gene pool of complex tropical rain-forest ecosystems is very vulnerable, and "EARTHCARE" means a concern for the genetic resources represented by this global resource.

Tropical rain forests are also extremely vulnerable because they very often occur on soils of low fertility and marginal structure. Most of the nutrients are in the living vegetation and not in the soils. They are cycled rapidly in a virtually "leak-proof" system. If the forest is removed, however, rapid oxidation, rapid leaching, and—if the soil is situated on a slope—rapid runoff, quickly remove the nutrients from an already nutritionally low budget. In some cases, problems of hardpan may occur, completing a chain of site degradation. Fortunately, unless the trees are burned after clearing and the land converted to agriculture, these sites are not permanently degraded, and become quickly revegetated with brush and secondary species. However, it may take centuries for anything approaching the original forest to return to an area, and some species may be lost from an area forever. Where man keeps an area open for his crops for more than two or three years, and where an area is larger than two or three acres

(4.8 to 7.2 hectares), the forest can be permanently altered. Clearly, only soils that are suitable for sustained agriculture should be cleared of their rain-forest cover. Ecological surveys must precede the removal of forests in the future.

The Assault on the Rain Forest

The unprecedented rate at which tropical rain forests are being cleared or drastically modified has aroused the concern of reputable scientists and organizations around the world. Two of these scientists (and myself as chairman) are on this panel. The United Nations Environment Programme (UNEP) has made it a priority item. It has been reliably stated that if present trends continue and conservation efforts are not greatly increased, "most of the existing forests of the tropics will be destroyed or replaced by seral communities by the end of the century." The International Union for Conservation of Nature and Natural Resources and the World Wildlife Fund (WWF) have initiated a strong campaign to save adequate and representative parts of the rain forest and have developed ecological guidelines for tropical lands. The Food and Agriculture Organization of the United Nations (FAO) estimates that in Latin America alone "between 12 and 25 million acres [between about 5 and 10 million hectares] of forests are being felled for agriculture each year." In Central America, two-thirds of the tropical rain forest has already disappeared. In Indonesia, 66 million acres (26.7 million hectares) of rain forest are officially classified as denuded by uncontrolled cutting. Similar kinds of reports come from many tropical countries. This growing concern is why the Sierra Club's International Program is conducting a study project on tropical rain-forest conservation. I am director of this project,* which is being carried out in Venezuela with Ing. Mondolfi's organization, the Rural Welfare Council, with funding from UNEP.

Several forces are responsible for the disappearance of the forest. Slash-and-burn agriculture, called in various places *milpa, conuco,*

---

*The results of this project have been published as *Tropical Rain Forest Use and Preservation: A Study of Problems and Practices in Venezuela.* March 1976. International Series Number 4. San Francisco: Sierra Club; and in an enlarged Spanish version as *Conservacion de los Bosques Humedos de Venezuela.* June 1976. Caracas: Sierra Club–Consejo de Bienestar Rural.

swidden, etc., can have either slight, temporary effects ("stable slash and burn") or more serious, long-term effects ("extended slash and burn"), depending on the size of the area cleared and on the length of the period of cropping and fallowing. Unfortunately, with population pressures, more and more of the shifting agriculture is of the unstable latter type. Roads penetrating the rain forests, for whatever purpose, are the means by which man's impact can be extended. The shifting cultivator, driven by land hunger, soon follows the road-building equipment. Government programs of "securing the frontier," "conquering the jungle," or "filling in the blank spots on the map" usually start with roads—witness the trans-Amazonia highway. Fire is certainly an important damaging agent—whether it is used deliberately in connection with land clearing or occurs inadvertently when it escapes during the burning of sugar cane. Even in a rain forest there may be drier seasons when felled trees become dry enough to burn or when fire can nibble at the edges of the standing forest. Grazing has been extended into rain-forest areas rather than intensified on savannah areas. Logging will change the primary rain forest. With primitive tools and harvesting of only one or two species, the impact is relatively minor, although the exploited species can disappear completely over time. There is now a trend toward greater utilization, even to the extent of an "all-tree, any-tree" approach, which would, of course, have a devastating effect on primary rain forests because secondary species would be favored. Intensive wood cropping on some areas, however, might relieve the pressure on some key areas of original rain forest and make their protection somewhat easier. Mining can have a direct impact, as can reservoirs that flood lowland rain forests, but it is the roads associated with these developments that are even more significant than the activities themselves.

Conclusion

There is no doubt that cutting and clearing of the world's remaining tropical forest is proceeding at an alarming rate. In a large area that I am studying in Venezuela, forest cover decreased by one-third during the past twenty-five years. Satellite monitoring by the Food and Agriculture Organization may help us determine just how fast it is disappearing on a global scale. The two panelists document some of the specifics of the problem. The scientific community may bewail, but the political decision-makers do not act, or do not act soon enough. Perhaps, through the urging of the international community at meetings such as EARTHCARE, national pride in a national and international heritage may move governments to action. It was largely by this means that the idea of national parks was spread globally. Perhaps this EARTHCARE Conference can support the program initiated by IUCN-WWF, which have designated this year (1975) as the "Year of the Rainforest."

33. Regeneration Problems of the Tropical Rain Forest

Arturo Gómez-Pompa
National University of Mexico, Mexico

The Roots of Ecology

The tropics support a great variety of ecosystems, which have attracted the attention of naturalists for a long time. They have been the scene of many important advances in the biological sciences. For example, Charles Darwin, the outstanding biologist of all time, obtained fundamental information in the tropics, and such basic concepts as evolution, species diversity, and speciation have deep roots there. It is astonishing to realize that all of these advances were made at a time when little knowledge was available about the biota of these regions.

I often wonder what our world would be like if development had been based in the tropics. How would human society have evolved if man had discovered and utilized the resources of the tropics to the same degree he has those of the temperate regions? Would we produce food as we now do? Would we manage land differently? Would our concepts of conservation and wildlife management be the same?

The Evergreen Rain Forest and the Process of Regeneration

The term "rain forest" is often generalized to include ecosystems of very diverse natures; I will use the term to mean evergreen rain forests of the tropical lowlands. These are without doubt the most diverse ecosystems on Earth and yet the least known—a paradox that concerns scientists and citizens all over the world. These ecosystems are threatened today in many regions of the world by the increasing demand for new land for the Earth's growing human population.

I would like to discuss one of the most important aspects of the tropical rain-forest ecosystems, one that may help us to see more clearly the problems due to the improper application of knowledge derived in temperate regions to tropical areas. I refer to the process of regeneration of the rain forest.

It should be kept in mind that the most significant characteristic of tropical rain forests is their inability to regenerate after heavy and extensive destruction. Yet many people take it for granted that a forest will regenerate after it has been destroyed. This belief is based on experience gained in heavily populated countries of the temperate regions, where forests have been reestablished successfully even after intensive exploitation and deforestation. But I would like to point out some of the differences between the regeneration of tropical rain forests and that of temperate forests.

Ecological Succession in Tropical Forests

When a tree falls in a forest due to wind, old age, and so on, it is replaced. This is "ecological succession," a process which is, in turn, the result of a long process of evolution; it is homeostatic: it helps maintain the stability of the ecosystem. Ecological succession has been known for a long time and is well accepted, though not all of its characteristics and specific variations are known yet. Because of their natural biological richness and absence of definite limiting factors, the hot, humid tropics in which the rain forests occur have one of the most complex successional processes known.

Let's go into more detail on this process as it occurs in the rain forest. When a tree falls, a "light gap" appears. Some of the seeds in the soil germinate, and some of the established small plants on the site grow faster, while others die. This means that regeneration occurs mainly from pre-established plants under natural conditions.

Regeneration Potential

We studied regeneration potential as the key to understanding the entire process of ecological succession in tropical rain forests. First, to determine the "floristic potential" of its soil, we collected soil samples from the rain forests to find out which species are represented by living seeds. After several years of studies, we discovered that the species present are mainly secondary species; primary tree species were not present in the samples. The obvious question came to mind: What is happening to the seeds of the primary-tree species? We conducted several studies and we found that they either germinate immediately or are soon eaten or decomposed.

That finding was extremely important, because it means that the only way for a primary-tree species to regenerate is for its seedlings to be growing in the understory of the forest when a light gap occurs. Temperate forests may have similar potential in seedlings, but a major difference between them and tropical forests is that the more viable seeds of primary species are stored in the soil of temperate forests. This permits many species to survive for many years, even while the forest is being intensively used. Through a survey of the scientific literature, we found that seeds of primary temperate-forest trees were viable for an average of ten years, while those of primary tropical rain-forest trees were viable for an average of only twenty-five days. Some of the findings we made in tropical Mexico are being confirmed by other studies conducted in Central America and Asia.

A primary tree of a rain forest has to have a large seed, with enough food reserves for fast growth into a large seedling well above the ground. This fast germination is a desirable adaptation because the large seed with its food reserves appeals to frugivorous animals. Fast germination reduces that problem. In addition, the seed is subject to the intense microbial activity in the litter of the tropical soil: a seed without any special protection decomposes rapidly. Immediate germination also helps prevent this problem. This adaptation was favored in evolution and is not a widespread characteristic of tropical trees. Also, primary-tree seedlings are adapted for photosynthesis under filtered light of low intensity.

Secondary species are also fast growing, but they are adapted to intense light. Their problem, then, is different from that of the primary species: because light is the most important trigger factor for germination, the seeds have to remain dormant and viable until an opportunity to grow arrives—which happens when a light gap occurs. These species have other adaptations that are different from those of primary species, such as their small, well-protected, dormant seeds.

Secondary species are extremely important because they rapidly restore the environment for the primary species. Several studies have been made by my research group on the evolution and adaptation of secondary species. The results of those studies verify the fundamental role that secondary species play in the natural regeneration process.

If we examine them from an evolutionary standpoint, we see that the regeneration system is well adapted to both the present ecological conditions of the tropics and to changes that occurred in the tropics before man appeared on Earth.

The Advent of Agriculture

Up to now I have been talking only of the natural process. With the early agricultural activities of man, the light gaps became larger, although primitive groups have unknowingly imitated the natural process of creating only small openings in the forest. After a few years of being used for agriculture, the fertility of soil in cleared forest patches declines, and various pests become a problem. The only solution has been to abandon the land, allowing time for the old gap to regenerate.

With this primitive, shifting agriculture, natural succession has operated quite well, and examples of a mosaic of land use produced by shifting cultivation can be seen all over the tropical parts of the world. This is a well-established system and, as long as the population density is low, it is an ecologically sound use of hot, humid tropical areas. Many different adaptations of this system exist in the tropics today.

Recently, however, the pressure of increasing populations and their increasing demands for agricultural land have posed new problems, compounded by the scarcity of the remaining primary forests. The fallow period of shifting cultivation has shortened, resulting in infertile soils in some areas.

In areas where no primary forests are left—even if we allow time for succession to occur—forests will never return to their original state because there are no more primary trees around to produce seeds, and there are no viable seeds in the soil. The larger the clearing and the more intensive and prolonged its use, the less the possibility of regenerating primary trees. In these places, a new and stable, but impoverished, ecosystem composed of secondary species (the only ones available for regeneration) has been established. If to this we add the ever-increasing phenomenon of extensive permanent agriculture, the possibility for future recovery of the forest is even less.

It is my impression that because of these factors, we have lost many hundreds of species of plants in the tropical world in recent times, even before we knew they had existed. This trend will accelerate in the years to come, because the loss of primary trees accompanies the loss of other plants of that ecosystem. This is also true for many of the animals that cannot adapt to the new, arrested-successional habitat.

There is much more to say in this regard, but I would like to leave it there in your minds. If you wish more information, you could read some of our publications and also the literature that has appeared recently on the subject. (See, for example, Gómez-Pompa et al. 1972.)

Solutions

Short-Term Solutions

I want to turn now to possible solutions. There are no simple ones; the problem is complex and of tremendous magnitude. I see two types of solutions, short-term and long-term. The short-term solutions are:

1. To increase the surveys of the biological diversity of tropical rain forests, especially those that are in immediate danger;
2. To increase the number and quality of botanical gardens, biological reserves, and national parks in regions occupied by rain forests;
3. To develop land-use planning methodologies in which ecologically sound considerations could be incorporated;
4. To study the regeneration process further, with the idea of

obtaining sound recommendations for determining the size and shape of biological reserves;
5. To study carefully the land-use systems of "primitive" agriculture, which may offer solutions to our problems in the future; and
6. To investigate the use of rain forests as a productive resource in itself, since the main reason the forest has been destroyed is that is is seen as an enemy of agricultural development.

It is evident that the best way to conserve tropical rain-forest ecosystems is by convincing the people with facts. Rain forests are not incompatible with man. They can provide useful products, and people can secure their livelihood from them. I believe that this approach is the only one that could effectively stop the destruction of these magnificent forests. Their purely aesthetic values should not be overstressed: romantic conservationism can do more harm than good.

The Long-Term Solution

The only long-term solution is to eliminate poverty, ignorance, and hunger among people in the tropics. The world cannot live in peace if there is too much wealth on one side and too little on the other. Selfishness and injustice should yield to cooperation and justice. My thinking may appear naive, but I foresee a dark future for the Earth's wild plants and animals if human beings cannot develop a widespread, well-understood, and scientifically based land ethic.

Man may try to destroy nature. And while there is a good chance that some of the millions of species of living things on Earth will adapt to man's disturbances, the human species might not be one of them. It is in our hands now to provide for the future.

References

Gómez-Pompa, A., C. Vazquez-Yanes, and S. Guevara. 1972. The tropical rain forest: A nonrenewable resource. *Science* 177(4051):762–765.

34. Scientific, Aesthetic, and Cultural Values of Rain-Forest Wildlife

Edgardo Mondolfi
Consejo de Bienestar Rural, Venezuela

Dr. Hamilton and Dr. Gómez-Pompa, with great knowledge and experience, have covered the vast and complex topic of the tropical rain-forest flora. I shall deal almost exclusively with the characteristics, values, problems, status, and potentials of the animal wildlife of the Venezuelan rain forest. These features of wildlife are common to all tropical countries with rain forests. The scarcity of scientific knowledge on this subject is reflected in the brief and sparse bibliography. For this reason, I shall present a general view, using the case of Venezuela, with which I have had firsthand experience.

The Diversity of Rain-Forest Wildlife

The different types of rain forests in the neotropical realm, as well as in West Africa, Southeast Asia, New Guinea, and the eastern portion of Australia, harbor a most outstanding fauna characterized by a great diversity of species, remarkable adaptations to different habitats, and a large number of endemic taxa. The tropical rain forest is the most favorable habitat for life on Earth. Temperature, humidity, and rainfall are constant within narrow limits and at a level nearly ideal for life. The long history of environmental stability, the abundance and variety of plants and animal life, the remarkable adaptations shown by many species of animals, and dynamic organism-to-organism interactions make the neotropical rain forest of major importance to biologists and to our understanding of nature in general.

No doubt, these characteristics gave clues about the mechanism of evolutionary change to Alfred Russel Wallace as he worked in the Malay Archipelago, and to Charles Darwin in his brief excursion to the Brazilian rain forest. For the taxonomist, ecologist, zoogeographer, geneticist, comparative anatomist, and in general to anyone who studies evolution, the tropical rain-forest fauna offers a rich and fascinating field. Expeditions conducted by museums and other institutions have gathered a wealth of vertebrate and invertebrate specimens from tropical rain forests around the world, allowing zoologists to classify the taxa and determine their geographical ranges. In spite of these efforts, however, inventories of tropical rain-forest animals are incomplete, and there are many gaps in the taxonomy and biogeography of several mammals, amphibians, reptiles, and so on. In the very large tracts of rain forest still unexplored or little explored by zoologists, as in Amazonia, there is no doubt that many animals, especially insects, remain undescribed. This is true for mammals, too, especially in regard to small rodents and bats, numerous species of monkeys, marsupials, and other groups.

Because of the extreme diversity of the bird fauna of the neotropical rain forests, there are many unusual species and perhaps even species to be discovered. It is noteworthy to mention the extraordinary ornithological wealth of Colombia: 1,556 species, which divide themselves into 2,640 subspecies—some 56 percent of all bird species found in South America (de Schauensee 1964). A great proportion of this diversified avifauna inhabits the lowlands and montane rain forests. In Venezuela, the Phelps collection has catalogued more than 1,300 species of birds. The rich avifauna in the rain forests of Surinam, Guyana, Brazil, Ecuador, and Peru shows the immense zoological importance of these areas. As Jürgen Haffer has said:

> The richest forest fauna of the world is found in the tropical lowlands of central South America. This fauna inhabits the vast Amazonian forests from the base of the Andes in the west to the Atlantic coast in the east, and its range extends far to the north and south of the Amazon valley onto the Guianan and Brazilian shields, respectively (Haffer 1969).

The outstanding scientific value of the Guyana-Amazonia region of Venezuela is enhanced by the *tepuis*, sandstone plateaus that, owing to the long isolation of the biota and the diversity of life zones, shelter a most extraordinary flora and fauna, rich in endemic forms.

In the Guyana-Amazonia region of Venezuela, eighty-three mammal species, including several subspecies, have been recorded. This number does not include the bats, a very diversified group, with more than sixty species in the area.

Especially important is the fact that the tropical rain forest, with a long history undisturbed by major geological events or climatical changes, contains a great number of primitive animal forms that are strictly dependent upon their special and fragile habitats. In regard to the monkeys, a most conspicuous and also a highly diversified group with sixteen genera in the South American rain forest, Philip Hershkovitz states:

> Living New World Platyrrhini are in general more primitively structured than known living and extinct Catarrhini. They fill and overlap the gap between the more monkey-like prosimians at one end and the more primitive catarrhines on the other. This broad span of evolutionary sequence makes living platyrrhines particularly attractive for studies of critical evolutionary stages not preserved in lines leading to modern Old World Primates (Hershkovitz 1972).

Citing other unique and primitive mammals of the Venezuelan Guyana-Amazonian forest, we may mention the giant, lesser, and silky anteaters *(Myrmecophaga tridactyla, Tamandua tetradactyla,* and *Cyclopes didactylus)* the two-toed and three-toed sloths *(Choloepus didactylus, Bradypus variegatus* and *Bradypus tridactylus);* the giant and naked-tail armadillos *(Priodontes maximus* and *Cabassous unicinctus);* several species of prehensile-tailed porcupines *(Coendou* sp.); the paca *(Agouti paca);* five species of squirrel *(Sciuridae);* several spiny rats *(Echimyidae);* the bush dog *(Speothos venaticus);* the Brazilian tapir *(Tapirus terrestris);* the giant otter *(Pteronura brasiliensis);* and two brocket deer *(Mazama americana* and *M. gouazoubira).*

Among the birds, the following groups are typical of rain forests in the neotropics and are considered primitive: several species of the tiniamous *(Tinamidae);* currasows and guans *(Cracidae);* the hoatzin *(Opisthocomus hoazin);* the trumpeter *(Psophia* sp.); the sunbittern *(Eurypyga helias);* potoos *(Nyctibius* spp.); the *Guácharo* or oilbird *(Steatornis caripensis);* toucans *(Ramphastidae);* jacamars *(Galbulidae);* mot-mots *(Momotidae);* antbirds *(Formicariidae);* gnateaters *(Conopaphyagidae);* oven birds *(Furiidae);* manakins *(Pipridae);* and cock-of-the-rock *(Rupicola rupicola),* umbrella birds, bellbirds, cotingas, becards, and fruitcrows *(Cotingidae).*

Venezuela has many peculiar snakes, among them the largest

snake in the Western hemisphere, the aquatic anaconda *(Eunectes murinus)*. There are many interesting frogs, such as the beautifully colored, poison-arrow frogs *(Dendrobates* sp.*)*, and a wealth of lower animal forms, especially invertebrates.

Tropical rain forests are genetic reservoirs of plants and animals of great scientific or economic value. To maintain these genetic resources, substantial samples of the habitats and of the animal populations must be protected. In this regard, Duncan Poore writes, "Evolution continues all the time and the process can go on only within the natural community. If species are planted in gardens [and we may add: if species of wild animals are captured and bred in captivity], they may be preserved, but the evolutionary process will be checked or diverted" (Poore 1974). He goes on to say that, since new and valuable uses for the organisms are constantly discovered in medicine, pest control, and in the breeding of plants and animals of commercial value, it would be blind and irresponsible to destroy the sources of so much potentially valuable genetic material.

In the IUCN's ecological guidelines for development in the American Humid Tropics (Anonymous 1975), as well as in those for the development of tropical-forest areas of southeastern Asia (Poore 1974), it is stated that "the conservation of natural ecosystems will provide a reservoir of populations of wild plants and animals, and of the variation within them under conditions which will enable evolution to continue under substantially natural conditions."

When dealing with the wildlife of the tropical rain forest, we should remember that this is a very fragile ecosystem, an ecosystem that, once disturbed, may need hundreds or perhaps thousands of years to reestablish its equilibrium. Man's activities can rupture the ecological equilibrium and endanger valuable and interesting species of wild animals. The enormous scientific and cultural values of the flora and fauna of the rain forests constitute a patrimony that belongs not to any particular country, but to all of humanity—a natural heritage that must be under the zealous custody of each government and its citizens. Scientists and technicians must assume their share of responsibilities and contribute to the preservation of this immense wealth for the enjoyment and well-being of future generations. Nations fortunate enough to contain still-large areas of tropical forests, Poore says, should be proud to be the trustees of a resource of the highest significance in the modern world.

Subsistence Hunting

Many species of wild mammals and birds, as well as turtles, some species of lizards, and even alligators, snakes, some species of ants, earthworms, and grubs are important sources of food for rural people in the tropics. Latin American Indian groups, as well as *campesinos* that live within rain-forest areas, practice a very primitive agriculture and usually do not have meat from domestic animals at their disposal. They have to supplement a deficient diet of cassava and other root crops and fruit with animal protein derived from subsistence hunting and fishing.

The most important hunted species of mammals in the Venezuelan rain forests, many of which are found in other Central and South American countries, are the tapir, collared and white-lipped peccaries *(Dicstyles tajacu* and *Tayassu pecari)*, brocket deers, pacas, capybara *(Hydrochoerus hydrochaeris)*, agoutis, acuchis *(Myoprocta* sp.*)*, several species of monkeys, and armadillos. Kinkajous *(Potos flavus)* and coatis *(Nasua nasua)* are also hunted.

Among the birds most commonly hunted are the guans (several species of *Crax, Mitu,* and *Pauxi)* and the rare *Nothocrax urumutum,* the nocturnal curassow; among reptiles, two species of land tortoises *(Geschepne carbonaria* and *G. denticulata)* and some of the river turtles, like the large Orinoco-Amazon turtles (the tartaruga and other species of the genus *Podocnemis),* and their eggs; iguanas and their eggs; and some lizards, such as the common tegú *(Tupinambis teguizin).*

The ecological impact of subsistence hunting has not been clearly established because there is not enough information on the number of animals and the frequency of hunting done by Amerindian groups. The most primitive Amerindians (like the Uaika, the Pieroa, and the Motilones of the Perija region of Venezuela) use the bow and arrow and the blow gun, with curare-tipped poisonous arrows. They know the jungle and are very good hunters, but they kill only when they need to eat and they live in equilibrium with the forest.

Subsistence hunting is probably not very harmful to the animal populations. But when people who live in communities, villages, missions, or mining and lumbering camps use modern weapons and dogs to track game, animal populations near rain-forest settlements can be diminished to a dangerous point. This can mean significant reduction of the species, not just of the local population. But subsistence hunting is not as bad as commercial hunting.

Commercial Hunting

The exploitation of the rain-forest fauna by commercial hunting is distressing. Some commercial hunting is for providing meat; dry meat to markets, fresh meat to restaurants. The more important and harmful commercial hunting is the killing of animals for their pelts and skins. This is very dangerous for the spotted cats of Latin America, as it has been for the leopard *(Panthera pardus)* and cheetah *(Acinonyx jubatus)* in Africa. Also threatened are the ocelot *(Felis pardalis)*, margay *(F. wiedii)*, and jaguar *(Panthera onca)* of the American tropics. As you know, the United States government has banned the commerce of the pelts of many endangered species. Pelts that still go to the fur trades, mostly in Europe, command high prices. Another mammal endangered in South America is the beautiful and large Brazilian or giant otter. Also endangered are some reptiles, such as the Orinoco crocodile *(Crocodylus intermedius)* and the American crocodile *(C. acutus)*, owing to commerce in their hides.

Another significant factor in the commerce in tropical wild fauna is the supplying of animals for pet shops. Many live animals are exported from the rain forests of South and Central America to be sold as pets in the United States and Europe. This is a big drain on animal populations. Hunters take monkeys of many species, as well as ocelots and margays, and toucans, parrots, and many other species of birds. The toll is large because many of these animals die awaiting shipment to other countries or in captivity with their owners; it is not easy to keep animals from tropical countries in captivity. Yet another big drain of live animals has been the demand for monkeys in biomedical research. As you know, several species of monkeys are important in studying human diseases, such as arteriosclerosis, viral diseases, and tropical diseases. The primate research centers in the United States, Germany, and Japan require a large number of South American primates for their research.

Destruction of Habitat

Now we may talk of one of the more indirect harmful factors, the disturbance or destruction of habitat. The tropical rain forest, as Dr. Gerardo Budowski has said, is complex, because there are many different types of rain forest. But it is a very fragile ecosystem. The diversified interactions among plant and animal species are destroyed by deforestation for grasslands and agriculture. Under these circumstances, many or all of the rain-forest animals will disappear because they are dependent on their peculiar habitats. If those unique habitats disappear, the animals will perish. And you cannot replace them. It takes thousands of years for the rain forests to build themselves up, as has been stated by Dr. Gómez-Pompa and Dr. Hamilton.

The destruction of habitat that is going on in many of the tropical rain forests and cloud forests in Latin America is endangering beautiful species of birds like the trogons, including the national bird of Guatemala, the quetzal *(Pharomachrus mocino)*. There are some odd species in Venezuela in the same group, and many other species that live only in cloud forests. Habitat destruction is also occurring in the lowland forests. The preservation of substantial tracts of high-quality habitat is indispensible to maintain the wildlife of the tropical rain forest of Latin America.

The Use and Preservation of the Rain Forest

What are the potentials for tropical rain-forest animals in the neotropics? Legally, their meat and other products can be provided for the rural population. This must be part of planned, multiple use of rain forests, based on a sustained-yield principle, as are forestry resources. We need more basic research and information. There has not been enough ecological research on the South American rain forests. Some work is being done in countries in Africa, South America, and especially Southeast Asia. Some is being done by local organizations, some by institutions like the New York Zoological Society and the Smithsonian Institution. The Smithsonian Institution now has a research program on wildlife ecology in Venezuela designed to compare the ecology of the animals of tropical rain forest in the Guatopo National Park with that of animals of the savannah ecosystem in the *llanos* (plains).

There is potential for tourism and recreation in the tropical rain forest. Many of you have had the privilege of being in tropical rain forests; they are of magnificent beauty, and it is pleasant to walk in them. Of course, one will see more plant life than animal life. Some people will be disappointed the first time because they do not see many animals, as they do in the rich African savannahs or even the savannahs of South America, which have large populations of ungulates and wildfowl.

But there is wildlife in the tropical rain forest. One sees the birds, and hears the strange voices of toucans, trogons, antbirds, and bellbirds. However, to be fortunate enough to see some of the mammals, one must know how to look for them because the forest is dense, and many of them are very shy, nocturnal, or arboreal. But one can see their tracks.

Thorough research has proved that the mammals of the neotropical humid forests are a very diverse group in form and function, and include many unique and primitive forms of considerable scientific interest. Hershkovitz (1972) recently reviewed the zoogeography and ecology of the Recent mammals of the neotropical region, which range in size from the tiny bristly mouse (*Neacomys* sp.), which weighs only 15 to 20 grams, to the tapir, which weighs between 225 and 300 kilograms.

The tropical rain forest is a magnificent world and a marvelous source of recreation that certainly will attract tourists. But tourism will have to be organized in a way that will not harm the beauties of the rain forests and their very fragile ecological balance.

What can be done? Dr. Gómez-Pompa has pointed out many important things. First, we must know more about the biota of the rain forest; in regard to wildlife, we know very little. We need basic ecological research; we need more biological stations. Venezuela is fortunate to have one in the cloud forests of Rancho Grande, situated in the heart of Henri Pittier National Park. It has been a center of research for many years. It was started by the famous Dr. William Beebe (Beebe 1949). We are planning to set up another station in the beautiful rain forest of Guatopo, one-and-a-half hours from Caracas, with the help of the Academy of Sciences and the Smithsonian Institution, and we are planning other stations in the Amazonian part of Venezuela.

One of the most important tasks is to preserve areas large enough to contain representative numbers of species of the varied and interesting rain-forest fauna. This can be done in national parks. In Venezuela, so far, we have some nineteen national parks (Table 1), many of them in cloud forests or lowland rain forests. Canaima (La Gran Sabana) National Park, which has been enlarged to 3 million hectares and is one of the biggest parks in the world (probably the second largest), is representative of ecosystems of the southeast part of Venezuela. We also need fauna refuges. We have two of these in Venezuela, one on the coast, very rich in waterfowl, and the other in the *llanos*. We need others in the Amazonia region. We also have

Table 1. National Parks in Venezuela

Park	Date Established	Area (in hectares)
Henri Pittier (Rancho Grande)	1937	107,800
Sierra Nevada de Mérida	1952	190,000
Guatopo	1958	92,640
El Avila	1958	85,192
Yurubi	1960	23,670
Canaima	1962, 1975	3,000,000
Yacambú	1962	14,580
Cueva de La Quebrada El Toro	1969	8,500
Archipiélago "Los Roques"	1972	225,153
Macarao	1973	15,000
Mochima	1973	94,935
Laguna de La Restinga	1974	10,700
Médanos de Coro	1974	90,280
Laguna de Tacarigua	1974	18,400
Cerro El Copéy	1974	7,130
Aguaro-Guariquito	1974	569,000
Morrocóy	1975	46,000
Alejandro de Humboldt (El Guácharo)	1949, 1975	15,500
Terepaima	1976	16,971
Total Area		4,631,451

Principal Sources: Galindo and Gabaldón 1974, Secretariat of ICUN 1975.

a sanctuary for the most unusual bird of the tropics, the *Guácharo* or oilbird, which nests in large caves in the eastern part of Venezuela and which was described by Alexander von Humboldt. A few days ago [27 May 1975—Editor], the forest surrounding the cave, including the bird's foraging areas, was set aside as the Humboldt National Park.

References

Anonymous. 1975. *The Use of Ecological Guidelines for Development in the American Humid Tropics.* New Series, no. 31. Morges, Switzerland: International Union for Conservation of Nature and Natural Resources.

Beebe, W. 1949. *High Jungle.* New York: Duell, Sloan, and Pearce.

de Schauensee, R. M. 1964. *The Birds of Colombia and Adjacent Areas of South and Central America.* Narberth, Pennsylvania: Livingston Publishing Co.

Galindo, Héctor, and Mario Gabaldón. 1974. *Parques nacionales y monumentos naturales.* Dirección de Recursos Naturales Renovables, Ministerio de Agricultura y Cría, Caracas.

Haffer, J. 1969. Speciation in Amazonian forest birds. *Science* 165(3889): 131-137.

Hershkovitz, Philip. 1972. Notes on New World monkeys. In *International Zoo Yearbook,* eds. Joseph Lucas and Nicole Duplaix-Hall, pp. 3-12. London: The Zoological Society of London.

Poore, Duncan, compiler. 1974. *Ecological Guidelines for Development in Tropical Forest Areas of South East Asia.* Occasional Paper No. 10. Morges, Switzerland: International Union for Conservation of Nature and Natural Resources.

Secretariat of IUCN, compiler. 1975. *World Directory of National Parks and Other Protected Areas.* Morges, Switzerland: International Union for Conservation of Nature and Natural Resources.

Editor's Note: Ing. Mondolfi has written a much more comprehensive discussion, in Spanish, of Venezuela's wildlife and its problems. Entitled, "Fauna silvestre de los bosques húmedos tropicales de Venezuela," it is published in Lawrence S. Hamilton, Julián Steyermark, Jean Pierre Veillon, and Edgardo Mondolfi, 1976, *Conservación de los Bosques Húmedos de Venezuela,* Caracas: Sierra Club-Consejo de Bienestar Rural.

Savannahs and Other Grasslands

A buffalo is coming they say.
He is here now.
The Power of the buffalo is coming;
It is upon us now.

I give grass to the buffalo;
May the people behold it,
That they may live.

  —Joseph Epes Brown,
  The Sacred Pipe:
  Black Elk's Account
  of the Seven Rites
  of the Oglala Sioux.
  Copyright 1953 by the
  University of Oklahoma Press.

These are the gardens of the Desert, these
The unshorn fields, boundless and beautiful,
For which the speech of England has no name—
The Prairies. . . . In these plains
The bison feeds no more. Twice twenty leagues
Beyond remotest smoke of hunter's camp,
Roams the majestic brute, in herds that shake
The earth with thundering steps—yet here I meet
His ancient footprints stamped beside the pool.

Still this great solitude is quick with life. . . .

  —William Cullen Bryant
  "The Prairies"

Terrestrial Ecosystems:
Savannahs and Other Grasslands

35. Conservation Problems of Savannahs and Other Grasslands

Kai Curry-Lindahl
United Nations Environment Programme,
Kenya

Introduction

As the white man's influence swept through the ranges of wild grazing animals all over the world—the prairies in North America, the *llanos* and *pampas* in South America, the savannahs and high veld of Africa, the steppes of Europe and Asia, the grasslands of arid Australia—indiscriminate land use and tremendous hunting pressure wiped out or greatly reduced incredibly productive natural resources. Through the centuries, man has increased the area of marginal lands all over the globe by changing fertile regions to semideserts and deserts. Today most grassland habitats of the world are in marginal land. But such land has a value, provided it is wisely used and managed instead of gradually destroyed. The great value of marginal lands is their productivity in the form of animals, many of which are large or medium-sized mammals with a great ability to convert poor vegetation to excellent meat. Many of these mammals do not drink water, or can do without it for long periods.

Man must learn to coexist with natural marginal lands and to restore the man-made ones to their former wealth, measured in terms of productivity. The conservatism of agriculturalists, who are unwilling to recognize that animals other than the few species hitherto domesticated are a potential resource, is distressing. In dealing with crop cultivation and livestock grazing on marginal lands, they always presume that wildlife is in conflict and must retreat, and that expansion of animal husbandry is always progress. This is not only nonsense; it is a highly dangerous philosophy that has already caused the loss of enormous areas of productive land. It is increasingly important to get rid of unrealistic and conventional use criteria for development of marginal lands. Effective utilization must include what these areas already produce: wild vegetation and wild animals for the production of meat, hides, and other products.

For Africa, this approach is particularly necessary because, with its diverse grassland fauna, it has a unique opportunity to develop the living natural resources in a constructive way. Savannahs were until recent times the principal food niches both for larger herbivorous and carnivorous mammals. In most African savannah areas, the great herds of grazers and browsers with their associated predators have been replaced by wandering cattle. This change has not only resulted in lower general productivity, expressed in protein quantities, but has also caused fundamental alterations in the vegetation and the soil, the disappearance of surface water, and the spread of diseases such as trypanosomiasis, rinderpest, and foot and mouth disease—in other words, a lowering of the environmental quality with long-term fatal consequences for man himself.

What Are Grasslands?

Grassland communities represent one of the major biomes of terrestrial environment, varying from tropical woodland savannahs to temperate steppes. Some are long-grass plains and others are short-grass prairies, independent of whether they are tropical or temperate. Grassland habitats include the savannahs of tropical Africa, the flood plains of Asia, the steppes of Ukraine, the high plateaus of Mongolia, the kangaroo grasslands of Australia, the *llanos, campos,* and *pampas* of South America, and the prairies of North America.

Grasslands pass by gradual stages into semideserts. In fact, many deserts all over the world are man-made; not long ago they were productive natural grasslands which have since deteriorated due to overutilization. Therefore, an important proportion of today's deserts and semideserts are still potential grasslands, and many grasslands are, in turn, potential forests; both can recover if they are not continuously overexploited.

Some shrublands and scrublands of the world may also be included among grasslands. This type of bush and brush country occupies a large portion of Australia and is also widespread in Asia, Africa, and South America. Large tracts of semiarid areas in North America support vegetation which is predominantly sagebrush (mainly *Artemisia tridentata*).

It is striking how productive grasslands once were; they were the basis for early civilizations. Unchanged savannahs, steppes, and prairies once supported biomasses that, in the form of animal protein, were the largest ever produced in terrestrial habitats. The Earth's largest array of large mammals converted green plants into quantities of meat that have never been reached since. This tremendous productivity was more or less constant and did not damage the vegetation or the soil. The grassland habitat was for millennia a perfectly balanced ecological system which offered optimal conditions for life.

Man himself has been intimately related to grasslands since he emerged from the forests of Africa about two million years ago. Despite this long experience, he has failed completely to understand the ecology of savannahs, steppes, and prairies. He has devastated few other biotic regions as thoroughly as these—changing the nature of the land through extermination, plunder, and the introduction of exotics; overcultivating and overstocking without regard for the carrying capacity of the environment. Even today he continues to convert large areas with productive soils and vegetation into useless deserts. In the Old World, the destruction has gone so far that in many regions it now appears to be irreversible. In the New World, man has had less time for destructive work, so damage there may be repairable through sound range management.

Today distinctive agricultural and/or pastoral economies characterize these grasslands, which are used for growing agricultural food plants and provide pastures for grazing animals. Has this change of land use led to an increased productivity on a sustained-yield basis and to the long-term benefit of man? The reply cannot be a general "yes" or "no," because not all grasslands respond in the same way to the influence of man and his livestock. Grasslands in temperate areas are less vulnerable to misuse than are those in subtropical and tropical regions. Also, the pressure on grasslands from increasing human and livestock populations varies in many parts of the world. However, where grasslands are continuously taxed by cultivation, fires, grazing, trampling, and other forms of utilization, they deteriorate at an accelerating rate. Unfortunately, this is what is happening on all continents and especially in subtropical and tropical regions. In these often arid and semiarid areas with erratic rainfall, a considerable proportion of the grasslands are marginal or submarginal. But they were, and in some places still are, extremely productive in the form of wild animals which convert the grasses and bushes to proteins.

In a world where food for human beings is increasingly scarce, it is a criminal waste to destroy the renewable natural resources of marginal grasslands, such as wild animals, and to replace them then with domestic stock, for example, which in a few years will change the land to useless deserts and lead to a collapse of former productivity. Tragically, this is what happens today on an increasing scale in the tropics and subtropics—precisely where the need for animal proteins is greatest. This does not mean that the temperate grasslands are without conservation problems, however.

Natural grasslands evolve where rainfall provides conditions that are intermediate between those of deserts and forests. Temperature and edaphic and biotic factors are also involved—for example, the grazing of ungulates. In subtropical and tropical regions, fire is an important factor in forming and maintaining grasslands. Natural ignition by lightning occurs, but is negligible compared to man-made fires. In Africa, man-made fires have swept the savannahs for thousands of years and have almost become a natural force. It has even been postulated that tropical savannahs are a product of man-made fires. It may be partly so at present, but the fact that Africa has evolved a well-adapted, specialized, and diversified savannah fauna, tremendously rich in species, clearly indicates that savannahs are an old biome that existed long before man. Therefore, his use of fire is in no way natural or positive.

Natural grasslands in both temperate and tropical areas represent a wide range of habitats and major regions, determined primarily by rainfall gradient, moisture, temperatures, soils, and herbivorous animals. Some areas are pure grasslands, without trees. Others are wooded savannahs and steppes, where, although trees visually dominate the landscape, grasses and bushes are still the most important food resource for its mammals. The soils contain large amounts of humus and are much richer than those in rain forests. As habitats, grasslands look simple, but in reality they are complex, with a number of limiting factors involved. If the ecology of a savannah, a prairie, or a steppe is not studied before man starts to exploit it, labor may be wasted, and the results of development schemes prove disastrous. This is a firm rule, too often ignored.

Secondary grasslands are mostly man-induced and, in the tropics and subtropics, usually maintained by fire. In temperate regions, secondary grasslands are often favorable to domestic herbivores, provided the rangelands are not overstocked; but in the tropics secondary grasslands never offer such a satisfactory environment to wild herbivorous animals as natural grasslands because the grazing spectrum of the former is unbalanced.

Africa offers a good example of how various natural grassland biomes succeed each other from north to south, crossing the continent from west to east in broad parallel belts, each responding to the different environmental conditions, chiefly rainfall, with its distinctive vegetation. Each of these vegetation belts succeeds the other by steady transitions. Usually there is no clear-cut border between them, not even at the Sahara Desert, which in the south slowly emerges as an arid bush steppe. But here also human activities have greatly changed the landscape by overutilization, which, north of the equator, presses the southern boundaries of the desert and of each type of grassland southward. This is the spread of deserts and arid lands in Africa. Climate alone cannot achieve the same effects.

Grasses of climax communities of natural grasslands are usually perennial, and some species live for many years.

Temperate and tropical grasslands often produce herbivorous mammals, from large-sized ungulates to small rodents, numerous birds, reptiles, and invertebrates. Many of these animals are not exclusively adapted to grasslands; they can enter and utilize forest habitats, an ecological flexibility that can be useful to man and should be taken into account in land-use planning.

Distribution of Grasslands

The areal extent of grasslands on the various continents is tremendous. Yet it is estimated that grasslands at present constitute about

27 percent of the world's vegetation, compared with 33 percent for forests and 40 percent for deserts. These figures are rough estimates because the exact proportions cannot yet be accurately quantified. Parts of some major deserts do carry vegetation but have been included in the desert percentage above. Vegetation types like tundra, montane plant communities above the timber line, and strips of chaparral (macchia), and mangrove have been excluded from the estimates. The surface of the Earth is three-fourths water. We have to safeguard the relatively small area available to terrestrial life. Half of the Earth's land surface is uninhabitable for human beings because it is too cold or too dry or too high.

More than half of Africa receives less than 40 inches of annual rainfall and more than one-third of the continent less than 20 inches. This means that a considerable part of Africa consists of semiarid, low-yielding rangelands which are marginal; in fact, at least half of its land area is marginal or submarginal for crop production. About 90 percent of Africa's savannahs are also marginal for livestock production, but, as mentioned above, they have a very high protein productivity when utilized by the indigenous herbivores which have evolved in these arid habitats.

Similar features also characterize grasslands of tropical and subtropical Asia, Australia, and South America, while other conditions prevail in temperate grasslands. The latter are less vulnerable to misuse.

The Deterioration and Desertification of the World's Grasslands

When looking at the present state of the world's grasslands and their functions within ecosystems, it is essential to bear in mind the important role the grassland vegetation plays in stabilizing the water cycle and preserving the soil, and that grasslands, in turn, cannot exist without either of these renewable resources. Grasslands facilitate evaporation and keep the soil porous. Humus and soil filter melting snow and rainwater, resulting in flows of crystal-clear streams and springs. Important salt nutrients both from the air and the rains, essential for vegetal and animal life, also flow to the soil and the subsoil water. These are just instances of a very complex pyramid of interrelationships. When man destroys one or several of the components, the pyramid collapses.

If the grassland vegetation is destroyed, the fertile soil layer disappears. When the vegetation dies, so do the animals. When the country thus grows sterile, the civilization existing there cannot survive. Many of today's deserts lie as silent memorials to once glorious civilizations that 2,000 to 3,000 years ago were based on extremely fertile lands supporting important animal and human populations.

The wild fauna are also a vital natural resource. Irreplaceable biological treasures—chiefly mammals and birds—are gone forever because their habitats were altered or completely destroyed, or the animals themselves were killed off without thought of their value as a natural resource.

It would lead too far afield to review here all the errors and crimes man is committing against the grasslands. Agriculture and livestock production have seldom been planned. For centuries, and around the globe, man has moved into virgin land, opened up the country, and despoiled the vegetation, which rapidly led to a ruining of soil, water, and animal resources; and this continues today, in an era when the main problem for humanity is the lack of proteins. It takes time to stop this destruction and even longer to reclaim what has been lost; but to restore many maltreated areas, the cost would be so large and the time involved so great that few politicians will find it worthwhile to take action and initiate a long-term commitment.

In Asia 2,000 to 3,000 years ago, many areas of the Middle East were fertile and supported flourishing civilizations, mighty kingdoms, and large populations. Today the same areas are pure deserts, where only the ruins of great cities provide reminders of former glory. In northern Syria alone, man-made deserts comprising more than a million acres (approximately 400,000 hectares) have buried more than a hundred cities. Nothing now but stone skeletons, these cities died because the land around them was killed by the inhabitants. Babylon met the same fate. So did the cities of ancient Mesopotamia and many other countries with a glorious past.

No natural deserts exist in India, but in recent times desertlike tracts have come into being—a consequence of human misuse of vegetation and soil through overgrazing by cattle. For the same reason, large areas of African semideserts and savannahs that formerly supported a wealth of wild animals have turned into sterile lands. There are many indications that the spread of the North African deserts is primarily due to the cutting of trees and overgrazing by cattle and goats.

The same phenomenon is to be seen in the southern Sahara, at the edge of the savannahs. Likewise, in many parts of east Africa, west Africa, southern Africa, and Madagascar, the deserts are spreading in previously fertile country with alarming speed. As soon as overgrazing by cattle, goats, and sheep occurs or farmers practice unwise agricultural methods, productive savannahs with comparatively rich vegetation and wild animal life rapidly disappear by erosion and give way to deserts. Dunes that have been kept in place by the vegetation for centuries revert to moving sand waves, killing everything in their path. River sources dry up, and so does the land along the dry river beds. The rains, no longer stored by the vegetation, sink deep down into the earth or run off, carrying fertile soils to the sea.

It is informative to recall what has happened to the North American grasslands. Many of the North American grasslands have also been transformed into agricultural deserts by erosion that followed cultivation practices performed without regard for environmental conditions. In addition, the cattlemen who invaded the western grasslands after the bison slaughters in the 1870s and 1880s made serious mistakes of overstocking. As early as 1885-1887, the combination of overstocking and drought depleted the grasslands; the ground cover was badly damaged by overgrazing and overtrampling, and 90 percent of the livestock was lost by starvation. The cattle-raising industry collapsed, but many cattlemen remained in the area. Then farmers invaded the Great Plains, encouraged by the homestead laws. Plows and cultivation damaged the grasslands even more, while the overgrazing effects worsened. The deterioration progressed, and after a few years the farmers gave up. The scars of their mismanagement are still to be seen. The natural regeneration of the Great Plains' grass fields is very slow, especially when they are today being grazed and trampled by cattle.

Misuse caused a serious depletion of most grasslands in the range country. During the 1940s, 55 percent of the range area was estimated to be so depleted as to have less than half its original grazing capacity. Another 30 percent was not so seriously depleted, but the forage was far less than the normal grazing value. An overall estimate gave 93 percent of the total range area as being depleted to some extent.

Intensified research coupled with conservation campaigns has led to an amelioration of rangeland and cropland conditions in the United States, but still the western plains are far from being as productive in proteins as they were a hundred years ago, when the white pioneers advanced westward. Today many ranchers on the Great Plains have realized the dangers of plowing up the short-grass prairie. They now understand that the economy of grasslands offers permanent values that do not exist in one-crop agriculture, with its economic ups and downs and vulnerability to parasites. After a hundred years of hard lessons, recognition has finally been given to the place of grass in the grasslands. Large areas of agricultural land have been converted back to grass country chiefly in the form of cultivated meadows and permanent pastures. This reconversion of plowlands to pastures is good conservation, a step toward restoration. An important animal industry can be built up if it avoids earlier mistakes. A complete restoration of the former tremendous productivity of the Great Plains would include the return of the bison. This species is less manageable than cattle, but there is space for both.

The pronghorn antelope in western North America provides an example of how productive rangelands once were. Since the first settlers arrived there during the nineteenth century, year-round hunting reduced the antelope population from an estimated 40 million to about 30,000 in 1924. Conversely, this same species has responded to conservation and management measures. From 1925 to 1955, the pronghorn recovered phenomenally and now numbers about 400,000.

The United States bases a considerable part of its economy on agriculture and is considered to be one of the richest agrarian countries of the world. By the end of the 1940s, most of the cultivated soils of the United States had lost more than half of their long accumulation of organic matter. Even worse is the loss of the soil itself. More than 100 million acres of the fertile top layer of cultivated soils in the United States have been destroyed by erosion. That is more than one-sixth of the country's agricultural land. Most of the American soils are ruined when six of the topsoil's approximately ten inches have been lost. This is happening over increasingly large areas. In many locations farmers are tilling the much less productive subsoil. This deteriorative process has taken but a few hundred years in the United States—less time than in any other area of the world.

During the dust bowl disasters in Colorado, Texas, Kansas, and Oklahoma in the 1930s, the loss of topsoil ranged between two and twelve inches. The windblown sand-dust damaged millions of acres of farms, and drifting dunes buried a vast area; land and people were ruined. It was a hard lesson, entirely caused by human misuse of the soil. The United States Soil Conservation Service estimates that at least 14 million acres of the Great Plains currently under cultivation should be returned to grasslands. Whether the plains have ever produced as much protein as when they were grazed by bisons and other wild animals is a moot question. In any event, one thing is sure: the soil did not deteriorate in that not-so-distant past, when millions of bisons roamed the North American plains.

In Europe, too, the productive life of the cultivated plains is slowly deteriorating. The spring floods have been canalized. They no longer nourish the meadows or provide storage water to the marshes or the soil, and the groundwater level is sinking.

South America's various types of grasslands have deteriorated through cultivation and pasture. There has been little or no effort to manage the grasslands, to let them periodically rest and recover. Lack of rotational grazing combined with overstocking has led to decreasing productivity; erosion; a sinking watertable; a drying up of springs, wells, and rivers; and the growth of man-made deserts.

In Australia, grasslands were grazed in a virgin state by wild mammals and birds for thousands of years and no erosion occurred; but when Europeans pastured sheep and cattle there, disastrous effects soon followed. Degenerated pastures and removal of surface soil by erosion now characterize all Australian grassland communities.

Great tracts of the world's arid and semiarid regions are covered by bushes, shrubs, and small trees. Some are evergreens or partially so, but the larger part is deciduous and often thorny. There is also a thin cover of grass or at least patches of grass. It is a matter of choice whether one refers to such bush and shrublands as semideserts or as arid savannahs, steppes, or prairies. But in many regions of the world, particularly in Africa, Asia, and Australia, and also in the Americas, this kind of biome is distinguished as a category in its own right. However, the dividing line between bushlands and semideserts or arid savannahs is not always easy to draw. Bushlands under the strong influence of fires and other human factors become transformed into semideserts in the same way that savannahs deteriorate to bushlands.

Natural bushlands have peculiar conservation problems and should be managed accordingly. They are relatively productive, which is clearly indicated by the larger mammals living there. However, they are extremely vulnerable to misuse, which easily upsets the interplay of moisture gain and loss. Evaporation is pronounced in regions with low relative humidity, dry winds, hot sunshine, and high temperatures. Therefore, it is important that the existing vegetation be able to make use of the rainfall in the right way. In natural habitats it generally does, but where the vegetation has been altered or partly destroyed, precipitation in the form of light showers may completely evaporate before it can sink into the soil and be used by plants; or torrential rains may wash away immediately in a surface runoff. In both cases the results are gradual impoverishment by desiccation and lowered productivity.

Many data indicate that semiarid grazing lands are being destroyed at an accelerating rate. This conversion of productive areas into deserts is pronounced around the edges of the Sahara and the deserts of the Middle East. It is also marked in other areas of Africa, Pakistan, India, Mongolia, and Australia. It can even be observed in western North America.

Desertification in northern Africa has been much discussed during recent years. There are two schools of thinking: one advocates that expanding aridity is induced by changes in climate; the other that it is mainly man-made, the result of overgrazing, overtrampling, overcultivation, wood-cutting and other removal of vegetation for fuel and shelter, and burning. More and more data from investigations in Africa during the last decades indicate that the latter view is in most cases correct. The recent climate may have contributed to this state of affairs but could not alone have achieved the same ecological degradation. The human and livestock misuse of semiarid grasslands has gradually led to a complete loss of the vegetation cover, leaving the soil bare to wind and water erosion, and dunal movement. Recent discoveries show that surprisingly large amounts of airborne soil dust from the Sahara and Sahelian zones are transported long distances into and over the Atlantic.

Grasslands of yesterday and deserts of tomorrow in various transitional vegetation stages exist in many areas of the world. For example, some authors believe that present North American deserts like the Sonoran and Chihuahuan used to be grasslands. Man-made pressures on grasslands increase due to the population growth. This is in itself a difficult socioeconomic ecological problem, but if climate were responsible for the disappearing grasslands, there would be no recourse.

In the Saharan desert and subdesert, the destruction of vegetation and wild animals has been going on for thousands of years; but in modern times automatic guns and motor vehicles have brought a number of mammalian species to the verge of extinction. Desert and subdesert antelopes, such as the addax *(Addax nasomaculatus)*, the scimitar-horned oryx *(Oryx tao)*, and four species of gazelles *(Gazella)* represent, in these harsh habitats, formidable resources converting the poor grasses to excellent meat without destroying the supporting environment—in contrast to the destructive cattle and goats, which tend to be the real desert-makers. It is not easy to establish efficient nature reserves for the nomadic antelope, which wanders great distances following the erratic rainfall. Therefore, the only way to save the wild desert herbivores must be through inter-African cooperation and strict hunting regulations, coupled with education and information schemes.

It is chiefly in Mauritania, Mali, Niger, and Chad that the last remnants of these desert antelopes are to be found. The scimitar-horned oryx, the addax, and the various types of gazelles are remarkably well-adapted to arid conditions. They can live for long periods without water and some of them are probably capable of existing without drinking water at all, although most of them drink surface water when it is available. Also, gazelles probably play an important role in the germination of acacia seeds and, therefore, help disperse vegetation and combat desertification. These features emphasize how productive these desert herbivores are as a protein resource, and how efficiently they make use of desert habitats. Yet they are gradually being exterminated and nothing else can replace them—an extremely unwise land use. It is necessary to take immediate and energetic measures to save the desert antelopes before it is too late; this is one of the most important conservation problems of Saharan Africa.

South of the Sahara Desert, various belts of dry thorn-scrub steppes and grassland and woodland savannahs stretch across Africa from the Atlantic Ocean to the Red Sea. For the countries in these vegetation belts—Mauritania, Senegal, Mali, Upper Volta, Niger, Chad, and the Sudan—the main conservation problem is desertification.

The so-called "drought catastrophes" which hit Sahelian and sub-Sahelian Africa are primarily a consequence of the environmental degradation caused by man and his livestock. The present land use in arid and semiarid Africa is an open invitation to drought calamities on the scale which we witnessed in 1973. This disaster has been labelled an emergency, but in reality it is the present overgrazing and misuse of water, soil, and vegetal resources, that is the real catastrophe. The drought which followed is merely the second stage of the same environmental change. Next time, Somalia or Kenya will be hit. Indeed, it seems clear that if the countries concerned do not stop the evil at the root, they will be continually stricken by drought disasters, with increasingly serious effects.

It may appear that I am advocating a philosophy based on the assumption that pristine or primeval conditions in nature always are the most productive and that they should therefore suit modern man; that is not the intention. I do contend, however, that an ecological land survey is a vital necessity before any decisions on land use and development are made. The long-range, maximum sustained yield of a given area and a development program directed toward obtaining that yield must be established. When seeking the answer for an area already exploited and therefore probably already destroyed, it is important to go back and retrace the area's former characteristics and the dimensions of its former productivity. Without doing so, its ecological potential will be unknown, and accurate conclusions cannot be drawn.

The Productivity of Marginal Grasslands

Africa serves as an instructive example of the grassland productivity of animals—specifically, large and medium-sized mammals. Only a century or two ago, the mighty savannahs of Africa provided scenes we today find difficult to visualize. Numerous species of large and small antelopes, mixed with giraffes, buffaloes, rhinoceroses, and elephants, lived on the various kinds of savannahs. These plains were regions of biological equilibrium—highly productive climax areas created by nature through countless ages.

In this highly specialized ecological system, each organism was a necessary element in the perpetuation of the region. Different animal species made efficient use of the whole array of available niches within the habitat. The various species of hoofed animals grazed and browsed together, choosing different kinds of bulbs, twigs, leaves, and fruits as food. Selection had eliminated competition, and in this way the effect of grazing was evenly distributed. With the exceptions of species such as the eland, the oryx, the impala, the gerenuk, Grant's gazelle, and Thomson's gazelle, which need little or no water, herds congregated on the savannah regions around lakes, rivers, springs, and small water holes—for not many savannah species can live for long periods without water.

During dry periods, most of these hoofed animals moved to other regions which had recently had rain and where the grass was plentiful and green. And so it went, all the year round. The hoofed animals followed the rain, changing pastures regularly, and as a consequence the grazing never destroyed the vegetation, which after the next rainy period was as fresh and green as ever.

The African savannah could support huge numbers of large meat-producing animals, which in turn were kept under control by the environment itself with its drought spells, predators, parasites, and diseases—the latter two operative only when the host animals are in a weak condition, often brought about by a lack of nutritious food due to extreme dry spells. Although the wild hoofed animals a century or two ago were far more numerous than the tame cattle on the plains today, they did not destroy the land even in the course of hundreds of thousands of years. But man, along with his goats and cattle, has in 200 years managed to transform large parts of the flourishing African savannah into a desert shadow of its former splendor and wealth. Cattle have probably existed in tropical Africa for 500 to 600 years, and in some coastal regions even longer; but it is only during recent centuries that the destruction of the land by man and his domestic animals has had such terrible consequences.

What has been said about the fertility of the African savannahs and their incredible number of wild animals may seem like a figment of the imagination, but it is not. Europeans visiting both the tropical African savannah and the South African veld have left eyewitness accounts of them. And in various parts of the savannahs, the national parks and nature reserves remain as oases, rich in wildlife. The savannahs in the Virunga National Park of Zaire, for example, produce more meat per square mile in the form of large hoofed animals than any other region in Africa, and still the land is not degenerating. Outside the parks and reserves, savannah fauna can still be found in some regions not yet destroyed by domestic animals, owing to the fact that African sleeping sickness ("nagana") spread by tsetse flies has prevented the breeding of tame cattle in those areas.

The Eurasian steppes too were inhabited in the past by wild herds of large mammals, mainly horses and asses, adapted to this special habitat. Much of the European steppe has been cultivated and the large mammals exterminated. On the Asian steppes they have been depleted by hunting. In both these areas this has meant a loss of a valuable resource, because many of the steppe mammals represented an important conversion of poor vegetation into meat and proteins.

The saiga antelope of the dry steppes and semideserts of the USSR is a spectacular example of how a species indigenous to an area can prove a source of revenue. After the saiga was almost exterminated in the 1920s, sound management raised the population to about 2 million. About one-quarter of these animals live between the Volga and the Ural rivers, where the population is about four times as dense as in Asia. In this region alone, professional hunters kill between 150,000 and 350,000 saigas annually, representing about 6,000 tons (approximately 5.4 million kilograms) and more than 2 million square feet (roughly 170,500 square meters) of hides, edible and inedible fat, and horns. Despite this killing, the number of saigas is increasing. In view of the sparse vegetation of the dry steppes, this productivity is astonishingly high. Production of saiga antelopes has proved to be the best form of land use in these arid regions. In the same way, the prairies of North America; the arid plains of Australia; the semideserts, savannahs, and forests of Asia; and many regions of Europe could profitably produce wildlife. With a conservation and management policy that accommodates the natural patterns of indigenous wild animals, many areas could certainly reach a higher annual productivity and economic yield than they achieve under present conventional forms of land use. Similar examples of high natural productivity could be drawn from marginal grasslands in other parts of Asia, in Australia, and in North and South America.

The productivity of marginal grasslands (as well as that of the grasslands themselves) is an ecologically fragile quantity. Therefore, it is of great importance that the planning and management of land use in marginal areas conserve natural productivity and not upset the conversion cycle of the system. In tropical rangeland, the present tendency to build up livestock populations artificially by providing water and controlling disease, without any consideration whatsoever to the land's carrying capacity and the delicately balanced interaction of rainfall, soils, vegetation, and wild animals, has led to numerous ecological disasters. They have repeatedly caused famines hitting not only domestic animals, but also man. Temperate grasslands react differently. For centuries, and in some areas even millennia, grasslands with fertile soils have sustained agriculture and animal husbandry.

Many former European rulers in Africa and Asia—and those in charge of present technical aid schemes to developing countries—assume that successful agricultural and stock-raising practices in their temperate homelands are also the best forms of land use for tropical soils and grasslands. This is a misfortune from which Africa, in particular, has suffered for more than 100 years. This mistake has led to disastrous failures, irreversibly ruining enormous areas of the continent and incurring the loss of great sums of investment capital.

The Problems of Nomadic and Seminomadic Pastoralism in Tropical Grasslands

There are planned and current projects for the settlement of nomads in Africa. Is this wise? The way of life of nomads in marginal, arid and semiarid areas is, in fact, an ecologically sound use of poor rangelands, as long as the livestock does not exceed the carrying capacity of the area. Originally, pastoral nomadism had little effect on the vegetation and wild animals in unspoiled habitats. There, the wild grazing animals changed pastures regularly and did not destroy the vegetation.

Pastoral nomadism of the past was a rational use of land, though less productive of animal proteins than the wild herbivores also on marginal lands. Serious damage to the habitat did not occur until veterinary practices increased their numbers so that the grazed vegetation could not recover despite the regular change of pasture. Ever since, this deterioration has continued, preparing the ground for desertification.

The important ecological lesson is that ranching on marginal lands is bound up with movement if habitat is to be conserved, and livestock has to be numerically adjusted to the carrying capacity of the land. Nevertheless, it is far inferior to the protein productivity of wild grazing animals which, despite a higher biomass than livestock, do not destroy the environment on which they subsist.

The African diet is in many parts of the continent deficient in animal protein. Domestic livestock cannot satisfy the protein needs of the continent's people because the animals are not adapted to the tropical environment and so destroy its productivity. Wild animals are not subject to environmental limitations; on the contrary, they are a product of the environment. Therefore, the utilization of wild animals as a food resource is the best and most rational use of African marginal lands, both for economic and ecological reasons. The results of experience all over tropical Africa clearly indicate that a new approach to land use planning, management, and utilization based on wild animals is necessary.

The Use of Fire

For millennia, fires have swept tropical grasslands, particularly in Africa, where white settlers have followed the example of the Africans. Also in Asia, New Guinea, Australia, and South America grasslands are regularly and deliberately burned. The use of fire in grasslands is a highly controversial method. Ecologists are, in general, against burning; agriculturists are for it. Before discussing this important aspect of grassland ecology, I think it necessary to say that the long-term environmental effects of continuous burning are usually impoverishment of species and increased erosion.

Farmers start bush fires regularly in order to burn off dry, old grass, to produce ashes valued for the nutrients they add to the soil, to kill weed seeds, or simply to clear the savannah for planting crops. Livestock owners burn dry grasses so that fresh new grass will shoot up in the ashes, providing pasture for cattle, or to cover their tracks when stealing cattle. Hunters also burn the dry grass because the new green shoots attract grazing antelope, which makes them easy targets for snaring. Or the hunters start fires in order to drive out animals for hunting. Honey collectors, or honey hunters as they are called in Africa, regularly set fires even in areas not inhabited by man. They use fire as a tool for collecting honey and just leave the fire burning after them.

African savannahs are usually burned once a year, in some areas twice. In rare instances some parts may escape burning for a year or two. The regular burning produces a kind of fire climax vegetation, where fire-resistant, thick-barked trees, bushes, and perennial grasses dominate. Normally, fires maintain savannahs, but even without burning there would still be savannahs because grazing animals continually crop the vegetation. Fires may be useful as a management tool locally, and occasionally their short-term effect on the vegetation may also be beneficial. But in the long run, annual burning is detrimental to the soil, the vegetation, and the animals. It reduces the nitrogen content and the organic matter in the topsoil, impoverishes the vegetation, and destroys or wipes out a number of animals in and on the soil. It also leaves many animals without food—browsing mammals, for example. In short, indiscriminate annual burning slowly reduces the general fertility of the area. Controversy about the effects of burning tropical grasslands is due to the lack of knowledge of fire ecology. Well-planned and executed research programs need to be carried out on all the various vegetation types over a period of years.

Europeans as well as Africans often claim that burning is necessary to remove dry grass and help fresh grass to grow. But seldom does dead grass inhibit the growth of its offspring. In reality, many factors other than fire destroy the mat of dead grass. It is trampled and crushed by many animals, eaten by rodents and termites. It decomposes, contributing to the humus, like all organic litter. As a part of the soil formation, it is, like dead trees, host to numerous tiny animals living on or inside the dead stems. The raging fires that annually strike African grasslands deprive the savannah habitats of many life forms that could exist there; since they destroy so many of the organisms that have colonized the area since the previous fires, the procedure must be regarded as unnatural—even if it has been followed for centuries. Still, burning continues to this day, and gradually affects larger and larger areas. Moreover, the fires are as a rule not controlled: they often occur at the wrong time of the dry season and of the day and, as a consequence, destroy indiscriminately.

Ecologically and biologically, annual burning is a waste of resources, a destruction of organic matter in many forms—decomposing material, stored seeds, eggs and pupae, growing plants, and animals. Fires prevent diversity of communities and habitats. At the least, large areas in nature reserves should never be burned artificially, but rather should be protected from the indiscriminate use of fires. Planned fires can stop accidental fires. Firebreaks can do that job too, but they are expensive. Moreover, regular burning at the edge of savannahs, semidesert areas, and deserts enables the latter to advance and conquer productive grasslands. Similarly, fires sweeping through the edges of forests force them to give way to savannahs; and fires in bushlands and thickets likewise create savannahs.

In Madagascar, 90 percent of the fires that sweep the country every year are set by cattle owners. Unfortunately they don't care if the fires spread uncontrolled over enormous areas and burn the few remaining natural grasslands and forests and even planted forests. The burning in Madagascar is the prime cause of the country's denudation and the rapid degradation of its natural resources. The repeated fires, combined with the trampling by cattle, have made the soil sterile and lifeless over very large areas.

It is impossible to generalize the pros and cons of burning in tropical areas. Each area has to be considered separately, relative to the kind of land use to which it is put. But there is no doubt that for natural communities repeated artificial burning is detrimental to productivity, particularly since most fires are excessive. The conclusion inevitably is that grass is more useful when grazed than when burnt.

Water Development in Marginal Grasslands

The creation of artificial water supplies usually creates more long-term environmental problems than it solves. Sinking boreholes and building dams to increase the water supply in arid and semiarid grasslands or to provide waterpoints in places where no permanent surface water exists are only short-term solutions; they do not solve environmental problems, but usually aggravate them by encouraging animals and people to remain in an area longer than the pastures can naturally maintain them.

Schemes for artificial water supplies are well-intentioned; they are often proposed, financed, and constructed by aid-giving agencies of developed countries as assistance to developing ones. These projects have usually been planned and carried out by engineers, and perhaps sometimes in consultation with geologists, with the aim of bringing water to semiarid regions. Advice from ecologists is rarely sought and little or no attention is paid to the ecology of the area concerned and to the project's long-term consequences.

Natural water and food supplies govern the distribution of grazing animals; in semiarid areas, they are in a delicate balance with the grassland environment and can forage in an economic, nondestructive say. Where surface water is not available throughout the year or not at all, the vegetation is also meager, but it nourishes many wild herbivores which do not need water; and during the rainy season, wild and domestic animals can feed on the grasslands. When an artificial water supply is installed, it functions as a magnet for people and domestic animals, but the pastures are as poor as before; they cannot support the increased use and so disappear. It happens too often, particularly in Africa, that water development schemes are planned so badly that large areas of grasslands experience lowered productivity followed by conversion to deserts. A network of water boreholes in a semiarid region already overgrazed by livestock is one of the most rapid means for creating deserts. It is better to leave the area as it is and to use water development funds for meaningful rangeland restoration programs.

Tsetse Flies as a Component in the Grassland Ecology of Tropical Africa

Tsetse flies and trypanosomiasis are an important part of the ecology in tropical Africa. Trypanosomiasis—usually called "nagana" in animals and "sleeping sickness" in human beings—is caused by microscopic flagellates *(Trypanosoma)* of several protozoan species, of which at least two and probably three are pathogenic to man. The trypanosomes are carried by the tsetse flies *(Glossina)* of which there are numerous species. Two of them (*G. palpalis* and *G. morsitans*) are vectors of the human disease. Wild animals are normally not affected by trypanosomes—although they serve as hosts for them—because of adaptation through natural selection; man and cattle are not immune, despite two million years of human presence in Africa.

Traditional tsetse-control measures include bush clearing to alter the habitats of the flies; wholesale or selective elimination of the hosts—wild mammals which carry trypanosomes but are themselves immune to nagana; or spraying enormous areas with toxic chemicals. Apart from the tremendous ill effects that these techniques have on the productive environment, they have not resulted in a permanent solution to the pest problem.

Usually, tsetse flies are looked upon as an evil. Yet, thanks to these insects, considerable quantities of wild mammals still exist in the tropical parts of Africa. The presence of tsetse flies on bush and woodland savannahs and in forests has been an obstacle to human settlement and to the introduction of livestock on marginal lands. This land has thus been saved from the destruction that usually follows in the wake of grazing and trampling by herds of cattle.

Since the marginal lands of Africa are unsuited to cattle raising, it is unprofitable to eradicate wild animals merely because they are hosts to trypanosomes; it is, in addition, inexpedient to clear the large areas that are their habitat. Only in areas with established agriculture may it be economically justifiable to reduce the tsetse. Elsewhere, a far greater economic return can be obtained from marginal lands by leaving the tsetse there and utilizing the wild animals as a food resource or as inhabitants of a national park or equivalent reserve.

It is often claimed that the tsetse is so effective at transmitting nagana that it invariably appears among cattle whenever they share a habitat with infested wild mammals. Yet little is actually known of the role wild animals play in the transmission of trypanosomes to cattle. In many areas, the expansion of the tsetse is connected only to the dispersal of domestic cattle. For example, the first decade of this century, the western Tanzania population of this fly started to expand northward into the savannahs of Ankole in Uganda, which were grazed by cattle. By 1958, the whole of the eastern half of Ankole was infested. In this case, it was not the wild ungulates that spread nagana to new areas.

The Interaction between Grassland Vegetation and Herbivores

The great array of ecologically well-adapted species of herbivores on the African grasslands merits attention. These animals function in naturally maintaining stable ecosystems and producing a high biomass, with more than thirty species coexisting in some areas. In most cases, the lands are marginal with poor soils and vegetation cover. Optimum conditions for most domestic livestock are narrow, and environmental conditions adverse to their production prevail over large parts of the world. For this reason, repeated attempts have been made to introduce or improve livestock production on ecologically unsuitable lands, but at great cost and with very little economic return. Artificial environmental changes made to meet the requirements of domestic livestock have usually overlooked the inability of these animals to adapt to adverse climatic conditions. Moreover, the detrimental effects of these changes on the land mean loss of productivity and destruction of habitats. Using animals well-adapted to the original conditions is economically more rational. As a rule, such animals already exist in the area involved; they have evolved as products of the environment.

Studies on productivity, biomass, and carrying capacity show that on many rangelands, particularly in tropical countries, wild animals make more efficient use of the environment than do domestic livestock, and that it is possible to maintain high production on a sustained yield basis without adversely affecting the habitats and the ecosystem. The protein potential of wild animals on vast areas of the globe is much higher than that of livestock. Several factors are responsible for this. The prime factor is the marvelous adaptation of wild animals to local conditions through an interplay with the environment which affects diet, nutritional efficiency, and water requirements. Other factors are the reproduction, mortality, and growth rates; average liveweight; meat yield in proportion to body size and percentage of fat content; resistance to drought, starvation, parasites, and diseases; habitat utilization, and so forth. In all these respects the wild animals are far superior and more efficient, both individually and collectively, than domestic livestock.

This means that African ungulates utilize poor natural grazing lands so efficiently that they reach a biomass five to ten times higher than that of cattle in excellent, heavily fertilized, artificial pastures. Moreover, the wild ungulates constantly produce a high yield of meat without destroying the soil and vegetation upon which they subsist, while cattle in similar habitats destroy the environment by overgrazing and trampling. In many African countries, the wild animals are the only, or the principal, supply of animal proteins. Despite the depleted wild animal populations, the consumption of so-called "bush meat" is remarkably high in a number of countries.

The foregoing shows that it is not unrealistic to envisage a rational exploitation of wild animal populations in Africa and Asia for protein production. Economically and ecologically, this is the best form of land use on marginal and submarginal lands. This emphasizes the great economic importance of wild animals as a source of protein. In addition, the value of their hides is high. It is tragic for Africa and Asia, and for mankind in general, that lands that could be so productive have been and are being rapidly destroyed by inappropriate use.

The greatest single danger to Africa's wild animals comes from the herds of ecologically unadaptable, tick-ridden cattle competing for grazing space on semiarid pastures. The cattle threaten the animals by threatening the land itself, and ultimately they threaten man. Will governments realize in time that a high production of proteins can be constantly maintained from wild animals on lands that deteriorate under other forms of use?

While Africa is exceptionally well-stocked with highly productive wild herbivores, other areas of the world also can produce wild animals as an additional source of wealth. Many parts of Asia, Australia, and South America, as well as the temperate regions of North America and Europe, can be used more rationally than at present.

In Australia wild animal cropping schemes have been economically successful. In recent years the country has exported weekly fifty-five tons (50,000 kilos) of meat from three species of kangaroos.

Only a few African countries have thus far initiated large cropping schemes, and relatively few species have been utilized. The potential of African ungulates is so high that cropping production can be increased many times when refrigerated transport facilities and other technical facilities become available.

Cropping of wildlife can also be carried out in the form of game ranching, and in some areas game and livestock ranging can be combined economically. In South Africa and Rhodesia game farming is a common practice. Many ranches produce higher quantities of first-class meat at lower cost after they have shifted from livestock to wild animal production. The costs of maintenance and harvesting are negligible, and the profits of the land owners increase considerably. Some ranches in Rhodesia base their production on sixteen or more species of wild ungulates—one reason for their success.

Ranchers have found that soil and vegetation grazed by wild animals have recovered from previous misuse. In Transvaal, a rancher discovered that 8,000 pounds (approximately 4,000 kilos) of cattle per square mile (259 hectares) had degraded his land; 17,600 pounds (approximately 8,000 kilos) of springboks and bushbucks on the same area enabled the land to recover and yield greater profit. In Transvaal about 3,000 game ranchers were operating successfully in the early 1960s, many of them on land previously depleted by livestock. In the 1970s this figure greatly increased. In other parts of Africa the potential for such game farming is even greater.

Cropping and game farming have produced both fresh and dry meat that have found markets overseas. Canned game meat too—like that of the antelope and hippopotamus—has been exported and accepted in Europe and America. However, it is essentially to meet the domestic protein needs of African countries that game meat is so invaluable.

In addition to the protein value of wild animals in marginal lands are the recreational and educational values so well displayed by existing national parks and equivalent reserves, which in some African countries are the basis for a flourishing tourist industry.

Management and Utilization of Grassland Ecosystems

Management problems of the environment are as varied and as changing as the habitats themselves; they depend on the type of land use and the pressure of its human population. Basically, the object of renewable resource management is to keep the areas concerned optimal, diversified, and harmonized with the environment in order to respond to man's needs. A sound management policy also requires some basic conservation concepts, including ecological and biological considerations and a respect for native plants and animals, which should always have priority over exotic species. The development and application of management measures to obtain the greatest public benefit from wildlife, or any other natural resource, should never be allowed to go so far as to threaten a species or subspecies with extinction.

Modern management of the environment must function by planning ahead, foreseeing the future's tremendous human pressure on habitat and wild animals. From centuries of human mistakes and land misuse, it can learn to reduce as far as possible man's detrimental impact on the environment and wildlife—much of it irreparable.

Nevertheless, management measures should include environmental restorations designed to put back the natural interactions that lead to wildlife fertility in a healthy landscape.

The ecosystem dimension is important in all natural resources management. It is desirable to prevent management techniques from conflicting with natural ecological processes. Simplification of grassland ecosystems and uniformity of vegetation are the results of monoculture, which may be initially profitable but in the long run often leads to degradation and loss of stability. Accelerating human population growth is an increasingly important factor influencing ecosystems negatively. It is imperative that this relationship is not overlooked in human ecosystems management and in planning the use of renewable natural resources at national or regional levels.

Ecosystem management is the antithesis of unplanned exploitation, which, so far in a reckless manner, has characterized man's "development" of the environment. An ecologically-based and well-explained land use policy will help the citizens of a country to understand their roles as a part of the world environment. If they appreciate this, they will also understand why renewable natural resources should be managed so that nothing irreplaceable is destroyed.

Grasslands and the Human Population Increase

In my opinion, all attempts to plan and manage grassland resources, as well as all other renewable natural resources, would not lead to man's progress as long as the human population continues to grow at the present rate. This is the basic conservation problem on which all other serious conservation problems depend. The key to the global population issue is the acceptance by each nation of the responsibility for sound national population policies by which the level, growth, and distribution of its population are related to its available resources, to its capacity to manage these resources according to ecological principles, and to the quality of life to which its people aspire. Equally important is the need for each nation to accept the corollary responsibility to assure that its population's demands on the natural resources and environment beyond its national boundaries do not impair the rights and interests of other nations.

An Action Program for Grasslands

We live in a hungry world, and we cannot afford to let our crops become ravaged through the destruction of the soil in which they grow. Yet this is just what we have been doing for centuries with wild animal crops all over the world. They have been wasted and plundered to the point where most of the once miraculously productive wild lands are now dead or dying. This is one of the darkest chapters in our history and has provoked an ecological crisis threatening man.

Nowhere has the prosperity of our planet been so convincingly illustrated as in its great plains. Unmatched among the Earth's terrestrial habitats, they continuously produced large herbivorous animals in quantities that today seem unbelievable. Yet this situation existed in some regions as recently as 50 or 100 years ago—a time when a man might have watched by the hour as a herd of bison thundered across the prairie of North America.

Few people then fully understood that the pristine landscape, with its tremendous number of large animals, was a dynamic expression of intact habitats that had evolved to ecological perfection and a maximum conversion rate—a kind of giant symbiosis between vegetation and animals. Where the landscape has not been irreversibly undermined by careless habitat destruction, this great symbiosis can be restored. The very idea is a challenge because it offers humanity the possibility of repairing for future generations what it has damaged. The strongest argument for such a scheme is the necessity to counterbalance the factors threatening human survival. The increasing demand for food cannot be satisfied unless governments and their people understand that wild grasslands and wildlife rank among the most important renewable natural resources.

Perhaps the symbiosis spoken of here in symbolic terms is not only a relationship between vegetation and herbivores; in the long run, man might well be involved also, whether he uses wildlife as a source of food or as recreation to relieve the tensions of life. National parks and equivalent reserves are essential components in any system of land use in the grassland biome as well as in other biomes, not only because they constitute a wise land use in the form of recreation, education, and research, but also as sample areas for comparison with regions which have been modified by man. Too frequently, genetic resources are completely forgotten in the planning and development of natural resources.

The simple fact is that large areas of the world's rangelands are overused. Many of them are continuously deteriorating due to overgrazing and overtrampling, and cannot sustain sedentary or semi-nomadic pastoralism. The time is at hand to consider a global plan of grassland restoration before deterioration has gone beyond the point of no return. Long-term programs are needed for repairing and rebuilding lands that have been seriously damaged by the mistakes of the past. These programs will be costly and will require cooperation among nations. In addition to resulting in a higher quality and quantity of resources produced, such programs may also lead to a better way of human life—one in harmony and cooperation with the environment.

36. Wildlife of Savannahs and Grasslands: A Common Heritage of the Global Community

Norman Myers
Nairobi, Kenya

Introduction

Tropical savannahs, especially their grasslands, are likely to be fundamentally modified before the end of this century. Some savannahs sustain the most spectacular array of large mammals left on Earth. African grasslands, in particular, support exceptional throngs of wildlife. Tanzania's Serengeti Park alone contains 2 mil-

---

This paper is presented entirely in the author's capacity as a private individual. It should not in any way be construed as representing the views of the Food and Agriculture Organisation.

lion wildebeest, zebras, and gazelles, many thousands of other herbivores, plus their associated predators. Additional large numbers of wildlife roam outside the protected areas. Yet present patterns of land use in Africa suggest that much of the wildlife outside the parks may be diminished by 1980, that many parks themselves may be impoverished by 1990, and that few parks at all will survive until the year 2000. Despite the best efforts of conservationists who have established an exceptional network of protected areas and are now doing a great deal to stem a tide of competing activities in wildland habitats, it is plain that development patterns and population pressures could eventually overwhelm many present measures for conserving wildlife in Africa. The situation is much the same in savannahs elsewhere in the world.

Savannah Zones

Savannahs may be taken to include grasslands of various forms in tropical and subtropical zones. Some of them are open plains, while in others trees and shrubs sometimes cover as much as 30 percent of the ground. Some savannahs are characterized as tall-grass woodlands, some as grassy plains with scattered wooded patches, some as short-grass steppes, some as dry-scrub or bush country, and some as floodplain savannahs. Some of them are edaphic in origin, some are degraded forest, some are fire-induced climax communities, and some derive principally from climatic determinants. Many of them arise from a mixture of these and other factors. Generally, savannahs do not receive more than 75 centimeters of rainfall and may receive as little as 25 centimeters. Thus, savannahs span a spectrum of phytoclimatic forms from moist to semiarid zones.

In South America, *llano* savannahs, or moist grasslands, exist mainly in Venezuela; *planalto campos*, or dry brush country, principally in southern Brazil, Paraguay, and Argentina; and fertile *pampas*, or flat grassland plains devoid of trees, in Uruguay and Argentina. In Asia, savannahs proper are confined chiefly to subtropical parts of India and China, while semiarid steppe extends across parts of Afghanistan and Iran. As in the Neotropical Realm, Asian savannahs have lost most of their wildlife through man's competitive use of their habitats and because of the inability of institutions to reflect mankind's interest in this common heritage.

It is in Africa that savannah wildlife retains a significant portion of its former diversity and numbers. There, savannah is the prevalent formation. It covers about 40 percent of the continent south of the Sahara and contains a great variety of biotic types, including the shrub belt of the Sahel, and semiarid sectors bordering the Kalahari in Botswana and Namibia; dry-wooded savannahs of the Guinea-Sudan belt north of the equator, matched by a similar zone south and southeast of the equatorial rain forest; thorn-bush country in eastern Africa; and open grasslands in Kenya, Tanzania, Uganda, and parts of southern Africa. Nine-tenths of this huge expanse of savannah constitute so-called marginal areas, which support a wide range of wild herbivores, although not in the diversity or quantity of the rich grasslands.

Films and books about African wildlife usually present the wildlife of the grasslands, since it is in the grasslands that the congregations of wildlife are most spectacular. For the same reason, it is in grasslands that most parks and reserves have been established and where most visitors go on safari. In grasslands too are found extensive habitats of elephants, rhinoceroses, buffaloes, giraffes, zebras, dozens of species of antelopes, lions, leopards, cheetahs, hyenas, wild dogs, and jackals. Yet these grasslands are far from typical of African savannahs: they cover a mere 5 percent or so of Africa south of the Sahara.

Modification of Savannah Grasslands in Africa

Many of Africa's grasslands are being changed, with progressive impact. To the extent that this trend facilitates the production of food and other development in Africa without depleting the biotic potential of grasslands, supporters of wildlife should not resist the changes. Provided unique and representative examples of wildlife can be safeguarded through the establishment of parks and other measures, conservationists should appraise the situation realistically. To regret the disappearance of any wildlife at all is a sure way to ensure the end of most wildlife within a few years: it implies that supporters of wildlife are indifferent to Africa's needs for development, which frequently entail basic changes in habitat. The most pronounced form of environmental degradation, as perceived by the citizens of emergent Africa, is impoverishment of the socioeconomic environment.

The question is how to reduce conflict and promote coordination between wildlife conservation and economic development. Put another way, the question is how much wildlife should be protected, in what areas, for what periods of time, and at what cost to whom? After all, just as there can be no long-term development without conservation (conservation in the broad sense, applied to all environmental resources), so there can be no conservation except within the context of development. If, through an expanded approach (as I propose later in this paper), wildlife conservation can reflect the socioeconomic aspirations of emergent Africa, enough wild animals will be left at the end of the century to occupy a whole year's safari.

At least three major activities are modifying grasslands in Africa: livestock husbandry, anti-tsetse measures, and the spread of cultivation.

Livestock Husbandry. Hitherto, man's activities in savannah grasslands have consisted largely of livestock husbandry, notably pastoralism, entailing no great conflict with wildlife except where overgrazing and overburning have reduced the carrying capacity of grasslands for both domestic and wild herbivores. Now, however, an abrupt change is overtaking the traditional forms of land use: stock-raising communities harbor not only growing numbers of people, but growing aspirations for better lifestyles. In their need for new grazing areas and in response to the shortage of meat in Africa and elsewhere, stockmen occupy rangelands that formerly were little utilized. At the same time, they work within increasingly narrow profit margins. These twin factors—to make livestock operations both more extensive and more intensive—cause the stockraiser to seek profit from every animal in his expanded holding. Thus, he is less likely to remain ignorant of—much less to ignore—competition from wild herbivores for such rangeland resources as forage and water. As the stockman gradually squeezes antelopes, zebras, and other herbivores out of existence, the natural prey of the wild predators disappear. Domestic herds then become increasingly subject to depredation by lions, cheetahs, and others.

The stockman's response is much the same, whether he is a commercial rancher in Kenya or southern Africa, a pastoralist (such as the Masai of East Africa or the Herero of Namibia) attempting upgraded husbandry, or a nomadic stockman (such as the Tuareg in the Sahel zone of West Africa) struggling to survive: namely, to eliminate whatever wildlife conflicts with his livelihood. In all of these regions, wildlife has declined markedly since 1960, and especially since 1970. Parts of Kenya Masailand have been cleared of much of their wildlife in just a few years. If present trends persist, wildlife in many areas could be pretty well eliminated by 1980 through competitive pressures from livestock alone.

In addition, land-use legislation in several countries now permits savannahs that were formerly regarded as tribal property to be demarcated and accorded individual or group registration. This means that fencing and other paraphernalia of modern ranching are widely installed, thereby disrupting the migratory habits of such wild herbivores as zebras, wildebeest, and gazelles. It also means that a herd of wild animals on the trek becomes private property, to be treated as the ranch owner sees fit, as soon as it steps onto his property. This legislation has proved a principal factor in the decline of wildlife in Kenya's Masailand.

Of course, the livestock industry in emergent Africa, as in developed countries, must remain a valid and viable sector of national economies. The stockman believes that his interests are just as important as those of the conservationists from developed countries who see the world's natural heritage at stake.

Anti-Tsetse Measures. Large expanses of savannah bush and woodlands remain little modified by man's activities because of the tsetse fly, which causes the disease nagana in domestic stock and sleeping sickness in human beings. Now, a large-scale campaign is to be directed at controlling and eradicating the tsetse fly. The World Food Conference in November 1974 specified the tsetse zone of Africa as among the principal regions in the world where food production could be greatly increased if ecological barriers were overcome. Present estimates suggest that improved measures to overcome tsetse could release almost 7.8 million square kilometers to livestock. Much of this area is in the miombo biome, with its thick woodlands, but extensive sectors are in bushy and lightly wooded savannahs.

The Spread of Cultivation. Generally, savannah grasslands have not proved to be suitable for arable agriculture, since most of them receive less than the 75 centimeters of rainfall hitherto considered the minimum for most crops. But cultivator communities now encounter increasing congestion in the small fertile zones. As a consequence, they are spreading into the next most favorable biotopes for human settlement, the grasslands. Since the savannahs are limited in extent, they are at a premium for both the expansion of human population and for the survival of wildlife.

In parts of Kenya the migratory surge causes human populations in some savannah zones to increase at rates between 10 and 35 percent per year. Similar trends are at work in Tanzania, where many savannah territories have recently experienced an exceptional upsurge in human numbers. In Uganda, with its greater population density and moister savannahs, the pressure to occupy wildlife country is of far longer standing; in 1929, elephants ranged across 70 percent of the country, but by 1959, the proportion had fallen to 17 percent, and by 1973, to less than 10 percent.

In Ethiopia, the control of malaria in savannah lowlands recently has permitted an influx of land-hungry people from the highlands. Parts of southern Sudan, with their immense stocks of wildlife, have been designated for broad-scale agricultural schemes. In Zambia, a country with only 4.5 million people in an area larger than Texas, 30 percent of whom are urbanized (the average for tropical Africa is less than 10 percent), similar pressures are directed at savannahs. The major aggregations of wildlife in Zambia occur, not in the miombo woodlands that cover most of the country, but in floodplains, with their open grassy habitats. The alluvial deposits of the floodplains have a high potential for cultivation; thus, rural migration concentrates on a small but significant portion of the country's total area. Similar pressures arise in Rhodesia, Angola, and Mozambique—all countries with notable savannah ecotopes and spectacular wildlife communities. While the per capita food production in Africa continues to decline, there is an urge to press into use every available piece of suitable land at a rate faster than that at which the population itself is increasing.

To gain an impression of these trends, the conservationist who flies into Nairobi should pause before he heads for the main wildlife areas hundreds of kilometers away. He might take a trip 15 kilometers north of the city, into Kikuyuland. There he would see people trying to make a living off the land at densities approaching 800 people per square kilometer, with plots one-third the size of the average peasant's holding in India. The conservationist would encounter a similar situation among the Buganda, Luo, and other tribes around Lake Victoria, in Uganda and Tanzania as well as in Kenya. On the slopes of Kilimanjaro, the Chagga people try to earn a living through horticulture rather than agriculture, at densities comparable to rural Holland, but without the capital inputs that sustain market-garden farming for the Dutch. African peasants will not stay in such congested conditions if they hear that they can earn a sufficient living off the land in savannah territories.

The conservationist could obtain an even more vivid impression of the population-resources-environment "crunch" in Kenya if he returned to Nairobi from Kikuyuland by way of a certain northern suburb. There, he would see communities at densities higher than those in Harlem (New York City) or Watts (Los Angeles)—but these people do not live in high-rise tenements: they live in single-story dwellings, forty to the hectare.

Whether he is the resident of a city or a peasant in a congested rural area, the man with limited prospects becomes all the more inclined toward savannahland farming when large tracts are opened up through technological innovation. For example, drought-avoiding strains of maize now can be cultivated in areas with only 13 to 18 centimeters of rain each season. This means that extensive sectors of Kenya's wildlife country soon will be available for cultivation and settlement.

Future Trends in the Use of Savannahs. These changes in the use of savannahs direct immediate pressures at wildlife outside such protected areas as national parks. In the long run, similar patterns will prove still more significant for wildlife across broad stretches of savannah Africa and could eventually threaten the parks and reserves themselves. The demographic, economic, and land-use projections are plain enough.

Tropical Africa contains human populations with a "younger profile" than those of any other region on Earth. The populations thus present potential for population expansion until well into the next century, regardless of what is done about family planning. The situation in Kenya gives us an idea of the outlook. Its populace already exerts considerable pressure on wildlife, including the ecosystems of parks and reserves, yet the present total of 14 million people is growing at a rate of 3.5 percent or more per year, one of the highest growth rates in the world. Despite the government's family-planning activities (one of the finest national programs in Africa), there will almost certainly be 30 million Kenyans by the end of the century, perhaps many more. Even if the fertility rate were to fall this moment from the present level of 7.6 (still growing) to 2.0, the population would continue to expand for several generations hence: the parents of the future are already born. Only 12 percent of Kenya is truly arable and fertile, another 6 percent moderately so. The average amount of such land now available throughout Kenya is 0.8 hectare per head, which will drop to 0.28 by the year 2000 if there is no fall-off in fertility, and to 0.38 if fertility rates can be brought down to 4. This means that at the end of the century at least 12.5 million people will be looking for land, over and above the number that can be accommodated in present-day arable zones (plus overloaded urban areas), if the average net cash income per farm family is to be increased to the equivalent of $300 (1972 value) plus subsistence. The main zones designated to absorb this landless populace are the savannahs, notably the grasslands.

A similar prospect faces Uganda, where at least four-fifths of the land area lends itself to agriculture. Of this agricultural potential, roughly two-thirds is now utilized through cultivation and grazing. Since many areas are only partially utilized, the great proportion of agricultural land is already under some form of human occupation. Within only another fifteen years, all unutilized and underutilized potential will be entirely absorbed by population expansion apart from increase in living standards. Most areas surrounding Ruwenzori National Park are considered now to be at or beyond the limits of carrying capacity. Within another ten years at most, most other parks and reserves will encounter similar pressure to put them to different use.

An analysis along these lines indicates potential shortages of land in most African countries that have extensive savannah grasslands. The writing is on the wall, and conservationists around the world should consider whether their present measures to safeguard their wildlife heritage in Africa are appropriate in view of the flood of competitive activities that will surely overtake grassland savannahs within the foreseeable future. More, they should consider whether current support would be sufficient even were it to be expanded several times within present strategies.

Parks and Reserves

In the meantime, a main hope may lie in parks and reserves. More than 390,000 square kilometers of savannah are now protected in eastern and southern Africa, with more in central and western Africa. This expanse, almost as large as California, protects the main wildlife concentrations, such as those at Serengeti, Ngorongoro, and Tarangire in Tanzania; Amboseli, Mara, and Tsavo in Kenya; Kabalega, Ruwenzori, and Kidepo in Uganda; Luangwa Valley and parts of the Kafue floodplains in Zambia; Gorongoza in Mozambique; Etosha in Namibia; and many others. But hardly a single one of these parks and reserves protects the entire ecosystem on which its wildlife depends; many protect half or less.

When the hinterland sector of a park's life-support system is given over to antithetical activities, such as intensified livestock husbandry or cultivation, the wildlife of the park itself becomes progressively impoverished, even though no destructive activities occur within the park.

Parks should be seen as no more than heartlands within broad ranges of supporting territory that provide for genetic exchange, protection against disease, and scope for the various dynamic and compensatory factors which may be viewed as what is ultimately unique about African parks. As Sir F. Fraser Darling has expressed it, "National parks are nuclei of cells in the body of the nation. The rest of the country must supply the cytoplasm, as it were, both to help in the renewal of the nuclei and to sustain the biological systems of the country."

Moreover, parks as isolated ecological islands become prone to a range of natural disasters, including shifts in make-up of animal communities, changes in vegetation configuration, pandemic diseases, and similar destabilizing processes of ecosystem dynamics. Tourism helps to maintain the economic viability of parks as a form of land use, but not nearly so much as is often represented. Game cropping, whether within parks or outside, could likewise reinforce the rationale of wildlife conservation as a medium for economic development, but, mainly because of marketing problems, not all cropping projects establish the capacity of wildlife to make its point in the marketplace.

The long-run prospect is that many parks and reserves may not survive the land-use changes which are soon to transform savannahland Africa, unless these protected areas can somehow be better integrated with their physiobiotic and socioeconomic environments. The topic is complex enough at local levels, and has been dealt with on various occasions elsewhere. The present paper considers the problem mainly within an international context, with emphasis on institutional deficiencies rather than ecological discontinuities. Not only do existing parks and reserves require better protection; the main need is for more parks, reserves, and other conservation units. Yet land-use patterns are moving in the opposite direction. In order to adapt to the situation, wildlife supporters should consider measures beyond appeals to governments to set aside additional areas. To reiterate a central point, the institutional underpinning of new parks and reserves must meet the development aspirations of local communities as well as the conservation needs of world society. This may mean a reorientation of approach to take account of socioeconomic constraints of conservation in Africa. The institutional environment is an integral part of life-support systems of parks, even though not always perceived as such.

Disparity in the Perception of Wildlife Values

This aspect of the problem is marked by the socioeconomic divergence between the advanced world where most wildlife supporters live, and the developing world where many wildlife spectacles are found. African countries in particular are scarcely in a position to subsidize the rest of the world through safeguarding mankind's heritage of wildlife. The United Nations list of the world's twenty-five poorest countries includes sixteen in Africa, several of which contain major wildlife.

Several African countries already conserve wildlife at considerable expense to themselves. For example, Tanzania, with a governmental budget less than New Yorkers spend on ice cream each year, devotes a larger slice of these national funds to Serengeti, Ngorongoro, and other wildlife spectacles than the United States spends on Yellowstone and its other parks. Tanzania does this in order to safeguard what outsiders are not slow to remind Tanzanians is also the world's heritage.

In addition to the socioeconomic gap, there is a cultural dissonance between Africa and the developed world. Citizens of emergent countries do not yet perceive wildlife conservation with the urgency of affluent nations. Rather, they have good cause to perceive parks as huge blocks of their countries set aside for the disportment of rich white foreigners on safari.

Thus, bioecological factors are not the only absolutes at issue. Conservation programs which entail the removal of substantial segments of land from human occupation should give attention to considerations of social equity and economic efficiency. To be sure, there can be no long-term economic advancement without attention to ecological imperatives: conservation and development are two sides of one coin. But the subsistence peasant of Africa has more immediate concerns than ultimate environmental stability. Unless he can take care of short-term needs, he may not be around to see how longer-term considerations work out. When he engages in land-use practices which are wasteful of local resources—resources which sometimes appertain to humanity's heritage—he is not generally acting in blind disregard of his own ultimate welfare; rather he is acting with rational regard for today's supper. If education is to play a role, it should be a two-way process. Many a conservationist could become better acquainted with the facts of staying alive as perceived by the savannahland peasant.

Wildlife as a Common Heritage

The second part of this paper focuses on the "common heritage" aspect of wildlife conservation. It attempts a preliminary exploration of the topic and seeks to ask some of the right questions rather than to provide conclusive answers.

There is little doubt that society at large would derive appreciable benefit from increased conservation of savannah wildlife. There is equally little doubt that communities in several parts of the world, notably among affluent countries, would ostensibly like to see more done. They might even be prepared to pay for it. After all, they are ready to pay large sums of money each year for books about elephants and lions and zebras, and they expend large amounts of another high value resource, leisure time, in watching television films on the same subjects. Only a trifling fraction of these expenditures are replicated in conservation measures on the ground in Africa.

Television programs are one way of evaluating people's interest. A typical program in North America dealing with African wildlife runs half an hour and attracts 12 to 35 million viewers, according to Nielsen ratings. If leisure time is assessed at an approximate worth of one dollar per hour (some economists put it higher), this means that one program presumes an evaluation of African wildlife by the viewing public of $6 million to $17.5 million. Of course people derive other satisfactions from viewing television besides their pleasure at elephants and leopards. But thirty such programs during one year argue as evaluation of at least $100 to 300 million as "ballpark figures." Book sales are an additional economic measure of people's interest. Yet private donations to United States conservation organizations in support for African wildlife, such as the World Wildlife Fund and the African Wildlife Leadership Foundation, total only around one million dollars per year. This suggests that people might be willing to pay much more to safeguard African wildlife if an efficient and equitable means could be devised to collect anywhere near the maximum that each person would be ready to pay.

Given this interest in African wildlife, why is broader-scale conservation not undertaken? To stimulate greater efforts, many wildlife supporters outside Africa urge a conservation rationale with stress on "environmental ethics" and "rational use of resources." They have urged this approach for years, and meanwhile the situation has deteriorated at accelerating rates. Perhaps the time has come for an appraisal of why this approach has not worked better and whether it stands a better chance in the future. There may be room for fresh strategies to complement the procedure of "same as before, only more so."

Wildlife in Africa as elsewhere is a common heritage. To be sure, a wild creature in the territory of an individual nation may be regarded, from the strict standpoint of national sovereignty over national resources, as the exclusive possession of that nation. But this is far from the entire story. (The Stockholm Conference on the Human Environment confirmed the rights of individual nations to exercise sovereign rights over their natural resources.) Wildlife species and communities are integral components of the planetary ecosystem. In significant senses, the spectrum of Earth's wildlife can be considered an indivisible part of society's heritage, now and forever. At least this is how many people perceive wildlife. They feel that the heritage of humanity, as well as that of Africa, has been impoverished by the loss of the quagga, for example.

In very limited manner, this situation is already marked by emerging concepts of joint responsibility for resources of global significance. A few institutional initiatives have been devised, notably the World Heritage Trust, the Convention on Wetlands of International Importance, and the Convention on International Trade in Endangered Wildlife and Wildlife Products, among others.

The World Heritage Trust seeks to protect outstanding examples of mankind's natural and cultural heritage. But the trust will assist only a few of the present parks and reserves in Africa and will certainly not broach the wide-scale conservation needs postulated in this paper. The same applies to UNESCO's Biosphere Reserve system. This proposal, derived from bioecological criteria, does not reflect the economic and institutional problems of conserving common heritage resources, even though these problems are, in the view of this paper, at the heart of the issue.

After two years or more, only one of these initiatives has received the minimum number of governmental ratifications to bring its provisions into force. Nevertheless, these devices serve as a measure of efforts on the part of the international community to take a few cautious steps toward communal responsibility for unique resources of exceptional value to mankind. With regard to African wildlife, however, these initiatives are on such limited scale that they would scarcely make marginal impact on overall needs of conservation.

Collective responsibility for the common heritage does not mean collective rights in particular resources; there need be no conflict between heritage concerns and sovereignty interests. In practice, the globalist perspective can be integrated with national considerations insofar as on-the-ground management of wildlife and wildlands remains with individual countries, while the international community occupies itself with devising methods to interpret the concept of common heritage through an institutional framework.

The present paper explores the conceptual background for more comprehensive measures. It adopts a functional approach to the problem. For example, it does not examine legal principles and techniques such as the public trust doctrine—methodologies which are in the course of intensive development in a number of countries and which afford protection for heritage resources. This paper's

Common Property versus Private Property

As described, wildlife in Africa is regarded as a heritage not only of Africa but of society at large. But everybody's heritage tends to be treated as nobody's business. However much society may value African wildlife, it has little opportunity to translate this interest into effective action. An individual person can own a house or a car—a situation which induces him to take care of it and which largely deters others from using or misusing it. But society can exercise no such property rights—much less the protection opportunities which accompany property rights—in wildlife. Society cannot "own" the cheetah in a manner to make certain the species survives. [For literature on ownership rights, a central issue in the economics of environmental conservation, see the bibliography at the end of this paper.—Editor]

This applies not only to society around the world, but to society within individual countries. A country can of course seek to exercise ownership by declaring legal protection for its wildlife (a measure not yet generally available at the international level). But this tends to work best in localized situations. The United States has instituted legal safeguards for the pupfish, which is confined to a few waterholes of western rangelands where livestock requirements have exercised overly competitive use of water supplies. This is in line with the important legal principle *Sic utere tuo ut alienum non laudas*—i.e., do not use your own property in a manner which harms another's. But, for many years, the United States had difficulty in enforcing laws in defense of bald eagles, which ranchers alleged take small stock. (Recent special measures have improved the situation.) The problem is still more difficult for African countries with limited funds for legal protection of widely and sparsely distributed species such as cheetahs and rhinos. The stockman who disposes of the cheetah to prevent commercial loss, or the poacher who kills rhinos for commercial gain, expects to get away with his illegal action since he knows that law enforcement is not widespread enough to safeguard wildlife over extensive areas. Still more to the point, the cultivator who digs up cheetah or rhino habitats can hardly be charged with any offence at all, either legal or moral.

The crux of the problem lies in the fact that public good does not always run parallel with private good. So divergent can the two paths become that it would not be easy to vest a public body with adequate authority to safeguard a resource of common heritage in conflict with private interests, whether within the context of an individual country or at a global level.

This assertion is qualified by the public trust doctrine and similar legal mechanisms in the United States, where community concern for community values has been extensively articulated. For example, much progress has now been accomplished to regulate private use of public resources such as the atmosphere and large water bodies. The same applies in a number of other countries, notably those in Europe. Countries of emergent Africa have not hitherto been in a position to develop legal procedures along these lines, nor have they perceived the same urgent need. Still less has the international community developed legal constraints to regulate use of natural environment resources.

For centuries, institutions of marketplace, law, and political systems have tended to formulate and consolidate the rights of private property. The needs of common property, such as the resources of society's heritage, have suffered in default. This has not mattered much until recently. The "community's estate" has required little protection as long as it was not threatened on a large scale. Now that common-property resources such as wildlife are diminished through misuse and overuse, societal systems to safeguard them are in short supply.

Strictly speaking, the term "common property" refers only to those resources for which institutional measures have been established to express the community's interest through mechanisms approximating to property rights. An arrangement of this sort usually facilitates a regulatory system that conserves and allocates resources. Examples of these resources include fisheries in territorial waters, and forests and rangelands held by public agency. Such resources are considered to be *res communis*, or resources for which no institutionalized regulatory process has been established. In this sense, African wildlife as a resource of global heritage is to be regarded as *res nullius* in principle; and in practice within individual countries (whatever the legal frameworks might say), it usually amounts to the same thing.

But resource economists frequently use the term "common property" to refer to resources for which would-be exploiters enjoy "open access." This means the resource is available for anyone to come and harvest it as he wishes (e.g., the man who hunts wild creatures for their products), or to dispose of as he thinks fit when the resource exists in competition with his exploitation of associated resources (e.g., the man who cultivates wildlife habitats). For purposes of this paper, the term is used in its latter, albeit loose, sense.

Common-property resources, then, are up for grabs by anyone who wants to make use of them or otherwise dispose of them. Worse, a private entrepreneur often finds it in his interest to overexploit these "free goods"—though they are free to only a few individuals, while of great scarcity value to society. He sees incentives in seizing as much as he can before other exploiters get at them. Worst of all, when an exploiter abuses a common property resource, or uses it up, society often finds it has lost a unique resource. Wildlife, for example, has no substitute. As a consequence, wildlife is regarded as beyond the normal estimations of value, i.e., as priceless. Yet, since there is generally no private owner of wildlife resources who can trade them in the marketplace and thereby establish a dollar price to reflect their economic value, wildlife is treated as if it is priceless, i.e., worthless.

Various efforts at wildlife management in Africa seek to tackle this problem through economic exploitation of wildlife products (trophies, meat, skins, tourist attractions). But these measures are not yet on a scale to meet the overall requirements of Africa's extensive wildlife resources. The upshot is that wildlife is progressively depleted because of its characteristics as common property.

The Bison as a Model of Grasslands Wildlife

The "irrational" dissipation of wildlife is illustrated by the story of a grasslands species, the bison. The bison's products could be exploited at a cost which did not reflect society's loss. Bison hunters thus created greater public detriment than private reward. But, as is usual when individuals destroy resources of common heritage, the advantages accrued to a small group of people concentrated in place and time. This allowed each member of the group to derive appreciable benefit. By contrast, the losses were spread among the collectivity of society. This caused each citizen's immediate perceived loss to be small, while future members of society had no say in the matter.

So millions of bison were killed for products which could be marketed at once, viz. tongues and hides. Even when it became apparent that within a decade bison meat could be shipped to market, thereby increasing private profit several times, the slaughter continued in disregard not only of society's long-term interests but of the exploiters' interests a few years hence. This was because the bison, as common property, did not lend itself to sustained-yield exploitation. The exploiter who planned a harvest year after year would find his competitors had left little to exploit. By the time the bison's demise was halted, its habitats had been taken for private purposes which would not have tolerated the bison's competitive use of grasslands. Thus the tragic outcome of a "common situation."

Something similar applies even to an exploiter who harvests a private-property resource in times of accelerating technology and sky-high interest rates. Many forms of natural resource exploitation now require that an entrepreneur recover his investment-plus-profit within a foreshortened time span, often only half a decade or less. This militates against considerations of long-term sustainable harvest for resources with lengthy self-renewal periods, such as elephants, when the costs of exploitation resources (capital, equipment, skills, time) allocated to the on-going harvest are not competitive with alternative uses for these resources. In this situation, it is rational to engage in a once-and-for-all operation. In broader perspective, present marketplace mechanisms are so out of kilter with society's needs that it would not be financially sound to invest in conservation of the oceans.

The experience of hunting tribes, which exploited common-property resources in the form of wild animals, indicates that they recognized the dangers of unregulated harvest. Their primitive cultures adapted to the constraints of scarcity through various forms of territorial jurisdiction and related measures which promoted limited ownership rights in segments of the resource. But these systems tend to break down when stable subsistence cultures come into contact with commercial considerations of traders from outside.

The bison's experience serves as a model for understanding what is now overtaking many wildlife species in Africa. The black and white rhinoceroses are overexploited merely for their horns, the spotted cats for their skins, the giraffe for its tail hairs, the elephant for its ivory. Nobody perceives sufficient incentive to maintain close, continuous protection over these creatures, though everybody loses as their numbers decline. Half a million wildebeest in Botswana and Namibia have been reduced by as much as 90 percent due to man's conflicting activities in their grassland habitats. Several gazelle species, the wild ass, and other herbivores have been reduced to threatened status as rangelands are given over to increasingly competitive use by domestic stock; the quagga and the blue buck have been entirely eliminated for this reason. Predators experience greater attrition than most species because they attack private property in the form of domestic stock; in parts of eastern and southern Africa, more lions, leopards, cheetahs, wild dogs, hyenas, and jackals have probably been accounted for during the past five years (when poison became widely and cheaply available) than during the previous twenty-five years.

In some cases, a species encounters more than one threat. The elephant is hunted for its ivory, but its numbers are declining principally through habitat loss, which now occurs in parts of its range at a rate of about 3 percent a year (not altogether coincidentally the rate of human population growth in many countries). When elephants are overexploited for their ivory, the exploiter is no more culpable than the markets which unwittingly stimulate excess demand for elephant products. (Of course, the exploiter who hunts elephants in breach of the law bears greater responsibility.) When landless individuals occupy elephant habitats in order to meet fundamental needs of daily subsistence, it is difficult to attach much blame at all to their actions.

If agriculture in Africa could be made intensive rather than extensive, there would be far less pressure on wildlands. But this implies greater development of cash crops such as coffee, which yield a high economic return from a small area. African peasants would be more inclined to stake their future in coffee if they could be assured a fair return for their efforts in international markets. But the trading price of coffee hardly rises in real terms. In this sense, the life-support systems of African wildlife extend to the American breakfast table, where coffee is the commodity which has shown least price increase since 1960. By the same token, a major way to safeguard African wildlife would be to increase Africa's rate of urbanization. If the United States were not 70 percent but only 10 percent urbanized, the pressure of people on American wildlands would be far greater. African countries are not likely to urbanize, however, except in response to the modernization of their economies. (People will crowd into shantytowns, but that is an altogether different process, and it will not draw sufficient numbers off the land to relieve natural environments.) Modernization means opportunity for manufacturing, which means access to markets in the advanced world—where African goods now encounter tariffs, quotas, and a range of other trade restrictions. Hardly any other factor would help Africa wildlife more (and perhaps nothing less will adequately work) than speedy economic modernization within a framework of "ecodevelopment," yet given the attitude of the developed world, hardly any other prospect seems less likely.

The Institutional Environment of Wildlife

In all these instances, the core factor is the conflict between public and private interests. The institutional environment of wildlife in Africa, as in most other regions, favor private interests. This is important in determining conservation strategies to counter destructive trends. Rather than engage in moral exhortations or appeals for education to stop people from seemingly cutting their own throats, wildlife supporters might look to the causes as well as the symptoms of the depletive process. When an individual exploiter overuses a common property resource (the case of the modern "bison hunter"), or exploits associated resources in implicit indifference to common heritage values at stake (the cultivator of wildlife habitats), he is not being unduly narrow-visioned or worse. He is pursuing his own self-interest within the rules of the game as laid down by society's inadequate institutions. It is not so much the exploiter who is to blame, as society for not devising better rules. Society's efforts to conserve its heritage should include incentives of a nature and on a scale to match the individual's incentive to look after his own property.

The destructive process occurs not only within the context of individual countries, but at the level of global society. One quarter of the world community consumes three quarters of the goods traded in international markets. This induces additional pressures on African savannahs, through, for example, demands for beef. But the process is better illustrated by a more acute example, that of tropical forests. These forests, especially in Southeast Asia, are overexploited primarily to supply affluent-world customers with paper, veneer, and other forest products. The entrepreneurs in question exploit their forest concessions for private purposes. At the same time, they cause loss of habitat for the orangutan and other species (at least half a million altogether, including insects) in Southeast Asia's forests. Many commercial organizations at work are multinational corporations based in North America, Europe, and Japan. Thus, several sectors of society contribute to the degradation of society's heritage. To relieve the pressure on these tropical forests, customers overseas would have to be satisfied with fewer forest products. This would mean higher prices for the reduced supply—not a favorable prospect in inflationary times. Moreover, growing demand could be met in part by increased exploitation of forest reserves in North America, but this would likely arouse opposition from local environmentalists.

The wildlife supporter in the developed world might recognize that what he does with his right hand he may undo with his half dozen left hands: he is both conservationist and consumer. The "spillover effects" of his actions are unwitting and generally remain unremedied. This relationship is likely to become a pervasive factor of exploitation patterns that overtake wildlife habitats in decades ahead.

The institutional constraint can be further illustrated by another grasslands species in the United States. The black-footed ferret, as a common-property resource in conflict with local interests, is among the most endangered species in the world. The farmers who poison prairie dog populations and thus eliminate the ferret's chief prey are not likely to forfeit a significant slice of private income for the sake of wild animals in their midst. To be sure, if they were to act with extreme public spirit (following education campaigns?), they would thereby safeguard society's interests. But that is precisely the point. The farmers do not see why they should accept uncompensated sacrifice on behalf of their fellow citizens of the United States, let alone of the rest of the world or of future generations. Much the same applies to several other species which the most advanced country on earth cannot safeguard beyond reasonable risk.

A farmer with prairie dog colonies where ferrets are known to exist receives government compensation totaling several hundred dollars to offset grass consumed by prairie dogs; in return, he is not permitted to engage in poisoning. But ferrets are so rarely seen that the arrangement breaks down: no compensation is received, and poison is used in areas where ferrets are subsequently discovered. As is so often the case with natural environment values, the asymmetry of institutional mechanisms promotes farming/marketplace interests against wildlife/nonmarket interests.

Within the context of African savannahs, resources other than wildlife are misappropriated. Soil, water systems—in fact many physical and biotic components of savannah ecosystems—are treated as common-property resources. Stockmen of all sorts and conditions overgraze rangelands for private benefit, thereby degrading plant cover and soil stability for all. Not only domestic livestock but wildlife suffers when the rangelands' natural productivity deteriorates. Individuals precipitate these destructive processes not because they are unusually stupid or selfish, but because their individual actions serve individual needs with implicit indifference to other considerations. Why should the individual concern himself with society's affairs if society does not reward him for it?

To use an extreme illustration of this point, suppose a conservationist finds himself stranded on a desert island, whereupon he sees a large creature haul itself out of the sea onto the beach. To the conservationist's surprise (and satisfaction?, since it represents a large stock of food), it turns out to be an individual of a species long believed extinct, Steller's sea cow. What will he do then—spare the animal and starve to death?

In short, destructive activities appear rational to savannahland citizens for the same reason that similar activities appeared rational to American farmers forty years ago when they brought on the great dust bowl. Those American individuals were persuaded to look out for the public good as well as their own private good only when the collectivity of society (the government) made it worth their while to do so. African individuals are likely to prove amenable to similar incentives, and hardly to anything else. Certain savannah resources are of interest not only to society at local levels (the African governments of the countries in question), but to society at large around the world. Global society should work out some sort of bargain with private interests in savannahlands to establish an equilibrium between forces tending to degrade the global heritage and forces seeking to safeguard it.

To the extent that people in many countries outside Africa derive satisfaction from the mere thought that African wildlife exists, they receive this benefit free. They have not had to pay for it; indeed they have had little chance to pay for it even if they had wished—other than through private organizations, which are subject to the problem of the "free rider," viz., the man who would contribute voluntarily to wildlife conservation if he could be sure that others would similarly contribute rather than take a free ride, or who refrains from contributing because he counts on taking a free ride by virtue of others' contributions. Just as people have not had to pay for wildlife's survival, they now find scant opportunity to pay to prevent its demise.

Expanding Pressures on the Common Heritage

The conflict between private interests and collective needs in Africa will grow more acute in years ahead. In part this is because private interests will increasingly encroach on the public domain. To date, savannah wildlands have remained little affected by human activities, but they will become progressively subject to competitive if not exclusive use by various forms of economic development. In part too the conflict will be exacerbated because of the increasingly integrated nature of society's activities. As economic interactions become more complex, they tend to produce more spillover effects. The accumulative impact of these spillovers tends to concentrate at weak points in the mechanisms by which society runs its affairs. Especially susceptible to spillover damage are common property resources, because they have no "owner" to look out for their welfare and protect them from harm.

The interdependent relationships of society's workings are now an accepted phenomenon. Poke one part of the economic system and it bulges somewhere else. Put a strain on one sector and the pressure is transmitted through the system until it shows up in vulnerable areas, such as society's heritage in wildlife. A falloff in cash crop production in a country's fertile localities can lead to a spread of subsistence maize cultivation in savannahlands, with repercussions for a park ecosystem hundreds of miles from the cash crop areas. When developed-world consumers look for low-priced beef from African savannahlands, they are unaware that the most "efficient" stock raisers, i.e., those who produce beef at the cheapest rates, have streamlined their operations by eliminating wildlife from their savannah territories. Some of these spillover threats to wildlife are traceable to a visible cause-and-effect sequence of events and so can be remedied comparatively easily; if industry, livestock watering installations, crop irrigation, and a hydropower project in one country reduce river flow through a park in another country, it should not be impossible (even if virtually unprecedented) to arrive at some compromise solution. But some spillover pressures arise because of dimly perceptible networks of diffused circumstances, making counter-measures less simple. When failing economies and unemployment problems in several countries of the developed world lead to trade restrictions against manufactured products of several African countries, thereby contributing to the spread of subsistence agriculture across many savannah territories of the continent, it is difficult to isolate causative factors—and impossible to treat them in isolation. In these circumstances, where multiple sectors of society contribute to the problem, the appropriate response lies in a comprehensive strategy on the part of society as a whole.

Most threats to African wildlife are now the collective consequence of unwitting actions by multitudes of citizens in many lands of different continents. These citizens go about livelihoods of various forms—all of them legitimate livelihoods as perceived within a local framework of aspirations and prospects, not always so acceptable as perceived from the standpoint of society. This situation requires a radical reorientation of conservation strategies. In the past, it was often sufficient to resist the direct and deliberate actions of individuals who were easily identified as being in conflict with wildlife. True, poaching is now worse than it has ever been, but it is far from the number one problem, namely, disruption of habitat.

Now there is a new scene. In 1960, there were plenty of savannah wildlands in Africa. Fifteen years later it is plain what is happening to them. In another fifteen years they could well be changed beyond recognition unless significant sectors can be conserved for wildlife. To achieve this objective, there is need for an expanded approach to complement conventional measures. When society around the world asserts an interest in Africa's wildlife, it should recognize a responsibility far greater than the gestures it has made so far. Practical measures lie outside the scope of this paper, but the principles at issue suggest that, for starters, we should direct more attention to the root causes of wildlife's present decline. This in itself would help to formulate a consensus about wildlife conservation as an objective of the community at large and would serve to establish the concept of common heritage.

As difficult as this fresh strategy would be, it would be appraised in the light of available options. Present measures amount to an implicit decision by society to permit its wildlife heritage to decline drastically, if not disappear.

Summary and Conclusion

The outlook for the wildlife of Africa's savannahs illustrates the need for a conservation strategy that reflects society's interest in this common heritage. The same applies to wildlife in savannahs of other continents and in other biomes.

Wildlife is now threatened for the most part because inadequate institutions encourage the misuse of wildlife and its habitats as common-property resources. Private needs conflict with community purposes, both of which are perceived of as rational interests by the respective participants. The institutional environment for conservation and development tilts the balance in favor of private considerations. In the face of this dilemma, there is need to identify and define new options for wildlife conservation, with emphasis on common heritage values. Society should expand its support for wildlife in recognition of the joint responsibility for a deteriorating asset of mankind's patrimony.

This means that conservation will reflect the increasingly interdependent character and needs of the community at large. This may well run counter to contemporary trends, where interdependency is a fact which not all members of the community wish to recognize. But whether the community perceives itself as a community or not, it will sooner or later be obliged to respond as a community to the problem of vanishing wildlife: either sooner, through conservation measures of sufficient scope, or later, when it finds that the decline of wildlife represents a loss through which the community is indivisibly impoverished.

Bibliography

The following materials deal with the issue of ownership rights and the economics of environmental conservation:

Demsetz, H. 1967. Toward a Theory of Property Rights. *American Economic Revue* 57(2):347-359.

Wunderlich, G., and W. L. Gibson, Jr., eds. 1972. *Perspectives of Property.* Pennsylvania State University: Institute for Research on Land and Water Resources.

For analysis of the whole field of common property resources in relation to environmental values, see:

Anonymous, ed. 1971. *Problems of Environmental Economics.* Paris: Organisation for Economic Cooperation and Development.

———. 1972. *Political Economy of the Environment.* Paris: Organisation for Economic Cooperation and Development.

———. 1974. *Problems in Transfrontier Pollution.* Paris: Organisation for Economic Cooperation and Development.

Barkely, Paul W., and David W. Seckler. 1972. *Economic Growth and Environmental Decay: The Solution Becomes the Problem.* New York: Harcourt Brace Jovanovich.

Freeman, A. Myrick, III, Robert H. Haveman, and Allen V. Kneese. 1973. *The Economics of Environmental Policy.* New York: John Wiley & Sons.

Kneese, Allen V., and Blair T. Bower, eds. 1972. *Environmental Quality Analysis: Theory and Method in the Social Sciences.* Baltimore: Johns Hopkins Press.

Kneese, Allen V., Sidney E. Rolfe, and Joseph W. Harned, eds. 1971. *Managing the Environment: International Economic Cooperation for Pollution Control.* New York: Praeger Publishers.

Krutilla, John V., ed. 1972. *Natural Environments: Studies in Theoretical and Applied Analysis.* Baltimore: Johns Hopkins University Press.

Krutilla, John V., and Anthony C. Fisher. 1975. *The Economics of Natural Environments: Studies in the Valuation of Commodity and Amenity Resources.* Baltimore: Johns Hopkins University Press.

Natural Resources Journal. 1972. The Human Environment 12(2).

Swedish Journal of Economics. 1971. Environmental Economics 73(1).

Tsuru, Shigeto, ed. 1970. *Environmental Disruption.* Proceedings of International Symposium [sic] on Environmental Disruption in the Modern World, Tokyo, 1970. Tokyo: International Social Science Council.

Deserts

... *The barren stretch in Africa—*
Where nothing can grow, because it does not rain,
And, by the same token, where no rain can fall, because
Nothing grows there....

—Frederik Paludan-Müller
Adam Homo (translated by William G. Mattox)

Cover my earth mother
four times with many flowers

Cover the heavens
with high-piled clouds

Cover the earth with great rains
cover the earth with lightnings

Let thunder drum over all the earth
let thunder be heard

Let thunder drum over all
over the six directions of the earth

—Zuñi
"Storm Song"

Terrestrial Ecosystems: Deserts

37. The Perplexity of Desert Preservation in a Threatening World

Roy E. Cameron
Argonne National Laboratory, U.S.A.

Deserts in Perspective

Deserts Past, Present, and Future

Although two "cradles" of western civilization—the Nile River valley and the valleys of the Tigris and Euphrates rivers—are situated in desert areas, deserts still are scantily populated, except at oases and along rivers or streams, where water is more likely to be available. (According to the usual definition, deserts are regions that receive less than 25 centimeters of precipitation water equivalent per year.) In places where crops cannot be cultivated through irrigation, grazing has been extensively undertaken by nomadic tribes and is continued by desertic peoples to this day.

Most people have never known or experienced the desert, and many think of deserts as vast, barren, and inhospitable arid regions characterized by sand dunes, rugged hills, mountains and plateaus, angular rocks, "open" soil, sparse vegetation, and few animals. In fact, however, most arid areas of the world, except those in Australia, are not flat and low (Perry 1970), and most contain mountain ranges that influence the climate.

Deserts are among the last ecosystems considered for preservation. Other terrestrial ecosystems and their interjacencies—savannahs and other grasslands; wetlands; temperate and tropical forests; rivers, streams, and lakes; islands and the coastal marine environment—receive more attention in this regard, largely because they are easier to settle and develop than deserts are.

The world's attention was drawn to the fact that deserts are expanding (a process called "desertification") by the recent advance of the Sahara in the Sahelian strip (Council on Environmental Quality 1974), which is as large as the conterminous United States and is inhabited by 25 million, mostly nomadic, desertic peoples whose numbers have been decimated by several successive years of droughts. [Deserts caused by desertification are to be distinguished from natural, extant deserts; desertification is a separate, although a very important, problem (Le Houérou 1970, Kassas 1970).—Editor]

Industrialization has affected deserts: industry has been established in centers of population, and minerals have been increasingly sought and extracted. More recently, technological advances and the "demographic upsurge," with their own consequent needs, have directed greater attention to the exploitation and "reclamation" of deserts. The establishment and development of the state of Israel is an outstanding example of this: not only is the land being reclaimed there, but even the climate itself is being changed (Lowdermilk 1971).

Pressures on Deserts

General Considerations. Most desert research and development is devoted to determining the best use of the desert for the greatest economic return and to halting and turning back desertification. ("Esthetic" qualities—including even unique wildlife and wildlife habitat—are generally subordinated to other purposes unless they can be "exploited" for recreation.) Man continues to change the physiography of the desert, both directly—by building dams, mining, and establishing suburbs—and indirectly—by causing soil erosion through the clearing of vegetation, cultivation, and overgrazing.

Sources of Energy. In October 1973, the major oil-producing nations shocked the world by quintupling the price of oil, which caused dramatic worldwide inflation, a recession, and a quadrupling of prices for fertilizer, an indispensible component of the so-called "green revolution" in developing countries. The increased price of petroleum has intensified pressure to explore the use of alternative energy sources. In the arid and semiarid western United States, reserves of fossil fuels (particularly coal) are likely to be exploited at an accelerating rate in the near future, with the greatest consequences for the deserts of North America since the onslaughts of settlers, miners, and—lately—off-road vehicles (ORVs).

Lebensraum. The need for *Lebensraum* is another source of real and potential impacts on desert regions: more space will be needed for a world population (now more than 4 billion) that is growing at an annual rate of 2.2 percent (Cook 1974). The expansion of human habitation into the approximately 20 percent of the Earth's surface that is desert is the "logical" solution to the problem: developers view the globe's vast, uninhabited deserts as the most logical places for the additional people to live in. "Better out than up; better spread than piled," is one opinion. (Mandatory zero population growth, euthanasia, eugenic control, dictatorial restrictions on numbers of children, and the selective allocation of food and living space portend as harsher, "Orwellian" realities.)

However, as "jack rabbit" homesteaders and "get-rich-quick" land schemers in the arid parts of the western United States discovered, the vast, uninhabited areas of the desert are not uninhabited without reason: there is no water, or very little of it when it is needed!

Preservation of Deserts

Deserts and Mankind. Unfortunately, deserts, whether in temperate, tropical, or polar regions, are considered "threats" to mankind. There is a strong belief that they must be "conquered"—wrested from the inhospitable forces of nature and put to the best economic use: they are not to be left undisturbed, *au naturel!* The prevailing attitude is that all of their potential for human habitation and use must be developed to the fullest, regardless of their beauty or the value of their natural state. It is now the desert's turn to be "threatened" by mankind, although it is mankind's use (and abuse) of deserts that in many areas has intensified or accelerated desertification beyond the natural response to the dynamic equilibrium among components of the environment—whether soil, climate, or living things.

Man is the most important factor to consider in any attempt to preserve deserts. This is especially so because it appears that man can even induce unfavorable changes in climate, initiating or perpetuating droughts in deserts (Charney et al. 1975). Man's thoughtlessness has also indirectly caused deserts to expand, as when he introduced rabbits, but not their predators, to Australia. Even Antarctica's cold polar desert has been contaminated with widely used pesticides!

Far too many inhabitants of deserts ignore the fact, recorded throughout the history, that the desert is a fragile entity, a beauty to be treated gently, a living *objet d'art*. Those who love the desert insist that it be admired, courted, and studied, treated gently, and bound up if wounded—not run over roughshod or desecrated. Even the objectivity of science yields to art in the description and appreciation of deserts.

The Desert Protective Council. One way those who love the desert can protect it is to band together for that purpose. Unfortunately, too many lovers of the desert are pacifists when it comes to protection and preservation; they depend upon governments to ensure its wise use and to rectify its mismanagement and abuse. As an alternative to their pacifism, they can form citizens' groups, such as the Desert Protective Council, Inc., centered in California. More than twenty years ago, the council foresaw the need for a force of concerned citizens devoted to averting and diverting pressures on deserts. Its major concern is to ensure that sound policies are adopted for the use of the desert's natural resources. Its representatives take an active part in the procedures of the several government agencies that formulate policies for the use of American deserts and mountain areas that are ecologically linked to deserts. The council publishes educational pamphlets and booklets and sponsors field trips that are open to all.

The council has stated its purposes simply:

    a.   To safeguard for wise and reverent use, by this and succeeding generations, those desert areas that are of scenic, scientific, historical, spiritual, and recreational value, and

b. To educate children and adults, by all appropriate means, to a better understanding of the desert, in order that the purposes of the organization may be attained.

The Desert Protective Council's purposes and activities are excellent examples to follow for others who have the initiative, resources, and freedom for similar undertakings.

Desert Experiences

Research and Travel in the Desert Biome

Since personal experiences, observations, and impressions that would be inappropriate in a purely technical article are acceptable in the context of the EARTHCARE Conference, I will share some of the experiences I have gained during twenty years of formal education, training, and research in the arid and semiarid western United States (not to mention my boyhood years there); in the arid or semiarid areas of such countries as Mexico, Ecuador, Chile, Peru, Morocco, the Soviet Union, India, New Zealand, and Australia; in the polar deserts of the Antarctic (where I have spent seven summers) and Arctic (where I have spent two); and, most recently, in the western United States again, where I have been working on the rehabilitation of strip-mined land.

I have collected and analyzed soil samples and taken photographs in many other arid and semiarid areas, such as those in Colombia, Argentina, Egypt, Israel, and Iceland. I also have visited and studied high-mountain and volcanic areas (in one sense, they, too, are "deserts"—i.e., places that are harsh for life, a concept that can be enlarged even further to include extraterrestrial environments, such as the moon and Mars).

Indeed, because of the apparent similarity between Earth's deserts and extraterrestrial environments, I spent some fourteen years amassing more than 20 tons of aseptically collected samples of desert soils, the largest such collection in the world (Cameron et al. 1966). The samples have been or will be used in baseline studies and for comparison with extraterrestrial soils and, possibly, microorganisms. To date, my studies on the samples have been reported in more than 100 publications, among them a generalized article on the North American Desert (Cameron 1974) and a paper on the conservation and preservation of Antarctica's terrestrial ecosystems (Cameron 1972).

The Search for Pristine Sites

Man's Incursions. I gained valuable insight travelling through the most desolate desert regions, where my associates and I sought out the most nearly "pristine" areas for our studies and traversed much desert terrain in the process, setting up tent camps in remote sites, or backpacking in the more rugged areas--sometimes after being set down by helicopter. The nature of the research demanded rigorous self-policing of trailbreaking, camping, and backpacking, and those who participated in the studies learned to appreciate the desert from both the scientific and the esthetic points of view.

As we sought undisturbed sites for our scientific work, it became evident to us that more and more of the desert was being subjected to incursion by man: in many of the places where we had hoped to find undisturbed sites, others had left their mark, sometimes only as tracks, crushed vegetation and animal burrows, or a dead animal or two, but at other times as indiscriminately dumped trash, rows of broken bottles or cans that had been used for target-shooting practice, and discarded automobiles.

Signs of Man and Beast. We became adept at detecting evidence of the presence of animals or people—not only through obvious signs, such as the mummified carcasses of animals dead for seventy-five years or more along the Old Bolivian Trail in the Atacama Desert, but also through telltale microorganisms—and we avoided them when possible. We learned that the slogan "Take only photographs, leave only footprints" did not apply in some desert areas: in extremely arid areas (for example, in the Mecca Hills of California and in the ice-free, so-called "dry valleys" of Antarctica), footprints were evident in the dry, impressionable soil after more than a decade. We saw sinuous, scar-like, centuries-old caravan, wagon, and herd trails. In short, we became expert at distinguishing "undisturbed" areas from "contaminated" areas.

In one of the "remote" sites in the Mecca Hills of the Colorado Desert that we chose in 1961 for its extreme aridity and its similarity

Deserts

to the Tanzerouft of the Sahara and to the Atacama Desert of northern Chile, we observed that even footprints left indelible marks on the barren clay hills. We had to abandon the site because the hills became crisscrossed with the tracks of those who had no obvious objective other than to leave their "signatures in the soil." The final insult came when a road developed in the hills, bringing us a mechanical corpse—a junked and stripped car!

Weiler (1969) estimated that it would cost $1 million to remove junked items from public lands in the southwestern United States, $10 million to remove the trash from some 2,000 unauthorized personal dump sites, and $3.5 million to clean up the litter only from the lands of the Bureau of Land Management—not to mention the astronomical cost of cleaning up public roads throughout the West on Forest Service, Park Service, Bureau of Reclamation, and other public lands.

On the following pages, I note some of the observations I have made in the desert biome, especially in North America, and will mention some of the unique problems of preserving desert ecosystems. In particular, I will discuss the recreational use of deserts—specifically the increased use of off-road vehicles (ORVs); military and Indian reservations; and mining. Finally, I will speculate on the prospects for preserving deserts in view of the world's probable future needs.

Recreation and Off-Road Vehicles

Mechanized "Nomads"

Although it is true in general that "all types of land use are centered around the use of available water" (Gerakis 1974), the recreational use of the desert in many cases does not depend upon a constant or large quantity of water: the modern "nomads" of the desert are mechanized, unlike those of the many arid and semiarid regions of Asia and Africa, where nomadism involving animals is still an intrinsic mode of existence.

I have seen an increasing use of the desert for recreation and have been especially impressed by the widespread use of a great variety of ORVs (motorcycles, jeeps, dunebuggies, four-wheel-drives, etc.), which have a high potential for damaging or destroying the vegetation, wildlife, and other fragile wilderness values of deserts (Figures 1, 2, and 3). The use of ORVs can conflict with other types of recreation, and while the number of people who use ORVs is relatively small, the damage they cause is far out of proportion to their numbers and is most serious in arid areas (Council on Environmental Quality 1974). Motorcycles are among the worst offenders.

Uncontrolled Use

The intensive use of some 6.5 million hectares of desert in southern California for recreation is almost completely uncontrolled. It conflicts sharply with the interests of stockmen, agriculturists, miners, and recreationists who do not use ORVs and is a serious threat to the

Figure 1. The effects of ORV Activity on the Mohave Desert, Dove Spring Canyon, California, in 1968. (Courtesy of L. Boll, Bureau of Land Management, Bakersfield, California.)

Deserts 419

preservation of desert ecosystems (AAAS Committee on Arid Lands 1974). The weekend exodus from the Los Angeles area to the deserts is a formidable sight!

The most destructive ORV activity is the annual Hare and Hound Motorcycle Race, which takes place on Thanksgiving Day over nearly 250 kilometers of fragile Mohave Desert between Barstow, California, and Las Vegas, Nevada. In 1968, there were 900 entries in the race; in 1971, 2,900 entries; and in 1972, 2,600 entries. In 1974, 3,000 motorcyclists ripped across the desert in the eighth annual race, even though a California archaeologist had pointed out to the Bureau of Land Management (BLM) that the race was cutting a 400-meter-wide swath through very early archaeological sites (Broadbent 1974). BLM's own environmental impact statement recommended against permitting the race (Anonymous 1974a).

The race caused devastation on a broad scale, and the predictions

Figure 2. The Same Site as in Figure 1, in 1970. (Courtesy of L. Boll.)

420   Natural Systems

the BLM scientists had made in their environmental impact statement were borne out. The "area of influence of the race course increased ... 1,632 acres [660 hectares] or 31%." Subsequent (and consequent) wind-caused erosion destroyed seedlings that had germinated before the disturbance, and it threatened other plants as well; the race increased thirty-day particulate levels in the air by 31 percent; many archaeological and historical sites were irreparably damaged; and the population of small mammals in the study area was reduced by 90 percent!

Vandalism

Vandalism of Archeological Sites. Vandalism is an historical and worldwide phenomenon in deserts, whether it be due to ORVs; the breaking up and hawking of pieces of the Dead Sea Scrolls; the

Figure 3. The Same Site as in Figure 1, in 1972. (Courtesy of L. Boll.)

robbing of ancient Egyptian tombs and the hauling away of the artifacts to European museums; or the blackmarketing of ancient, pocket-sized clay Inca figurines. In the deserts of the southwestern United States, the wealth of fossil sites is exploited by "pothunters," and archaeological sites have sustained substantial damage from ORVs (Anonymous 1975a). Only a small proportion of archaeological sites has been found or studied by scientists (Broadbent 1974, and personal communication). Intaglios (religious artforms cut into the desert floor by prehistoric Indians) near Blythe, California, clearly show deliberate vandalization by ORV users (Council on Environmental Quality 1974). Stebbins (1974) has estimated that if the present rate of ORV use continues, 80 percent of the artifacts will be destroyed within ten to fifteen years. (This does not include the vandalism of archaeological sites by ORV "pothunters.")

Vandalism of Plants and Animals. Man frequently operates on the sand-castle theory: one person makes a sand castle, and for no obvious reason, another destroys it. This principle also operates in regard to nature's "sand castles," not only landforms or terrain, but plants and animals also. In arid southwestern California, relict Washingtonia palms have been burned by vandals. Sometimes the destruction is not deliberate, but merely thoughtless—as when people cut open barrel cacti to see whether they *really* do contain water. In other cases, commercial dealers or private individuals remove plants for window pots or yards, with the same result.

As for the animals, why not have a desert tortoise in the backyard? Rattlesnakes are deadly, so why not kill them? That jack rabbit, skunk, squirrel, or coyote may be carrying rabies, so better shoot it. Tarantulas aren't too dangerous after all, so let's take one home!

Threatened and Endangered Species: The Case for Biotic Diversity

It is an unspoken maxim, but one not everyone recognizes, that all animals ultimately depend on plants for their food and survival—a fact that is very obvious to anyone who travels extensively in deserts: no plants, no animals. One becomes aware of the precarious existence of desert animals, even those that are "adapted" to the harshness of the desert environment. Too many people do not realize that once the vegetation is gone the animals must go too; nor do they realize that the gene pool of a species gone to extinction is lost forever. It is not easy to convince people that not only is a wide variety of plant species necessary for the long-term welfare of human beings, but that the dependent fauna is also necessary for ecological stability.

There are now a number of threatened and endangered species of desert plants. As in other biomes, they are usually narrow endemics—that is, they occur only in restricted ranges and in fragile, threatened habitats. ("Endangered species" are in more immediate danger than "threatened species" and may become extinct unless they are protected or the threats to their existence removed.)

The United States. There are several major centers of endemism in the arid and semiarid areas of Texas, California, Nevada, Arizona, Utah (Anonymous 1975b), and—perhaps—New Mexico (Dick-Peddie 1975, and personal communication). Endemic species often exist in low numbers and inhabit quite small geographic ranges. Hence, it is not surprising that a report (Anonymous 1975b) on the endangered and threatened plants of the United States, prepared by the Smithsonian Institution pursuant to the Endangered Species Act of 1973, reveals that the largest number of endangered species are succulents from arid or semiarid areas, which are commercially exploited or privately collected. Fortunately, it already is illegal to take such plants in some states.

On the international level, the United States has signed the International Convention on International Trade in Endangered Species of Wild Fauna and Flora, which prohibits trade in all species of the Cactaceae (Order Opuntiales). However, as Dr. Robert Jenkins of the Nature Conservancy and others have rather cynically remarked, "I never heard of anyone arrested for taking plants."

In the deserts of the southwestern United States, fish are the most severely affected group of animals; in Arizona, half or more of the species are depleted or extinct (according to a personal communication from Jenkins). This loss and endangerment of species is attributed to the indiscriminate use of water, the pumping dry of streams, and soil erosion. Obviously, both education and laws with adequate provisions for enforcement are necessary if endangered animals and plants are to be protected and preserved.

Other Arid Areas. Grazing by domestic animals (cattle, sheep, and especially goats), considered the primary cause of desertification, has not only left scars and barren areas, but has cut deeply into populations of wildlife. For example, in the arid Galápagos Islands, the cradle of Darwinism, the depredations of man and beast (goats, dogs, and rats) have eliminated, or threaten to eliminate, several species of birds and reptiles, including species of giant tortoises.

Other endangered species of arid or semiarid areas are the Arabian ostrich and the West African ostrich *(Struthio camelus syriacus* and *S. c. spatzi,* respectively); the Andean condor and the California condor *(Vultur gryphus* and *Gymnogyps californianus,* respectively); various species of eagles, hawks, and falcons (Order Falconiformes); Attwater's prairie chicken *(Tympanuchus cupido attwateri);* and the masked bobwhite quail *(Colinus virginianus ridgwayi).* More than thirty species of marsupial are endangered in Australia, as are some fifteen species of rodent in the United States and Australia. Over thirty species of even-toed ungulate (Order Artiodactyla) are endangered in arid areas around the world, among them several species of deer *(Cervus* spp., subspecies of *Odocoileus virginianus,* etc.), gazelles (primarily *Gazella* spp.), ibex *(Capra walie),* oryx *(Oryx leucoryx),* Shou *(Cervus elaphus wallichi),* stags *(C. e. barbarus* and *C. e. hanglu),* and yak *(Bos grunniens mutus)* (U.S. Department of the Interior 1974). [For more detailed information on rare and endangered species of animals, see *Red Data Book,* vol. 1 *(Mammalia),* vol. 2 *(Aves),* vol. 3 *(Amphibia and Reptilia),* and vol. 4 *(Pisces),* Secretariat of IUCN, compiler, Morges, Switzerland: International Union for Conservation of Nature and Natural Resources, 1972 and after. Flowering plants are treated in vol. 5 *(Angiospermae),* 1970.—Editor]

The Soils of Deserts

Soil Ecosystems

Much of my research in deserts has dealt with soil ecosystems. While soil may not be "pretty," except to the soil scientist, it is a prerequisite for the existence of life in the desert: not only is it a source of nutrients, but it is substrate and a habitat as well. Just as animals are dependent upon plants, so plants are dependent upon the soil. All life, in fact, is ultimately dependent upon the soil. Even the oceans are nourished and renewed by the soil and soil constituents that wash into them.

Two principal soil-forming factors are time and climate. While a soil may have formed many thousands of years ago, under different climatic conditions, it is in continuous "dynamic equilibrium" with its environment. Shifts in the physical environment, and hence in the soil-environment equilibrium, are normally slow and best measured on a geological scale of time.

In arid and semiarid regions, stability of the soil is extremely important: when the soil is disturbed, it erodes rapidly. Since a desert preserve is in essence a soil preserve, desert soils must be treated as gently as any animal or plant.

Disturbance of Desert Soils

Foci of Infestation. The soil ecologist considers disruption of the soil to be the most serious threat to desert ecosystems as a whole. Thus, there is reason for great concern when the soil is disrupted over wide areas by overgrazing or the indiscriminate passage of ORVs, which break the soil "pavement." Not only is the soil loosened thereby, permitting the wind to pick it up and carry it away, but it is also exposed to the impact of rain and to the transporting power of running water, which in deserts may come as a deluge.

When small areas of soil and ground cover are disrupted in the desert, they become "foci of infestation" for much more than their immediate environs. There is a far-reaching "ripple" or "domino" effect, both in space and in time, much like that of a disease which spreads through an organism—in some cases resembling a malignancy, in others forming benign and localized "blemishes." Wind transports clays and other soil "fines," salts, and trace metals throughout the drainage system, depositing silt and salt on vegetation and in reservoirs, leading to destructive floods.

Soil Crusts. In deserts, soil-stabilizing crusts of microphytes (algae, lichens, mosses, liverworts, and so on) are common, forming series of distinguishable climax communities along moisture gradients. They are significant sources of nitrogen and carbon for desert ecosystems, and hence of soil fertility, and they carry out other indispensible transformations. Not only trees, shrubs, and grasses, but these very important crusts may be disrupted. As you would expect, this causes the subsequent disruption, alteration, dispersal, or destruction of a wide variety of dependent animals, from microscopic protozoans to insects, birds, fish, amphibians, reptiles, rodents, and larger mammals. It may take five or more years of favorable solclime (soil-climatic conditions) for a disrupted crust to become reestablished.

Dust Clouds. Because crusts of soil-stabilizing microphytes are so important to desert ecosystems, and the consequences of disrupting them so dire, it is difficult to justify any use of the desert's surface before the impact has been assessed. In the Hare and Hound Race of 1974, more than 7 million cubic centimeters of soil cover were removed just before and after the race in a 100-meter by 100-meter study area; 2 weeks later, 4.7 million cubic centimeters more had been lost (Anonymous 1974b). These values, multiplied by the 250-kilometer by 0.4-kilometer racing trail, yield an astronomical figure!

The 600-meter cloud of dust raised during the race significantly increased the turbidity of the atmosphere, intercepting sunlight and—perhaps—intensifying the "icebox effect" and furthering the worldwide expansion of glaciers (even if only infinitesimally).

Who can estimate the real damage done to the desert by ORVs? The indiscriminate use of ORVs in the desert is neither right nor privilege—it is sacrilege.

Military Bases

Warfare and Deserts

So far as I can determine, nearly all countries that contain arid areas either maintain military bases or conduct military exercises and testing in them. Nuclear devices have been tested in remote arid areas of the United States, the Soviet Union, China, and India, and in the Sahara by France. Other countries also wanting to be members of the "nuclear club" will test similar devices; this undoubtedly will occur in the Middle East.

The effects of nuclear testing on arid regions were well

documented by studies of the Nevada Test Site in the United States (Wallace and Romney 1972), and studies (especially of fauna and flora) are still being undertaken, decades after the first underground detonations.

Though armies have ravaged deserts for centuries by occupying or traversing them, modern warfare (as in the North African campaigns of World War II and lately in the Middle East) can devastate fragile desert environments. The land, its inhabitants, archeological sites, animals, and plants often lie in the path of military operations.

It is well known, for example, that Ottoman troops eliminated, for several hundred years, virtually all forested areas of Biblical fame by cutting them for firewood, with no thought of replacing them. They thereby eliminated many of the original animals and increased the extent of desert. Truly, it is an unusual individual, such as our panelist, General Avraham Yoffe, who will move an entire army encampment during wartime to save a stand of rare wildflowers (Rothstein 1972).

Practice Maneuvers

Vast areas of the southwestern United States are occupied by military reservations, much of them "off limits" to nonmilitary personnel and therefore in a sense "protected." However, certain areas—bombing and gunnery ranges—used for target practice are dangerous and will be unusable for other purposes until they have been cleared of duds, some of which remain from military exercises carried out during World War I.

During World War II, General Patton's armored divisions chewed up an area along the Arizona-California border near Yuma. The damage is still evident. The military base occupies a zone of transition between the Sonora and Yuma Deserts—an area invaluable for the study of transitions in landscape, soil, fauna, flora, and microbiota. My own research team was ejected from the area while it was studying desert ecology there. To avoid future "invasions," the military shortly thereafter fenced and posted the area as off limits to the public. Unfortunately, that desert-transition ecosystem is unique—found nowhere else in the world. I cannot predict whether it will continue to be "preserved."

More deplorable, perhaps, was "Desert Strike," a month-long

practice maneuver staged by the military in the middle and eastern Mohave Desert in 1964. Approximately 100,000 men and many kinds of heavy equipment took part (Jaeger 1965). The maneuver destroyed desert vegetation, wildlife, and wildlife habitat over an enormous area. I doubt that any thought was given to preservation of the area, and I know of no attempt to evaluate or repair the damage. Smaller such operations continue to this day, but there appears to be no requirement that environmental impact statements (EIS) be filed on this kind of abuse.

Research Budgets

The use of military lands within the United States concerns environmentalists because more than 6 million hectares of land are under the jurisdiction of the Department of Defense, a large proportion of it in the western states (Bureau of Land Management 1973). Some 40 percent of the world's research money is spent on weapons systems, with 70 to 80 billion dollars being spent annually by the United States Department of Defense. By comparison, an insignificant amount is spent on desert research, of which only a "drop" is spent to protect and conserve natural desert areas!

Indian Reservations

The stewardship of American Indian lands is seldom mentioned, although much is said about the plight of the Indians and the injustices they have suffered. I will not discuss the details of causes and effects, lament the present lot of many Indians, nor compare the exploitation of their resources with that of the Third World's (see, for example, Brom 1974).

Despite their economic and social ills, the American Indians are the custodians of the more than 22 million hectares of land under the jurisdiction of the Bureau of Indian Affairs, a substantial proportion of it in arid and semiarid western states.

For example, in Arizona, with a population of more than 100,000 Indians, there are almost 8.1 million hectares of Indian land and nearly 7.7 million hectares of tribal land occupied mostly by the Navajo Nation—in reality, a nation within a nation. The next largest expanse of tribal Indian lands, 2.4 million hectares, also is situated in the arid southwest, in the state of New Mexico.

Since they cannot return to their historic nomadic way of life, the American Indians have had to adopt a sedentary existence, irrigation-assisted agriculture and the raising of sheep and cattle being their major occupations. As in the Middle East and the Sahara, the land has suffered from overgrazing and the long-term use of poor-quality water. This is also true, of course, of non-Indian lands in the United States, but the Indians often lead more of a marginal existence than do non-Indians and therefore may cause more harm to their land, which often is not suited to agriculture or grazing.

There is a great need for education on the tribal level. The Indians should realize that the ecological concerns of the non-Indians are now theirs also. "Laissez-faire" is no longer an acceptable way of life for anyone.

Recently, the Indians (among them, the Crow, Sioux, Apache, Shoshone, Cheyenne, Ute, Zuñi, Navajo, and Hopi) have become concerned about the intensifying impact on their land of increased national and international requirements for energy. The rate of disturbance has accelerated as the need has grown.

The Navajo lands contain the largest coal deposit in the United States—or in the world, for that matter—and by 1972, surface mining had disturbed approximately 1,200 hectares of their land (Box 1974). As you might expect, the Indians are concerned about these threats to their land, culture, and historic sites: not only will surface mining take their land out of production for years, destroy or displace their historic artifacts, and change their mode of life, but it could affect their water resources, possibly intensifying drought.

The establishment of coal-gasification complexes creates a need for water (which competes with the Indians' need), causes noise and air pollution, and requires the disposal of fly ash. New towns, which will have sociocultural impacts of concern to the Indians as well as to others, will have to be constructed (Baldwin 1975). For example, it is expected that 15,000 residents will be drawn to a new community by 1978 if plans for a gasification plant on the New Mexico Navajo reservation are realized.

Although the mined land may be rehabilitated or reclaimed to productive use, it will not be "reassembled" in its original state: the Indian will never again see the land as his forefathers saw it. On the other hand, the mining companies are concerned that the Indian will attempt to graze the reclaimed land prematurely, thereby delaying or inhibiting its "full" use. Needless to say, the impact on the desert ecosystem in these areas will be tremendous, and preservation and protection of the desert may be a subordinate, after-the-fact concern again.

An updated version of Custer's Last Stand has been taking place as the Indians' lands have been progressively leased for mining purposes: these desert denizens may find themselves included, along with the land, in "National Sacrifice Areas" (Brom 1974). Any apparent victories, as that over General Custer, will have longer-ranging effects, also to their disadvantage.

Strip Mining

Impacts in Arid Regions

Open-pit and strip mining are proliferating in arid lands. The large pits, such as the copper mines at Chuquicamata in the Atacama Desert of Chile and the 790-meter-deep pit at Bingham, Utah, in the Great Basin Desert of the United States, are outstanding proof of that fact. Surface mines, while they are less expensive, more efficient, and safer than underground mines, create other problems, the major one being that they remake and essentially destroy the landscape until it can be reclaimed or rehabilitated to productive use. (In mining terminology, the land is not "restored," that is, returned to its original, undisturbed, pre-mined state; it is "reclaimed" or "rehabilitated.")

The inevitable pits and piles of overburden, or spoils, which the general public may consider as only waste and an eyesore, are not the most important problems created by surface mining; the damage occurs as existing natural communities and the structure of the soil are destroyed; aquifers disrupted; land rendered unproductive for years; and the quality of water degraded by the consequent turbidity, salinity, and high levels of toxic trace metals. Also, the competition for water from other interests aggravates the problem.

Coal

Half of the world's known coal reserves are situated in the United States, most of them in the states west of the 100th meridian (Figure 4). It is low-sulfur coal and occurs in beds 15 or more meters thick. Nearly 52 million hectares of land in the western United States are underlain by coal, and approximately 600,000 hectares

(25 billion tons of coal) can be surface mined by current methods (Figure 5) (Box 1974).

In the United States, the surface mining of coal will be a very important industry in the near future. It has already begun in Montana, Wyoming, New Mexico, the Dakotas, Arizona, Utah, and Colorado (see Figure 6). It will have a significant impact on the environment of large areas of the arid and semiarid United States. Even though mining interests must face the fact that public lands

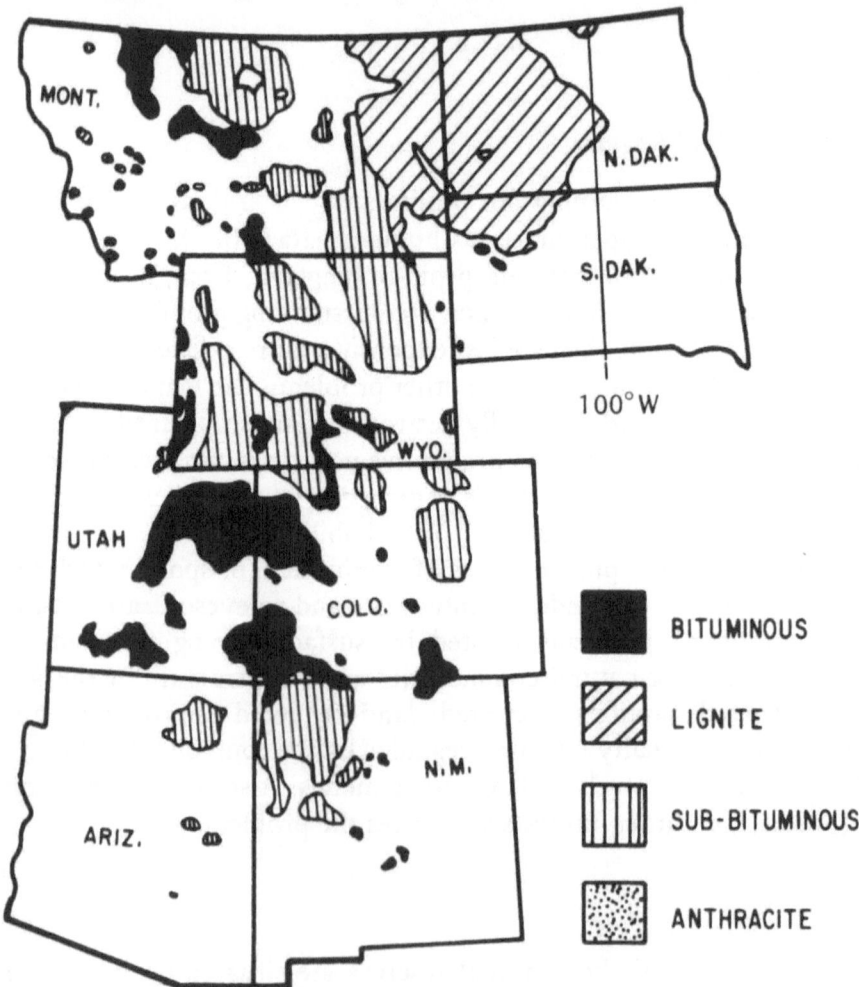

Figure 4. *Coal Resource Regions of the United States West of the 100th Meridian.*

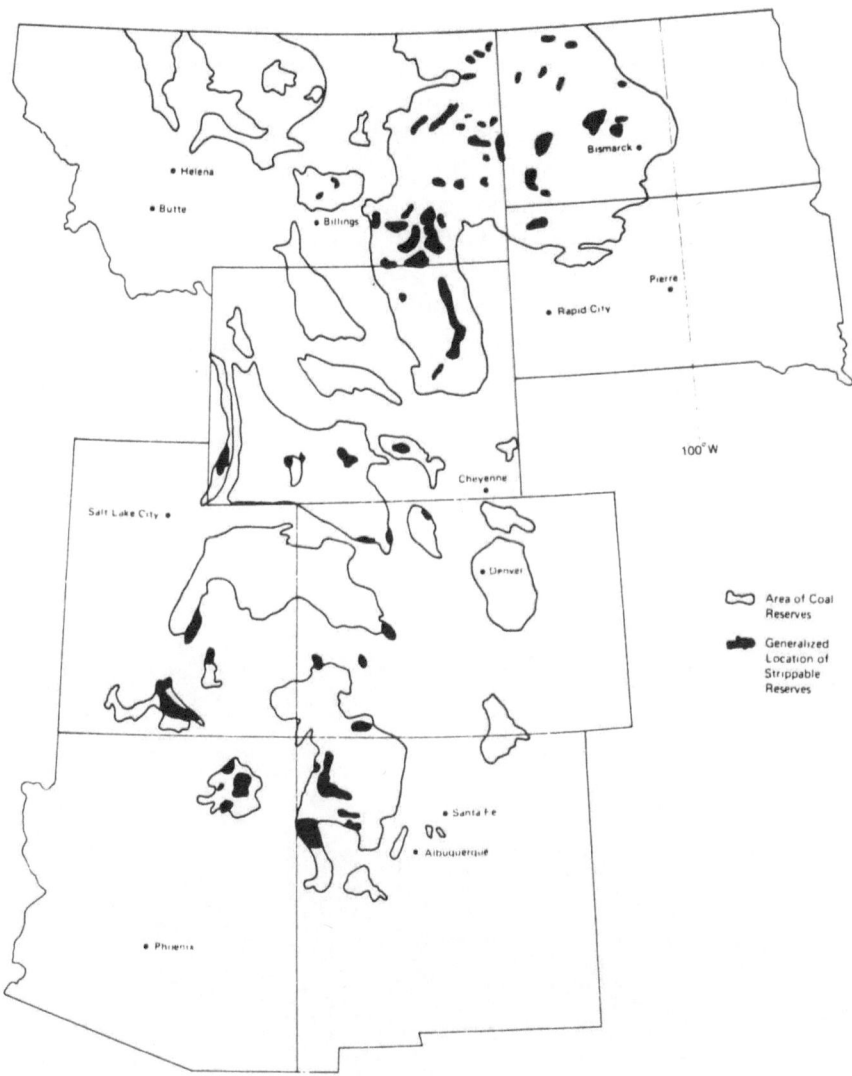

Figure 5. Sketch Map of Strippable Coal Reserved West of the 100th Meridian (after Box 1974).

are to be shared with recreationists (Weiler 1969), the need for energy is now crucial, and coal-mining companies, as well as uranium companies, have been very active in the West.

The three largest coal-mining companies in the United States—Kennecott Copper, Continental Oil, and Occidental Oil—are owned by copper and oil companies. They and other companies have obtained long-term lease agreements on public and Indian lands, for the most part on a "preferential-right" basis, wherein an exclusive prospecting permit was exchanged for a government lease at a cost of a $10 filing fee (Brom 1974). Other leases were given (for a nominal single low bid of $3.31 per acre [1 acre is equivalent to 0.4047 hectare]) for 330,000 hectares of public land underlain by billions of tons of western coal (Brom 1974)! In addition, power companies are obtaining crucial water rights (more than half the coal produced goes to utility companies) (Miller 1975).

Other companies, taking advantage of archaic mining laws, are

Figure 6. A Strip Mine in the Western United States. Notice the pit in the foreground and spoilbanks in the rear and at the right. (Courtesy of Marvin Singer, Council on Environmental Quality.)

exploring and staking claims for uranium. Very little is known about the reclamation of land mined for uranium (Walchek 1975), or the concurrent impact of uranium-milling operations (personal observations).

Coal Gasification

In addition to strip mining, large-scale coal-gasification plants are also being considered. Such plants, which convert coal into a synthetic gas similar to natural gas, offer the only workable solution to crippling curtailments of gas supplies in the immediate future.

The construction of thirty-six coal-gasification plants is planned by the Federal Power Commission by 1990. It has been calculated that if they are built in the western United States, each one of them will use 8 million tons of coal and from 25 million to 37 million cubic meters of water a year and could make necessary the rehabilitation of 1,000 square kilometers of land (Malde 1975).

Of utmost concern to many people is the fact that the mined land in the West can not be restored—can not be put back the way it was or, in many cases, put back to its former use. Rehabilitation—a return to productivity—may be the best that can be accomplished for some of the land. The term *potential use* implies more than the mere establishment and ecological succession of plant communities: it also implies a complex array of geologic, pedologic, hydrologic, biologic, climatic, and micrometeorological factors. In addition, a less easily defined array of social, political, economic, and esthetic factors reflects the impact of mining. I very much doubt that "95 percent of the disturbed land can . . . be successfully reclaimed" (Anonymous 1975c)—if the lands concerned are arid and semiarid western lands.

Water and Topsoil

Water will be the key to mining in the West, yet 20 percent of the United States west of the 100th meridian receives an average of less than 25 centimeters of rainfall a year, 25 percent receives between 25 and 38 centimeters, and 30 percent receives between 38 and 50 centimeters; thus, 75 percent of the region receives less than 50 centimeters of annual precipitation (May 1967). A committee of the National Academy of Sciences found that areas receiving 25 centimeters or more of rainfall per year could be rehabilitated only if the landscape were properly shaped and if successful techniques were demonstrated. None of the committee's estimates were based on comprehensive, long-term, controlled experiments, however (Box 1974).

A knowledgeable environmental geologist has pointed out that the soils of the Great Basin are relict soils, formed during two short periods between about 1,500 and 5,000 years ago. They are similar to relict surface soils in Utah, Colorado, Wyoming, New Mexico, Arizona, and California (Curry 1973). The claim has been made that topsoil removed from mined sites can be stored and then replaced; the claim does not take into account, however, the fact that the structure of the soil has been destroyed, organic matter lost, and the microflora crucially altered (Miller and Cameron 1976).

The amount of topsoil removed varies from a few centimeters to a meter or so in depth. In this case, the topsoil has been treated more in a mining or soil-engineering context, rather than in a pedologic, agronomic, or ecological context. No one yet knows to what extent the topsoil changes when it is stored, nor whether it is even possible to put the soil back into place effectively.

Obviously, the strip mining of coal will have a monumental impact on each site; but coal will nevertheless be mined, for it is needed—especially since the oil-exporting nations continue to have such a tight grip on the world's supply of fuel.

Oil Shale

In addition to coal, oil-bearing shale deposits in the United States are being considered as a source of synthetic fuels. It is estimated that 100 billion barrels of oil [1 barrel is approximately equivalent to 159 liters.—Editor] could be obtained from shale in Colorado, Utah, and Wyoming (Seamans 1975). If the oil-shale deposits are mined, even more of the land and its associated biotic communities will be disrupted.

Needless to say, the arid and semiarid western United States face huge impacts from impending mining activities. Procedures for carrying out the recovery of mined sites have been developed (Box 1974, Curry 1973, Malde 1975), but they offer few grounds for optimism. Efforts to preserve deserts also face severe challenges from the impending mining. Legislation that prohibits mining in easily damaged arid areas remains to be tested, as do the recovery procedures, since it is "unlikely that a stable plant community comparable to undisturbed areas can be expected in less than ten to twenty years" (Box 1974). The use of native species will be emphasized in recovery, but introduced species will also be used (Box 1974). While they are useful for reclaiming desert areas, introduced species do not contribute to the preservation of desert areas, and there are few commercial sources of native seeds so far.

Prognosis

Prognostication has become acceptable even in the most conservative of scientific circles. Permit me, therefore, to speculate on the likelihood that efforts to preserve the world's deserts will succeed—especially deserts considered to have a high economic potential. In this regard, I must state that the preservation of deserts undoubtedly will be subordinated to the pressing global need, due to increased world population, for food, energy, relief from pollution and wastes, and shelter.

The Importance of Water

Fortunately (or unfortunately, depending upon one's point of view and upon one's wish to preserve deserts), water will continue to be the crux of both the problems and their solution in arid and semiarid regions: its use, management, supply, and recycling will become increasingly more difficult.

At present, Earth's deserts have limited "carrying capacities" for human beings, determined primarily by the effectiveness of water—that is, by its quantity, quality, duration, seasonality, and physical availability. Antarctica, for example, is an ice-bound desert having more than 90 percent of the world's fresh water, but it is presently both economically and physically unavailable.

In some areas, such as the Tucson Basin in Arizona, subterranean water-bearing strata have been "mined" in the sense that their fresh water has not been replaced. It would be impossible, within a geological scale of time, to satisfy the area's immediate or future needs.

Some people claim that there is "plenty" of water in deserts—

as in those of Asia—albeit water of poor quality and situated at great depths underground (Petrov 1970).

But access to sufficient water will not eliminate intense solar radiation or strong winds from deserts, nor should it be forgotten that irrigation does not solve the problem of saline or calcereous soil, a fact often demonstrated within historic time. The so-called domino effect applies to irrigation, in that subsequent problems develop (Dregne 1970), problems that may be worse than their precursors.

Changes in local or regional climate that would increase humidity might occur in some desert areas, as they have in extensively revegetated areas of Israel. However, a worldwide change in climate, whether natural or as a consequence of man's activities, may cause either a new ice age or the converse (Kellogg and Schneider 1974): the desert will "bloom like the rose" or intense desertification will set in, and the preservation of deserts may then be beside the point.

Desert "Islands"

One indirect way to preserve and protect deserts, especially in the face of humanity's need for food and shelter, as well as for *Lebensraum*, is to create desert "islands"—artificial, manmade "oases" that are self-sustaining, or nearly so, in otherwise naturally waterless areas. Such islands or oases would be essentially closed life-support systems like those in spaceships and space stations and proposed for use in extraterrestrial colonies.

Hydroponics, desalinization, the use of solar or nuclear energy, and the recycling of water, wastes, and even air, along with stringent conservation measures, could be undertaken in these artificial oases at a level previously unknown on our planet. Such "experiments in desert living" would be useful for extraterrestrial colonization if population pressures on Earth should continue to increase.

The Soviets, for example, consider their Antarctic bases training grounds for extraterrestrial colonies (Quartermain 1966). The "Autonomous Living Unit" (ALU-1), designed and constructed at the University of Arizona, is a compact-living module that fits the natural arid environment and is a sign of the future (Paylore undated).

If the difficult problems of harvesting plankton can be solved, human beings may have an energy supply equivalent to that presently derived from the combustion of oil and natural gas. The mass culturing of algae may someday become the desert's major industry, reducing the spectre of worldwide hunger. There are already large-scale, plastic-enclosed power-, water-, and food-producing plants in arid regions of the United States, Mexico, Abu Dhabi, and Iran (Paylore undated). Respect for the desert, for its preservation and protection, could be a welcome "spinoff" from such endeavors.

Recreation in the Desert

The undisciplined recreational use of the desert will spread like a blight unless man's activities in these areas are regulated. If the costs of fuel or vehicles increase sharply, then recreational use of the desert may decrease, as it did during the curtailment of oil in late 1973. Similarly, there must be legislation to control the numbers of vehicles and people in the desert, and the legislation must be enforced.

BLM estimated the "visitor-days" to California's deserts as 4.3 million in 1968 and 13 million in 1973, and predicts it will reach 50 million by the year 2000, 70 percent of it involved with ORVs (Anonymous 1974b). At present, most BLM land (over 48 million hectares in the western states) is open to ORVs. In a current lawsuit, plaintiffs want BLM lands closed until they are declared open and until an adequate EIS has been published (Rehm 1974). However, BLM can ignore the recommendations of an EIS.

Concerned-citizens groups are far outnumbered by those seeking relief from urban pressures and needing open space, silence, and the recreational amenities of deserts. Their need for escape is manifested on a grand scale in the desert "backyard." (In the United States, a prevailing attitude is that what goes on in one's backyard is nobody else's business.) Government organizations and bureaus (BLM, for example) are currently adequate neither in number nor authority to be regulatory and enforcing powers in the desert, and the ORV groups, whether organized or not, are powerful because they have considerable economic clout. (Witness, for example, the U.S. television commercials showing the allure of using ORVs to chew up the desert landscape.)

Mining and Related Activities

Mining and utility interests will not be dissuaded from their goals in desert areas so long as there are pressing national and international demands for minerals and energy, although environmental protection is of increasing concern in the surface mining of coal (Grim and Hill 1974). The government and the general public will back them up: greed and comfort are strong motivating forces.

An episode that actually occurred proves that they will not be dissuaded. In Montana, a rancher who refused to part with his land for a large-scale commercial operation was given this parting shot by the official who had approached him: "We'll get you in the end" (Miller 1975). Threats of this kind eventually become realities, raising the question of whether personal freedom still exists (Toole 1976).

The rehabilitation or reclamation of mined lands in very arid regions has been declared impossible (Curry 1973); certainly, it is improbable that it will become stable or "permanent" unless there are technological breakthroughs—climate, proper use of the soil, and effective moisture, of course, being key factors. Data that portend continued droughts are not encouraging for arid areas (Hyder 1975).

If the land continues to be mined in the absence of any means to rehabilitate or reclaim it, "National Sacrifice Areas" (a new Appalachia) may soon be evident in a "Western" setting (Brom 1974). Mining will take precedence over the production of food in desert areas, at least until the mining is essentially over. Nevertheless, the fact that deserts are mankind's future "land bank" must not be forgotten (Weiler 1969).

I predict that desert preserves—if they are even recognizable at some future time—will not be significantly enlarged, although attempts are presently being made to do so. Most likely, they will become shrinking "islands," as have wilderness areas in other biomes—natural zoos within an overpopulated human zoo.

Population

The preservation and protection of deserts will be complicated by the uncontrolled growth of Earth's human population. Half a decade ago, it was stated at an international conference on arid lands that the unanswered question is "whether mankind will permit the population to increase indefinitely at the current rapid rate" (Dregne 1970). Unfortunately, the answer has been "yes," since in five years the rate has increased 0.2 percent beyond the predicted 2.0 percent (Cook 1974).

There is only one "spaceship earth," and, as scientists have observed, not even rats care about other rats when they are too crowded and don't have enough to eat. Simply stated, on the human level: callousness will increase as population multiplies. It may be more than speculation to say that it all began with a bang, and that it will end with a "bust" unless mankind comes to its senses and agrees on a common way out of this world of increasing perplexities.

Epilogue

Man puts his hand to the flinty rock and overturns mountains by the roots.

He cuts out channels in the rocks, and his eye sees every precious thing.

He binds up the streams so that they do not trickle, and the thing that is hid, he brings to light.

—Job 28:9-11

References

AAAS Committee on Arid Lands. 1974. Off-road vehicle use. *Science* 184: 500-501.

Anonymous. 1974a. Barstow to Las Vegas: A license to destroy. *The ORV Monitor* [Environmental Defense Fund, Berkeley, California] (December), pp. 3-5, 8.

———. 1974b. ORV's and the California desert: A case study. *The ORV Monitor* [Environmental Defense Fund, Berkeley, California] (June), p. 2.

———. 1975a. Desert's quiet users heard. *The ORV Monitor* [Environmental Defense Fund, Berkeley, California] (February/March), p. 4.

———. 1975b. *Report on Endangered and Threatened Plant Species of the United States.* Presented to the Congress of the United States of America by the Secretary, Smithsonian Institution. Serial Number 94-A. 94th Congress, 1st Session, House Document Number 94-51. Committee on Merchant Marine and Fisheries. U.S. Government Printing Office, Washington, D.C.

———. 1975c. Hathaway is endorsed by Sawhill who cites "balance." *Weekly Energy Report* 3(16):8.

Baldwin, T. E. 1975. WESCO Environmental Impact Statement Hearings in Window Rock, Arizona (12 March 1975), and Farmington, New Mexico (13 March 1975). Summary. Argonne National Laboratory.

Box, T. W., ed. 1974. *Rehabilitation Potential of Western Coal Lands*. National Academy of Sciences and National Academy of Engineering. Cambridge, Mass.: Ballinger Publishing Co.

Broadbent, S. 1974. Scavengers on wheels. *Sierra Club Bulletin* 59(4):9-11.

Brom, T. 1974. The Southwest: America's new Appalachia. *Ramparts* 13(4): 17-20.

Bureau of Land Management. 1973. *Public Land Statistics*. Bureau of Land Management, U.S. Department of the Interior. Superintendent of Documents, U.S. Government Printing Office, Washington, D.C.

Cameron, R. E. 1972. Pollution and conservation of the Antarctic terrestrial ecosystem. In *Proceedings of the Colloquium on Conservation Problems in Antarctica* (Blacksburg, Virginia, 10-12 September 1971), ed. B. C. Parker, pp. 267-306. Lawrence, Kansas: Allen Press.

———. 1974. North American desert. In *The New Encyclopaedia Britannica. Macropaedia*. 15th ed., vol. 13, pp. 203-204. Chicago: Encyclopaedia Britannica.

Cameron, Roy E., Gerald B. Blank, and Doris R. Gensel. 1966. *Desert Soil Collection at the JPL Soil Science Laboratory*. Technical Report No. 9, pp. 32-77. Pasadena, Calif.: Jet Propulsion Laboratory.

Charney, J., P. H. Stone, and W. J. Quirk. 1975. Drought in the Sahara: A biogeophysical feedback mechanism. *Science* 187(4175):434-435.

Cook, R. C. 1974. *World Population Estimates, 1974*. Washington, D.C.: The Environmental Fund.

Council on Environmental Quality. 1974. *Environmental Quality*. Fifth Annual Report of the Council on Environmental Quality. Washington, D.C.: Superintendent of Documents, U.S. Government Printing Office.

Curry, Robert R. 1973. Reclamation Considerations for Arid Lands of Western United States. Hearing of the U.S. Senate Public Lands Subcommittee, Committee on Interior and Insular Affairs, Record of 15 March 1973.

Dick-Peddie, W. 1975. A Proposal for National Science Foundation Support of a Natural Area Workshop for New Mexico. Workshop, New Mexico State University, Las Cruces, 23 May 1975.

Dregne, H. E. 1970. The changing scene. In *Arid Lands in Transition*, ed. H. E. Dregne, pp. 7-12. (International Conference on Arid Lands in a Changing World, University of Arizona, 1969). Publication No. 90. Washington, D.C.: American Association for the Advancement of Science.

Gerakis, P. A. 1974. Environmental consequences of land use in arid and semi-arid areas. *Journal of Soil and Water Conservation* 29(4):160-164.

Grim, Elmore C., and Robert D. Hill. 1974. *Environmental Protection in Surface*

Mining of Coal. National Technical Information Service Report PB-238 538. Cincinnati: National Environmental Research Center, Office of Research and Development.

Hyder, Charles L. 1975. *Water Supplies and Water Consumption.* Testimony Regarding the WESCO Coal Gasification Plant Proposed Draft Environmental Statement prepared by the Bureau of Reclamation (INT DES 74-107), presented during March 1975 in Window Rock, Arizona.

Jaeger, Edmund C. 1965. The preservation of deserts. In *The California Deserts.* 4th ed., pp. 189-193. Stanford: Stanford University Press.

Kassas, M. 1970. Desertification versus potential for recovery in circum-Saharan territories. In *Arid Lands in Transition,* ed. H. E. Dregne, pp. 123-142.

Kellogg, W. W., and S. H. Schneider. 1974. Climate stabilization: For better or worse? *Science* 186(4170):1163-1172.

Le Houérou, H. N. 1970. North Africa: Past, present, and future. In *Arid Lands in Transition,* ed. H. E. Dregne, pp. 227-278.

Lowdermilk, Walter C. 1971. The reclamation of a man-made desert. In *Man and the Ecosphere: Readings from Scientific American,* eds. Paul R. Ehrlich, John P. Holdren, and Richard W. Holm, pp. 219-227. San Francisco: W. H. Freeman and Co.

Malde, H. E. 1975. Recovery and Rehabilitation of Coal Lands. Paper read at Coal Energy Symposium, 18 April 1975, at New Mexico State University, Las Cruces, New Mexico.

May, M. 1967. Mine reclamation in the western states. *Mining Congress Journal* 53(8):101-105.

Miller, A. 1975. The energy crisis as a coal miner sees it. *Montana Outdoors* 6(2):7-14.

Miller, R. Michael, and R. E. Cameron. 1976. Some effects on soil microbiota of topsoil storage during surface mining. In *Papers Presented Before the Fourth Symposium on Surface Mining and Reclamation,* Louisville, Kentucky, 19-21 October 1976, pp. 131-139. Washington, D.C.: National Coal Association, Bituminous Coal Research, Inc.

Paylore, Patricia. No date. The University of Arizona, Arid Lands Research Office. Tucson: The University of Arizona Press.

Perry, R. A. 1970. Productivity of arid Australia. In *Arid Lands in Transition,* ed. H. E. Dregne, pp. 303-316. (International Conference on Arid Lands in a Changing World, University of Arizona, 1969. Publication No. 90. Washington, D.C.: American Association for the Advancement of Science.

Petrov, M. P. 1970. Deserts of Asia. In *Arid Lands in Transition,* ed. H. E. Dregne, pp. 279-302.

Quartermain, L. B., editor. 1966. Testing ground for the moon. *Antarctic* [New Zealand Antarctic Society] 4(6):285.

Rehm, George. 1974. EDF files suit. *The ORV Monitor* [Environmental Defense Fund, Berkeley, California] (June), pp. 5-6.

Rothstein, R. 1972. General Yoffe's biggest battle. *Audubon* 74(2):64-67.

Seamans, Robert C., Jr. 1975. A unified approach to energy self-sufficiency. *Information from ERDA* 6(8):2-4.

Stebbins, Robert C. 1974. BLM desert plan found inadequate. *The ORV Monitor* [Environmental Defense Fund, Berkeley, California] (June), pp. 4-5.

Toole, K. Ross. 1976. *The Rape of the Great Plains: Northwest America, Cattle and Coal.* Boston: Little, Brown and Co.

U.S. Department of the Interior. 1974. *United States List of Endangered Fauna.* Office of Endangered Species and International Activities, Fish and Wildlife Service, U.S. Department of the Interior. Washington, D.C.

Walcheck, Ken. 1975. The long pines. *Montana Outdoors* 6(2):15-19.

Wallace, A., and E. M. Romney. 1972. *Radioecology and Ecophysiology of Desert Plants at the Nevada Test Site.* Biology and Medicine TID-25954. Washington, D.C.: Office of Information Services, U.S. Atomic Energy Commission.

Weiler, F. J. 1969. Public land management in the arid Southwest. In *Future Environments of Arid Regions of the Southwest,* ed. Gordon L. Bender, pp. 23-32. Contribution of the Committee on Desert and Arid Zone, Research No. 12. Tucson: Bureau of Publications, Arizona State University.

38. The Nature of Desert Ecosystems

Avraham Yoffe
Nature Reserves Authority, Israel

We must understand deserts in order to preserve them and benefit from them. To understand them, we first must characterize them. This we can do because each kind of ecosystem or biome has its own unique combination of distinguishing characteristics. The salient characteristics of desert ecosystems, in comparison with those of ecosystems in more humid regions, are as follows:

1. The biomass of desert ecosystems is lower than that of all other known terrestrial ecosystems, whereas their turnover rate is among the highest.
2. Desert ecosystems are highly unstable, but at the same time they are highly resilient. They are unstable in that their plant biomass (phytomass) and production fluctuate greatly from year to year, in direct relation to wide annual variations in climate. Desert ecosystems are highly resilient in that they are able to "bounce back" from low to relatively high levels of phytomass and production, in response to changes in environmental conditions.
3. The diversity of species in desert ecosystems is very low, but the diversity of control mechanisms (i.e., of mechanisms that permit them to adapt to their environment) is very high.
4. The net production, or biological yield, of mature or climax desert ecosystems, like that of most other mature ecosystems, is very low. Stated another way, as a desert ecosystem approaches successional climax, or maturity, most of its energy is diverted from the production of new biomass to the maintenance of existing biomass.
5. A consequence of the foregoing characteristics of deserts is that a very delicate equilibrium exists between the biota of a desert and its harsh environment, the main limiting factor being water.

The Actual and Potential Benefits of Desert Ecosystems

Most of the Earth's once natural ecosystems have been either destroyed or radically changed by man. Only in rain forests, deserts, and tundra do extensive areas of climax or near-climax biotic communities still exist. It is our duty to preserve and enhance the world's remaining natural, undisturbed desert ecosystems; only then can we derive benefits from them, benefits such as opportunities for research, education, recreation, agriculture, and the prevention and treatment of disease.

Research

Since many desert ecosystems are still close to their natural state, they afford an excellent opportunity for us to understand ecosystems in general. The following characteristics of ecosystems can be studied conveniently and to good advantage in deserts:

1. The structure of ecosystems—i.e., the species and numbers of plants, animals, and microorganisms that constitute ecosystems.
2. The dynamic relationships among the various biotic components of ecosystems.
3. The relationships between the biotic and the abiotic components of ecosystems—i.e., the interplay between plants, animals, and microorganisms on the one hand and soil, geological structure, and climate on the other.
4. The physiological function of each biotic component of ecosystems, in relation to the main limiting environmental factors—for example, water, salinity, and temperature.
5. The adaptive control mechanisms that permit the various biotic components to survive in the harsh environment.

Education

The understanding of past and present desert cultures is of high educational value. Ancient and modern desert civilizations have succeeded in adapting to their arid environment, which has profoundly affected their religions, philosophies, ways of life, and individual moral values.

It is worth noting in this regard that all three monotheistic religions originated in the desert.

Study of mechanisms by which plants and animals have adapted to their desert environment helps us understand our own possibilities and limitations in relation to nature. From the aesthetic point of view, the desert, virtually untouched by man, is unique in its primeval landscape and features. It influences contemporary poetry, art, philosophy, and moral values just as it affected human cultures in the past.

Recreation

Because their climate, geology, fauna, and flora often are completely different from those of areas in which most human beings live, deserts have recreational value. The mere presence of a completely different biota is rewarding in itself. The uniquely colored geological formations of deserts are another great attraction. Also, the many remains of ancient desert civilizations are of considerable general interest.

Agriculture

The combination of a desert climate and the availability of both fresh and brackish water in some desert areas permits the flourishing of an agriculture that can produce special kinds of products at unusual times, for both export and local use. The rift valley between the Red Sea and the Dead Sea is a classic example of such an area. Fifteen years ago, it was completely barren; today, it supports a thriving agricultural community that annually produces thousands of tons of vegetables and flowers.

Another possible use of the desert for agriculture involves self-contained greenhouses in which highly saline water, purified with solar energy, is used in a closed system or circuit to raise such crops as tomatoes, cucumbers, and flowers. The first experiments with such greenhouses promise unusually high yields for a comparatively low cost, since the driving force of the system is solar energy.

Medicinal and Curative Values of the Desert

The dry climate, pure air, cloudlessness, and intense sunshine of deserts have medicinal value against asthma and other respiratory disorders. Mineral springs, whose waters have special temperatures, occur in many deserts. They are useful for the treatment of a great many skin and other diseases. In addition to their medical value, the pure air and cloudless skies of deserts are valuable for such scientific pursuits as astronomy.

Threats to Desert Ecosystems

At least four human activities pose a threat to deserts and therefore to the special benefits people derive from them. The threats are overgrazing, the use of marginal areas for agriculture, mining and quarrying, and colonization.

Overgrazing

The paramount threat to deserts is overgrazing, which is the tendency of local human populations to permit animals to graze freely, without regard to the carrying capacity of range or pasture. This tendency already has led to the complete destruction of vegetation in some places, which in turn has caused severe economic loss to nomadic populations. In addition, overgrazing has turned nondesert areas into "deserts," a process termed "desertification." [The distinction between natural, extant, deserts, which deserve to be preserved in their natural condition, and "deserts" produced unintentionally by human action, which ought to be prevented, is an important one.—Editor] Desertification is increasing at an asymptotic rate.

Yet even when the natural vegetation has been destroyed through overgrazing, water and food are supplied to animals and the imbalance maintained. In the past, the numbers of grazing animals were kept in check by natural factors—for example, by rainfall and hence by the availability of water. Today, however, water can be shipped into an area by mechanical means or shallow wells can be bored, and food can be supplied, saving from death animals that otherwise would have succumbed to drought.

Overgrazing has undesirable secondary effects. For example, it harms wildlife by disrupting the natural equilibrium between native plants and wild herbivores, and it leads shepherds to use scarce trees and shrubs for fuel. In Israel, this poses a special threat to the remaining few hundred pistachio trees *(Pistacia atlantica)*, which are hundreds of years old. Also, shepherds shoot partridges, gazelles, and other wild animals for food.

Agriculture in Marginal Areas

The tremendous increase of human populations in many semiarid regions has compelled people to try to grow field crops in them. This is successful in good, rainy years, which occur only once every four to five years; but in dry years fields turn into completely barren dust bowls, which accelerates desertification.

Mining and Quarrying

In deserts, open mines and quarries often turn scenic landscapes into vast areas of potholes, dust bowls, and slag heaps. People tend to ignore this kind of threat to the environment when it occurs in a desert; in populated, nondesert areas, such activity is not tolerated. Because some people stand to gain economically from such activities, it is difficult to fight this practice.

Colonization of Deserts

In deserts, most human population and activities are concentrated within small, intensively used patches devoted to irrigated agriculture, industry, mining, and cities which are surrounded by large expanses of desert in which there is little or no human population. These enclaves of human enterprise may occupy only 1 to 5 percent of a region, yet they contribute 95 to 100 percent of the region's economic production. Nevertheless, the surrounding, unoccupied desert contributes significantly to the quality of life of the people who live in desert towns and villages: for example, it has recreational and educational value for them. Their attitude toward that larger environment—its expansive qualities, its landscape, natural inhabitants, and habitats—may significantly influence their attitude toward living in the desert settlement itself.

In countries that have only limited open space for the inhabitants of their nonarid regions, large areas of uninhabited deserts are of great recreational and touristic value. For this reason, it is important to protect large expanses of natural and seminatural desert ecosystems from the inadvertent depredations of human activities.

In spite of their special character, the effects on the surrounding desert of the concentrated human activity of intensive settlements and the reciprocal effects of the desert on the settlements have hardly been studied. The nature of these interactions originates in the sharp contrast between the two systems. On the one hand, the desert ecosystem is extensive; it is poor in water and food sources and supports very sparse populations of specially adapted plants, animals, and people. The human "oasis" ecosystem (i.e., the settlement), on the other hand, is small and intensive; is rich in water and food; and supports dense populations of plants, animals, and human beings that are not specially adapted to the desert environment and that are maintained by the continuous import of water and food from the nonarid regions.

Several kinds of ecological interactions may take place across the sharp boundary between the natural desert ecosystem and the human oasis ecosystem. For instance, the people and domestic animals of the settlement may endanger the sparse plant and animal populations of the surrounding desert by grazing, gathering, and hunting. Animals or diseases inadvertently introduced by man may have similar effects. Also, the concentration of food and water resources in the settlement, as well as the settlement's more favorable microclimate, may attract animals from a wide area of desert. In some cases, this may lead to the rapid and explosive increase of desert populations in and near the settlement (e.g., the explosion of the partridge population in the Negev), with harmful consequences for the agricultural systems of the settlement and for the surrounding desert ecosystem.

Solid and liquid wastes discharged from the settlement into the desert may have similar biological effects, since they are an additional source of food and water for desert animals. In some cases, of course, they may permit animals to survive without at the same time damaging the oasis ecosystem, but in other cases they disrupt the normal mechanisms of population regulation. Also, activities that go on in human settlements may directly or indirectly modify the physical environment of the surrounding desert (for example, by causing soil erosion), with harmful consequences for desert populations and sometimes even for the settlement itself (for example, by creating a source of dust).

The following questions may be asked about each ecological interaction between oasis and desert:

1. For how great a distance from the settlement is the desert ecosystem significantly affected?
2. Does the presence of the oasis increase or reduce the stability and resilience of the natural ecosystem?
3. What repercussions could settlement-induced impacts on the natural ecosystem have for the settlement itself?
4. How can undesirable colonization of the desert be prevented or held to a minimum?

The Negev Desert, below 200 meters in elevation, is a classic example of a desert that has been—and that probably will continue to be—colonized mainly by intensive irrigated and urban enclaves. The situation in the Negev provides many opportunities to study the relatively neglected ecological problems I have described, problems that will increase in importance as the Negev itself and as the arid parts of other countries are developed.

Measures for Preserving Desert Ecosystems

Desert ecosystems will be preserved only if people become aware of the delicate balance in which such systems exist. In other words, it is a question of perception.

Where possible, states that possess desert areas that are still little affected by man should set up planning bodies or agencies specifically for those areas. The chief aim of these agencies—which should consist of scientists (ecologists, pedologists, and so forth), landscape planners, economists, and administrators—should be to formulate plans for determining which areas will be protected and which areas will be open for such human activities as agriculture and industry.

The functions of the planning agency should be:

1. To survey completely the areas in question.
2. To decide which areas are suitable for human activity, and which areas should be set aside for complete protection. (The protected areas cannot be small because

desert ecosystems need vast areas for complete natural development.
3. To formulate a set of rules for people who already live in protected areas that will both ensure the people's livelihood and also prohibit damage to the biotic communities within which they live.
4. To formulate a set of strict rules for the complete protection of geological formations, landscape, flora, fauna, and the remains of any ancient civilizations within protected areas, as well as of existing civilizations.
5. To manage and develop, as well as to protect, the reserve areas, including the proper protection of all of their existing elements and the restoration of plants and animals extirpated or reduced in numbers through mismanagement during the last few decades or centuries.
6. To allocate, according to objective criteria, the areas best suited for agriculture, recreation, the extraction of raw materials, and so on, which should be strictly controlled by a set of rules and laws set up to prevent even the slightest damage to the protected areas.

After the planning agency has devised a plan that is acceptable to all interested parties, the government in question will have to pass legislation designed to protect the areas proposed for complete protection as nature reserves, national parks, wildlife reserves, wildernesses, and so on, the precise name depending upon the country involved.

There are at least two good examples of the types of restoration described in functions 4 and 5 above, namely, the restoration of the ancient runoff farming systems in the Negev, which date to the first millennium before Christ, and the restoration of wildlife inside a fenced area of 3,000 acres near Elat. The success of the former effort is evident in the ancient farm reconstructed near the ancient Nabataen town of Avdat. The farm, which is already sixteen years old, is run by Professor Michael Evenari of Hebrew University. In the effort near Elat, animals mentioned in the Bible but extinct in Israel for several hundred years are being brought from all over the world to create breeding herds that in the future will populate the desert. Some of the animals being restored in this effort are the Nubian ibex, Dorcas gazelle, addax, Persian wild ass, Somalian wild ass, and a population of ostrich from the Danakil Desert of Ethiopia.

39. Mareotis: A Productive Coastal Desert in Egypt

M. Imam
Cairo University, Egypt

Egypt, a Desert Country

Egypt is part of the great Sahara Desert, a hot desert. It is generally accepted that the Sahara as we know it is a relatively modern geographical phenomenon, the most recent stage in a more or less progressive shift from parkland to steppe and from steppe to desert. The Paleolithic people who lived in the now arid regions of Egypt never knew desert conditions; all evidence points to rainy conditions during their existence.

Desert conditions became acute in the northern part of Egypt fairly late in Paleolithic time because of decreasing rainfall (de Cosson 1935, p. 17; see also Selected Bibliography). As a consequence, in the face of the continual constriction of the great interior tracts of grassland, the human population began to congregate in the few, limited areas moistened by rain or by water from the Nile River, making it necessary for them to cultivate crops.

The Nile River moistens only about 3 percent of Egypt; the rest of the country, except for a narrow coastal belt, is an uninhabitable desert. Egypt's problem of overpopulation increases every year because its population of almost 40 million is concentrated in the 3 percent of the country near the Nile. The vast expanse of desert could provide plenty of space, but few life-supporting resources, for human settlement. The only promising part of the desert is the narrow coastal belt along the Mediterranean, which extends west of Alexandria nearly 600 kilometers to Salum, and inland about 15 kilometers. It is considered the richest part of the rain-fed area because of its relatively high rainfall. The average annual precipitation of the north coast of Egypt is 190 millimeters between Alexandria and Lake Burullus (Buhayrat al-Burullus) and 73 millimeters between Damietta (Dumyat) and Port Said. The coast's orientation relative to the direction of moist currents influences rainfall to such an extent that precipitation in one place may be twice that in another (Ali

1952). The average annual rainfall along the coast west of Alexandria is 150 millimeters.

The whole coastal strip consists of Recent and Pleistocene formations, which are delimited on the south by the Miocene limestone of the desert plateau. Tadros (1956) recognized three parts to the strip: a middle part, a western part, and an eastern part. The middle part of the strip extends westward from El 'Alamein to Fuka. It is a plain with fairly deep sandy-loam soil and is famous for its relative fertility. In the western part of the coastal strip, the relief is less regular and the country becomes rather hilly, dissected by many wadis and ravines. (Wadis are river valleys that are dry in summer, and filled with flowing water in the rainy season.) Hills in this part of the coastal strip are 40 to 150 meters or more above sea level. The eastern part, between El 'Amiriya and El 'Imayid is a narrow strip of white dunes of more or less regular relief that extends along the whole coast, and of a series of ridges and depressions. This part of the coastal strip, the eastern part, is the subject of this study (see Figure 1).

Mareotis

The eastern part of the belt between El 'Amiriya and El 'Imayid (29°E) is known as Mareotis—also spelled Mariot and Maryut. (The history of this area was recently reviewed by Kassas [1972]; an older publication on Mareotis is that of de Cosson [1935].) It has been under cultivation since before Graeco-Roman times. Rome regarded it as an important grain-producing area. At present, dry-farming agriculture, based upon the average annual precipitation of 150 millimeters, is practiced. The main crop is barley. In several scattered spots there are fruit—primarily olive—orchards, which must be irrigated during the long summer months until they have grown sufficiently for their roots to reach down to permanently wet layers. Within the last 2,000 years, the climate has not undergone "any very sensible change for the worse since Graeco-Roman times" (Ball 1942, p. 120). Yet Weedon (cited by Kassas 1972) begins his report on the Mareotis District by observing that "one of the most striking anomalies of modern Egypt consists in the fact that the whole district of Mariout, which in ancient times was, as I shall endeavor to show, famous for its fertility, should at the present day

Deserts

be for the most part barren and waste." The extraordinary number of structural foundations throughout the coastal area proves how great was the development of the Nome of Mareotis in Graeco-Roman times, when cultivation of the vine and olive was at its height (de Cosson 1935, p. 37).

Before any regeneration and development programs can be drawn up for the area, it would be very useful for us to investigate the basis of its prosperity for about 900 years during Graeco-Roman times and for its subsequent decay. Here we will discuss the main characteristics of the area, its potential resources, its land-use patterns, and the principal changes that have occurred there during the last 2,000 years.

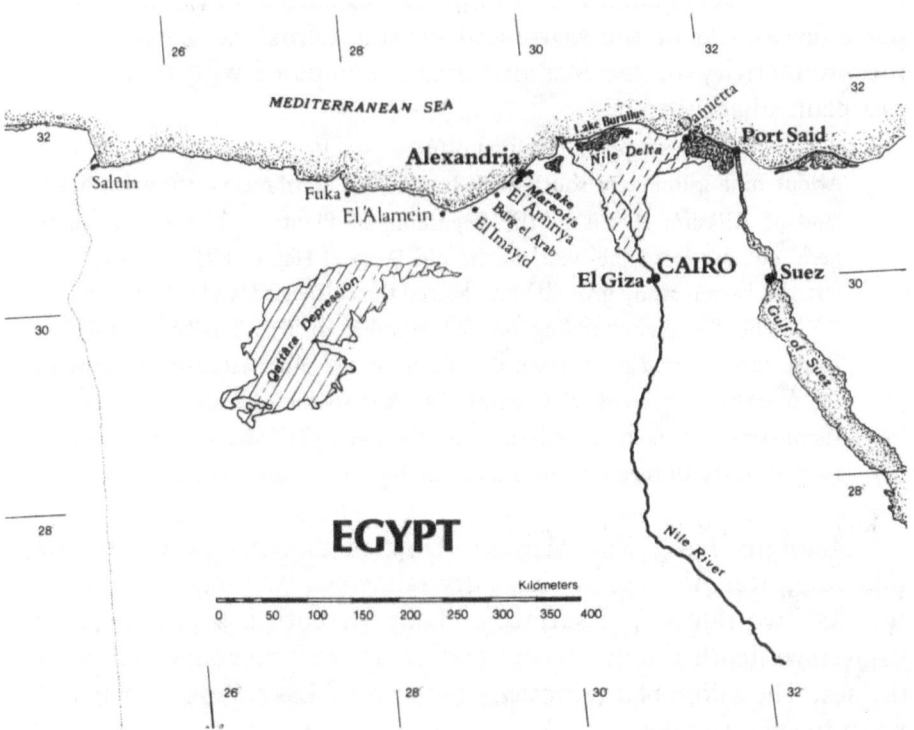

Figure 1. Egypt's Coastal Belt from Port Said to Salum in the Sahara Desert. The topography of this strip varies from hills dissected by wadis (alternately dry and wet river valleys), to a relatively fertile plain from Fuka to El 'Alamein, and dunes and depressions in the vicinity of Lake Mareotis.

The Topography of Mareotis

Along the coast there is a narrow strip of white dunes made up of loose, coarse, calcareous sand grains. Inland from the dunes, there is a series of nearly consolidated dunes (rocky ridges or bars) that alternates with a series of depressions. The system of bars and depressions runs parallel to the shore and extends east-west for varying distances. The ridges are closer together in the east than they are in the west; gradually, they fade away. They seem to represent a sequence of Pleistocene-shoreline bars (Shukri et al. 1956).

The ridges constitute an effective means of redistributing rainwater. Forms are usually placed at the bases of these ridges, where runoff water accumulates. The value of this natural magnification of the effectiveness of the sparse and variable rainfall is obvious when the productivity of the Mareotis area is compared with that of the vast plain adjacent to it.

> About nine kilometers south of El-Imayid . . . lighthouse is the great headland of Khashm el 'Eish, "the Beginning of Plenty," so named because here, coming from the west on the old Darb el Haj, or Pilgrim's Way, one left the barren stony ground and entered the cultivated lands of the Maryut.
>
> The traveller standing on the summit of this headland, 300 feet above the sea, will look down upon one of the most astonishing views in the Maryut; westward all is emptiness, but to the east are the houses of Hammam, the factory chimney of Gharbaniyat, and the building of Borg el Arab, all foreshortened into one big town (de Cosson 1935, p. 120).

Mareotis Lake. The Mareotis District contains Lake Mareotis (Maryut), near Alexandria, and a dry saline bed (salt marsh) between the first two ridges. The surface of Lake Mareotis is below sea level; its average depth is only 80 centimeters. There is no connection with the sea. The saline bed represents the arm of Lake Mareotis when it was a freshwater lake.

De Cosson (1935, p. 70) speaks of "the lake as it was—a deepish fresh-water lake, rather than the shallow, weedy sheet of water, mud flat as it is now." More recently, the lake's area has diminished from 38,500 to 6,900 hectares, only 2,400 of which are suitable for fish.

During each of the last ten years, the annual catch of fish amounted to 9,000 tons. At present, however, because of pollution by industrial and sewage wastes and continuing desiccation (Ghobrial and Abbas 1972, p. 446), the annual catch hardly exceeds 2,000 tons.

Besides being used for pleasure by the public, Lake Mareotis was an important route for waterborne transport to and from the interior of Egypt. It was a basin into which some of the waters of the Nile found their way, and its fresh water was used for irrigating the surrounding fields. De Cosson (1935, p. 71) adds, "There seems to be little doubt that two thousand years ago the lake was of greater extent than in modern times, and in addition, the long Abu Sir inlet ran westward towards El Imayid. . . . The fine jetties at Mareotis are a striking example, and they prove that the average depth of water in Graeco-Roman times was considerable." As the Canopic Nile and the channels that fed the lake gradually became filled with silt, the waters of the lake receded, and inland water communication, water supply, and irrigation all suffered. In consequence, the city of Alexandria and the District of Mareotis slowly declined. Finally, about the twelfth century, the Canopic branch disappeared and the former lake became a salty swamp, or *sabbaba*.

The Egyptian rulers of the twelfth and thirteenth century tried to save the ancient water communication link between Alexandria and the District of Mareotis. They failed, however, and Alexandria—without its freshwater lake and communication—deteriorated rapidly (de Cosson 1935, p. 80). Thus, the natural silting of the channels that had fed the lake contributed to the economic decline of the area.

The Climate of Mareotis

The general climate of the Mareotis area is Mediterranean: there is a long, dry, fairly warm period and a shorter, somewhat rainy, moderate winter. Rainfall, though low (150 millimeters per year), is an annually recurring phenomenon, but it is distributed irregularly from year to year, during the season, and from one part of the region to another. Köppen (cited by Draz 1954), in his classification of humidity provinces, considered the Egyptian deserts along the Mediterranean shore to lie within the limits of the steppe regions.

Tadros (1956) calculated the pluvithermic quotient for two sites along the coast, Alexandria and Mersa Matrouh. He showed that they fall within two levels (22 and 17, respectively) within the zone of warm, semiarid Mediterranean climate. The entire area, with an average annual rainfall of 150 millimeters, lies within the "rainfall desert" (Kassas 1975), in which the precipitation of 100 to 200 millimeters is still too low for the sustained production of crops. Along the coast, dew and mist frequently occur, especially in summer and autumn, contributing considerable moisture.

Management of Water in Mareotis

Wells. Shallow wells in the coastal area west of Alexandria derive their water from a comparatively thin layer of fresh water that floats on top of the salty water (Paver and Pretorius 1954). Most of the wells are used only for drinking water, which is raised by hand in buckets and poured for animals to drink. Windmills have been installed on some of these wells, for there is always a breeze from the sea.

The luxuriant fig groves west of Alexandria to Borg el Arab have been and still are irrigated by these wells until the trees' roots reach the water table. Care must be taken to avoid soil salinity problems created by the misuse of saline well water for the expansion of cultivation. Abdel-Samie (1960) concluded that the use of saline water to supplement irrigation during dry seasons leaves a residual salinity effect that is not necessarily corrected by runoff or rainwater during the rainy season.

Cisterns. Ball (1935) described cisterns as large, underground chambers that were excavated in limestone ridges during Roman times and that serve as reservoirs of rainwater. The site of a cistern is carefully chosen so as to collect runoff from a considerable area. Small drainage channels conduct water to an orifice that leads to the underground chambers. The water stored in the cisterns is usually fresh, and is used mainly for drinking. The importance of drinking water in the Mareotis District is obvious and need not be stressed.

Shafei (1952) relates the story of an old Arab sheik in the Bahig area:

> He told us that this year's winter rains (1949-1950) were good, the rain gauge at Burg el Arab recorded a total of nearly 300 mm., and the pastures were more than usually excellent to a distance of over thirty kilometres from the sea. But also, they could not benefit from this! There was no

fresh water, and it had been necessary to take the flocks every two or three days to the wells near the shore to drink, a distance of twenty kilometres and the same distance back.

Shafei (p. 82) reports the repair of sixteen cisterns near Bahig. There are many more of them in the area. Now the Egyptian government is constructing a water pipeline, which will replace the old one, to provide the Bedouins with drinking water.

Karms. Artificial hillocks, situated up to 30 kilometers inland, mark the limits of cultivation in Graeco-Roman times (de Cosson 1935, p. 27). These hillocks are karms, earthen banks built 3 to 4 meters high around cultivated fields. The flanks of karms cause runoff water to flow to either side, concentrating the water in areas where plants (mainly vines) are grown (Kassas 1972).

Cisterns are no longer functioning in the area. Shafei (1952) reported that he is working on repairing the karm of Borg El Arab. He added, "I saw hundreds at Gabes and on the way to Gojsa (Tunisia). They function in zones of 150 mm. rainfall" (p. 84).

Land Use in Mareotis

Agriculture. Dry farming has been the principal type of agriculture in the Mareotis area since Graeco-Roman times. Barley is the grain crop. Fruit orchards dot the area. Figs *(Ficus carica* L.) do best on the sand dunes close to the seashore and in the nearby area of ridges and depressions. Olives *(Olea europea* L.) do better inland.

Nowadays, with the peaceful settlement of the native Bedouins, more land is devoted to figs and olives. Almonds *(Prunus amygdalus* Batsch.) are also grown, and almond cultivation is spreading into areas in which water is available for irrigation. Vine *(Vitis vinifera* L.), date palm *(Phoenix dactylifera* L.), carob *(Ceratonia siliqua* L.), and pistachio *(Pistacia vera* L.) are among the orchard crops of the area, but these are not grown on a large scale. The fig and olive plantations constitute the replacement of a natural climax vegetation (open scrub) with an artificial flora of comparable structure and ecological relationships.

There is a serious disadvantage in depending on an annual crop in an area with notable variation in annual rainfall (Kassas 1972).

Therefore, the economic future of this region will depend largely on the cultivation of trees.

The Bedouins are no longer dependent on barley for bread-making; however, they now use wheat flour that reaches them by way of the asphalt coastal road. They are now encouraged to grow more trees because they are better paid for land if it is under cultivation with trees. Land devoted to barley is considered natural land, and the Bedouins are not paid for it.

Grazing. The land that is not cultivated, mainly the higher ground, is open for grazing by goats and sheep. Grazing is prohibited in barley fields and fruit orchards, within the framework of social traditions. According to Arab law, an indemnity of 5 piasters (1 Egyptian pound equals 100 piasters) must be paid for any twig nibbled by a goat or camel (Shafei 1952, p. 68).

Recently, Libya has become a good market for the sheep of the Egyptian coast. Controlled, sound grazing would certainly add to the income of the local Bedouins, but intensive grazing and the nomadic life common in the semidesert areas of Egypt have played a great role in the destruction of the vegetative cover.

Mining. Gypsum was extracted from the area in Graeco-Roman times and is still mined today. The loamy soil is removed for the manufacture of cement. The calcareous rocks of the ridges are used in the manufacture of sodium carbonates and bicarbonates. An oil field has been discovered near El 'Alamein.

Tourism and Recreation. Desert tourism is growing and will continue to expand, and the prospect of Mareotis becoming a summer resort is great. Egyptians are seeking the open country and pure air of the coasts as relief from the crowded cities, and Europeans and people from nearby highly industrialized countries are seeking the fresh, pure, and healthy air of Egypt's coastal deserts. The developments planned for the coast should keep the area in as healthy a state as possible.

The Decline of Mareotis

During ancient times, "there was only one incident to disturb the peace in this period. . . . Peace was guaranteed in Mareotis, and this was the golden age of prosperity which lasted nine hundred years..." (de Cosson 1935, p. 38). But in the early seventh century, the country was invaded three times within thirty-four years: armies swept from the west and the east, and then wandered for decades with their sheep, camels, and goats, robbing the soil of its cover.

More destructive invasions and uncontrolled grazing occurred during the tenth and eleventh centuries. In the words of de Cosson, "Thus did neglectful government and Bedouin lawlessness replace the patient planning and planting of the thousand years of Graeco-Roman civilization. The wind, rain, and sand completed the destruction, leaving only the foundations of countless buildings to tell the role of this once prosperous land" (p. 63).

The Regeneration and Development of Mareotis

After the Second World War and the decisive battle of El 'Alamein, people began to resettle in the Mareotis area. The government provided the area with an irrigation canal and a new pipeline for drinking water, and is continuing to develop the area with three major new projects—an oil pipeline, the desalinization of seawater, and the Qattara Depression project.

The Oil Pipeline. An oil pipeline between Suez on the Gulf of Suez and Sidi-Kreir on the Mediterranean Sea, 20 kilometers west of Alexandria, is near completion. Pollution of the seawater by oil seems inevitable. If a spill does occur, currents probably will spread the oil and pollute the beaches east of Sidi-Kreir. The economic benefits of the pipeline and the requirement of clean beaches and coastal waters for recreation and commercial use make it imperative that these apparently conflicting needs be reconciled.

The Desalinization Project. In coastal deserts, the shortage of drinking water can be alleviated by desalinizing seawater. Thus, an atomic powerplant, which will be used to desalinize seawater, is being installed near Sidi-Kreir.

The Qattara Depression Project. The Qattara Depression is about 90 kilometers wide and 200 kilometers long, and has an elevation of 130 meters below sea level. Its northern edge is situated some 70 kilometers from the coast of the Mediterranean Sea. A canal will be cut across the "isthmus" between the Depression and the Mediterranean, and a large hydroelectric power station will be built at the precipice of the Depression. [See J. Ball. 1933. The Qattara Depression of the Libyan Desert and the possibility of its utilization for power-production. *Geographical Journal* 82(4):289–314.—Editor]

Much of the electricity will be used on-site for local industry; some of it will be used to desalinize seawater for irrigation purposes. Atomic power will be used to dig the canal.

The environmental consequences of this large-scale project should not be underestimated, in spite of the pressures created by overpopulation and development programs. To assure that people will be able to enjoy the economic benefits of this project, studies that focus upon means for maintaining a healthful living environment must be carried out.

The high-paying salaries associated with industrial projects will attract Bedouins; in fact, many of them already work as drivers, transporting passengers and goods between Egypt and Libya. The impact of this shift on agricultural productivity will be serious.

Wildlife and the Conservation of Genetic Resources

Because of its relatively high rainfall, the Mediterranean coastal belt is floristically the richest part of Egypt: about half of Egypt's 1,800 to 2,000 species of flowering plants grow there (Täckholm and Täckholm 1941). Also, endangered species of plants and animals inhabit the area.

The region's flora constitutes a genetic reservoir, the potential source of a number of cultivars. There is a pressing need for a reserve to keep these genetic resources available for future breeding-improvement programs. The idea of a network of such natural reserves in Egypt is being discussed, and the coastal belt stands first in priority.

References

Abdel-Samie, A. G. 1960. Observations on the seasonal variations in the salinities of some coastal zone soils. *Bulletin de l'Institut du Désert d'Egypte* 10:95.

Ali, F. M. 1952. Outstanding variations in rainfall on the north coast of Egypt. *Bulletin de l'Institut du Désert d'Egypte* 2:5.
Ball, J. 1935-1937. *The Water-Supply of Mersa Matruh.* Survey Department Paper No. 43. Survey of Egypt, Cairo [?]. Cairo [?].
Ball, J. 1942. *Egypt in the Classical Geographers, by the Late Dr. John Ball,* ed. G. W. Murray. Survey of Egypt. Cairo: Government Press.
de Cosson, A. 1935. *Mareotis; Being a Short Account of the History and Ancient Monuments of the Northwestern Desert of Egypt and of Lake Mareotis.* London: Country Life, Ltd.
Draz, O. 1954. Some Desert Plants and Their Uses in Animal Feeding; *Kochia indica* & *Prosopis juliflora. Publications de l'Institut de Désert d'Egypte,* no. 2. Heliopolis.
Ghobrial, S. G., and M. M'Abbas. 1972. [The effect of pollution on the Egyptian fisheries] [in Arabic]. Presented at the symposium "Progress in Man–Environment Development" at Khartoum, organized by the Arab League Educational, Cultural, and Scientific Organization.
Kassas, M. 1972. A brief history of land-use in Mareotis region, Egypt. *Minerva Biologica* [Turin] 1:167.
——. 1975. A. Arid and semi-arid lands. I. An overview. In *Review of the Environmental Situation and of Activities Relating to the Environment Programme,* p. 21-23. Report of the Executive Director, United Nations Environment Programme. UNEP/GC/30; Na 75-153.
Paver, G. L., and D. A. Pretorius. 1954. Report on reconnaissance hydrogeological investigations in the western desert coastal zone. *Publications de l'Institut de Désert d'Egypte,* no. 5. Heliopolis.
Shafei, A. 1952. Lake Mareotis, its past history and its future development. *Bulletin de l'Institut de Désert d'Egypte* 2:71.
Shukri, N. M., et al. 1956. The geology of the Mediterranean coast between Rosetta and Bardia. Part II: Pleistocene sediments. *Bulletin de l'Institut d'Egypte* 37:395.
Täckholm, V., and G. Täckholm. 1941. Flora of Egypt. Vol. 1, Pteridophyta, Gymnospermae and Angiospermae, part Monocotyledones: Typhaceae, Gramineae. *Bulletin of the Faculty of Science,* no. 17. Fouad I University, Cairo.
Tadros, T. M. 1956. An ecological survey of the semiarid coastal strip of the western desert of Egypt. *Bulletin de l'Institut de Désert d'Egypte* 6:28.

Selected Bibliography

Source materials on the ecological history of the Egyptian Sahara and adjacent Africa, compiled by the editor.

Huzayyin, S. 1956. Changes in climate, vegetation, and human adjustment in the Saharo-Arabian belt with special reference to Africa. In *Man's Role in Changing the Face of the Earth*, vol. 1, ed. W. L. Thomas, Jr., pp. 304–323. Chicago: University of Chicago Press.

Monod, Theodore. 1963. The late Tertiary and Pleistocene in the Sahara. In *African Ecology and Human Evolution*, eds. F. Clark Howell and François Bourlière, pp. 117–229. Viking Fund Publications in Anthropology, no. 36. Chicago: Aldine Publishing Co.

Street, F. A., and A. T. Grove. 1976. Environmental and climatic implications of late Quaternary lake-level fluctuations in Africa. *Nature* 261(5559):385–390.

Wendorf, F., R. Schild, R. Said, C. V. Haynes, A. Gautier, and M. Kobusiewicz. 1976. The pre-history of the Egyptian Sahara. *Science* 193(4248):103–114.

It's wonderful to see
the caribou come down
from the forest,
and start pouring north
over the white tundra,
anxiously avoiding pit-falls in the snow.
Jai-ja-jija.

And it's wonderful to see
the short-haired caribou
in the early summer
start wandering.
Jajai-ja-jija.

. . .

It's wonderful to see
the caribou with their winter fur
returning to the woods,
anxiously avoiding us, the little men,
and following the ebb-mark of the sea,
with a rustle
and a creak of hoofs.
Oh, it's wonderful.
Jajai-ja-jija.

—Netsit (Copper Eskimo man, Musk Ox Folk)
"Song about the Reindeer, Musk Oxen,
Women, and Men Who Want to Show Off"

The Polar Regions

40. The Polar Regions and Human Welfare: Regimes for Environmental Protection

Gerald S. Schatz
National Academy of Sciences, U.S.A.

Introduction

When the United Kingdom of Great Britain and Northern Ireland (U.K.) took its Antarctic territorial dispute with Argentina and Chile to the International Court of Justice in 1955, both Argentina and Chile contended that the Court lacked jurisdiction; in 1956, the Court removed the Antarctica cases from its list. Recently, in the aftermath of the "Cod War," the U.K. brought proceedings in the International Court against Iceland's proposed extension of fisheries jurisdiction. In 1974, the Court ruled that Iceland could not unilaterally exclude U.K. fishing vessels from certain areas, that Iceland's assertion of exclusive fishing rights did not apply against the U.K. in these circumstances, and that Iceland and the U.K. were under mutual obligations to continue their negotiations toward an equitable solution of their differences. This is not emphatic rule-making or rule-application of the sort found typically in domestic law—in legislation or in case law. International law is not a set of statutes, although there are statutes; it is a disciplining procedure in international politics. In their voluntary associations, nation-states have accepted the principle that in most instances they must be prepared to argue for their actions. This is a splendid accomplishment in the

---

Opinions expressed in this paper do not necessarily reflect the views of the National Academy of Sciences.

development of human society. The antedating principle of concern here is that nation-states legitimately may protect their vital national interests and security. It follows that:

1. To propose international regulations in disregard of national interests and international law and politics is naive and may be perceived as a threat, thus upsetting existing accommodations; and,
2. When states share security concerns and are sensitive to each other's national interests, they strengthen the effectiveness of international accommodations.

These caveats should guide our consideration of environmental protection in the polar regions.

Consider now what specific national, human interests are shared in the polar regions. Consider developments and pending developments that appear to threaten these interests. Consider the adequacy of existing international accommodations to cope with threats to these shared interests. Consider possible courses of corrective action consistent with these shared interests.

In its importance to the human environment, one set of circumstances overrides all other polar problems. This is the military "balance of terror" between the United States and the Soviet Union. Both nations share, with all humanity, an indisputable interest in ensuring that nothing is done to cause either one to fear that its military-security interests in the Arctic are threatened and that nothing is done in the Arctic to preclude the deescalation of the armaments race. All humanity shares with the Antarctic Treaty countries an indisputable interest in maintaining the demilitarized status of the Antarctic. That these two points are not inconsistent is a measure of the horrifying absurdity of catastrophic weaponry.

Humanity shares an interest in ensuring that polar geophysical, biophysical, and biological processes of importance to human welfare on a global scale are not perturbed adversely and significantly. The polar regions are not isolated, quaint geographical phenomena: they help to govern global natural processes that dictate conditions of life on Earth, and they are influenced by environmental change elsewhere on Earth. Precisely how this happens is not known—which means that humanity shares a particular interest in the polar regions and an attitude of prudence towards them. The significance of the polar regions in the global environment has become clearer over years of polar exploration and polar science, and the increased ability of science to detect, identify, and appreciate global processes. Thus, Richard E. Byrd reflected on the rationale for his establishment of Bolling Advance Weather Base, which he manned alone on the Ross Ice Shelf in the Antarctic night of 1934:

> The farmer whose livelihood comes from crops, the people whose stomachs are kept full by these crops, the speculators who gamble in them, the industrialist whose factories depend upon the farmer's purchasing power, the sailor on the seas—all these and others, even to the casual holiday tourist, have a vital stake in weather. But few of them appreciate the extent to which the poles enter into their local schemes. (Byrd 1938)

As rapid population growth and its exacerbation by food scarcities have made themselves felt, society has become acutely aware of the social and economic significance of weather and climate. Apparent changes in climate and their effect on agriculture and fisheries have shown beyond doubt humanity's current sensitivity to the modification of climate. The physical and biophysical relationships between climate and the polar regions have been under study by several panels and committees of the National Research Council, with an eye toward determining where to direct polar studies. As summarized in the National Academy of Sciences' *News Report*, in a discussion of Southern Ocean studies:

> The relationships appear to be such that it is not exaggerating to say that a geophysical change such as a glacial shift on the Antarctic continent or an unusual warming can affect shelf ice, changing the distribution of sea ice, altering the process by which waters recirculated to the south become cold again, sink, and travel outward in the abyssal circulation, changing the locations of waters of a given temperature interacting with the atmosphere very far to the north, shifting the breeding regions of marine life and the growing regions of weather-vulnerable crops. (Schatz 1974b)

The Joint U.S. POLEX [Polar Experiment] Panel, a National Research Council panel concerned with the scientific program of the evolving international polar experiment in meteorology, has observed:

> There is little question of the importance of polar regions as key areas for climatic variation consideration. . . .
>
> For the Arctic . . . the importance of polar processes to midlatitude weather is not yet fully known, and it would seem unwise to permit the polar regions to remain an unknown element of global weather. . . .
>
> There is . . . no consensus as to the physical basis of climate change. . . .
>
> It is clear that the thermal and dynamic properties of arctic sea ice are unique and of great importance to meteorological processes. It is also clear that catastrophic changes in the Arctic could, in principle, be triggered by lesser events. . . . (National Research Council 1974c)

Because of the tremendous influence of Antarctic and Arctic ice-sea-air interactions upon oceanic and atmospheric circulations and upon world climate, and because each polar region responds to major glacial changes in the other, geophysical perturbations in the polar regions may have large-scale effects elsewhere on the planet. It is possible for human enterprise—whether purposeful or inadvertent—to affect polar ice-sea-air interactions, with consequences for the polar regions themselves and for climate-dependent agriculture and fisheries.

The perturbation could come with the best of intentions: it could be the minimal consequence expected of an insufficiently assessed northern water-management project; it could be the result of a well-intended effort to engage in weather modification in a small way to reverse what are perceived as adverse weather trends; it could be the outcome of an attempt to make shipping safer. Joseph O. Fletcher, at the time working for the Rand Corporation and later head of the National Science Foundation's Office of Polar Programs, told a symposium in 1968:

> [H]owever objectionable some of these schemes may be for political, economic, or environmental reasons, there is good reason to believe that we probably do possess the engineering capability to influence the extent of, or even eliminate, the [Arctic] ice should we really need to do so. . . .
>
> The primary factor limiting our ability to influence climate is our poor understanding of how our climatic machine operates—no matter what we are physically capable of doing. We cannot apply our technological capabilities in a purposeful way until we can predict the consequences

for the entire ocean and atmosphere, which is a single, interacting, physical system. (Fletcher 1972)

The state of science concerned with the polar ice and climate has improved considerably since then; but, as we have seen, it has not improved to the point where "we can predict the consequences" of an act affecting the polar ice "for the entire ocean and atmosphere." In these circumstances, it is at least reasonable to ask what the consequences might be of oil spills in polar waters. How and to what extent would a spill of crude oil in icy waters affect the albedo? The rate of formation and degradation of sea ice? The air-sea interaction?

The United States has mounted an extensive program to examine the environmental aspects of possible oil exploitation off the north coast of Alaska, in advance of a projected Arctic offshore-leasing program. The State of Alaska has shown signs of eagerness to begin leasing in the Beaufort Sea; yet a National Research Council panel has found that subsea permafrost, which can exert great pressures on structures, is insufficiently mapped. And the state of knowledge of the effects of a polar oil spill is indicated by an exchange between R. C. Ayers, Jr., and H. O. Jahns, both of the Exxon Production Research Company, and J. L. Glaeser of the Exxon Company; and W. J. Campbell of the U.S. Geological Survey and S. Martin of the University of Washington. In this exchange, the Exxon group takes issue with portions of an earlier paper by Campbell and Martin (1973) whose subject was the interaction of oil with the Arctic Sea ice. The Exxon group declared:

> Our comments should not be construed as an attempt to minimize the importance of oil spill prevention in Arctic operations or the need for effective means of dealing with an oil spill, should one occur. On the contrary, the Arctic environment will require special precautions to minimize the risk of an accidental spill and special techniques for cleanup in ice-infested waters.
>
> Continuing research will be required to improve our understanding of the Arctic environment. However, concern about the possibility of a significant alteration of the heat balance of the Arctic Ocean from a major oil spill appears to be unwarranted. (Ayers et al. 1974)

"Appears to be unwarranted" is hedging one's bets. Martin and Campbell replied:

> In summary, we stress that no data exist on the dispersion of a medium-to-large oil spill in pack ice and that the data for small spills obtained from either accidents in shorefast ice or controlled spills in pack or laboratory ice do not allow for an accurate assessment of the extent of the albedo decrease caused by the cumulative effects of oil spills in the Arctic Ocean. The question of scale is an important one, and we believe that, until more relevant studies are carried out, a cause for concern still exists. (Martin and Campbell 1974)

There is also an important set of biophysical relationships to be considered in connection with the extraction of minerals from the seabed in the polar regions, should it be pursued. Obviously, changes in polar temperatures and currents as a result of changes in the ice and the extent of open water will affect marine life in the immediate vicinity. A less direct relationship ultimately may be more important to human welfare: as the Antarctic plays a major role in the abyssal circulation, it influences the waters very far to the north. Antarctic waters are among the world's most biologically productive, and the Antarctic Bottom Water that flows northward carries nutrients with it, essentially feeding fisheries on which economies in many parts of the world depend. Increases in turbidity due to seabed mining in the Antarctic, should it come to pass, could well affect the production in the Antarctic of marine nutrients on which commercial fisheries thousands of miles away depend.

The foregoing are some of the polar geophysical, biophysical, and biological processes in which humanity shares interest and dependence, as surely as human welfare depends upon the continued demilitarization of the Antarctic Treaty area and upon extreme care regarding nuclear and thermonuclear weapons so long as we curse ourselves with their existence. What legal and political frameworks accommodate considerations of the polar environments in their relation to global processes and human welfare? What has been done and what, if need be, realistically should be done to protect humanity's shared interests in the polar regions beyond the clear bounds of national jurisdiction?

The Antarctic Treaty

Neither of the polar regions is a commons, as the term is generally understood—for political reasons in the Arctic and for legal and political reasons in the Antarctic. Both polar regions present situations in which the resolution of issues of national jurisdiction beyond generally recognized territorial seas and continental-shelf interests is deferred—essentially indefinitely. In the south polar region, this is national self-restraint: a Norwegian-Soviet difference over respective rights in Svalbard is being pursued by the two countries diplomatically, not militarily.

The Antarctic Treaty provides a measure of environmental protection for Antarctic lands and shelf ice, and it provides a consultative forum for the evolution of protective steps in connection with the treaty area (south of 60° south latitude). The treaty does not impinge upon rights under international law in the water column. Antarctic waters and their living resources, as elsewhere in the common sea, are protected only to the extent provided in miscellaneous unilateral requirements and policies, bilateral agreements, and international conventions, where there is enforcement; at least there exist political arrangements within which protective measures may be considered.

The status of the Antarctic seabed with respect to the Antarctic Treaty is unclear. While the treaty's consultative machinery is privileged to worry about it, there is a possibility that an emerging United Nations seabed authority may be privileged to exploit it.

It is important therefore to take a closer look at the Antarctic Treaty, at what it represents, at what the treaty relationship has accomplished and can accomplish, and at the cost of disrupting that relationship.

A Closer Look

The Antarctic Treaty is a multilateral agreement, open to accession, whose parties include claimant states (some with conflicting claims), nonclaimant states, and states that have taken the position from time to time that Antarctica is not susceptible to claim except under very elaborate procedures of qualification.

Before the Antarctic Treaty regime was established, the Antarctic was the scene of occasional violence between parties of different nations. An informal agreement to defer territorial issues in Antarctica during the 1957-1958 International Geophysical Year was followed by the signing of the treaty in 1959 and the treaty's entry into force in 1961. Nineteen states are now parties to the Antarctic Treaty (see Appendix C).

Claims. The treaty does not do away with claims issues; rather, it obligates parties to the treaty not to act on claims issues. In this regime, claims are not to be asserted, enhanced, or derogated, and no new claim can be asserted nor can any act or activities constitute the basis of future claims. The parties to the treaty are obligated "to exert appropriate efforts" consistent with the United Nations Charter "to the end that no one engages in any activity in Antarctica contrary to the principles or purposes" of the Antarctic Treaty. Procedures for the settlement of disputes—ultimately leading to the International Court of Justice—are provided in the treaty.

Military Activities. The treaty prohibits "any measures of a military nature," including military maneuvers and training, weapons tests, nuclear explosions, and military conflict in the treaty area. Nuclear explosions, whether for military or civilian purposes, and the disposal of radioactive wastes are also prohibited within the treaty area.

The language of the Antarctic Treaty is expressly consistent with "the purposes and principles embodied in the Charter of the United Nations," which encourages "regional arrangements or agencies for dealing with such matters relating to the maintenance of international peace and security as are appropriate for regional action." The Antarctic Treaty is a regional arrangement for the maintenance of international peace and security; therefore, a failure of the treaty or a very serious external challenge to it would be in order for consideration by the United Nations Security Council rather than any other United Nations body, save the International Court of Justice in appropriate circumstances.

Consultation. The treaty provides a consultative mechanism, whose participants are the signatories that are active in the area, for "exchanging information, consulting together on matters of common interest pertaining to Antarctica, and formulating and considering, and recommending to their Governments, measures in furtherance of the principles and objectives of the Treaty," including measures regarding a number of matters, including the "preservation and conservation of living resources in Antarctica." Recommended measures become effective if they are approved by all of the countries entitled to send delegates to the consultative meetings.

The Scientific Committee on Antarctic Research (SCAR) of the International Council of Scientific Unions acts as a nongovernmental coordinating body for Antarctic science and as adviser to the several treaty governments on Antarctic matters of scientific concern. Antarctic Treaty consultative meetings have shown themselves quite receptive to SCAR's advice, which has included a variety of conservation measures. Curiously, and possibly because some of them feel insufficiently assured by the language of the treaty and want no Antarctic option precluded, the treaty governments have not been altogether receptive to recommendations by the consultative meetings.

Thus, SCAR has recommended to the consultative machinery, which in turn has recommended to the treaty governments, the adoption of "Agreed Measures for Conservation of Antarctic Flora and Fauna," which provide for the designation of so-called "Specially Protected Areas." The consultative powers have yet to give the agreed measures the unanimous approval necessary to make them internationally binding, but have accepted them as "interim guidelines," pending their unanimous adoption. With rare exceptions, these guidelines are followed in practice. The strength of informal agreement among the Antarctic scientific community and its sponsoring governments in this area is considerable. (See Appendix C, part 2.)

Resources and Ownership. The treaty is silent on all ownership-resolution matters, and as a consequence is silent on the disposition of Antarctic resources. When the prospect of Antarctic sealing came before the consultative powers, the consultative machinery developed a separate Convention on Conservation of Antarctic Seals, now before governments for ratification. The sealing convention is outside the treaty, and it was therefore unnecessary to consider ownership issues within the treaty framework. This procedure may serve as useful precedent for facing the problem of the exploitation of Antarctic krill (primarily *Euphausia superba*). Whaling conventions long have existed; their effectiveness long has been in dispute, but that is a question of resolve and not of regime. Clearly it is possible to deal with resource matters in the Antarctic without damaging the treaty relationship.

Oil has been the subject of many news stories on the Antarctic recently, and because a drive for oil in the Antarctic would clearly threaten the treaty relationship, as well as the Antarctic environment, that situation deserves clarification in this context. First, the records of United States geological exploration in the Antarctic are open and are in the international scientific community's appropriate world data centers. Second, a United States Geological Survey paper by N. A. Wright and P. L. Williams (1974) declares that "Antarctica seems to have some petroleum potential, but lack of information precludes any real appraisal." Wright and Williams reported that "Antarctica now has no known economically recoverable resources of any category, nor does Antarctica have any known mineral districts."

The Antarctic Treaty is nonexpiring: it can be opened for renegotiation and for unilateral withdrawal in 1991, but it need not be. At least until then, any Antarctic Treaty country that renounces the treaty unilaterally or simply flouts the treaty relationship, with its political requirement of staving off ownership questions, likely will find itself enmeshed in diplomatic dispute. It is doubtful that financial institutions asked to invest in resource exploitation in those circumstances would respond with confidence.

Exploitation of Resources

Questions of resource exploitation are currently before the Antarctic Treaty powers. Under the existing regime, there is neither rush nor need to reach definitive settlement. Indeed, the political characteristic that has made this relationship work is acceptance of the idea that some matters need not be settled.

What of the possibility of a conflict of laws between the emerging international oceans regime and the special situation of the Antarctic Treaty? The draft provisions that have emerged from the most recent law-of-the-sea negotiations in Geneva appear to provide no machinery comparable to the Antarctic Treaty consultative machinery to deal with environmental protection and questions of protection of natural processes. Is it worthwhile, then, to try now to set aside the Antarctic Treaty and go for something else?

Fortunately, the very high cost of Antarctic operations makes active exploration for oil in the Antarctic unlikely; similarly, it makes the mining of nodules on the ocean bottom in the Antarctic unlikely. Fortunately also, if a 188-nautical-mile [348-kilometer] zone of resource trusteeship beyond a 12-nautical-mile [22.2-kilometer] territorial sea becomes the law of the sea, no such zones will reach as far south as the Antarctic Treaty border—60°S latitude.

If events warrant the resolution of a conflict of laws in this situation, the International Court of Justice is the proper venue for resolution without damage to the treaty relationship.

Nevertheless, it is evident that the Antarctic Treaty consultative powers must be prepared to deal with a resource challenge if it arises. It is within the authority of the consultative machinery to suggest the separate circulation among the Antarctic powers of an interim agreement to this effect: that, while reserving all rights, the Antarctic powers agree that the disposition of profit from any exploitation of resources in the treaty area be managed by an appropriate body of the United Nations, most likely the emerging United Nations seabed authority, according to formulas set by that body, subject to the condition that any such exploration and exploitation be conducted only if environmental protective regulations established either as interim guidelines or as approved measures by the Antarctic Treaty consultative powers are satisfied fully. In simple terms, any exploration for and exploitation of resources in the Antarctic Treaty area— no matter by whom—ought to be conducted only with appropriate environmental protections that are acceptable to the Antarctic Treaty consultative parties. The Antarctic Treaty consultative parties could begin to consider now the desirability of instituting measures within the treaty regime to protect not only the land area, but the ice and the sea in the Antarctic as well.

The Arctic

No special protective regime exists for the Arctic seaward of acknowledged national jurisdiction. There have been unilateral assertions of presumptive rights to Arctic sectors, but these have not been accepted generally, and the historical basis initially cited for one major sector assertion has been found faulty, resting on a mistranslation. The sector principle as asserted in the polar regions has not been accepted in international law. If, as appears likely, extended resource-trusteeship zones become the law of the sea, it is evident that extensive portions of the Arctic will be directly in the care of the Arctic nations.

The climatic importance of the Arctic suggests that Arctic

powers have particular responsibility in protecting the Arctic, both within their territorial seas and where they may be entrusted with mineral-extraction privileges, as well as in their Arctic lands, rivers, and airspaces. Directly to the point here is Principle 21 of the Declaration on the Human Environment, Stockholm, 1972:

> States have, in accordance with the Charter of the United Nations and the principles of international law, the sovereign right to exploit their own resources pursuant to their own environmental policies, and the responsibility to insure that activities within their jurisdiction or control do not cause damage to the environment of other States or of areas beyond the limits of national jurisdiction. [See Appendix A.—Editor]

Accordingly, those who would use the Arctic and those who would protect the Arctic from significant, hazardous geophysical perturbation share an interest in increasing knowledge of the Arctic environment—particularly in relation to trans-Arctic natural processes; this is reflected in growing international cooperation in Arctic studies.

The current Arctic Ice Dynamics Joint Experiment is primarily a United States study program, but Soviet scientists are cooperating in it, and United States and Soviet scientists have worked recently on each other's Arctic scientific stations. The Soviet scientists E. P. Borisenkov and A. F. Treshnikov, in spelling out the Soviet proposal for a Polar Experiment (POLEX) as part of the Global Atmospheric Research Program (Borisenkov and Treshnikov 1971), have noted that "participation of organizations and institutions of the Soviet Union and other countries is very desirable" and that the POLEX program "can be much improved if it is augmented by the research of other nations." Arctic processes, they have pointed out, "directly influence the nature of the weather and of the climatic conditions over the Soviet Union and adjacent territories" and generally exert "a considerable influence on the development of macroprocesses."

There is a need for a regional arrangement among the Arctic powers that, without prejudice to the rights and claims of any, would provide for mutual consultation on activities that affect, or are likely to affect, the Arctic Ocean and ice in relation to weather and climate and other natural global processes, and to provide for continuing, cooperative, scientific studies in this connection, with the free and open exchange of information therefrom. The arrangement would be in addition to existing bilateral and multilateral study arrangements and in essence would be an Arctic environmental forum and environmental information-exchange service, which would be concerned with the Arctic environment in relation to global natural processes. Seabed activities carried out in the Arctic under the authority of any international seabed regime ought to be conducted only in accord with measures for environmental protection that are held necessary by the parties to such an Arctic regional arrangement.

Conclusion

Man's ability to perturb the polar regions surpasses his technical knowledge of the scale of the expected effects of such perturbation, and the legal and political arrangements for governing the use of the Arctic and Antarctic lag disturbingly behind the human ability to affect these regions and their roles in the natural processes of the whole Earth. Nevertheless, as the Antarctic experience has demonstrated, and as informal scientific cooperation and bilateral and other study programs in the Arctic suggest, there is among both the Arctic powers and the Antarctic powers a willingness to pay attention to shared environmental concerns. It is tempting to try to rewrite existing law and political relationships, but perhaps the most useful task of this forum—as it considers regimes for the protection of the polar environments in their relation to human welfare—is to expose areas of shared interests.

Again, to propose international regulations in disregard of national interests and international law and politics is naive. Such proposals may be perceived as a threat, thus upsetting existing accommodations; but shared security concerns and the willingness of states to be sensitive to each other's national interests strengthen the effectiveness of international accommodations. In international law, as in our relationship with our environment, the Golden Rule applies with a vengeance.

References

Ayers, R. C., Jr., H. O. Jahns, and J. L. Glaeser. 1974. Oil spills in the Arctic Ocean: Extent of spreading and possibility of large-scale thermal effects. *Science* 186(4166):843–845.

Borisenkov, E. P., and A. F. Treshnikov. 1971. The polar experiment: Arctic ice dynamics joint experiment. Trans. S. M. Olenicoff. *AIDJEX Bulletin* (November).

Byrd, Richard E. 1938. *Alone.* New York: G. P. Putnam's Sons.

Campbell, W. J., and S. Martin. 1973. Oil and ice in the Arctic Ocean: Possible large-scale interactions. *Science* 181(4094):56-58.

Fletcher, J. O. 1972. Ice on the Ocean and World Climate. In *Beneficial Modifications of the Marine Environment,* symposium sponsored by the National Research Council and the U.S. Department of the Interior, 1968. Washington, D.C.: National Academy of Sciences.

Martin, S., and W. J. Campbell. 1974. [Untitled reply to Ayers et al.] *Science* 186(4166):845-846.

National Research Council. 1974a. *Priorities for Basic Research on Permafrost.* Report of the Ad Hoc Study Group on Permafrost, Committee on Polar Research, National Research Council. Washington, D.C.: National Academy of Sciences.

———. 1974b. *Southern Ocean Dynamics: A Strategy for Scientific Exploration, 1973-1983.* Report of the Ad Hoc Working Group on Antarctic Oceanography, of the Committee on Polar Research, National Research Council. Washington, D.C.: National Academy of Sciences.

———. 1974c. *U.S. Contribution to the Polar Experiment (POLEX): Part I, POLEX-GARP (South).* Reports of the Joint U.S. POLEX Panel to the U.S. Committee for the Global Atmospheric Research Program; Committee on Polar Research, Committee on Atmospheric Sciences, and Ocean Science Committee, National Research Council. Washington, D.C.: National Academy of Sciences.

Schatz, Gerald S. 1974a. International scientific concerns in the Antarctic: XIII SCAR. *News Report* [National Academy of Sciences] 24(7):4-5, 9.

———. 1974b. Polar research: Emphasis on understanding world climate. *News Report* [National Academy of Sciences] 24(6);1, 8-10.

———. ed. 1974c. *Science, Technology, and Sovereignty in the Polar Regions.* Lexington, Mass.: Lexington Books, D. C. Heath and Co.

Wright, N. A., and P. L. Williams, eds. 1974. *Mineral Resources of Antarctica.* U.S. Geological Survey Circular 705. Washington, D.C.: U.S. Government Printing Office.

41. Oil and Gas in the Canadian Arctic: Exploitation, Issues, and Perspective

Douglas H. Pimlott
University of Toronto, Canada

Introduction

The Arctic is a challenging place: it is beautiful but austere, simple yet complex, harsh though sometimes gentle, fragile but durable, vulnerable yet sometimes resistant to the impact of our technology. The challenge to industrial colonizers is to recognize the need for extreme care and to have the patience to emulate the aboriginal people, who developed an elaborate technology and a way of life that enabled them to cope with a harsh environment while living in harmony and in balance with it. The challenge to the aboriginal people is to adopt the elements of European culture they find desirable without becoming totally subservient to the gadgetry of its technology, without being overwhelmed by its apparent sophistication, and without losing their capability to return to a subsistence economy should the world suffer economic and political disintegration. The challenge to scientists is to adapt their roles in society so that, while continuing their efforts to increase knowledge and develop understanding of polar ecosystems, they can exert much more influence on government and industry to apply existing knowledge in the exploitation of renewable and nonrenewable resources.

The Arctic challenges us too because less than 5 percent of the surface of its land will ever contribute to the world economy in ways that would be measured today in the calculation of gross national product. We are just beginning to question the value of growth for its own sake—but now, in the infancy of that questioning, the Arctic challenges us to learn instantly the difference between the cost of a resource and its value. I am convinced that we will not be able to preserve the beauty or maintain the integrity of the Arctic unless we develop the ability to make that distinction: unless we learn quickly to formulate benefit-cost equations in which intrinsic values are as important as the resources that are rated so important to the economy.

Dr. Douglas Clarke worked in the Canadian Arctic for several years and developed a strong sense of identification with the land and a keen awareness of the intrinsic values of its resources. He reflected on the value of its wildlife resources in these terms:

> The last area of values related to wildlife is the national heritage—certainly the most important. The national image of the North is of a vast land, but not an empty one. There are bears and wolves, caribou and muskoxen, geese and ptarmigan. The fact that Canada should ever be known as a land that failed to take due account of its wildlife in its haste for development is totally unacceptable to our people. We must be held responsible. How much is it worth to be proud? How much does it cost to be ashamed? These are the intangibles that we must evaluate when we make decision. (Clarke 1973)

Dr. Robert B. Weeden stated the case for intrinsic values in Leopoldean terms that express the philosophy of many ecologists and naturalists toward the Arctic:

> In a very real sense what I am proposing is not only a milieu for Alaskans but an opportunity for the world. The world needs an embodiment of the frontier mythology. The sense of horizons unexplored, the mystery of uninhabited miles. It needs a place where wolves stalk the strand lines in the dark, because a land that can produce a wolf is a healthy, robust, and perfect land. (Weeden 1970)

As Dr. William A. Fuller (1970a) said:

> Let us hope that we can introduce into northern development enough science, humanity and common sense that, when all the sound and fury has died away there will still be wolves and strand lines for them to stalk.

But these words were spoken by men with finely attuned sensitivities for animals and land—men with strong ecological consciences who had developed the ability to "think like a mountain." They were not spoken by the politicians who establish the terms of reference on which development will be based, nor by the industrialists who invest the capital in development schemes and who will expect to get a substantial return on it over a relatively short term.

In this paper, I will trace the history of petroleum developments in the Canadian Arctic during the past two decades. My objective will be to show how Canada is approaching the development of its "last frontier." In so doing, I will hope to provide some of the information, some of the perspective, to answer the question: What are the chances that intensely industrialized, growth-oriented societies can exploit the nonrenewable resources of the Arctic without repeating all the mistakes made in the frontier areas of 100 or 200 years ago? Where it seems appropriate to do so, I will draw comparisons between Canadian and United States approaches.

The Arctic

What Kind of Place?

I was introduced to the Arctic in the summer of 1966. It had a powerful impact on me. Although I had "bushwhacked" a great deal in the boreal forest, and in the tundra-like regions of Newfoundland, I was not prepared very well for the Arctic of central Baffin Island, where we were at least 1,100 kilometers north of trees. The sense of shelter, the ability to use fire as a source of comfort and warmth, had obviously been very important to me, and I missed them very much. When first we travelled on foot in search of wolves and storms blew up, I felt very naked and at the mercy of the elements.

But the endless summer days were a source of great delight to me. The opportunity to be active in the field at any time of the day was a unique experience, and one that I still enjoy a great deal. At the opposite end of the spectrum, the long winter nights seem to slip by quickly and are much more than compensated for by the positive features that I associate with the other seasons of the year.

The simplicity of the plant and animal communities was striking, but it appealed to me. In making the transition from Algonquin Park in Ontario, where I had also studied wolves, I had gone from a place that had forty-three species of land mammals to one that had only seven. The differences were just as striking whether one compared birds, insects, fish, or plants. But since we were seeking to understand the phenomenon of the predator, I was delighted to be in a situation where the number of variables I had to consider was less.

Five years later, I visited the western Arctic, near the mouth of the Mackenzie River and the Beaufort Sea. I gained the feeling that I was in quite a different place: the northern swoop of the treeline and of permafrost, the admixture of animals of the boreal forest, and the presence of a great, north-flowing river made a tremendous difference. I realized then that the Arctic defies simple, generalized descriptions in spite of its relatively simple biotic communities.

Approximately 3.2 million square kilometers of North America lie north of the treeline and can be called "Arctic." That part of it in Canada was entirely glaciated, so it had shallower soils and more bouldery landscapes than either Alaska or Siberia, which have large unglaciated areas. The southern extent of continuous permafrost, the northern limit of trees, and the characteristics of the various seasons are quite variable too. In short, the ecological experience associated with crossing the Arctic Circle can be a very different one, depending on where you are in the world. The length of day or night will be the same at comparable latitudes, but a lot of other things are likely to be different.

When you really try out winter by moving out from the imported tropical hothouses we live in, you come to respect a side of the Arctic that is important to the new colonizers. In the Arctic islands, the monthly mean temperatures in January and February are not far from -40°C. A modest wind soon "subtracts" an additional 5° to 10°C. Temperatures at this part of the scale create problems: brittle fracture in metals and mental fracture in men. If you are dealing with a substance like oil, the problems may be very difficult to solve.

How "Fragile" Is It?

When ecologists began to discuss their concerns about the impact of resource development on Arctic ecosystems, they spoke of it as a "fragile" area. The term stuck and it is used—perhaps overused—widely in television documentaries and in articles on the Arctic. Whether or not "fragile" is the right word, there can be little argument with David R. Klein's description of the situation:

> Arctic and Subarctic ecosystems, however, have unique characteristics which set them apart from the ecosystems of more temperate regions and which necessitate a new pattern of behavior by technological men if they are to survive as viable entities into the future. (Klein 1973)

There certainly is no doubt that Arctic ecosystems are sensitive and vulnerable to pressures and stresses that resource development can place on them. Their vulnerability is due to the presence of permafrost, particularly permafrost in fine-grained soils that have a high ice content. When the vegetative cover is disturbed, the permafrost tends to melt. If there is high ice content, the soil slumps and in many situations may wash away. Gullies can form quickly and cause deep scars on the landscape.

The simplicity of Arctic ecosystems, which I referred to earlier, seems to be due to the slow growth of plants, which have to live with a limited supply of nutrients, the long winter, and the low angle of incidence of the sun's rays even during the long days around the summer solstice.

The concept of fragility has been thoughtfully reviewed by Dr. Max Dunbar (1973). I have thought particularly about the vulnerability of many Arctic animals to human activities. For example, many species mature slowly and have low reproductive rates. Arctic char on Baffin Island may be twelve or more years old before they spawn and may spawn only every second or third year (Grainger 1953). The tendency of some Arctic animals to aggregate in particular areas or to return to specific areas at predictable times results in another element of vulnerability to human activities. Some of the animals that exhibit this characteristic are caribou, muskoxen, polar bears, walruses, narwhals, white whales, several species of geese, and seabirds. The aggregations usually occur in response to some particular aspect of the environment. The result is that the animals are very easily killed by hunters, are very susceptible to oil spills, or are easily dislocated or disturbed by seismic operations or by a multitude of other human activities.

To sum up, ecologists worry about the Arctic because of the way human activities can scar and degrade the landscape through erosion caused by melting of the permafrost; because of the uncertainties about what will happen if catastrophic events, such as a major oil spill in the Mackenzie Delta, the Beaufort Sea, or on the Arctic islands, occur; because of the sensitive nature of its fauna; and because so many animals behave in ways that make them disaster-prone when they come in contact with humans or human activities. The range of potential environmental impacts that may result from petroleum development in the Arctic were reviewed by Klein (1973) and by Bliss and Peterson (1973).

The Approach to Development

During the past twenty years, the government of Canada has made a very intensive effort to encourage the multinational resource corporations of the world to undertake exploration programs for oil, gas, and minerals in the Northwest Territories and the Yukon Territory. The effort to find and exploit these resources was based primarily on the growth-progress syndrome, not—at least until very recently—on a sense that they were required to meet Canada's needs.

Although the pace of northern development began to pick up immediately after World War II, in retrospect, many Canadians think of 1958 as an important benchmark: it was an election year, and the Conservative Party was making a strong bid to convince Canadians that it was the party of the future. The campaign focussed strongly on what a Conservative government would do to promote resource development. "The Vision of the North" was a prominent campaign slogan of the party. It is often referred to now as "Diefenbaker's Vision of the North," after the venerable politician who led his party to victory in the election and who is still a member of the House of Commons.

Subsequent events show clearly that the stage was being set for a program of intensive development. Land and geological surveying was greatly intensified, and in 1960 greatly improved geological maps were published. The maps showed the location of formations likely to bear petroleum and minerals. Before 1964, the average number of claims staked north of $60°N$ was less than 6,000 a year; during the next five years, there were five major staking rushes. In 1968, a peak of 52,000 claims were staked, which represents a tremendous level of activity in exploration and development (Passmore 1971).

In 1964, a railway was completed to Hay River on Great Slave Lake and to Pine Point, where a large lead-zinc mine was being developed. Late in the decade, intensive studies were conducted to determine the feasibility of establishing a deepwater port at Herschel Island on the Beaufort Sea.

In 1966, a Northern Exploration Assistance Program was established. It provided up to 40 percent of the exploration costs of Canadian citizens, or of companies incorporated in Canada. Added to this were many speeches in which the virtues of northern development were extolled and the rapid evolution of the "petroleum play" was referred to in glowing terms.

In the case of petroleum, the only field in production was a small oil field at Norman Wells, which had been discovered in the 1920s. However, the area held by oil companies under exploratory permits increased rapidly after the geological maps were published in 1960. Much of the interest was of a speculative nature, and there was only little exploration of permit areas in the early part of the decade.

The announcement of a major discovery of oil and gas at Prudhoe Bay came just after the Canadian government had joined a consortium to form Panarctic Oils Limited, in which it held 45 percent of the equity. The new company immediately began an active exploration program on its holdings of approximately 18 million hectares in the Arctic islands. It has constituted the most successful operation in northern Canada. To date, it has discovered approximately 15 trillion cubic feet [425 billion cubic meters] of gas and undisclosed quantities of oil, but at the same time, Panarctic achieved an unimpressive record in the conduct of its operations. It had made two major gas discoveries out of the first eight wells drilled; both—one on Melville Island and one on King Christian Island—had blown out. Fortunately for the environment, neither discovery was of crude oil (Woodford 1972).

After drilling wells on the Arctic islands, the company began to explore its offshore holdings in 1974 and 1975, drilling wells to determine the extent of one of the major gas fields it had discovered (Pimlott 1974c).

Many of the large multinational corporations also became interested in the potential of the Arctic; by 1971, 186 million hectares were under permit (Yates 1972). Earlier, the oil industry had been invited to draft the oil-leasing regulations so they would provide maximum incentives for exploration. The leasing arrangements were much more favorable than those that prevailed in Alaska, and in terms of petroleum speculation they had certainly accomplished their purposes (Thompson and Crommelin 1973). The areas held under exploration permits represented virtually all the potential oil-bearing formations north of the 60th parallel. They also included large offshore holdings, which covered the Beaufort Sea to the edge of the polar pack and virtually all of the waters enclosed by the Arctic islands, including the Northwest Passage. It even included McLure Straits at the northwest end of the passage, which is normally free of ice only one year in ten.

In the post–Prudhoe Bay climate, expenditures on exploration increased from $24 million in 1967 to $170 million in 1971. They remained at a high level through 1974, but are said to have dropped during 1975 because of uncertainty due to the government's failure to clarify royalty and other taxation policies.

There have been several discoveries of gas and oil in both the Mackenzie Delta region of the western Arctic and in the Arctic islands. In the case of gas, the combined total of all the finds still falls short of that discovered in the Prudhoe Bay reservoir. Although no announcements have been made on the size of the oil discoveries, it is believed that they are still far below the volume required for commercial production.

At present, much of the focus on exploration is shifting from the land to the sea. It appears that the results of exploration programs have indicated that "elephant-sized" discoveries are unlikely to be made on land. However, the industry remains hopeful that offshore regions contain reservoirs of the magnitude of Prudhoe Bay's.

Offshore operations will be quite diverse in nature. Up to the present, they have included drilling from artificially constructed islands in shallow water at the mouth of the Mackenzie River delta; the use of land rigs on thickened-ice platforms in the Arctic islands, and the use of a semisubmersible rig in Hudson Bay. In 1976, two drill ships were scheduled to begin operations far off the land in the Beaufort Sea. Approval in principle had also been given for the operation of a drill ship in Lancaster Sound in 1975; however, the company involved decided not to proceed with the operation that year. The well there was to have been drilled in 750 meters of water.

Soon after the Prudhoe Bay discovery, the Canadian government became intensely interested in the possibility of constructing both oil and gas pipelines from Alaska through the Mackenzie Valley to markets in southern Canada and the United States. A network of interlocking committees was formed; in 1970, as a result of their efforts, the government promulgated a generalized set of guidelines for the construction of pipelines.

Progress toward a rapid and smooth integration of the energy resources of the Canadian Arctic into a "continental pool" was profoundly disturbed by the voyage of the *Manhattan* into and through the Northwest Passage in 1969 and 1970. One result of the voyage was that many Canadians realized the importance of the

Arctic for the first time. However, the voyage created a chaotic situation in the echelons of the federal government. Although they were never publicly admitted, there were strong fears that Canadian sovereignty over the Northwest Passage was being challenged by the United States. The government reacted by enacting the Arctic Waters Pollution Prevention Act in 1970. Its enactment was primarily a matter of international politics. It was a clear case of using an expression of concern for the environment as a means of strengthening Canada's jurisdiction over the ice-bound waters enclosed by the Arctic islands. The United States objected strongly to some of the provisions of the act, which, it claimed, were contrary to international law. However, the bill was passed unanimously by the House of Commons in June 1970 and came into force in 1972, when regulations were promulgated under the act (Pharand 1973).

The "qualified" success of the *Manhattan* in traversing the Northwest Passage, the legal barriers that were being posed to the trans-Alaska pipeline by United States environmental groups, and the fears in British Columbia that major oil spills from tankers operating between Valdez and the state of Washington off the coast of British Columbia encouraged pipeline proponents in Canada. The two departments (the Department of Indian Affairs and Northern Development and the Department of Energy, Mines, and Resources) that were promoting the construction of both oil and gas pipelines moved rapidly to take advantage of the promising situation. Early in 1971, Canadians became aware of the fact that the federal government was offering a Canadian route as an alternative to the trans-Alaskan route. The offer was being made in a series of speeches by ministers of the federal cabinet. However, Canadians also believed that it had been made directly at an informal meeting with high-ranking members of the United States administration in Washington by the Minister of Energy, Mines, and Resources. Significantly, these activities took place shortly after the trans-Alaska pipeline hearings in Washington and Alaska, at which the United States Department of the Interior's preliminary environmental impact assessment was reviewed and strongly criticized.

However, environmental concerns about the development of oil and gas had also surfaced in Canada over the construction of an oil pipeline. Under pressure in the House of Commons and in the press, the government felt it necessary to qualify statements that had been made about Canada's willingness to accept an oil pipeline. Top executives of multinational oil companies jetted in and out of Ottawa. By the late spring of 1971, it had become evident that the industry would continue its fight to gain approval of the trans-Alaska pipeline. The federal government recognized that, while there would probably be a gas pipeline through the Mackenzie Valley, there was no immediate hope for an oil pipeline. The government then went on and attempted to ensure that nothing would stand in the way of the construction of a gas pipeline.

In the meantime, two consortia, Arctic Gas Limited and the Northwest Project Study Group, had been developed as competing organizations to prepare feasibility studies, engineering plans, and social and environmental impact assessments. The government brought strong pressure for the amalgamation of the two consortia, and this was achieved in 1973. The new organization, Canadian Arctic Gas Study Limited, continued the Environment Protection Board but deemphasized its role. The job of preparing the environmental data for the public hearings was given to the more traditional environmental section, which had come into Canadian Arctic Gas Study Limited from the Northwest Project Study Group.

These hearings are now going on before Justice Thomas Berger of the British Columbia Supreme Court, who was appointed in January 1974 to study and make a recommendation on the application for the right-of-way of a gas pipeline through the Northwest Territories and the Yukon Territory. [See editor's note on page 510.]

Environmental Issues and Problems

The Environmental Issues

During the 1960s, few Canadian conservationists outside of those in the federal government realized the extent of the developments that were shaping up in the Arctic. Virtually no questions were being raised about the potential impact of exploration on Arctic ecosystems in Parliament, in the press, or even in the publications of conservation organizations. In retrospect, it seems evident that Canada simply did not have either the individuals or the organizations that could or would take the initiative to alert the country to the dangers posed by the massive thrust by government and industry in the Arctic.

The development of powerful national conservation organizations has proven to be difficult in Canada. During the 1960s, the Canadian Audubon Society (now the Canadian Nature Federation) and the Canadian Wildlife Federation were the principal national conservation groups. Both operated at submarginal levels throughout the decade; bankruptcy was an ever-present threat. As a result, neither had much capability to monitor the policies, programs, or activities in the North. As will be seen, the Canadian Wildlife Federation did an important job in 1970 of bringing the problems into focus in spite of the limitations imposed on it by the shortage of money and human resources.

In the case of governmental agencies, the Canadian Wildlife Service was the principal one interested in the protection of Arctic ecosystems. Members of the Canadian Wildlife Service were actively engaged in studies of Arctic foxes, caribou, muskoxen, waterfowl, seabirds, and other species in the postwar period; however, the main thrust of these studies was over by the early 1960s. Local government was evolving in the Yukon Territory and in the Northwest Territories. Game management was one responsibility assumed by the game branches when territorial governments were formed. Thick walls soon developed between the two game branches and the Wildlife Service, and the Service encountered strong resistance to its request for funds for Arctic research.

During the critical predevelopment period, the Wildlife Service had little influence on the formulation of environmental policies for the Arctic. I became aware of this in 1969 as leader of a study for the Science Council of Canada (Pimlott et al. 1971). We attempted to analyze the situation for our report, but the Wildlife Service was extremely sensitive about its relationships with the territorial game branches and refused to give our research team access to the files that would have provided us the knowledge we needed to review the situation and to report on it to the government and the public.

Responsibility for the aquatic ecosystems of the North came directly under the Department of Fisheries (which developed into the Department of the Environment in 1972). In this case, there were no jurisdictional conflicts with either territorial government. However, the department was very ineffective in representing Arctic environmental interests. The Fisheries Research Board established an Arctic biological station in 1955; it was formed to be responsible for

"Canadian research on marine and anadromous fish and coasts." In terms of the two territories alone, its area of responsibility constituted 30 percent of the area of Canada; however, up until 1972, "the full-time scientific staff of the station has never exceeded nine!" (Sprague 1973). The interest of the department and the Fisheries Research Board was primarily in the commercial fisheries of the two coasts. As a result, it did even less than the Canadian Wildlife Service to promote the development of an ecological conscience among the agencies that were so actively promoting the development of oil, gas, and minerals in the Arctic.

The result of the lack of environmental purview began to be evident to Canadians in 1969, the same year that Panarctic Oils' first well on Melville Island blew out of control. It turned out that virtually nothing had been done to enact environmental legislation to control any phase of exploration activities. As in the case of Naval Petroleum Reserve Number Four in Alaska, the heavy equipment used in seismic operations had been allowed to work on the tundra during the summer, causing severe erosion of the tundra on the Tuktoyaktuk Peninsula and in at least one other area of the Northwest Territories.

Dr. William A. Fuller's interest in Arctic conservation began in 1963 when he had "observed cat tracks on the North slope of Alaska west of point Barrow" (Fuller 1970a). This caused him to become concerned about the potential effect of exploration activities on the Arctic. He worked for more than five years within the International Union for Conservation of Nature and Natural Resources and succeeded in convening a "Conference on Productivity and Conservation in Northern Circumpolar Lands" in October 1969 at the University of Alberta (Fuller 1970b).

The conference was a significant event. The activities leading up to it stirred interest at many levels, including the Department of Indian Affairs and Northern Development. Nine days before the Conference, Mr. Chrétien, the minister of the department, gave his first "We will protect the Arctic" speech. His department made a modest grant to support the conference. He sent a telex message that was read on his behalf at its opening; in it, he stated that his department would establish land-use regulations, support a broad program of research, and propose water-conservation legislation to Parliament (Chrétien 1970). A member of his department presented a paper that filled in some of the detail (Naysmith 1970). The conference passed a resolution welcoming his statement, but continued by stating that, while "endorsing the program, deep concern is expressed that it should be implemented on a scale and with a speed and determination that will adequately cope with the thrust of development now penetrating the north."

The conference was conducted as a very circumspect, unemotional, scientific activity; however, it served as a focus for Canadians to develop perspective on what could happen to the Arctic if exploration crews were allowed to run rampant in the North.

The new perspective was sharpened by the course of events in both the United States and Canada in 1970. A preliminary injunction against construction of the trans-Alaska pipeline was granted in April 1970 under the mandate of the National Environmental Policy Act; the tanker *Arrow* sank off the coast of Nova Scotia, bringing out how little was known about the effect of oil and how nearly impossible it would be to clean up a spill in the Arctic (McTaggart-Cowan 1970); Panarctic's second natural gas well blew out in October and burned with a deafening roar for ninety-one days; and the Canadian Wildlife Federation sent out a "Crisis in the North" letter in February, stating a succinct case on the nature of the hazards and arguing a strong and rational case for a partial moratorium on oil exploration until 1974.

The Wildlife Federation's proposal did not go far, but it and many other events of 1970 intensified discussion among the public, which speeded up the enactment of environmental legislation and made it evident to government that people were finally aware of what was happening in the Arctic. Before the end of 1970, the Arctic Waters Pollution Prevention Act and the Northern Inland Waters Act were passed by Parliament. The Land Use Regulations were also promulgated under the Territorial Lands Act, which had been revised by Parliament so that land-use activities could be controlled under it.

The events and expressions of concern for Arctic environments during the early 1970s are too numerous to review in an account of this nature. They are documented in the proceedings of a federal-provincial wildlife conference, a conference of the Royal Society of Canada, a special report by Pollution Probe at the University of Toronto, and many articles and editorials in newspapers and in television documentaries. To a large extent, the focus was on the construction of pipelines, but other types of activity occasionally were involved. The period was epitomized by very sharp contrasts. Background studies for assessing the environmental impact of the construction of the gas pipeline were quite intensive. They were conducted on behalf of industry by an environmental section of the Northwest Project Study Group and by the Environment Protection Board. The Department of the Environment had many scientists involved in research along the route, and the Arctic Land Use Research Program provided grants for a variety of studies.

Offshore drilling in the Beaufort Sea and elsewhere in the Arctic provided an example of the contrast. In this case, the Department of Indian Affairs and Northern Development had been encouraging exploration for years. However, in 1973, when the cabinet was requested to give approval-in-principle for offshore drilling operations, no specific research had been undertaken, and the proposal to the cabinet made no mention of the need for an environmental-impact assessment. In addition, the plans for the operation were shrouded in secrecy. In order to report on those plans for a Native organization, the Committee for Original Peoples Entitlement (COPE), I had to search out restricted, confidential, and secret reports and other documents within government agencies and industry (Pimlott 1974c, Pimlott et al. 1975).

A decision was finally made to conduct a research program in the Beaufort Sea. It was an entirely inadequate program, conducted in a single season during one of the worst ice years in a decade. In other Arctic areas, offshore drilling has been conducted without even the pretense of research programs. In one case, approval was based on an environmental-impact assessment that made a mockery of scientific procedures (Canadian Arctic Resources Committee 1974a, 1974b).

In another case, which involved the construction of artificial islands, an environmental-impact assessment was made of the effect of construction of an island on beluga whales; however, the terms of reference for the study specifically excluded the consultant from considering any other impacts of the exploration process, the most obvious of which was the effect of oil spills (Pimlott 1974c). This topic would have been particularly relevant because the island was constructed at the interface of the Mackenzie River delta and the Beaufort Sea, an area that is very critical to many species of birds as well as to whales.

One of the most controversial events was the announcement by

Prime Minister Trudeau in April 1972 that an immediate start would be made on the construction of a Mackenzie Valley Highway, which would terminate at Tuktoyaktuk on the Beaufort Sea and open up the entire valley to traffic and people from the south (Trudeau 1972). When the announcement was made, the preliminary engineering surveys had not been done. In addition, no research had been conducted, and none ever was undertaken, to determine either the potential social or environmental impact of the project—this in spite of the fact that by then the government had required the pipeline consortia to embark on such a program in preparation for public hearings on the pipeline.

On the side of industry, one of the most positive initiatives was taken by Arctic Gas Limited, which contracted to have its environmental studies and its environmental impact assessment prepared by a specially constituted body, the Environment Protection Board. It seemed like an important innovation because the board included some of the best-known ecologists in Canada; most important was the fact that the results of all studies were to be made public at the discretion of the board. However, the board barely survived the amalgamation of the consortia. In addition, it has become clear that the petroleum industry considers that the Environment Protection Board was a dangerous experiment.

Unfortunately, there seems to be no likelihood that the endeavor will be considered a precedent for the industry to follow in the future. Rather, it appears that the model will be the Arctic Petroleum Operators Association, an organization that has hosted annual Arctic environmental conferences and sponsored studies on various topics of interest to exploration and production systems. These studies are, however, held confidential among the sponsors and are unavailable to environmental organizations, and even to many government agencies.

Perhaps the most significant development on the public-interest side was the formation of the Canadian Arctic Resources Committee (CARC) in 1971. CARC sponsored a workshop in early 1972 on the social, environmental, and legal aspects of northern development (Pimlott et al. 1973). I stated our reasons for forming CARC in these terms:

> By the end of March [1971], we were convinced that Canada badly needed an organization which could provide a pair of eyes to look in on the North

in a more perceptive way than any existing citizens' organization was capable of doing; which could act in an Honest Broker capacity to attempt to ensure that the things that needed to be done in advance of development of whatever type, got done; which could help to bring to the surface the question of what was to be done about the claims of the native people; and which could help to overcome the barrier to factual information existing between the Canadian people and the Government on matters that pertained to development, the native people and the environment. (Pimlott 1973)

We solicited the financial support of both the government and the oil companies for the workshop. It was our objective to emulate the low-key approach of the Conference on Productivity and Conservation in Northern Circumpolar Lands, while at the same time dealing frankly with the problems. But, even though we were very thorough in our approach to fund raising, no oil company nor any department of government would give us financial assistance for the workshop. This is the way I spoke of CARC and the situation it faced in my talk at the opening of the workshop:

In a memorandum to the Committee, I described CARC as a social experiment. In retrospect, I think that the description was an apt one.

In the first place, it was an attempt to form an organization which would further public participation in quite different ways. Those of us who conceived of the organization had often taken adversary positions on government policies on resource and environmental matters. But we recognized that significant things appeared to be happening to attitudes and approaches towards both the environment and the native people. We reasoned, if attitudes in government and industry had changed, perhaps it was possible to achieve results through processes of reason rather than of confrontation.

In the second place, there are not many organizations where people with a wide diversity of backgrounds work together to further the cause of the native people or the environment. We hoped to add a little to the development of that type of organization.

But the experiment goes beyond that. It seemed to us that there are at least two fundamental problems which must be faced if Canadian society is to come to terms with socio-cultural and environmental problems. One is the need for much greater day-to-day participation by people in decision-making processes. The other is honouring the right of people to know. (I apologize for the cliché—I know of no phrase that states the case better.) Government in Canada is not finding it easy to come to terms with these problems. Part of the experiment has been to determine if CARC could do anything to speed up the process of letting the people in on decisions in the North. As far as I am concerned, that is what the Canadian Arctic Resources Committee is mostly about.

It is difficult to be certain whether or not the Workshop will make a significant contribution to the cause. We were not very successful in getting financial support from the government or the industries most directly involved in the North. Both, however, will be ably represented in the working sessions, and Mr. Chrétien will speak to us on the opening night. So I express pleasure about the latter and reserve judgment on the former. I am sure that the success of the Workshop will be judged by what we accomplish, not by who pays for it.

Sometimes I describe myself as a realistic idealist. Assuming that role for a moment, I have to admit that it was not very realistic to have expected the petroleum-centered industries or the Federal Government to welcome CARC with open pocket books. After all, the companies we solicited money from have blood brothers, or fathers, who were badly stung by rather similar mosquitoes in Alaska. We claimed that we were not after as much blood. But who would be certain what might happen if we were given the chance to sting? And what would the reflections of Standard Oil be if Imperial Oil contributed funds to an organization that had the remotest similarity to The Wilderness Society or the Sierra Club—two of the Alaskan mosquitoes.

In terms of realism, some similar things can be said about our attempts to obtain financial support from the Federal Government. The Department of Indian Affairs and Northern Development and the Department of Energy, Mines and Resources have been active in promoting the development of the North. Several members of our Committee have been critical on many occasions of northern policies and programs. It was not very realistic to expect that the Government would support an organization which included such people. We ran into opposition within the Civil Service, in areas where we least expected it. It caused us to reflect on the strength of aggressive instincts when territorial boundaries are being transgressed. It was another element of realism to which we should have been more attuned.

But in terms of realism, there was another side too. In 1968, a government was elected that had included participatory democracy as a plank in its election platform. In retrospect, I consider that it provided the license for some of our lack of realism.

Time lends perspective. It is interesting to look back over events of the past three years and see how environmental activism has shaped events in Canada. There has been confrontation over issues on several occasions. Many people consider that it is a poor way of doing things: others have argued that there was no other way. Time will lend perspective— even if it will not be able to provide a black or white answer.

But, to return to the North, I feel certain that some things will not change. One of these is the increasing interest of people in the south for the place of the people who were there first. Another is an increasing awareness of the need to protect its fragile environments. One of the unanswered questions is whether people will be encouraged to help shape events or not. David Anderson said, "Let the people in." Is there a reasonable alternative? Time will tell. Eventually people and the North will know. (Pimlott 1973)

CARC has been very active on Arctic problems and issues since 1972. It has sponsored a series of publications including a regular letter, *Northern Perspectives,* a technical report on land-use regulations, and the proceedings of two additional conferences. It is a far cry from the doldrums of the 1960s.

In terms of the evolution of environmental policies and programs for the Arctic, the greatest disappointment has been the role of the Department of the Environment. A second disappointment has been the inability of the Department of Indian Affairs and Northern Development to overcome the conflicts of interests due to its dual role of protecting the environment and promoting the development of nonrenewable resources. This is the way I reflected on the problem at a CARC conference a year ago:

> The formation of Environment Canada delighted many conservationists. At last, we said, there is going to be a department of Government which has a broad mandate to protect the Canadian environment. Its formation seemed to bode particularly well for the Arctic since there were no competing governments in the North to develop confusion and jurisdictional squabbles. But since 1970, the Government has increasingly removed the responsibility of protecting the Arctic environment from the jurisdiction

of Environment Canada and placed it under the Department of Indian Affairs and Northern Development which is "seeking the exploitation of non-renewable resources". Environment Canada was not named as an administrative or a cooperating agency for the enforcement of either the Arctic Waters Pollution Prevention Act or the Northern Inland Waters Act. In addition, during the past winter I have realized that its prerogative to enforce the Fisheries Act in the NWT [Northwest Territories] is being progressively reduced and taken over by DIAND [Department of Indian Affairs and Northern Development]. The Environmental Protection Service of DOE [the Department of the Environment] has been, and is, operating at less than a marginal level both in terms of staff and budget. It even has unwritten orders to clear all of its actions to enforce Section 33 of the Fisheries Act with DIAND. EPS [Environmental Protection Service] is so ineffective that it might as well be withdrawn from the territories. (Pimlott 1974b)

To further widen the "credibility gap," the responsibility for environmental protection has not been given to the Conservation Branch of the Department of Indian Affairs and Northern Development, but to the branch that until recently was called the Northern Economic Development Branch. Of course, if the job were being done in an adequate way, it would not matter what agency or department did it. But there is much evidence that the Department of Indian Affairs and Northern Development has not resolved the conflict of interests between development and protection. The examples are legion.

The land-use regulations have been authoritatively criticized on many occasions by Andrew Thompson and others. More recently, Peter Usher and Graham Beakhurst have brought the weakness in implementation of the regulations into sharp focus. But in terms of perspective on environmental policy, one of the most significant developments was the virtual exclusion of mineral exploration from the regulations—by carefully worded definitions.

The approaches of the departments of Indian Affairs and Northern Development and of the Environment to preparation for offshore drilling in the Beaufort Sea are another example of the government's inability to bring environmental protection into perspective with the development of nonrenewable resources. That case history has been documented in some detail in a report which I wrote for COPE (Pimlott 1974a) and for the last issue of

Northern Perspectives (Canadian Arctic Resources Committee 1974a, 1974b), so I will not detail it here. A few points will serve to bring it into focus:

1. The Department of Indian Affairs and Northern Development has been issuing permits for petroleum exploration in the Beaufort Sea since the early 1960s. It made no attempt whatsoever to foster or encourage a program of research to assess the potential impact of exploration and development on the environment prior to submitting a memorandum to the cabinet to obtain approval-in-principle for offshore drilling.
2. The memorandum to the cabinet failed to give cabinet members any appreciation of the environmental hazards and risks that would be involved in petroleum operations in the sea. The Department of the Environment had two representatives on the committee that reviewed the memorandum, so it must share some of the responsibility for its inadequacies.
3. The Department of Indian Affairs and Northern Development allows the petroleum industry to "shelter" potentially important reports on environmental topics, such as meteorological and ice conditions, under proprietary-interests arrangements, so that they are not even reasonably available to members of government. It has even allowed these proprietary interests to dictate that a major conference on offshore drilling in the Arctic be held *in camera* and the distribution of the report restricted to participating oil companies and government agencies.
4. The government tried very hard to prevent information on the project from becoming publicly known. In fact, all the important documents associated with the plans were rated either "restricted" or "confidential." In retrospect, the most obvious effort at subterfuge was the steps taken to disguise the construction of artificial islands as normal extensions of land-use operations rather than to identify them as the first stage of offshore-drilling operations.

There are many more examples of the Department of Indian Affairs and Northern Development's inability to incorporate environmental considerations in decisions on resource development. Even a short list would include the limited research undertaken to determine the impact of seismic and other exploration work on the tundra and animals in the Mackenzie River delta and the Arctic islands; approval of Panarctic's first offshore drilling operation after only a crude and limited study had been made of environmental considerations; the recent approval-in-principle by the cabinet of major expenditures to support the development of mining operations in the Strathcona Sound area of Baffin Island—once again, before the Department of the Environment had conducted any environmental research; the decision to promote the development of the hydro potential of Great Bear Lake before any program of environmental assessment had been undertaken; the failure of the government to undertake even preliminary studies of much of the animal-resource base of the people of Victoria Island and Resolute Bay, even though oil companies were granted exploration permits for large areas on the island some time ago. The role of the Department of Indian Affairs and Northern Development in protecting Arctic environments is, of course, seen in a much more positive way by members of the department (e.g., Chrétien 1973, Naysmith 1971, 1973a, 1973b).

In retrospect, experience has shown that the concerns reflected in the resolution passed at the Conference on Productivity and Conservation in Northern Circumpolar Lands in 1969 and by the Canadian Wildlife Federation in 1970 were justified. However, the concern shown by Canadians over the possible impact of Arctic development has as yet had only a modest influence on government policies and programs for the Arctic.

The Rights of the Native People

The aboriginal rights of the Indian and Inuit (Eskimo) peoples of the Northwest Territories and the Yukon were ignored during the thrust for the development of Arctic petroleum resources in the 1960s. The most striking documentation of this statement is provided in the report of a study of the people of Banks Island. Usher (1971) traced how the government had issued petroleum-exploration permits for the entire island (71,700 square kilometers) and the adjacent waters without informing the Bankslanders of what was happening. In June 1970, representatives of two oil companies visited Sachs Harbour to tell the people that they held exploration permits and would conduct seismic exploration during the winter of 1970-1971. There appeared to be no recognition of the fact that the people, whose livelihood depended almost entirely on trapping, would feel that the wildlife and, as a result, their interests were threatened by the operations.

The case history gives a thorough account of the callous way the Bankslanders were treated. Exploration did come to the island, despite the objections of the people. However, as a result of the confrontation, they have been able to maintain some control over the time of year when exploratory operations will be conducted.

There have been some improvements in the approach of government since 1970, but the changes have not been profound ones. In the early 1970s, when government and industry were intricately involved in planning for offshore-drilling operations, Native people were not consulted and were not told about plans until they had reached the *fait accompli* stage. For example, the Department of Indian Affairs and Northern Development held a "Northern Canada Offshore Drilling Meeting" in December 1972 as one stage in the preparation of a request to the cabinet for approval-in-principle for the project. Officials from oil, gas, and consulting companies were included, but no Native organizations were invited to send representatives. COPE reflected on the situation in one of its news releases (Committee for Original Peoples Entitlement 1974). The rationale for the situation that existed through the 1960s was perhaps given by Prime Minister Trudeau, who, in referring to Indian land claims, stated that the government "would not recognize aboriginal rights" (Trudeau 1969).

The negative attitude of the government, the moratorium on development in Alaska until Native claims were settled, the rapid expansion of exploration activities in the North, and the discovery of oil at Atchinson Point on the Tuktoyaktuk Peninsula in 1970, caused great concern among Native people. One result was the formation of six organizations to represent their interest between 1969 and 1972: COPE, Indian Brotherhood of the Northwest Territories, Inuit Tapirisat of Canada (a national Eskimo organization), Yukon Indian Brotherhood, the Métis Association of the Northwest Territories, and the Federation of Natives North of Sixty.

The pressures on the government from these organizations and from public opinion in southern Canada gradually forced the government to adopt a more flexible policy. All the organizations eventually received financial support from the government to support their programs; by mid-1971 the Minister of Indian Affairs and Northern Development had begun to make statements such as, "I am ready at any time to sit down to discuss the settlement of their treaties" (Chrétien 1972). On August 8, 1973, the minister issued a communiqué on the "Claims of Indian and Inuit People," which stated that the government had changed its stance and was willing to negotiate a land settlement.

It remains to be seen how profound the change in policy really is. In the first place, the government has never considered a moratorium on development, such as was imposed in Alaska, while claims are being negotiated; in the second place, the communiqué appeared to exclude the possibility that the Native people would gain major control of land and resources. The key statement seemed to be, "An agreed form of compensation or benefit will be provided to native peoples in return for their interests" (Department of Indian Affairs and Northern Development 1973). Thirdly, a proposal for a settlement of Indian claims in the Yukon, made in a white paper earlier this year, appears to fall far short of the hopes and expectations of the Native people of that territory. The legal basis of land claims and the historical record of settlements has been reviewed by Cumming (1973).

But there are also things to be positive about. The government provided quite liberal funding for land-use and occupancy studies to establish the basis for negotiation of both Indian and Inuit claims. In the case of the Northwest Territories, the Inuit studies are virtually complete; the Indian studies for the Northwest Territories will probably be completed within a year.

Another positive action was the appointment of Justice Thomas Berger to conduct the hearings to determine the social and environmental impact of a gas pipeline through the Mackenzie Valley. When in law practice, Justice Berger had represented the interests of an Indian band on a land claim in British Columbia; as a jurist, he has a reputation of being an able arbitrator.

In retrospect, an action taken by the Indian Brotherhood of the Northwest Territories was probably of paramount importance in stimulating the government to put out the communiqué stating that it was willing to negotiate on land claims. The action was the filing of a caveat in the Territorial Court early in 1973. It warned developers to beware of going ahead with major programs, such as pipeline construction, because the Native people had a legal interest in the land and resources. In September 1973, Justice William Morrow found that the Indian people had sufficiently established their case to give them the right to file the caveat (Wah-shee 1974).

Perspective on Man and Resources in the Arctic

Each year since 1962, when I ended the last of four periods of my life spent in government service, I have become increasingly involved in the public-interest side of conservation and environmental affairs. For some years, my attention, like this series of biennial wilderness conferences, focused on the preservation of natural areas and endangered species. My interest in these is still very strong; however, my energies are devoted more and more to broader areas of environmental concern, such as the one I am talking about today. The evolution was prompted by the realization that the preservation of bits and pieces of species, parks, and wilderness will have little meaning unless they are part of a beautiful, diversely patterned fabric that is represented by a whole, healthy Earth. The increasing level of my activities has resulted from a growing conviction that democratic societies will not be able to come to terms with problems of environmental degradation unless "people" become much more involved in the day-to-day decision-making processes of government.

Intricate, symbiotic processes involving government and industry have developed over the years as the growth-progress syndrome has evolved. At the same time, the growth of massive bureaucracies has put a greater distance between politicians and the people they represent.

Events documented in many environmental case histories demonstrate that the system is not working well. In terms of EARTHCARE, it has led us into deep trouble. It is clear that something drastic must happen if we are to get clear of the quicksand that is tugging at us. In terms of the environment, a situation in which industry has relatively direct access to government, and people at the grassroots virtually none, contains some very important weaknesses. One is that the time perspective is too short: politicians have difficulty looking beyond the next election, and boards of directors have a problem looking beyond the time required to recover the capital invested in a particular project. EARTHCARE requires a much longer perspective.

Many events associated with the energy equation have shown that the activities of public-interest groups provide some of that perspective. But they must be willing to come to terms directly with issues in a way that keeps the issues before the public, turns them into political issues, but does not turn too many people off. Environmental groups must also, somehow, convince governments and society at large that they are of great potential importance in redressing and preventing environmental problems. That is the premise I will argue in this part of my paper.

The events associated with the move to exploit the oil and gas resources of the Arctic indicate both the nature of the problem and the need for grassroots involvement.

In Ontario, where I now live, the petroleum industry has been a leader in pollution abatement. It has been far ahead of other major resource industries, such as mining and pulp and paper. Sixteen years ago I recall seeing full-page advertisements featuring a pollution-treatment plant. The plant treated the wastes of a refinery so thoroughly that a wide variety of fish lived in a tank that was fed from it. I often cited the performance of the industry in talks I gave on environmental issues and problems.

I became aware during the 1960s that some of the major oil companies were developing environmental organizations. In the early 1970s, at least one Canadian subsidiary of a multinational corporation had a larger environmental staff than the Ontario Hydro Commission, the publicly owned corporation that supplies electricity to Ontario. But I am sure it is evident that my impression about the performance of the petroleum industry in the Arctic has not been nearly as positive. Except for activities pertaining to the gas pipeline, many things are being "swept under the rug"—by, for example, the "big-name" company that keeps its sense of environmental concern before the public, which has a strong environmental staff, but which specifically avoided having an assessment made of the really significant hazards posed by its operations—that is, the effects of an oil spill if one of its offshore wells were to blow out; or the company that has pressed very hard to drill wells from the ice with land rigs and that has submitted inadequate assessments of the potential impact of these operations on the environment in support of its applications. The examples are legion.

But no Canadian example measures up to the inadequacies involved in the trans-Alaska pipeline case. In that case, a consortium of multinational corporations prepared to build a pipeline across hundreds of kilometers of tundra, using a "cookbook approach" to the application of technology. Even members of oil companies will now admit that if the pipeline had been built as originally planned it probably would have been disastrous for Alaska.

There are many examples which indicate that when the investment chips are down it simply does not seem to be possible for industry, and particularly one working in remote areas, to give environmental protection adequate consideration in the scheme of things. It is also often the case that the priorities and conflicts within governments result in either inadequate laws and regulations or inadequate enforcement of those that exist.

In comparing the performances of the petroleum, mining, and pulp and paper industries in Ontario, I became convinced that image and location were very important in the complex of factors involved in decision-making processes. The petroleum refineries were close to or in centers of population: it was evident to a lot of people whether they were being "good citizens" or not. They were also selling their products in the same neighborhoods where the products were being produced. Image was important.

On the other hand, a great many pulp and paper mills and mines are in remote areas—often in company towns where the people are completely dependent on the mill or the mine for their livelihood. They see some of the effects of pollution, but they are employees, and keeping their job is all-important. The company's products are being sold on the markets of the world; environmental citizenship, and the image associated with it, is not a very important consideration.

The situation with many of the "dirty" industries did not start to improve in North America until "people pressures" began to influence government policies during the past decade. But, it is very difficult under present circumstances in Canada for these pressures to work effectively as far as the Arctic is concerned. Few people can see the problems or be aware of the dangers, so it is easy for the mass of the population to be lulled by sophisticated advertising campaigns.

The result is that many cumulative, chronic problems are missed completely. Once the initial flurry of interest has passed, only major problems will come to public attention.

In Canada, the pipeline situation is a positive one because public-hearing processes are involved. In the case of offshore drilling, the potential environmental hazards and risks are probably greater than in the case of the pipeline, but there is no requirement for hearings; thus, the processes involved have been very shoddy ones: a great deal of dust is being "swept under the rug."

In the United States, the public process is much more firmly established, so it would seem probable that a better balance could be achieved. However, I am aware that limited financial resources always pose serious problems to the groups involved. Organizations with million-dollar annual budgets are pitted against corporate giants that spend millions of dollars on advertising and public-relations programs every day.

My conclusion from all of this is that in Canada we must continue to fight for a much more open process than we now have. I think that in both countries we must try very hard to convince governments that they should support environmental organizations so that they can prepare for and participate effectively in public hearings. Professor Andrew Thompson, present chairman of CARC, has been a consistent proponent of such an approach for Canada (Thompson 1973).

The Mackenzie Valley pipeline hearings have established some precedents that I hope will have meaning for the future in Canada. In this case, the Native organizations have received funding "through the front door" (directly from the Department of Indian Affairs and Northern Development), while the environmental groups have been funded "through the back door" (by the Berger Commission).

However, the government demonstrated recently that it does not like the process at all, at least as far as the environmental organizations are concerned. The Northern Assessment Group (a consortium of environmental and conservation organizations) had received money from the Berger Commission to review the application which Canadian Arctic Gas Limited had made for the pipeline right-of-way. But when the Northern Assessment Group, as directed, applied to the government for additional funds to cross-examine and to present its evidence at the hearings, it was twice refused by the

Treasury Board, which is the court of last appeal when federal funds are involved. The problem was finally resolved by some intricate juggling that permitted the Berger Commission to again fund the group's operations. The implication seems very clear: the federal government does not want to establish the direct precedent of providing the money so that public hearings can become more meaningful processes. It is obvious that we still have a lot of convincing to do in Canada.

Conclusion

A sense of perspective on the use of the polar regions should surely be based on a sense of humility that our species has seldom achieved. But we must somehow find both the humility and the perspective if we are to change the frontier approaches that we seem compelled to adopt in commandeering resources for our use.

The sense of perspective could arise logically from an appraisal of the needs and wants of mankind and of our relationship with all other living things. In the past, we have found it difficult to gain perspective from appraisals of this kind. Could it be that things have changed enough in the last decade to make it possible now? If EARTHCARE has real meaning, perhaps this gathering can make a substantial contribution to the perspective that will allow us to utilize the resources of the polar regions without degrading those areas, without significantly affecting all the other species that make up these austere, beautiful, and tremendously important of Earth's ecosystems.

But I am convinced that wishing will not make it so. Our governments will not protect the Arctic and the Antarctic unless we demand it of them, unless those of us who are interested extend our range of vision and develop our capabilities to operate as environmental politicians. Platitudinous statements and nods of agreement are not enough. The gentle, nature-loving people of the world, and the multitude of scientists who offer society their wisdom but not involvement, must take off their gloves and participate much more directly in the affairs of men.

That is the only way that the concept and symbolism of EARTHCARE can really be given any meaning.

References

Bliss, Lawrence C., and E. B. Peterson. 1973. The ecological impact of northern petroleum development. *Fifth International Congress, Arctic Oil and Gas: Problems and Possibilities.* Le Havre: Fondation Française d'Etudes Nordiques.

Canadian Arctic Resources Committee. 1974a. Offshore drilling in the Beaufort Sea. *Northern Perspectives* 2(2):1-12.

———. 1974b. Arctic offshore drilling. *Northern Perspectives* 2(4):1-8.

Chrétien, Jean. 1970. Statement from the Honourable Jean Chrétien, Minister of Indian Affairs and Northern Development. In *Productivity and Conservation in Northern Circumpolar Lands.* Proceedings of a Conference (Edmonton, Alberta, 15-17 October 1969), eds. William A. Fuller and Peter G. Kevan, pp. 9-10. IUCN Special Publications (New Series) Number 16. Morges, Switzerland: International Union for Conservation of Nature and Natural Resources.

———. 1972. Mackenzie Corridor: Vision becomes reality. Keynote address to the 18th Annual Convention of the Pipeline Contractors Association of Canada (Montreal, 11 May 1972). Mimeographed.

———. 1973. Northern development for northerners. In *Arctic Alternatives: A National Workshop of People, Resources and the Environment North of '60* (Ottawa, 24-26 May 1972), eds. Douglas H. Pimlott, Kitson M. Vincent, and Christine E. McKnight, pp. 26-35. Ottawa: Canadian Arctic Resources Committee.

Clarke, C. H. Douglas. 1973. Terrestrial wildlife and northern development. In *Arctic Alternatives: A National Workshop of People, Resources and the Environment North of '60* (Ottawa, 24-26 May 1972), eds. Douglas H. Pimlott, Kitson M. Vincent, and Christine E. McKnight, pp. 194-234. Ottawa: Canadian Arctic Resources Committee.

Committee for Original Peoples Entitlement. 1974. *Drilling for Oil and Gas in the Beaufort Sea.* Press release of 8 February 1974. Inuvik, Northwest Territories: Committee for Original Peoples Entitlement.

Cumming, Peter. 1973. Our land, our people: Native rights north of '60. In *Arctic Alternatives: A National Workshop of People, Resources and the Environment North of '60* (Ottawa, 24-26 May 1972), eds. Douglas H. Pimlott, Kitson M. Vincent, and Christine E. McKnight, pp. 86-110. Ottawa: Canadian Arctic Resources Committee.

Department of Indian Affairs and Northern Development. 1973. Claims of Indian and Inuit people. Communiqué 1-7339, 8 August 1973.

Dunbar, Maxwell J. 1973. Stability and fragility in Arctic ecosystems. *Arctic* 26(3):179-185.

Fuller, William A. 1970a. Problems of northern resource development. Paper read at Mid West Wildlife Conference (Winnipeg, December 1970).

———. 1970b. Opening address. In *Proceedings of the Conference on Productivity and Conservation in Northern Circumpolar Lands* (Edmonton, Alberta, 15-17 October 1969), eds. William A. Fuller and Peter G. Kevan, pp. 7-8. IUCN Special Publications (New Series) Number 16. Morges, Switzerland: International Union for Conservation of Nature and Natural Resources.

Grainger, E. H. 1953. On the age, growth, migration, reproductive potential and feeding habits of the Arctic Char *(Salvelinus alpinus)* of Frobisher Bay, Baffin Island. *Journal of the Fisheries Research Board of Canada* 10(6): 326-370.

Hemstock, Russell A. 1970. Industry and the Arctic environment. *Transactions of the Royal Society of Canada* (Series 4) 8:387-392.

Klein, David R. 1973. The impact of oil development in the northern environment. In *Proceedings of the Third International Petroleum Congress* (Rome), pp. 109-121.

McTaggart-Cowan, P. D. 1970. *Operation Cleanup.* Report of the Task Force to the Minister of Transport. Ottawa: Ministry of Transport.

Naysmith, John K. 1970. Conservation in Canada's north. In *Proceedings of the Conference on Productivity and Conservation in Northern Circumpolar Lands* (Edmonton, Alberta, 15-17 October 1969), eds. William A. Fuller and Peter G. Kevan, pp. 320-324. IUCN Special Publications (New Series) Number 16. Morges, Switzerland: International Union for Conservation of Nature and Natural Resources.

———. 1971. *Canada North—Man and the Land.* Ottawa: Department of Indian Affairs and Northern Development.

———. 1973a. Management of polar lands. In *Meeting, International Union for Conservation of Nature and Natural Resources,* pp. 295-317.

———. 1973b. *Toward a Northern Balance.* Ottawa: Department of Indian Affairs and Northern Development.

Passmore, R. C. 1971. Environmental hazards of northern development. *Transactions Federal-Provincial Wildlife Conference.* Ottawa: Canadian Wildlife Service.

Pharand, Donat. 1973. *The Law of the Sea of the Arctic with Special Reference to Canada.* Ottawa: University of Ottawa Press.

Pimlott, Douglas H. 1973. People and the North: Motivations, objectives and approach of the Canadian Arctic Resources Committee. In *Arctic Alternatives: A National Workshop of People, Resources and the Environment*

North of '60 (Ottawa, 24-26 May 1972), eds. Douglas H. Pimlott, Kitson M. Vincent, and Christine E. McKnight, pp. 3-24. Ottawa: Canadian Arctic Resources Committee.

———. 1974a. Offshore Drilling in the Beaufort Sea: Report to the Committee for Original Peoples Entitlement (COPE). Ottawa: Canadian Arctic Resources Committee. Mimeographed.

———. 1974b. Delta gas: Time and environmental considerations. In *Gas from the Mackenzie Delta, Now or Later?*, pp. 93-107. Ottawa: Canadian Arctic Resources Committee.

———. 1974c. The hazardous search for oil and gas in Arctic waters. *Nature Canada* 3(3):20-28.

Pimlott, Douglas H., C. J. Kerswill, and J. R. Bider. 1971. *Scientific Activities in Fisheries and Wildlife Resources: Background Study for the Science Council of Canada.* Special Study Number 15. Ottawa: Information Canada.

Pimlott, Douglas H., Kitson M. Vincent, and Christine E. McKnight, eds. 1973. *Arctic Alternatives: A National Workshop on People, Resources and the Environment North of '60* (Ottawa, 24-26 May 1972). Ottawa: Canadian Arctic Resources Committee.

Pimlott, Douglas H., Dougald Brown, and Ken Sam. 1975. *Oil under Ice: Offshore Drilling in the Canadian Arctic.* Ottawa: Canadian Arctic Resources Committee.

Sprague, John B. 1973. Aquatic resources in the Canadian north: Knowledge, dangers and research needs. In *Arctic Alternatives: A National Workshop on People, Resources and the Environment North of '60* (Ottawa, 24-26 May 1972), eds. Douglas H. Pimlott, Kitson M. Vincent, and Christine E. McKnight, pp. 168-189. Ottawa: Canadian Arctic Resources Committee.

Thompson, Andrew R. 1973. *Public Hearings for Northern Pipelines in Canada.* Montreal: Arctic Institute of North America.

Thompson, Andrew, and M. Crommelin. 1973. Canada's petroleum policy: A cornucopia for whom? In *Proceedings of a Conference on Canada's Petroleum Leasing Policy.* Ottawa: Canadian Arctic Resources Committee.

Trudeau, Pierre E. 1969. Transcript of remarks at a Liberal Association dinner in Vancouver, 8 August 1969. In *The Violated Vision: The Rape of Canada's North*, ed. James Woodford. Toronto: McClelland and Stewart.

———. 1972. Press release of notes for remarks by the Prime Minister to a public meeting, 28 April 1972, at Edmonton, Alberta.

Usher, Peter. 1971. *The Bankslanders: Economy and Ecology of a Frontier Trapping Community. Volume 3: The Community.* Ottawa: Department of Indian Affairs and Northern Development.

Wah-shee, James. 1974. A land settlement—What does it mean? In *Gas from the Mackenzie Delta: Now or Later?*, pp. 83-92. Ottawa: Canadian Arctic Resources Committee.

Weeden, Robert B. 1970. Man in nature: A strategy for Alaskan living. In *Proceedings of the Conference on Productivity and Conservation in Northern Circumpolar Lands* (Edmonton, Alberta, 15-17 October 1969), eds. William A. Fuller and Peter G. Kevan, pp. 251-256. Morges, Switzerland: International Union for Conservation of Nature and Natural Resources.

Woodford, James. 1972. *The Violated Vision: The Rape of Canada's North*. Toronto: McClelland and Stewart.

Yates, A. B. 1972. Energy and Canada's north: The search for oil and gas. *Nature Canada* 1(3):9-14.

Editor's Note: A 245-page report based on the hearings held by the Berger Commission was published on 7 May 1977. Entitled *Northern Frontier, Northern Homeland,* the report is available for $6.00 from Printing and Publishing, Supply and Services Canada, Ottawa K1A 0S9 ($5.00 in Canada). On 8 September 1977, Prime Minister Trudeau of Canada and President Carter of the United States jointly announced their preference for the so-called "Alcan" route for the pipeline, which would pass through the southern Yukon Territory. Their choice was based in large measure on the Berger Commission's report.

HUMAN SYSTEMS AND INSTITUTIONS

42. Environmental Laws and Conventions: Toward Societal Compacts with Nature

Nicholas A. Robinson
Sierra Club, U.S.A.

Principle 21

Assembled at Stockholm for the United Nations Conference on the Human Environment, the nations declared as a common principle that:

> States have, in accordance with the Charter of the United Nations and the principles of international law, the sovereign right to exploit their own resources pursuant to their own environmental policies, and the responsibility to ensure that activities within their jurisdiction or control do not cause damage to the environment of other States or of areas beyond the limits of national jurisdiction.[1]

This is the now well-known Principle 21, included in the Declaration on the Human Environment and endorsed by the U.N. General Assembly.[2] [See Appendix A.—Editor] It speaks in terms of a nation's duty to protect the environment because of the obligations each state owes to other nations; it stops short of acknowledging the common need of every nation to protect the biosphere out of a due regard for the rules of nature and the well-being of people within nature.

Nonetheless, Principle 21 restates the only duties that our legal systems today provide for protecting natural areas. A discussion of its text allows us an understanding of the still evolving terms of our social compact with nature.

The well-being of each type of ecosystem on our globe is now imperilled. Although our legal system has been slow to respond, there are trends—such as those within Principle 21—that need encouraging. If nurtured, these trends may develop into a realistic duty of EARTHCARE. This essay will explore the ways in which different nations and the international community are adjusting their rules and sanctions, laws and enforcement, to reflect the scientific realities so ably set forth in the other essays contained in this volume.

Public International Law

The international law that guides relations among nations has paid little heed to environmental protection. It is founded either upon treaties or upon expression of consensus being "general principles of law recognized by civilized nations."[3]

Principle 21 is primarily a restatement of a general principle, although it can also be founded in specific treaties as well. As a principle, it is drawn from the common law maxim, *"Sic utere tuo ut alienum non laedas,"* which means, "Use that which you own or control in such a way as not to harm others." This maxim, derived from England's legal evolution, has its civil law analogue in the French concept of *droit de voisinage* and the German concept of *Nachbarrecht,* or the law of good relations among neighbors.

Prototypes and Precursors

These principles have been relied upon by international tribunals in circumstances similar to those involving environmental harm. In the "Corfu Channel Case,"[4] two British ships were damaged in crossing a minefield in Albanian territorial waters. Albania did not lay the mines, but it knew of their existence and yet did not warn the ships or remove the mines. The International Court of Justice, in awarding damages to Britain, stated that it is "every State's obligation not to allow knowingly its territory to be used for acts contrary to the rights of other States." The predecessor tribunal, the Permanent Court of International Justice, had characterized the role of damages in the "Case Concerning the Factory at Chorzow"[5] as follows: "Reparations must, as far as possible, wipe out the consequences of the illegal act and reestablish the situation" that existed before the injury occurred.

Yet nature is ill served by awards of money damages after injury has occurred. In the "Trail Smelter Arbitration," for instance, the tribunal held Canada liable to the United States for damages to farms and land in the state of Washington. The damages were caused by sulfur dioxide fumes emitted by a smelter in the province of British Columbia across the border in Canada. Ruling in 1941, the tribunal observed that "no State has the right to use or permit the use of its territory in a manner as to cause injury by fumes in or to the territory of another. . . ." The real issue here, insofar as protection of natural areas is concerned, is not whether owners of land were ultimately paid some money for losses in value or income, but whether the land was restored. The ruling did not examine this aspect, for the tribunal was not structured to extend its purview that far.

In the "Lac Lanoux Arbitration"[6] between Spain and France, the same principle was noted by the tribunal. Spain objected to a French plan to divert the Carol River, which flows across the border from France into Spain, for a French hydroelectric plant. The tribunal did not find it necessary to order France to change its plans or to explore liability, since it found that France replaced the water diverted with water of equal quality and quantity.

France was less forthcoming when New Zealand and Australia sued the French Republic in the International Court of Justice, seeking an order from the court requiring France to cease atmospheric nuclear-weapons testing in the southern Pacific. In the proceedings at The Hague, France refused even to appear before the court, citing reasons of national defense.[7] The tests were, however, terminated voluntarily thereafter.

The Carol River and French nuclear-weapons testing issues illustrate how difficult it is to apply Principle 21 *before* injury is done. Nonetheless, only action before injury occurs truly and wholly protects both the complaining party and nature.

Protective Duties Under Principle 21

Principle 21 has been construed to emphasize these protective duties to prevent harm. Professor Louis Sohn of the Harvard Law School contends that the language on exploitation set forth in the first half of the principle cannot mean that a nation has unlimited sovereignty to exploit in a manner inconsistent with the spirit of the U.N. Declaration on the Human Environment as a whole; the declaration, by its terms, embraces protection of the biosphere.[8] With respect to the second half of the principle, as the Canadian delegation to the Stockholm conference stated:

> The second element made it clear that those rights [to exploit cited in the first half] must be limited or balanced by the responsibility to ensure that the exercise of rights did not result in damage to others.[9]

Such restatements of one nation's responsibility to protect others ignore, of course, the issue of whether a nation has any duty under international law to protect the environment situated wholly within its jurisdiction and having no immediate adverse effect elsewhere. Clearly, many seemingly harmless acts will actually have unperceived ramifications elsewhere through interaction of ecosystems, e.g., through interference with a food chain or contamination with disease vectors.

However, even where no adverse impact is discerned, there is a growing body of treaty law now extending the duty to protect nature into the domestic or municipal forum. Two prime examples may be cited.

Convention on Trade in Endangered Species

The first is the Convention on Trade in Endangered Species of Wild Fauna and Flora.[10] With its initial ten signatories, this convention entered into force on 1 July 1975. It took twelve years of negotiations after such a convention was first proposed in 1963 at the Eighth General Assembly of the International Union for Conservation of Nature and Natural Resources (IUCN) in Nairobi. IUCN acts as secretariat for the convention, which bans the trade in animal and plant species specified in the convention's three appendices.

Appendix I to the convention lists species that are or may be threatened. Trade in such species must be strictly regulated. IUCN's Threatened-Plants Committee, for instance, estimates that 10 percent of the world's flowering plants (some 25,000 to 30,000 species) are threatened with extinction. As many as 5,000 species are found in the Philippines alone.[11] Appendix II lists species, trade in which must be controlled if the still somewhat abundant numbers of the given flora or fauna are to be preserved.

While the restrictions of the first two appendices do restrict nations within their borders, it is largely through Appendix III that municipal obligations directly impact on the internal natural area protection. Appendix III includes a ban on trade in any other species that any nation ratifying the convention already protects and needs the assistance of other States to make its protection effective.

The importation of any species listed in the three appendices may be permitted only if the given national scientific authorities determine that it would not be detrimental to the survival of the species. No primarily commercial uses are allowed for such imports. Export is permitted only if the species was gathered legally.

IUCN has identified practical problems with implementing the convention:

> Many of the major exporting countries, particularly in South East Asia and Latin America, are not yet equipped to impose the stringent controls required of them by the Convention. Much of their difficulty is geographical; it is clearly impossible for Indonesia to monitor each of its thousands of islands. Since a specimen exported illegally from Indonesia is more than likely to pass through Singapore, it is obvious that accession of the Convention by both countries is necessary if it is to be at all effective in either. Indeed, it has been suggested that loop-holes will remain unless Malaysia and Thailand, as well as Singapore and Indonesia, accede together.
>
> Another problem is the lack of public awareness of the value to the nation of retaining its full inventory of species, or of how vulnerable some of those species are. This means that the only control is that of the law, in countries where the law is relatively thin on the ground, or regards the protection of endangered species as less pressing a duty than the prevention of crimes against person and property. For as long as there is a market for animal and plant souvenirs, endangered species will be shot, snared or picked. And it counts for little if the tourist, ignorantly buying such souvenirs, should have them wrested from him at the frontier, when the animal or plant is already dead. . . .
>
> Entrepreneurial centers like Singapore and Hong Kong earn a considerable revenue from trade in souvenirs derived from endangered species.

Similarly, there are powerful lobbies in Europe, a notable example being the skins and leather industry in Italy. By the same token, the International Fur Trade Federation's voluntary bans on the sale of the skins of various spotted cat species have been widely disregarded in France, Italy, Spain, Scandinavia and Japan, where the demand is greater than ever before.[12]

The Endangered Species Convention, to be effective, thus imposes duties on nations to protect the listed species within their own jurisdiction. The states act as custodians for the common global resource represented by such species.

World Culture and National Heritage

This concept is made more explicit by the Convention Concerning the Protection of the World Culture and National Heritage,[13] which was adopted by the Seventeenth Session of the UNESCO General Conference and is awaiting ratification.

The "Heritage Convention" requires protection for natural and cultural monuments that are determined to be "of outstanding universal value."[14] States party to the Heritage Convention are to identify and delineate properties within their territories having such protection. An inventory, constituting the "World Heritage List," is then to be compiled.

While each state ratifying the treaty "recognizes that the duty of ensuring the identification, protection, conservation, presentation and transmission of the monuments . . . belongs primarily to the State" where the monument is located, the same states also "recognize that heritage constitutes a world heritage for whose protection it is the duty of the international community as a whole to cooperate."[15]

Thus, states party to the Heritage Convention agree to take no deliberate measures that may damage heritage situated in other states. A committee of experts is to be formed, pursuant to the convention, to allocate funds to safeguard monuments and to designate monuments that become endangered, whether or not the government of the nation in which monuments are situated requests their designation. Public pressure could then be brought to bear to protect the monument, and, if the state requests it, international aid could be made available.

In view of the international judicial and arbitral tribunal decisions described earlier, it may legitimately be asked how these conventions are to be enforced if they are violated. The extinction of a species or ruin of a monument is not repairable: damages awarded after the fact cannot ever compensate such a loss. The reluctance of nations to submit to the jurisdiction and orders of the International Court of Justice—the United States remains a prime recalcitrant in this regard—unavoidably incapacitates these conventions, and even Principle 21.

The Ghosts of Justice Holmes

Reluctantly, one is forced to conclude that international public law for the protection of nature is so embryonic in application as to be of little practical utility. The two treaties described above are still too new to afford relief. The concepts and restatements of duties in texts such as Principle 21 are crucial and well evolved, but more attention is needed for determining how the duties can be enforced and what real protection can be obtained for natural areas.

One is reminded of a succinct observation by Mr. Justice Oliver Wendell Holmes of the United States Supreme Court. In ruling on an admiralty law case entitled "The Western Maid,"[16] Justice Holmes remarked: "Legal obligations that exist but cannot be enforced are ghosts that are seen in the law but are elusive to the grasp."

If applications of international law for the protection of natural areas are few and ephemeral, we must turn our inquiry to the duties which nations undertake through their municipal laws. A state's internal lawmaking may effectively enforce international duties. Perhaps even citizens from different lands can seek to enforce some of these international duties in the political systems of the offending or negligent state.

Municipal Laws for Protection of Specific Natural Objects or Areas

The more an environmental harm affects the welfare of the community, the more likely it is that the community will understand legal-protection obligations. As early as 1273 for instance, Dagobert I, king of the Franks, decreed that if a person "corrupted by filth" a fountain, he had to pay a fine and clean up the mess.[17]

At the same time, if the welfare of only a few is harmed,

experience teaches that the state may ignore even the worst environmental harms. The methylmercury poisonings in Minamata, Japan,[18] or the harm caused by discharges of lead, cadmium, polychlorinated biphenyls (PCBs), and other toxic chemicals in Japan, North America, and Europe, are only recently coming under regulation. The banning of toxic emissions is not yet effective and certainly comes too late for the persons and animals already poisoned, crippled, or dead.

While the U.N. Commission on Human Rights' deliberations on the EARTHCARE Petitions, described elsewhere in this volume, reflect an attempt to refine the scope of a state's human rights duties under international law, effective action must be studied in the national forum.

Many currently effective environmental laws at the municipal level could be noted. UNESCO has urged creation of biosphere reserves throughout the world for nature study and preservation; many states have created such reserves.[19] The two world conferences on national parks have assessed and pressed for effective implementation of such municipal laws.[20] Iran sponsored the Convention Concerning the Protection of Wetlands of International Importance[21] in Ramsar, and coincidentally many nations have been adopting wetlands protection laws.[22]

The trend toward the protection of cultural heritage—a trend more developed than that toward the protection of natural areas—illustrates how the same type of law can emerge in a number of different jurisdictions. It is now common for many nations to have registries of monuments with governmental sanctions to assure their maintenance and protection. As precedents for the Heritage Convention, these laws merit close study, in connection with parallel laws for protecting natural areas and open spaces.

For instance, in the United Kingdom the ancient monuments acts of 1913 and 1931, together with the Historic Buildings and Ancient Monuments Act of 1953,[23] provide for a registry of "monuments," whether unoccupied or still in use, and require governmental authorization before any alteration of these scheduled monuments may take place. Grants are sometimes available to help the owner of such a site to adequately preserve it.

Local government in the United Kingdom is authorized to contribute to the repair of structures adjacent to scheduled buildings through grants and interest-free or low-interest loans;[24] it may also classify areas of special historic or agricultural value as "conservation areas,"[25] in which it will regulate all repairs, alterations, and new construction. In these areas, owners must undertake necessary maintenance and repairs or be subject to fines and even imprisonment; and improperly maintained property is subject to expropriation in order to guarantee its protection.

In England, as separately in Scotland and the Isle of Man, a National Trust for Places of Historic Interest or Natural Beauty has been established by law.[26] Significant buildings or regions, with a maintenance endowment, are contributed to the trust, to be preserved and open to the public.

In France, cultural monuments of historic or artistic value are registered under the *classement* system with the Ministry of Fine Arts. Adjacent buildings or edifices with archaeological value are listed in the *inventaire supplementaire.* No work is permitted without the permission of the Historic Monuments Architectural Service. Grants of up to 40 percent of maintenance or restoration costs are available to private owners, and the state can intercede to preserve a monument if the private owner does not act.[27]

The "Malraux Law" of 4 August 1962[28] extended this basic monument-by-monument preservation to include regional planning and the full rehabilitation of entire, historically valuable towns. The town councils and the ministers of culture and of construction jointly designate a "protected sector." If the town opposes, the designation can be done by order of the Council of State. Restoration of the site is then planned, supervised, and often executed by the state. The Marais quarter of Paris is a striking example of how an entire area has been restored.[29]

In 1970, the Byelorussian Republic in the Soviet Union enacted a law concerning the artistic and historic heritage. It provides for an inventory of historical monuments eligible for government protection regardless of ownership (e.g., ownership by the state, a collective farm, cooperative, or an organization of individuals) and classifies them according to their local, national, or union-wide interest. Collections of buildings with special value will be declared state reserves. Geographically, Byelorussia was a natural crossroads of the main trade routes linking eastern and western Europe, and Scandinavia and Byzantium.[30] Its own collections, archeological sites, and monuments—such as the Cathedral of Saint Sophia, the Cathedral of the Spaso-Efrosinievskogo Monastery in Polotsk, and the Byelaya Vezha (White Tower) at Kamenets—are all protected and restored from the ravages of World War II. Public support for cultural preservation is evidenced by the 400,000 membership in the Byelorussia Society for the Preservation of Monuments. In 1961, the Byelorussian Law on Nature Conservation was enacted, and by 1963, the ordinance On Protection of Natural Monuments and rules on protection, registration, and management of natural monuments had been promulgated.[31]

Canada has taken steps, both federally and in the ten provinces, to preserve historic buildings and natural regions. The eighteenth-century French port of Louisburg on Cape Breton Island has been restored along with the surrounding town. The British citadel in Halifax, Nova Scotia, is preserved, as is the battlefield of the Plains of Abraham in Quebec. North of Toronto, a Huron Indian village and Jesuit mission have been restored by the Province of Ontario.[32] The national parks and provincial parks have set aside and preserved systems of representative natural resources.

In the United States, as in Canada, laws have emerged at both the federal and state levels for natural and cultural preservation. The first national park was established a century ago;[33] cultural preservation came later. The New World lacked a wealth of ancient monuments, but it possessed a wealth of natural wonders that merited protection in the wake of resource exploitation and settlement.[34]

In 1966, the United States expanded the preservation of its natural resources with the creation of a National Wilderness System for preserving primeval forests and mountains in their natural, pristine state.[35] That same year, the United States enacted its National Historic Preservation Act,[36] which does not go as far as its European counterparts. Under the Fish and Wildlife Conservation Fund, federal grants have been combined with local funds to create thousands of local or state parks. Most states have extensive state park systems paralleling the national park system.

Not surprisingly, cultural preservation in the United States follows principles similar to those in Europe, though the National Historic Preservation Act, enacted in 1966, does not go as far as its European counterparts. Initially administered by the National Parks Service,[37] historic sites are now preserved by means of a National

Register of Historic Places, grants in aid, and an Advisory Council on Historic Preservation, empowered to review and evaluate the impact of any actions which are permitted or undertaken by the federal government affecting registered property. Historic preservation is defined to include "the protection, rehabilitation, restoration, and reconstruction of districts, sites, buildings, structures, and objects in American history, architecture, archaeology or culture."[38] Sites are listed upon nomination by the states and by federal agencies.

Similar developments can be cited in the USSR. Before the October Revolution, specimens of the natural heritage were protected by rich landowners. Thereafter protection continued, although most laws in effect were promulgated during the past two decades.

Protection for areas of unique natural features or cultural, historic, or aesthetic value is provided by law in each of the fifteen republics of the USSR. As in the laws of nature conservation, the basic principles include "the legal definition of objectives and goals of protection; the establishment of regimes of protection of the procedure of reserving tracts of land; and the obligations of executive bodies of the State power concerning such tracts."[39]

The USSR has vigorously preserved her historic cultural monuments. After World War II, a program was undertaken for the restoration of war-damaged monuments; the restoration of the Summer Palace near Leningrad required the training of artisans in such forgotten skills as parquetry and damask weaving.[40]

As a final example of state action, the comprehensive Japanese laws can be cited. Law 214 of 1950 governs the protection of "cultural properties." These consist of "tangible" buildings and works of art, "intangible" arts, "folk culture," and "monuments," including natural and historic sites, urban quarters, castles, gardens, and geologic sites.[41] Japan requires owners of cultural properties to maintain and conserve them and, to assist this process, makes grants to cover part of the costs. If a property is termed a "national treasure," the National Commission for Cultural Properties may instruct a private owner to act and will itself pay for expenses.

The use of cultural preservation as a precedent illustrates the range of protection laws that have gradually developed in municipal law. A consensus of state practice for protection appears to be developing in the cultural sphere, which may presage a parallel evolution in the natural sphere. An international appreciation for the views of the UNESCO Convention concerning the preservation of the world cultural and natural heritage will accelerate the protection of unique natural areas. The coming years may witness a needed extension of protection for unique natural areas in each country.

Two countervailing tendencies will affect any such extension. Natural areas, unique and worthy of protection as monuments, may be saved in their own right and also for their commercial value as tourist attractions. Yet at the same time, many natural areas are "ripe" for exploitation. Coastal wetlands, for instance, will be preserved in their natural state only if society fully understands how valuable and essential they are for human well-being. Our concept of "ripeness" must yield to the concept of a dynamic, ongoing partnership with nature. We cannot write effective laws to protect nature unless we understand and observe this underlying policy premise.

Municipal Laws for General Protection of Natural Areas

No matter how important it may be to preserve specific species, monuments, and other isolated objects, it is even more important to preserve the integrity of entire natural systems wherever they may be found. Few species or natural areas can survive if entire ecosystems are destroyed. If the air is polluted, ancient frescoes and carvings disintegrate. In short, the municipal protection of a specific heritage is undermined.

The fundamental premise for creating comprehensive laws to protect natural systems was ably put by René Dubos: "Unwise management of nature or of technology can destroy civilization in any climate and land, under any political system."[42]

This reality was recognized at the 1972 Stockholm Conference on the Human Environment and subsequently by the United Nations General Assembly. In 1973, the assembly emphasized that "the complexity and interdependence of such problems require new approaches." The world body stated that it was "convinced of the need for prompt and effective implementation by Governments and the international community of measures designed to safeguard and enhance the environment for the benefit of present and future generations of Man." At the same time, the assembly recognized "that responsibility for action to protect and enhance the environment rests primarily with Governments and, in the first instance, can be exercised more effectively at the national and regional levels."[43]

Yet even with "good" environmental laws enscribed among a nation's codes and regulations, many questions remain as to whether or not the laws will be well administered or enforced. Just as global laws are too weak to be effective, so local laws may cover too small an area or serve too limited a function. John Muir, the founder of the Sierra Club, recognized this in 1909 when he deplored the weakness of laws designed to protect forests in the United States:

> All sorts of local laws and regulations have been tried and found wanting, and the costly lessons of our own experience, as well as that of every civilized nation, show conclusively that the fate of the remnant of our forests is in the hands of the federal government and that if the remnant is to be saved at all, it must be saved quickly.
>
> Any fool can destroy trees. They cannot run away; and if they could, they would still be destroyed.... Few that fell trees plant them; nor would planting avail much towards getting back anything like the noble primeval forests.... It took more than three thousand years to make some of the trees in these Western woods.... Through all the wonderful, eventful centuries since Christ's time—and long before that—God has cared for these trees, saved them from drought, disease, avalanches, and a thousand straining, leveling tempests and floods; but he cannot save them from fools, only Uncle Sam can do that.[44]

Muir deplored the failure of "Uncle Sam," or the American federal government, to act effectively on a regional and interstate basis to protect forests. Since Muir's day, such regional or nationwide protection has become the norm. The Clean Air Act[45] in the United States recognizes this reality, although little has yet been done to implement some of its provisions, such as the duty to prevent any degradation of existing air quality.[46] The Sierra Club, having fought for the adoption of the law, still must fight to have it implemented.

The role of municipal laws is clear also in the sphere of water-quality protection. Recognition of the principles of international law, such as Principle 21, should require every government to curb any water pollution that cannot be assimilated by a river, lake, or estuary without harmful effects. The U.S. Water Pollution Control Act amendments of 1972 set forth the goal of eliminating the discharge of all pollutants into navigable waters (those subject to federal control under the U.S. Constitution) by 1985.[47] Whether or not this deadline can be met on time, the goal is consistent with the duty, specified in international law, to prevent the pollution of oceans through coastal waters or rivers which empty into them, and with respect to water resources shared with neighboring states. It should be noted that land-based discharges are the single largest source of oceanic pollution.

Within the guidelines of the USSR's Fundamental Principles of Water Legislation, the Water Code of the Russian Republic (Russian Soviet Federated Socialist Republic) describes its task as one of regulating "water relationships with a view to ensuring the efficient utilization of water for the needs of the population and the national economy; and to protect water from pollution, obstruction and depletion."[48] The experience with such water-pollution laws parallels that of the United States and other countries.

With respect to boundary areas, there exists the 1964 agreement between Poland and the USSR, Concerning the Use of Water Resources in Frontier Waters.[49] Article 4(2) of the agreement defines pollution as the "introduction into the waters, directly or indirectly, of solid or gaseous substances and heat in such quantities as may cause physical, chemical and biological changes which limit or prevent the normal utilization of the said waters for communal, industrial, agricultural, fishery or other purposes." The agreement then sets forth measures to keep the waters free of pollution.

The 1964 treaty between Finland and the USSR created a joint commission to oversee the implementation of the treaty's pollution-control measures on frontier waters.[50] The United States has similar treaties with Canada[51] and Mexico.[52]

Such laws are remedial and preventive: they are designed to avert pollution and other types of environmental degradation, not to assess liability after harm has been done. The prototype of this sort of preventive law is the comprehensive tool of "environmental impact assessment," pioneered in the United States through the National Environmental Policy Act of 1969 (NEPA).[53]

NEPA's impact on governmental decisionmaking in the United States has been profound. Its purpose is to require every agency of the federal government to evaluate and consider environmental protection *before* it acts. The law has been copied by over half the states of the United States,[54] and several foreign countries have adopted similar laws.

Much has been written about NEPA,[55] and it need not be

repeated here. Suffice it to say that NEPA requires all federal agencies to use interdisciplinary methods in decisionmaking and to study the impact on the environment of any major action or development proposal as determined by the agency.

Rather than comment on how NEPA has actually operated to date, a subject carefully examined in Oliver Thorold's excellent essay in this volume, it will be useful here to speculate on how NEPA may be used internationally to advance protection of natural areas and whole ecosystems.

It should be noted in passing that environmental impact assessment, such as NEPA requires, may be seen as national compliance with the rules of international law embodied in Principle 21. Moreover, effective environmental assessment would help fulfill U.N. General Assembly Resolution 2995.[56] That resolution, supplementing Principle 21, recognized that technical data on environmental control and the use of natural resources should be provided "relating to the work to be carried out by States within their national jurisdiction with a view to avoiding significant harm that may occur in the environment of the adjacent area." Such data "will be given and received in the best spirit of cooperation and good-neighborliness...."

How, for instance, might NEPA protect Canadian citizens and natural areas from potential harm due to action permitted by an agency of the federal government in the United States? NEPA's language, which speaks in terms of protecting the biosphere without regard to national political boundaries, requires each agency to evaluate its environmental impact anywhere on the Earth,[57] as I have argued elsewhere. Thus, if the Nuclear Regulatory Commission were to license a nuclear-powered electricity-generating facility near waters in the state of Washington, it would have to consider the potential environmental consequences for the resources and ecosystems of British Columbia and perhaps the Pacific Ocean—not merely the impacts likely to occur within U.S. territory.

This interpretation of NEPA was accepted by the Legal Advisory Committee to the President's Council on Environmental Quality.[58] Nonetheless, the Department of State initially rejected the interpretation and by doing so encouraged other agencies to take unrealistically narrow views of their environmental impact assessment duties under NEPA. The State Department reversed itself, but other agencies are still recalcitrant.

Conservationists proceeded to bring suits against such agencies. In *Sierra Club* v. *Atomic Energy Commission* [AEC],[59] the AEC conceded that NEPA required an environmental impact assessment of the export program under which the United States sells nuclear power plants and fuels to other countries. In *People of Eniwetok* v. *Laird*,[60] legal-aid counsel representing the inhabitants of a trust territory successfully contended that NEPA required the Department of Defense to study the impact of military cratering tests on the environment of the U.N. Trust Territories.

These cases have been capped to date by the decision of Judge Bryant in *Sierra Club* v. *Coleman*.[61] In this suit, conservationists contended successfully that the Department of Transportation had a duty under NEPA to study the impact of its aid given in construction of the last overland link of the Pan-American Highway through the Darien Gap in Panama and Colombia. Specifically, the court found that the department had proceeded in violation of NEPA by: (1) not asking the advice of the Environmental Protection Agency about the highway project's impact; (2) failing to study thoroughly the possibility that hoof-and-mouth disease (aftosa) could be introduced into North America by vectors carried along the highway; (3) failing to adequately examine alternatives to the route chosen or to the entire project; and (4) failing to adequately discuss the impact which the road would have on indigenous Indians; the study of alternatives necessarily will encompass an evaluation of the project's impact on the tropical rain forest and flora and fauna in the area.

The "Darien Gap" suit was brought by conservationists largely in the United States. It could just as easily have been brought by a Colombian environmental society or a Panamanian Indian. In this potential rests the utility of NEPA as a transnational tool for the protection of natural areas.

In our Canadian hypothesis, a Canadian conservationist could commence suit in a federal court in the United States to require a U.S. agency to consider protecting Canadian natural areas. The direct precedent for this course is a procedural ruling in the trans-Alaska pipeline litigation, *Wilderness Society* v. *Tom Morton*.[62] Canadian conservationists were granted permission to intervene as plaintiffs in that suit, to require the Department of the Interior to conduct a proper environmental impact assessment of the effects of building the trans-Alaska oil pipeline. The Canadians availed themselves of a very real right to force the United States under NEPA to protect Canadian natural areas.

This application of NEPA to events outside the territory of the United States is not entirely novel in law. An analogous series of cases exists in the seemingly unrelated sphere of commercial law regulating the offering and sale of securities, such as common stocks. Rule 10(b) of the Securities and Exchange Act of 1934[63] makes it unlawful for a person or corporation to defraud or mislead a purchaser of stock by means of U.S. mails, interstate commerce, or the services of the stock exchange. Since 1968, this law has been applied to fraudulent or misleading conduct even if it occurs in other countries. Thus, in *Schoenbaum* v. *Firstbrook*,[64] the Court of Appeals applied Rule 10(b) to sales of a Canadian corporation sold on the American Stock Exchange in order "to protect domestic investors who purchased shares on American exchanges and to protect the United States securities market from improper foreign transactions in American securities." This application is like ordering the hoof-and-mourth disease study to protect American cattle and other livestock and animals. It is called the "objective territorial principle" in international law and allows a state to act to protect itself in advance from harm inflicted from abroad. Canada employed this principle in extending preventive jurisdiction over the Arctic areas to curb the transport of oil by tankers in that region.[65]

In *Securities and Exchange Commission* v. *Gulf Intercontinental Finance*,[66] Rule 10(b) was applied when U.S. residents went to Canada to use a Canadian corporation to defraud investors in Canada through misleading advertisements in Canadian papers. Unfortunately for them, the papers had a small circulation in the United States; even though the harm occurred to Canadians in Canada, since some of the fraudulent acts occurred within United States jurisdiction, the court applied the law. This is recognized as the "subjective territorial principle" in international law. Thus, if a corporation in the United States authorizes the clear-cutting of a tropical rain forest in Borneo or South America, and secures a U.S. government loan guarantee to do so, the federal loan agency must analyze the environmental impact of its action, which in this case might bring about the destruction of a rain forest abroad. The prosecution of a U.S. citizen, under the Endangered Species Convention and U.S. criminal law, for importing to Italy an endangered species of Honduras in violation of that country's laws would be a further example of this subjective principle at work.

All this securities law precedent takes on tremendous environmental importance in view of the court ruling in *Natural Resources Defense Council* [NRDC] v. *Securities and Exchange Commission* [SEC].[67] NRDC sued SEC to require it to comply with NEPA. The compliance sought successfully to have SEC require all corporations which register stock and file reports with SEC to disclose the significant environmental impact of their activities. SEC is now considering what kind of disclosures of such information to require.[68]

If a company's activities might have a substantial adverse effect, it will have to make a disclosure to that effect; if its disclosure is not accurate, it would be subject to prosecution under Rule 10(b). Thus, if a Texas corporation were to develop, directly or through subsidiaries, a pulp and paper mill on the Caspian Sea, it would have to disclose its intention and its plans for protecting the sea and its jeopardized sturgeon, and the valuable hardwood forests used by Iranian craftsmen for their exquisite inlay work.

Private corporations, which enjoy the privilege of selling stock to raise investment capital in the United States, would have to become environmentally sensitive or risk a Rule 10(b) suit in a United States Court. Multinational corporations—now called more fashionably "transnational"—would be held accountable at least to report upon their environmental depredations wherever they occur. Protection of the biosphere would become the necessary standard of conduct, and the tendency of a company to flee from stiff environmental protection jurisdictions into "pollution havens" might be checked.[69]

Conservationists could use such information themselves in a transnational way. The reportedly sloppy and negligent offshore oil operations by Shell Oil Company in Nigeria[70] might be forced to use the state-of-the-art oil protection devices and safeguards now required in U.S. coastal waters. This could happen through pressure from Nigerian conservationists in Nigeria, while North American and Nigerian conservationists sue in a U.S. court to require Shell Oil to comply with SEC's environmental disclosure rules; under NEPA the company would have to reveal the harmful impact of its operations in Nigeria and the off-shore Atlantic waters by their failure to use the same care it uses in its operations in the waters of the Gulf of Mexico or the coastal waters of southern California.

If these speculations seem a trifle unrealistic or futuristic, think back on the "overnight" evolution of laws in other areas—for example, those which regulate working hours. In 1900, New York State enacted a law to restrict labor to ten hours a day, sixty hours a week. The law was invalidated in *Lochner v. New York*,[71] with Justice Holmes dissenting. That dissent later became the majority opinion when Lochner was reversed. No one today argues that such a regulation, or the elimination of child labor, is a violation of an employer's property rights. A similarly rapid evolution is possible in the laws governing the environment of today's interdependent world.

The Fabric of International and Municipal Laws Protecting Nature

The foregoing analysis suggests that at present the only effective protection of nature is provided by a nation's internal or municipal laws, when well administered. Unfortunately, such a system is too incomplete to protect the biosphere.

Consider the example of fluorocarbon emissions, which may adversely affect the Earth's protective ozone layer in the stratosphere. The United States accounts for 45 percent of the annual worldwide emissions of the two fluorocarbons most likely to deplete ozone— fluorocarbons F-11 and F-12—and western Europe accounts for 40 percent.[72] New York State has adopted a law to ban such emissions.[73] Clearly such a law even by a major state would be only marginally useful. Even if the federal government acted, only half the problem would be solved. Muir's complaint would rear its ugly head: "All sorts of local laws and regulations have been tried and found wanting."[74]

A nation that endangers the ozone layer threatens all others with environmental harm. Should not the rule of Principle 21 require it to immediately stop the emission of fluorocarbons? What of the nations defiling the oceans with contaminants? Or the nations which do not choose to join or enforce the Endangered Species Convention? What of Nigeria which allows oil pollution of the ocean and its coasts by not requiring the same stiff environmental protection laws that Canada and the United States have adopted?

The answer to these questions is discouraging. Any fool can destroy a tree and any collection of foolish nations can irreparably harm the biosphere and common natural heritage of the Earth.

Unless the consensus develops to create strong national laws reflecting adherence to more than just Principle 21, environmental degradation will continue in the foreseeable future.

To be sure, each nation will gradually act to protect one or another facet of the natural environment. A crazy-quilt or patchwork of different laws will emerge. Some species will be saved, others will be lost. Some rain forests will be cut as clear as the primeval woods of Europe or the eastern United States. In his paper, Dr. Gómez-Pompa speculates that perhaps the Northern Hemisphere nations destroyed much of their natural environment so that the world would reach a level of development and experience that would enable the equatorial and Southern Hemisphere states to achieve social justice in harmony with nature. This vision states a noble aspiration, and law can help its realization.

As we come to understand the functioning of natural systems, we can adapt the laws of society to the laws of nature. We can seek to protect natural areas for their own sake and not simply because they fall within the political bounds of a particular state.

Our times are a period of tremendous flux. This essay points to a few tendencies of law which may evolve into a kind of compact between our human society and the natural areas which nurture us. The evolution is not at all certain. Short term gain still offers a more expedient, and therefore a more attractive, choice to a majority of national leaders than a long-range, comprehensive EARTHCARE policy.

Law—municipal or international—can be fashioned to advance the protection of natural areas—the precondition for global well-being. The United Nations Environment Programme's Governing Council has called for the study and promotion of environmental law.[75] Little has been done to implement the call. Indeed, little has been advanced since Ambassador Alan Beesley's vigorous espousal of environmental protection in international law at the 1972 United Nations Stockholm conference. Spokesmen for environmental protection at the Law of the Sea Conference have been woefully few and far between.

Much remains to be done before environmental protection and respect for maintaining natural areas becomes the norm. Just as humane rules for the conduct of warfare were slowly developed over a century by the International Committee of the Red Cross, so rules for environmental protection in wartime may also slowly develop.[76] The ban on biological weapons and atmospheric nuclear testing, and the treaty obliging belligerents to safeguard cultural monuments[77] are examples of such developments.

As the other essays here have described the health fund threats to representative ecosystem types, it becomes apparent that new laws are needed to avert threats and maintain the respective natural systems. Specific laws, such as the model law prepared by Ruby I. Compton for preservation of wild, scenic, and scientifically important rivers,[78] are needed for each kind of natural area. Comprehensive laws, such as NEPA, are just as important. Binding international rules must be promoted also, probably on a resource-by-resource basis since it is unlikely that global, comprehensive law reform will be possible in the near future.

Several approaches are very likely to produce good laws for protecting natural areas around the world. Lack of space prevents my discussing them here, but merely listing them can be instructive nonetheless:

1. Identifying endangered natural areas and explaining to people why they are important and what should be done to protect them;
2. Making available laws from several jurisdictions that have undertaken such protection. Model laws also serve this function;
3. Encouraging political—but nonpartisan—conservationist, professional, and scientific advocacy of adopting the necessary laws, suitably adapted to the jurisdictions involved; and
4. Ensuring strict scrutiny by the same conservationist professional, and scientific elements of the laws' implementation.

Where regional law reform is needed, the case for protecting a natural system must be made regionally, as was done with the protection of the Baltic Sea. In instances where preservation was achieved, e.g., in Antarctica, vigilance must continue because vested interests may wish to exploit the resources without appropriate regard for environmental protection.

The early adoption of the Migratory Bird Convention by states in North and South America[79] is a preeminent example of how a specific resource was preserved by coordinated action among diverse states. Could the same be done for fluorocarbon emissions? It doubtless should be.

If the United Nations Environment Programme or some other instrumentality were to compile and promote as a basic common denominator the best laws protecting each natural area, perhaps a uniform pattern of nature protection could evolve independently of the cumbersome and slow treaty-negotiation process. States that fail to adopt such laws could be singled out and the educational process directed there by conservationists, scientists, and others.

Some global tools are now available. The *Earth Law Journal*,[80] published quarterly as a scholarly legal forum for discussing recent developments in international and comparative environmental law, together with occasional articles on environmental law in other law reviews, constitutes an immediately available tool. The magazine *Environmental Policy and Law*[81] and the newspaper *Development Forum*[82] provide a more general lay discussion of the administration and growth of environmental law.

What is needed now is a unanimous effort among the dedicated conservationists, scientists, and others in each land to forge a global law for protection of natural areas.

Conclusions

Despite the strictures of Principle 21, most states today violate it and also the international law it restates. Their violations most frequently result from ignorance of the harm that actions under their control inflict on other states. The interdependence of natural systems is not widely understood by administrators, legislators, or operators.

Even in states with extensive environmental laws at the municipal level, such as the United States, the sound administration of these laws requires vigilant conservationist pressure. As national laws grow, it may be possible for citizens of other nations to use them to enforce Principle 21 and protect natural areas elsewhere. The avenues for such action exist either in treaties requiring new municipal law, such as the Endangered Species Convention, or in securing through judicial or administrative rulings the extraterritorial application of relevant environmental safeguards, as with NEPA.

What is required ultimately is an evolution beyond the nation-state confines of Principle 21. The law must adapt human conduct to the limits fixed in natural systems and not contort those systems to fit human preconceptions influenced by technological considerations. We respect these limitations out of necessity with telecommunication through the air waves.[83] We are learning that the same necessity will require us to enforce measures to stop ocean pollution and ozone depletion, and to preserve wetlands and tropical forests, just to mention only a few of the issues that demand our attention.

The need for a global system of environmental laws and law enforcement is obvious; whether we can fashion such a pact between society and nature remains to be seen. The struggle to achieve such a goal, however, is the mission of EARTHCARE—one of the ends to which these proceedings are dedicated.

Notes

1. For the full text of the Declaration see *Report of the United Nations Conference on the Environment,* United Nations Document A/CONF. 48/14, Annex II, at 72 (1972). See the annotations to the Declaration in Louis Sohn, "The Stockholm Declaration On The Human Environment," 14 *Harvard Int'l L. J.* 432 (1973). [See also Appendix A, this volume.—Editor]

2. See General Assembly Resolution 2995 (XXVII) of 15 December 1972, and General Assembly Resolution 2996 (XXVII) of 15 December 1972. The text of the Stockholm Declaration constitutes Appendix A of this volume.

3. Article 38, Statute of the International Court of Justice. The recognition of other common principles relevant to current environmental issues was described briefly in my essay "Forging Global Law For Environmental Protection," in V. Fauerbach, ed., *EARTHCARE Program/Journal* (New York: National Audubon Society/Sierra Club, 1975), pp. 159-163 (hereinafter "Forging Global Law"). It may be useful to note the following as an historical footnote:

> Jurisprudence, even since Justinian's Digest in 533 A.D., has strived to achieve the 'science of the just and unjust' as well as the 'art of the good and the equitable.' These roots produced the legal systems in civil and common law nations around the world.
>
> While Roman law paid little heed to environmental problems as such, some basics were recognized. 'Air, running water, the sea, and consequently

the seashore' were 'common to all.' It was customary among nations mutually to abstain from the use of poisoned arrows and to safeguard diplomats.

In a sophisticated world some fifteen centuries later, substantially the same Justinian principles are being marshalled anew to define what is just and equitable in protecting the environment. Protection of the common heritage of shared natural resources is now pressed as a right. Mutual self-interest in avoiding 'poisoned arrows' has at least banned the atmospheric testing of nuclear weapons.

"Forging Global Law," p. 159. See Justinian, *Institutes* 1, 2.2, 2.3, 2.10 (4th ed., J. B. Moyle, trans., 1889).

4. "Corfu Channel Case," (1949) *I.C.J. Reports*, 4.

5. "Case Concerning the Factory at Chorzow" (1928) P.C.I.J. Ser. A., No. 17, 4. See also the "I'm Alone Case" (Canada v. United States), 3 *U.N.R.I.A.A.* 1616 (1935).

6. *L'Affaire du Lac Lanoux*, 24 I.L.R. 101 (1957).

7. ("Australia and New Zealand v. France"), reported and described unofficially in the *Journal of The American Society of International Law, Resolution Concerning The Nuclear Test Cases*, XIII (no. 3) *Int'l Legal Materials* 613 (May 1974).

8. Sohn, op. cit. *supra*, note 1.

9. United Nations Document A/AC 138/S. C., 111/S. R. 20 at 4 (1972).

10. Convention On Trade In Endangered Species of Wild Flora and Fauna, U.S.T., T.I.A.S., U.N.T.S. (1975).

11. Robert Allen, "Flower Smuggler, Drop That Pistil," *The New York Times*, 31 May 1975, p. 27.

12. IUCN, "Environment Outline XX, Convention On International Trade in Endangered Species of Wild Fauna and Flora," III *Development Forum* (no. 4), p. 10, cols. 3-4 (May 1975).

13. XI *Int'l Legal Materials* 1358 (1972).

14. Ibid., Articles 1-2.

15. Ibid., Articles 4 and 6.

16. 257 U.S. 419, p. 433 (1921).

17. James Marshall, "Pollution," 57 *Amer. Bar Assoc. J.*, p. 23 (January 1971).

18. Jun Ui, ed., *Polluted Japan*, Reports by Members of The Jishu-Koza ("Open Forum") Citizens Movement (Tokyo 1972).

19. See Resolution 2.313 Adopted By The General Conference of UNESCO

Human Systems and Institutions 537

at Its 16th Session, and Director-General's Document 16 C/78. The progress of the Biosphere Preserves program may be followed in the Final Reports of the International Co-ordinating Council of the Programme on Man and the Biosphere: SC/MD/26 (1972); SC/MD/33 (1973); SC. 74/Con F. 203/2 (1974). To a marked extent, the program does not increase the scope of already preserved natural areas; see, e.g., the current list of Biosphere Reserves designated by the United States:

1. Coweeta Hydrologic Laboratory and Experimental Forest, North Carolina (deciduous forest)
2. Fraser Experimental Forest, Colorado (coniferous forest)
3. H. J. Andrews Experimental Forest, Oregon (coniferous forest)
4. Three Sisters Wilderness Area, Oregon (coniferous forest)
5. Luquillo Experimental Forest, Puerto Rico (tropical forest)
6. Pawnee National Grassland, Colorado (high-plains grassland)
7. Central Plains Experiment Station, Colorado (central plains grassland)
8. Jornada Experimental Range, New Mexico (desert)
9. Amchitka Island, Alaska (arctic tundra)
10. Noatak National Arctic Range, Alaska (arctic tundra)
11. Olympic National Park, Washington (temperate rain forest)
12. Great Smoky Mountains National Park, Tennessee/North Carolina (deciduous forest)
13. Mount McKinley National Park, Alaska (arctic and alpine tundra)
14. Rocky Mountain National Park, Colorado (coniferous forest and alpine tundra)
15. Yellowstone National Park, Wyoming (coniferous forest)
16. Glacier National Park, Montana (coniferous forest and alpine tundra)
17. Virgin Islands National Park, Lesser Antilles (tropical forest)
18. Everglades National Park, Florida (semitropical ecosystems)
19. Organ Pipe National Monument, Arizona (desert)

20. See, e.g., Sir Hugh Elliott, rapporteur-general and ed., *Second World Conference on National Parks* (Switzerland, 1974).

21. The full text of the "Ramsar Convention" is given in Appendix B of the present volume.

22. See, e.g., The Freshwater and Tidal Wetlands Acts in New York State, U.S.A., Article 24 and 25, Environmental Conservation Law, 17 1/2 McKinney's Consol. L. of N.Y.

23. Ancient Monument Act of 1913; and Ancient Monument Act of 1953.
24. Local Authorities (Historic Buildings Act) of 1962.
25. The Civic Amenities Act of July 26, 1967.
26. See for England, the National Trust Acts 1907-53.
27. See generally Robert Brichet "Protection juridique des Villes anciennes," 8 *Monumentum* 115, 124-127 (1972).
28. Law 62-902 (1962).
29. See Hiroshi Daifuku, "National Legislation—France" in XIV *UNESCO Preserving and Restoring Monuments and Historic Buildings*, p. 32.
30. No. 568 *UNESCO Features* February (I) 1970, p. 5.
31. Vladimir A. Borissoff, "Some Aspects of Legal Protection of Natural Areas in the USSR," a paper submitted in a meeting of U.S. and USSR Specialists on Legal and Administrative Aspects of Environmental Protection, U.S. Council on Environmental Quality.
32. No. 636 *UNESCO Features* January (I) 1973, p. 10.
33. Yellowstone National Park, 16 U.S.C. 21, established 1 March 1842; Act Mar. 1, 1872, c. 24. §1; 17 Stat. 32.
34. See generally Roderick Nash, *Wilderness and the American Mind* (1967).
35. The Wilderness Act of 1966, 76 Stat. 890, 16 U.S.C. 1131 *et seq.*
36. The National Historic Preservation Act of 1966, 80 Stat. 915, 16 U.S.C. 470.
37. Historic Sites Act of 1935.
38. Ibid., §102(b) (3).
39. Borissoff, *supra* note 31.
40. Daifuku, *supra* note 29, p. 264.
41. Ibid., 34-35.
42. René Dubos, *A God Within* (1972).
43. *Supra*, note 2.
44. John Muir, *Our National Parks*, pp. 364-365 (1909).
45. The Clean Air Act is derived from the Air Quality Act of 1967, Pub. L. 90-148, 81 Stat. 485 (1967), and from the Clean Air Act Amendments of 1970, Pub. L. 91-604, 84 Stat. 1676 (1970), are as set forth in the United States Code, 42 U.S.C. §1857 *et seq.* (1970).
46. Fri v. Sierra Club, U.S. (1973); see N. A. Robinson, "Purging Pollution from Clean Air: Supreme Court Fiat in Fri v. Sierra Club," 169 *N.Y.L.J.*, p. 1, col. 1 (June 26, 1973).
47. Federal Water Pollution Control Act Amendments of 1972, P.L. 92-500, 86 Stat. 816, 33 U.S.C. 1251 *et. seq.*

48. Article 1, Law of 30 July 1972; Vedomosti RSFSR no. 27, item 692.

49. See generally G. Gaja, "River Pollution In International Law" in Alexandre-Charles Kiss (ed.) *Colloquium,* p. 354 (1975), Proceedings of the 1973 Colloquium of the Hague Academy of International Law.

50. Ibid.

51. Boundary Waters Treaty with Great Britain (on behalf of Canada), U.S.T.S. 548.

52. Treaty between United States and Mexico, U.S.T.S. 994.

53. NEPA is codified at 42 U.S.C. 1424 *et seq.*

54. Thaddeus C. Tryzna, "A Comparative Review of State Environmental Impact Laws within a Federal System," I *Earth L.J.* 133 (1975).

55. See, e.g., Malcolm F. Baldwin, "Environmental Impact Statements: New Legal Technique for Environmental Protection," I *Earth L.J.* 5 (1975); and Frederick Anderson, *NEPA in the Courts,* (1973). For similar laws in other nations, see "Forging Global Law," p. 162:

> In the U.S.S.R., the Decree of 29 December 1972 requires that all ministries in the Soviet Union add environmental protection to existing agency duties. Special offices in each agency require compliance with this comprehensive mandate.
>
> The Quebec Environmental Quality Act of 1972 authorizes prohibition of any act impairing human welfare or the quality of "soil, vegetation, wildlife, or property". In France, land development in crucial areas such as the Queras is strictly regulated, and the Ministry for the Protection of Nature and the Environment coordinates the many air and water and other programs started in the early 1960's.
>
> Japan's Diet enacted the "Basic Law for Environmental Pollution Control" in 1967 providing a framework for fixing environmental standards and enforcement. In Mexico, a federal "Law to Prevent and Control Environmental Pollution" of 1972 empowers the Minister of Health to fix rules minimizing urban and industrial pollution.

56. General Assembly Resolution 2995 (1972):

> ... Cooperation between States in the field of the environment, including cooperation towards the implementation of principles 21 and 22 of the Declaration of the United Nations Conference On The Human Environment, will be effectively achieved if official and public knowledge is provided of the technical data relating to the work to be carried out by States

within their national jurisdiction, with a view toward avoiding significant harm that may occur in the environment of the adjacent area. . . .

Note also that the United States, through the National Aeronautics and Space Administration, has reached an expert agreement with the USSR through the Academy of Science on scientific exchanges of earth resources information. In February of 1973, the USSR proposed that the United Nations establish a center for collection and distribution of data from Earth-orbiting satellites. Such a center could provide environmental exchanges of data as well as Earth resources data.

57. Nicholas Robinson, "Extraterritorial Environmental Protection Obligations of the Foreign Affairs Agencies: The Unfulfilled Mandate of NEPA," 7 *N.Y.U. J. Int'l L. & Politics* 257 (Summer 1974); see the comment on this article in A. Dan. Tarlock, "The Application of The National Environmental Policy Act of 1969 To The Darien Gap Highway Project," 7 *N.Y.U. J. Int'l L. & Politics* 459, p. 469 (1974).

58. Principal arguments for and against were treated initially by The Legal Advisory Committee to the President's Council on Environmental Quality, in *Report of the Legal Advisory Committee* 13-17 (December 1971). This report is not generally available and is now out of print; accordingly, it is useful to provide the relevant pages, including committee's official legal recommendation, here:

> One of the ironies in the new national awareness of the environmental damage that can be done by industry and large-scale economic development is that the United States itself is currently engaged in a number of foreign aid programs which involve many of the same factors, and which may be doing equal environmental damage abroad. By way of example, environmentalists point to the adverse ecological impact of Russia's foreign aid in the construction of the Aswan Dam in Egypt, which has produced a disruption in the flow of fertilizing minerals to the sea, a decline in the sardine fishing industry, a threat to grain crops, and increased human ravages from the explosion of the fluke parasite in irrigation canals.
>
> Discussions between representatives of the Department of State and representatives of the CEQ disclosed that those who were responsible for our foreign aid program, while proclaiming their sympathy with environmental goals, did not consider themselves bound by the reporting requirements of the National Environmental Policy Act. Section 102 of that Act requires Federal agencies to provide assurances that the environmental

impact of their program is kept to a minimum. The State Department simply refused to comply with that provision. In short, there was ground for concern that the United States was not practising what it preached about concern for the environment when it came to its activities in foreign countries.

Accordingly, the Legal Advisory Committee at its meeting on April 19, 1971, unanimously adopted the following resolution, submitted by its Subcommittee on International Environmental Law, chaired by Professor Strong:

> Whereas, the Department of State, in a legal memorandum and comments dated 4 May 1970, contended that §102(2) (C) of the National Environmental Policy Act of 1969, requiring submission of an environmental policy statement, does not apply to State and AID [Agency for International Development] actions carried out within the territorial jurisdiction of another nation; and
>
> Whereas, the Legal Advisory Committee of the Council on Environmental Quality believes that the language and legislative history of the National Environmental Policy Act of 1969, the law of the United States, and the administrative procedures of the Council on Environmental Quality support the conclusion that §102(2) (C) does apply to State and AID actions carried out within the territorial jurisdiction of another nation;
>
> Whereas, the Department of State through a representative meeting with the International Legal Advisory Subcommittee, has made five further representations on behalf of its position to wit:
>
> (1) The Act is ambiguous, the legislative history of the Act on this matter is unclear, as noted by Congressman Dingell, and the detailed steps required to implement Section 102(2) (C) cannot readily be applied to a situation where the action occurs in a foreign country;
>
> (2) many developing countries would regard the imposition of explicit environmental criteria on U.S. AID funding as an infringement on their sovereignty and as a pretext to delay their advance towards industrialization with what they see as its concomitant pollution;
>
> (3) unilateral action by the United States to influence environmental results in other nations is unwise because it would encourage further such unilateral actions by other nations in environmental and

other areas and would impede international solutions to these worldwide problems;

(4) specific environmental conditions like other economic or political conditions, should not be made mandatory in State and AID funding of foreign programs since experience has shown this complicates the administration of aid and impedes the development of U.S. relations with developing countries;

(5) the information which would be provided under a §102(2) (C) statement is available already.

The Legal Advisory Committee responds to these representations as follows:

(1) the legislative record stands as a public document and indicates an intention to include all Federal agencies not specifically excluded; further, the procedures for preparing 102(2) (C) statements *can* be applied within a foreign context without undue administrative difficulty;

(2) while the State Department's assessment of the attitude of other nations is generally correct, the attitudes of the developing nations are changing—in part because of U.S. diplomatic efforts to this end—and can be expected to continue to change in the direction of increased concern for the environmental effects of development; moreover, the §102(2) (C) statement can be a useful educational device to further this objective;

(3) and (4) the effect of the application of the §102(2) (C) statement to State and AID would not be to impose specific conditions on United States' spending abroad but to inform United States citizens and the citizens of foreign nations of the anticipated environmental effects of such spending, and to establish guidelines for the extension of aid which are in the long-range interests of the recipient countries;

(5) while the information to be obtained under the §102(2) (C) statement may in fact be available elsewhere, the purpose of the National Environmental Policy Act of 1969 is to bring all such information together in a common place to simplify the task of people seeking information about the environmental impacts of all programs of Federal agencies.

Now, therefore, the Legal Advisory Committee resolves that the Council on Environmental Quality hereby is urged to seek full compliance by State and AID with the provisions of §102(2) (C).

The subsequent response by the Department of State reflected a sharp reversal in position. On September 1, 1971, AID advised the Chairman of the Council on Environmental Quality that the agency had now adopted new procedures to insure that environmental factors would be identified, analyzed and considered in the decision-making process on capital projects in foreign countries in conformity with the interest and objectives of the National Environmental Policy Act.

59. Civil No. 1867-73 (D.D.C., August 3, 1974), F. Suppl.

60. People of Enewatak v. Laird, 353 F. Supp. 811 (D. Ha. 1973); In People of Saipan v. Interior, 356 F. Supp. 645 (D. Ha. 1973) *aff'd as mod.* F. 502 F. 2d 90 (9th Cir., 1974), the Court similarly noted in *dicta* that NEPA applied extraterritorially.

61. Civil No. 75-1040 (D.D.C., October 17, 1975), F. Suppl.

62. David Anderson and The Canadian Wildlife Federation, intervenors; 463 F.2d 1261 (D.C. Cir., 1972).

63. The Securities and Exchange Act is codified at 15 USC 78a, *et seq.* For a thorough discussion of the 1934 act's application extraterritorially, see Richard Mizrack, "Recent developments in the extraterritorial application of Section 10(b) of the Securities and Exchange Act of 1934," 30 *The Business Lawyer* 367 (January 1975).

64. 405 F. 2d 200 (2d Cir.), *rev'd on rehearing on other grounds*, 405 F.2d 215 (2d Cir., 1968), *cert. den.* 395 U.S. 906 (1969).

65. Canadian Arctic Waters Pollution Prevention Act, 18-19 Eliz. 2, C. 47 (Can. 1970); see also Richard B. Bilder, "The Canadian Arctic Waters Pollution Prevention Act: New stresses on the Law of the Sea," 69 *Mich. L. Rev.* 1 (1970).

66. 223 F. Supp. 987 (S.D. Fla., 1963).

67. 389 F. Supp. 689 (D.D.C. 1974).

68. See N. A. Robinson, "SEC announces new rules for disclosures," 175 *N.Y.L.J.*, p. 1 (25 November 1975).

69. Leonard Woodcock, writing in *The New York Times*, p. 34F (7 January 1973), quotes Carl A. Gerstacker, the chairman of Dow Chemical Company

in 1972, as follows: "I have long dreamed of buying an island owned by no nation, and of establishing World Headquarters of the Dow Company on the truly neutral ground of such an island, beholden to no nation or society."

70. E. I. Nwogugu, "Law and environment in the Nigerian oil industry," I *Earth L.J.* 91 (1975).

71. Lochner v. State of New York, 198 U.S. 45 (1905).

72. U.S. Environmental Protection Agency Press Release No. R-278, "EPA Receives Report on Economic Impacts of Control of Potential Ozone-Depleting Chemicals" (10 December 1975), the report by Arthur D. Little and Company is entitled "Preliminary Economic Impact Assessment Of Possible Regulatory Action To Control Atmospheric Emissions Of Selected Halocarbons," EPA-450/3-75-073 (1975).

73. Article 38, Chlorofluorocarbon Compounds, Environmental Conservation Law, 17 1/2 McKinney's Consol. L. of N.Y.; L. 1976, c. 713, McKinney's 1975 Session Laws, p. 1131.

74. Muir, *supra* note 44. Difficulties in implementing environmental laws exist in the Court as well as among bureaucrats and administrators. See "Forging Global Law," p. 163:

> It is not easy today to persuade a judge or the head of a ministry that nature protection should be a prerequisite before "business as usual" continues. Nigerian fishermen, suing in 1973 to restrain damage to fishing areas in Agge from oil exploitation, confronted a judge's ruling that it "will not be just" to stop the unintentional oil pollution since the work was licensed and, if discontinued, trade and employment might be disrupted. The fishermen's effort to protect nature failed, and they obtained only money damages.
>
> Intentional harm frequently is halted in national courts. As early as 1610, in *William Alfred's Case* the King's Bench in England held against a landowner who intentionally placed a piggery next to his neighbor's orchard. In 1970, on the other hand, in *Boomer* v. *Atlantic Cement,* New York's highest court refused to stop a cement plant from spewing forth dust and thereby destroying neighboring farms; the court concluded that the employment afforded by the cement plant and its contribution in local taxes out-balanced protection of natural areas. Like the fishermen, all the farmers got was compensation for their lost property. Nature lost.

See *William Alfred's Case,* 9 Co. Rep. 576, 77 *Eng. Rep.* 816 (King's Bench 1610); see *Boomer* cases: 55 Misc.2d 1023, 294 N.Y.S.2d 452, *rev.* 26 N.Y.2

219, 309 N.Y.S.2d 312; on remand 72 Misc.2d 834, 340 N.Y.S.2d 97, *aff'd Kinley v. Atlantic Cement Co.*, 42 A.D.2d 496, 349 N.Y.S.2d 199.

75. UNEP's Legal Task, An Interview with Dr. Hassan Ahmed, I *Environmental Law & Policy* 50 (no. 2, October 1975). See also draft U.N.G.A. Resolution by Australia and thirteen states, U.N.G.A. Doc. A/C. 2/L. 1975, "Conventions and Protocols in the Field of Environment," 13th Sess., 2d Committee, Agenda Item 59, authorizing UNEP to help develop further environmental law, urging ratification of existing treaties, and requesting regular report on the status of such laws.

76. See R. Russell, ed., *Earth, Air, Fire, and Water* (1974).

77. Treaty on the Protection of Artistic and Scientific Institutions and Historic Monuments (Roerich Pact), open for signature 15 April 1935, 49 Stat. 3267, U.T.S. 899. Hague Convention for the Protection of Cultural Property in the Event of Armed Conflict, 14 May 1954, U.N.T.S.

78. Ruby I. Compton, "Scientific and Scenic Rivers: Commentary and Model Statute, I *Earth L.J.* (Issue 3) 241 (1975).

79. Convention Between The United States and Great Britain For The Protection Of Migratory Birds, August 16, 1916, 39 Stat. 1702; see Migratory Bird Treaty Act, 16 U.S.C. 703; Missouri v. Holland, 252 U.S. 416 (1920).

80. *Earth Law Journal*, published quarterly by A. W. Sijthoff International Publishing Company, B.V., Post Office Box 26, Leyden, The Netherlands; subscriptions or single issues may be obtained from Academic Book Services Holland/Journal Department, Post Office Box 66, Groningen, The Netherlands, Vol. I, 1975; Vol. II, 1976.

81. *Environmental Policy and Law*, published quarterly by the International Council of Environmental Law, Bonn, Federal Republic of Germany; subscriptions or single issues may be obtained from Elsevier Sequoia, S.A., Post Office Box 851, 1001 Lausanne 1, Switzerland. Vol. I, nos. 1-2 1975.

82. *Development Forum* is published by the Office of Information and Center for Economic and Social Information, U.N., Palais des Nations, 1211 Geneva 10, Switzerland; subscriptions are free, one per person. Published monthly. Volumes 1-111, 1973-1975.

83. See the convention establishing the International Telecommunications Union (ITU) and its regulatory authority.

Human Settlement and Natural Areas

43. EARTHCARE in France

François Gros
Grande Traversée des Alpes Françaises,
France

Introduction

The notion of protecting the environment from human intervention arose in France only about a century ago. Before then, no one worried about nature. Nature was everywhere, and two-thirds of the French lived in the country and were of the country.

As long ago as the Middle Ages, princes had taken measures to protect forests and game. Such measures were not intended to protect the environment itself, but rather, to meet the needs of the princes: hunting and forests were their possession and privilege. Later, in 1669, King Louis XIV issued a genuine "Charter of the French Forests" on the advice of his minister, Colbert, to ensure a supply of timber for building his ships. These early measures for the protection of nature were taken only with some advantage in mind.

Scientific, aesthetic, and touristic purposes arose much later, when the famous French painters of the Barbizon School asked Napoleon III, as early as 1853, that the State protect part of the forest of Fontainebleau. Napoleon III heard their claim, and issued an imperial decree on 13 August 1861 preserving 624 hectares for artistic reasons, the first protection extended to this beautiful woodland.

In France then, one must wait until 1928 for the creation of the natural reserve of Camargue, when, thanks to private initiative, 10,000 hectares (including the Etang de Vaccarès) came under the supervision of the National Society for the Protection of Nature. The Camargue area is notable for its rich bird life, including flamingos; the Etang, the largest lagoon, is a bird sanctuary.

The first act of Parliament dealing with the protection of monuments and sites dates from 2 May 1930. Much later, the law of 1 July 1957 was enacted.

Types of Reserves in France

Since industrial and other human pressures on the environment are increasing every day, the idea of a reserve has become more important, and many are to be created. There are several kinds of reserves in France.

Official Natural Reserves

According to the law of 1 July 1957, the creation and boundaries of "official natural reserves" are fixed by cabinet decision if the owner agrees; if not, by a decision of the *Conseil d'Etat* (the equivalent of a "supreme court").

There are few official natural reserves in France: in 1972, there were only four; in 1974, sixteen. Their purpose is to protect species. A team of specialists in various disciplines conducts a scientific survey before an area is designated an official natural reserve.

Natural Reserves with a Specific Purpose

There are three primary types of "natural reserves with a specific purpose:" "zoological reserves," "botanical reserves," and "geological reserves." In some cases, two or more specific purposes may be served by one reserve.

Other Reserves Classified According to Their Legal Status

Three other types of reserves may be classified according to their legal status: "game preserves," "waterfowl reserves," and "survival reserves." Game preserves are numerous, efficient, and deal more with the breeding of species than with protecting the environment. Survival reserves will be discussed below in the section on national parks.

Special Reserves

"Special reserves" are limited in purpose. They exist to protect whole elements of aesthetic, historic, or educational value, or to comply with biological necessities. There are five types of special reserves: "natural sites reserves," "natural monuments reserves," "woodland reserves," "game preserves," and "waterfowl preserves."

Special reserves are officially declared when important natural sites or species are threatened to such an extent that their existence is endangered. Previously, protective measures were difficult to enforce; often it was too late to create a reserve, or, interest in such reserves and free shelters being mostly "psychological," there was little to guarantee their continued security.

Independent Natural Reserves with a Free Status and Free Shelter Reserves

"Independent natural reserves with a free status" are organized by the owners or tenants of tracts of land, without any specific legal action. They are numerous, fairly old (for example, Camargue), and of various sizes. Owners of this type of reserve can be the State (for example, Brittany's Gulf of Morbihan), local authorities (the Dombes swamps), public agencies such as the National Hunting Board, and private bodies. More frequently, such reserves are managed by specialized associations. Other reserves, called "free shelter reserves," are protected by the rules of charters adopted by their owners and with the assistance of specialized associations. There are more than a thousand of these today of various sizes.

National Parks in France

Aware that France—a leader in EARTHCARE since 1861—had no national parks, the Parliament enacted Law No. 60-708 of 22 July 1960, creating the French national parks. The law went into effect with Decree No. 61-1195 of 31 October 1961, which promulgated regulations for the application of Law No. 60-708.

In France, the State does not own the national parks; rather, they remain the property of local people. Their right of property is guaranteed, but hunting rights, which in France are linked with landed property, are not: they are abolished, but with financial compensation.

The success of this arrangement depends upon a special legal status that is conferred upon a part of the national territory. The limits of each park are determined by the *Conseil d'Etat* after preliminary studies and consultation with the local people.

In France, the aim of a national park is to protect and to organize—for scientific and educational purposes—an area whose fauna, flora, and other natural features are valuable or otherwise significant, without harming the traditional agriculture or forestry that have protected it to our day. The park is managed so as to preserve its original characteristics.

Structure of French National Parks

To accommodate all interests, three zones are determined, the national park itself, the neighboring zone, and "survival reserves." Fauna, flora, and geologic sites are protected in the national park itself, but agricultural and traditional pasture lands are maintained. In the neighboring zone *(la zone périphérique)* adjacent to a national park, people maintain their traditional way of life. To this end, various administrations regulate social, economic, and cultural development. This prevents the local populace from abandoning the area adjacent to the park, yet assists them in welcoming and accommodating visitors.

Priority zones called survival reserves can be established by law in one or several parts of a national park. In survival reserves, animals and plants receive better protection than elsewhere; however, human inhabitants and their customs are explicitly accommodated. At this time there are no survival reserves in France.

There are now [1975] five national parks in France (Table 1). Together, they occupy some 275,000 hectares. Besides the five existing national parks, a sixth (Mercantour) is planned in the southeast, in the Alps near Nice, and a seventh (Haut Ariège) in the eastern Pyrénées.

Table 1. National Parks in France, as of June 1975

Park	Area (hectares)	Year Established
La Vanoise	52,520	1963
Port-Cros	690	1963
Pyrénées Occidentales	45,715	1967
Cévennes	84,220	1970
Les Ecrins	91,740	1973

The process of classifying an area as a national park, thus guaranteeing its protection, unfortunately involves only fairly big regions whose exceptional fauna, flora, and landscape justify that designation.

Management

The Management of national parks is in the hands of a public agency consisting of a board of trustees made up of the various administrations involved, scientists, national sports and tourist clubs, and local authorities. Its decisions are carried out by a director and on-site personnel called "guards" or "rangers" *(gardes moniteurs)*, whose job it is to protect the parks from the public. Everything—financing, equipment, and maintenance—is entirely taken over by the state.

The significance of protecting areas as national parks was demonstrated in the "Vanoise Scandal," in which 25,000 hectares were proposed for withdrawal from the park to create a ski resort. Despite the lure of profit and pressure from some locally influential people who declared that it would be "dreadful to oppose the necessity of EARTHCARE to the economic and social priorities of the region," the park was saved. This amputation could have been a severe blow to other French national parks, but it was finally avoided.

The Regional Natural Parks of France

Between 1960 and 1965, advocates of EARTHCARE and national parks in France were not taken seriously; they were considered idlers who wasted their time watching butterflies or collecting flowers. Gradually, the idea of EARTHCARE was accepted and "regional natural parks" came to life with the decree of 1 March 1967. Regional natural parks must not be mistaken for national parks. Behind the regional natural park is an original concept of EARTHCARE which promotes rural space in areas of varied size (but not less than 5,000 hectares) forming a whole with precise boundaries. Rural territories revived by the regional natural parks are exceptional because of their natural riches or their historic and artistic heritage, or because of their proximity to a town, offering city dwellers easy access to nature.

Regional natural parks can only exist through the good will of the local inhabitants. Initiatives arise from the local authorities, who submit their plans to the cabinet-level Commission of Regional Parks, which then proposes them to the government. If the idea is accepted, it is studied by a board under the supervision of a *préfet* (the administrative head of a *département*). A region is not classified according to the decree of 1 March 1967 until a "constitutive charter" has been accepted by all concerned parties.

After the efforts and the good will of the local people and after the state recognizes a charter, the most important task is to protect the site, to safeguard the balance of the natural environment. Once these goals are achieved, the task is to give new strength to rural life and to allow the farmers, acting as ecology wardens, to rebuild the rural environment—encouraging not only cultural activities but also regional and traditional crafts. The flora, the fauna, and the rural life all have to be respected, promoted, and brought to life again; but above all, it is a way of life which must be saved. The regional natural park must be a meeting point between city and country.

While the procedures for creating a regional natural park are much simpler than those for creating a national park, certain rules, though they are not enforced by law, must be voluntarily respected. This explains why regional natural parks have had varied success. Financial problems may also occur, because financing is not entirely supported by the State and especially since these regions are poor and their level of self-financing is absurdly low. A solution to this problem will have to be found rapidly so that the generous impulse that gave birth to these parks is not stifled because of money matters.

Ultimately, eighteen regional natural parks (nine of which already exist) will cover 2 million hectares, some 4 percent of France's total territory.

Traversing the Alps: *La Haute Route*

Integral to the mechanism for establishing reserves and parks in France is the importance of man and his fundamental and beneficent role in safeguarding nature. Preserving species or sites is primary, but cannot be severed from social and cultural considerations. There are no precise limits to preservation or development. Therefore, the aim must be to harmonize the interests of man, animals, and plants.

Everybody and nature must profit by it. The State, with its financial help, can offer shelters and other facilities to hikers at a lower cost—by using what is already existing, not by creating hotels and refuges. Year-round maintenance is assured by the owners living in the villages. The mountain dwellers are ski-instructors in winter; guides, farmers, and craftsmen in summer. In a word, they protect life in all its aspects.

They are responsible for their own future and environment. While their incomes may be far lower than those of wage earners, it is the quality of life that matters to them. They are proud of their mountains and intend to stay there, keeping their homes.

Thus, the notion of "peasant," "mountain gardener," or "paid civil servant" is discarded and the protection of nature finds its true *raison d'être*. Man and his activities are dependent on nature. This must give new confidence. Otherwise, it is to be feared that the existence of the reserves and parks will only be a legal and easy pretext to do anything, anywhere else. If species are no longer in danger only in such reserves, the essential thing—namely, the balance between man and nature—still would be precarious.

Selected Bibliography

Bourlière, François, et al. 1975. Spécial: Réserve de Camargue. *Le Courrier de la Nature* 35:1-65 (January and February).

Fischesser, Bernard. 1973. *Richesses de la Nature en France: Réserves et Parcs naturels*. Paris: Horizons de France.

Weber, Karl, and Lukas Hoffmann. 1970. *Camargue: The Soul of a Wilderness,* trans. Ewald Osers. Berne: Kümmerly and Frey; London: George G. Harrap and Co. (Also published in German by Kümmerly and Frey as *Camargue: Seele einer Wildnis* and in French, also by Kümmerly and Frey, as *Camargue: L'Ame d'un Sol sauvage.*)

44. Nature and Human Settlement

Roland C. Clement
National Audubon Society, U.S.A.

The protection of nature depends upon man's ability to make cities livable, so that they will satisfy most of his needs and thus make unnecessary the widespread destruction of the environment that has characterized recent history. Because cities are where most people want to live, the task of human settlement is to build and maintain livable cities. This task is ignored until our cities have suffered so much abuse—as in our day—that people are torn between their fascination for the city and their frustrations with its problems.

The city is exciting and thus desirable as a human habitat, because it provides a maximum of interaction potential by bringing many different people together. It has long been the focal point of all social innovation, of all human progress.

To provide these advantages, cities must possess a degree of structural and functional unity, since this is what gives them a sense of personality and makes them livable. Dr. Constantinos Doxiadis has suggested that a 10-minute communication distance—by foot, bicycle, or taxi and public transport—is the optimum size for a city. Within that area, a proper city will satisfy all needs for social interaction—except, of course, that today our intellectual interactions may be very widespread.

We need unity with balance. "Balance" means a mixture of social groups and implies an equitable allocation of opportunity or access to the city's advantages by all its citizens. If the number of people in a city interferes with the 10-minute-contact criterion, they are too numerous, at least for the existing degree of social and technological organization: the city has become too big.

Even urban man lives in two worlds. Too many people have mistakenly assumed, or have been falsely taught, that the world man has built himself—the city—is a world sufficient unto itself. However, as we are beginning to realize, man's world is built within and is dependent upon the world of nature, not only for food and water but for a multitude of other human needs as well. Therefore, people have "the best of both worlds" only when their circumstances allow them to alternate, easily and frequently, between the city and the countryside. These circumstances can be purposely planned and provided for during the development of a city and its natural environs.

We need to be aware also of certain behavioral problems that stem from our attitudes toward ourselves, our environment, and nature in general. One of the most difficult of these is the aberrant notion that man is, or must become, the master of nature. This widespread attitude has resulted in a master-slave relationship, wherein man has demeaned other creatures; but, even worse in the short run, this has caused man to neglect environmental realities that affect him directly and unfavorably, often as serious losses that could have been avoided by realistic adjustments.

For example, we generally fail to allow for random but probable (and therefore to be expected) events in nature. Floods, droughts, hurricanes, tornados, and earthquakes are all natural events we cannot prevent, although we can learn to mitigate their effects if we will take them into account in our planning. Instead, we make believe they are unlikely to happen to us, or pretend we can control them if only we will invest more in the effort. Our only response to the flooding of large rivers in the United States has been to spend more money to control floods, with hardly any thought to adjusting our uses to the floodplains. The result is that, despite the expenditure of over $8 billion in flood-control works on the Mississippi River alone, each new flood is more destructive than the last because people, having been given false assurances of safety, crowd into the floodplain in increasing numbers, or build increasingly valuable works there.

We should, of course, study natural events so as to predict them more accurately, but instead of vainly trying to control what is uncontrollable, we need to adjust our behavior. Adjusting our behavior will minimize both the impact of natural events on people and the social costs of futile control investments. We must think of natural phenomena in scientific (probabilistic), not absolute, terms. (Absolutists may retort that this proposal is simply a negative form of absolutism; I invite them to wait until man has indeed demonstrated his control over natural elements before they base social plans upon promises that remain unfulfilled generation after generation. That would make intelligent allowance for uncertainty.)

To strike a balance between man's use of the Earth and its natural ecosystems is an ecological challenge, but ecology is such a new science that there is little knowledge of urban ecology as yet. What feedback processes, positive and negative, are set in motion by the urbanization process? We know that many general principles of ecology are applicable to urban environments, but few of them have been translated into tested guidelines for engineers and planners. This is an urgent task for ecologists.

The Quality of Our Environment

Meanwhile, however, we know some of the ingredients of a quality environment, and the several sciences—though they are not yet integrated ecologically—can document their importance and how their mismanagement turns natural assets into liabilities. What are these ingredients? And what management guidelines must we set ourselves if they are to contribute to that intelligent balance between man and nature we seek for our joint benefit?

The basic ingredients for human survival are air, water, food, shelter, and a sense of community. The first four ingredients serve the physicochemical needs of all organisms. Only the higher forms of life require the fifth: it serves our psychic needs.

Air

Air pollution is not only an annoyance; it damages our health and economy, and modifies climate. Particulate emissions, gases, toxins, radiation, and heat from industry, automobiles, and other sources, must therefore be controlled. Our technology must increasingly be directed at designing "leak-proof" cities.

Currently, urban parks play a crucial role in helping disperse and thereby dilute air pollutants. In New York City's Central Park, which is 1.7 kilometers wide, ambient levels of sulphur dioxide in the air are less than one-half those in the surrounding city. In addition, wooded parks temper local climate, reduce noise, accommodate groundwater, furnish habitat for interesting animals, and provide open space and a variety of recreational opportunities for people.

Air pollution is due to either ignorance or irresponsibility: it is an example of "dumping in the Commons" and thus of distorting the real costs of industrial production. Some people obviously derive short-term advantage from such dumping, but their impositions on the rest of society must be countered by collective action—that is, by law and its enforcement.

It should be noted that pollution hurts the poor most of all: it has been shown that, in almost any large city, zones of high pollution are also zones of greater poverty and higher morbidity. Polluted areas become "ecologic traps" for the people who cannot afford to escape them. We have institutionalized poverty, in part, by polluting the environment and social institutions.

Water

Two of the most widespread constraints on urban development are (1) the availability of water and (2) the contamination of water by inadequately treated wastewater or from excessive numbers of cesspools.

If annual precipitation is about 40 inches (100 centimeters), well distributed seasonally, a region can sustain three people to the acre (0.4047 hectare), provided the water is adequately collected and distributed. Where cesspools are the principal form of domestic-waste treatment, however, an acre of land may support only two people at the typically high rate water is used in the United States. Contamination is more likely at higher human densities. Less profligate use of the water would provide leeway for more people. Further, engineering skills make it possible to supply the water needs of many more people by importing water from other regions, at varying distances. However, I know of no city that has had sufficiently intelligent management to recognize that every drop of imported water must also be treated after it has been used. This separation of engineering and sanitation, of politics and economics, is responsible for the sad state of urban water supplies almost everywhere, and the even worse water-pollution problems of urban regions.

Almost everywhere, also, governments have overlooked the fact that development greatly alters regional hydrology. As more and more area is made impervious "black-topping," water runoff is magnified after rain. Since drainage patterns evolved before development, the result is increased local flooding, with all of its attendant misery, health hazards, and other social costs.

We now understand these hydrological relationships, but they have hardly begun to be built into public-works programs, even in the United States. Urban open spaces—in the form of water-catchment areas of adequate size along every stream—provide a means of mitigating these problems. The task, it seems, is to develop political processes that will serve as a "social memory" and prevent ignorant individuals and unscrupulous developers from neglecting these hydrological realities.

Carrying Capacity—The Key to Urban Planning

Providing adequate food and shelter and so organizing ourselves that a sense of community prevails among us are vast subjects in themselves. But we have probably overcomplicated them by failing to see them as problems of environmental carrying capacity. The ecological concept of carrying capacity is familiar in the fields of wildlife management and animal husbandry; however, the false notion that human beings are not part of nature has deterred us from applying the scientific insights of population biology to the solution of our own problems.

So strong has been our determination to see man apart from nature that we have developed a fixation on the Malthusian concept of limits, inverting the concept, of course, by first denying that people are limited by resources, and then proceeding to overexploit the resources, in a vain attempt to maintain excessive human numbers in dignity.

It is time to look beyond Malthus and to see that the structures of our societies have suffered from the effects of large populations and rapid growth and from the parallel error of developing beyond optimum levels. It is these flaws in our outlook that have made human community so rare that we have had to invent a heaven after death, to console people for the miseries of their earthly existence.

Neil W. Chamberlain (1972) has analyzed these relationships and has shown that urbanization involves positive-feedback (runaway) mechanisms that must be regulated if the city is not to be ruined by its own early success. We all know the evidence, but he was the first to systematize it. Urbanization first creates a large demand for food and this leads to the mechanization of agriculture. Mechanization displaces farm labor, which then migrates to the city, magnifying both the demands and the responses. Within the city, higher population stimulates invention and innovation, which leads to increased specialization of effort as a means of escaping the competition of numbers. The clever few soon take advantage of this specialized activity to command extra privileges for themselves; this deprives others, leads to alienation, and makes the city ripe for political demagogues.

Replacing the original sense of participation and community, many subpopulations develop, and the more determined minorities among them fight the system. The only way to maintain order in this increasingly complex and tenuous social system is to repress the malcontents or to provide increasingly expensive concessions. Either way, more and more centralized power is required. But, since not all claims can be denied, the community succumbs to growing welfarism as growth continues and the standard of living deteriorates for almost everyone. The city, which earlier provided so much excitement and satisfaction, gradually becomes a "concrete jungle." This is the social price of unregulated growth.

Under crowded conditions, individuals must also pay a high price in terms of personal health. Health is a product of a creative way of life, not of medical science, as so many have naively assumed. The opportunities for creative living are, in turn, by-products of social organization, since we do not live alone. Biologists have been aware of these implications at least since John J. Christian (1950) showed, nearly a generation ago, that the individuals in a population are healthiest when that population exists at 50 percent of the environment's carrying capacity. Unfortunately, perhaps, Christian's experimental subjects were mice, so most people have assumed that his evidence was irrelevant to human affairs. But our approaches are simplistic: because we can, by disregarding costs, produce more food, we have thought that we were outwitting the phenomenon of Malthusian limits. And because people are more adaptable than mice, we overlook the possibility that we may simply be abusing people more, not circumventing Christian's 50-percent rule. This simplemindedness may be our ultimate limit because, if we cannot solve our problems, nature will solve them for us.

As for development, we already know that the costs of providing services for either a mechanical or a social unit increase as the square of the number of units to be serviced ($C = N^2/2$). This is why, once an effective "critical mass" has been achieved in the developing of any system, further growth becomes overdevelopment and imposes exorbitant costs. If these added costs are allocated evenly (democratically), the existing system is degraded. The only way to avoid this disorganizing pressure—as we are now learning in trying to cope with growth in energy demands and costs—is to impose all the costs of new services upon those who demand them. Economists call this marginal-cost pricing, and it is an essential constraint on overdevelopment.

New, computerized studies of these relationships by Eugene P. Odum and Howard T. Odum (1972) point in the same direction as Christian's studies of biological stress. They show that we must value three sets of environmental components: (1) cultural creations, including the arts, technologies, social advances, and ethical and intellectual insights; (2) the natural ecosystems that serve as the substrate of our very existence, providing most of the essential services and resources; and (3) the opportunity for interaction between the "worlds" of man and nature.

We must prevent overdevelopment and overpopulation because they occur at the expense of natural ecosystems and ultimately of human society. This is why the Odums conclude that it would be "prudent for planners everywhere to strive to preserve 50 percent of the total environment as natural environment." They conclude that the minimum per capita area required for quality existence is 5 acres (2 hectares).

The preservation of wilderness, of species, and of ecosystems depends on our ability to solve the problems of development. We do not know how to build livable cities, but we have at last learned that to keep them livable we must prevent their overdevelopment. This is at least half the task—perhaps the more important half.

References

Chamberlain, Neil W. 1972. *Beyond Malthus: Population and Power.* Englewood Cliffs, New Jersey: Prentice-Hall.

Christian, John J. 1950. The adreno-pituitary system and population cycles in mammals. *Journal of Mammalogy* 31(3):247-259.

Odum, Eugene P., and Howard T. Odum. 1972. Natural areas as necessary components of man's total environment. *Transactions of the 37th North American Wildlife and Natural Resources Conference* (Mexico City). Washington, D.C.: The Wildlife Management Institute.

Interdependence:
The Interaction of Societies and Ecosystems

45. Interdependence: Society's Interaction with Ecosystems

Gerardo Budowski
International Union for Conservation of
Nature and National Resources, Switzerland

Interdependence is one of those interesting and appealing words that sums up for ecologists a series of concepts reflecting, perhaps better than anything else, the idea of a holistic approach to the environment. According to their degree of sophistication, computer-minded ecologists speak about "loops," "retrieval," and similar terms, while the peasant in a developing tropical country intuitively senses the changes that will occur in his day-to-day life as a consequence of new elements introduced into his environment.

Today, we will hear about three approaches to interdependence (probably from different parts of the world), and we will examine some of the negative and positive consequences. They may be only case studies, but I hope they will point the way to workable general solutions.

Interdependence involves much more than an awareness of problems and the search for solutions to them. I believe that interdependence is likely to dominate our policies and lifestyles in the next few years, simply because it is an inescapable and ever-growing trend in a shrinking world. One finds it in the industrialized world when it comes to the supply of raw materials or the cultural effects of mass communication, and one finds it in the most remote communities, such as in the mountains of Papua New Guinea, where hundreds of different tribes still practice their traditional ways of life but are confronted with such concepts as statehood, sovereignty, and "responsible" government.

I have just come back from Papua New Guinea, where I attended a seminar organized by the University of Waigani, at Port Moresby, on the appropriate theme, "The Melanesian Environment: Change and Development." As many of you know, Papua New Guinea will become independent in the next few months and a new government is in the process of being constituted. [Papua New Guinea gained its independence on 16 September 1975.—Editor] Several leaders of the new government attended the meeting at Port Moresby and made valuable contributions. While the word *interdependence* was not often pronounced, it was at the root of the deliberations about the mistakes that should be avoided, about the lifestyle that should be promoted, and about various facets of a lifestyle aimed—not at uniformity—but at adapting to the great diversity of that country. It was a fascinating experience to hear speaker after speaker—whether born in the United States or Australia or another westernized area, or whether a native of Papua New Guinea—seeking the best ways of achieving a harmonious future and a full comprehension of biological, physical, and cultural implications of interdependence.

Should Papua New Guinea strive for economic development to the utmost, or should it devise very careful development plans that will take account of the customary rights of its present population, mostly found in village settlements? There was no consensus, and it was unrealistic to search for one, but there was a clear indication that traditional ways of life should not be lost.

Perhaps some of the problems connected with this search for the future—taking into account all of the interdependent influences—was best expressed by the Honorable Stephen Tago, minister for environment and conservation in Papua New Guinea, when he said:

> We have generally taken our environment for granted. This makes the task that faces me as Minister for Environment and Conservation a difficult one, for what is generally taken for granted is not likely to be a policy issue. This is changing, however, since we increasingly realize that there needs to be a deep and lasting commitment of our people to maintaining and improving, but certainly protecting, our resources base.

The resources base, which is all we have, is what has given us the capacity to survive and to live and must give us the potential to develop. That, and our communities, together with the magnificent variety of our cultural heritage and its natural background, is the true wealth of Papua New Guinea. But we must be careful not to forfeit this wealth for short-term economic gains nor to ignore that such gains may be acquired at the expense of our people and their expectations.

However, we must have economic growth. But economic growth means changing the environment. As we know, it is at this point that injury to the environment can take place. If Nature is abused beyond its limits, its revenge is inevitable. The question is not whether we should have economic growth. There must be! Nor is the question whether the impact on the environment must be respected. It has to be. The solution to the dilemma clearly revolves, not about whether, but about how we are to manage our development.

Our needs should be weighed against the needs of our children, their children, and the generations which follow them.

We have a series of unique and complex ecological systems in this country, and their conservation is of great importance for the present and the future. Without trying to determine future generations' priorities, we collectively have a responsibility to keep the options open and prevent the unnecessary loss of unique species and natural areas.

Early consultation with the owners of the area must be started to find out about local land-use and development ambitions and to carefully explain the concept of national parks, wildlife reserves, or management areas. Local committees may be set up, and where this has occurred a significant participatory aspect has been created right through to the establishment of regulations for control. Once a protected area is established, it is essential that the education be carried through, particularly among the children, as they will be the future citizens and leaders.

In the past, our informal education was necessarily ecological in its content. We lived within our environment, and, if we transgressed its thresholds, by bitter experience we discovered the results. But this has changed. One of the major tasks of our present education system today is to recreate this awareness—an awareness of our responsibility to our common property—air, land, water, the other natural resources, and our cultural heritage. And this awareness must be extended throughout the community—to our children, our elders (perhaps the most knowledgeable already), my fellow politicians, and our administrators. Only in this way will we be able to respond to the challenges which face us.

> We have also been careful to include in our guidelines not only the physical and health aspects, but cultural considerations as well. We are concerned that a development project does not adversely affect our culture, but supports and enhances it. Hence, at all stages during the decision-making process for a project, the people should be consulted. The recognition of people's rights is likely to produce a more acceptable pattern of development. This means that the advantages and disadvantages of a particular project must be explained to the community and its wishes respected.
>
> The challenge is to us, our public servants, teachers, village elders, local government councillors—indeed, to all Papua New Guineans—to discipline our desires for quick, exploitative development and temper it with concern, so that we do not slowly destroy what is unique in our country.

These wise words, I feel, provide an appropriate background to our meeting.

I would like to conclude these introductory remarks with an axiom we should never forget, an axiom that was used by Minister Tago in his speech: "Good intentions are no substitute for good performance."

I do hope that our discussions this morning will provide us with the tools for performance based on ecological understanding, where the recognition of interdependence may lead us to a better and more meaningful relationship with our environment.

46. The Interaction of Ecological and Social Systems in the Third World

M. Taghi Farvar
Centre for Endogenous Development Studies,
Iran

Introduction

Is there an environmental crisis in the Third World? This paper examines the question through a number of case histories in Asia, Africa, and Latin America, and concludes that the crisis is in many respects even more acute in the developing countries than it is in the developed ones.

Agricultural development, industrialization, public-health campaigns in the tropics, nutrition-aid programs, urbanization, and population policies are shown to be following a self-destructive and environmentally disruptive path. The existing strategies have not only failed most of the time to meet their declared objectives, but they have caused much damage to the human environment, often in an irreversible way.

Humid-tropical and arid ecosystems are very fragile, yet we are increasingly pushing them beyond their limits of tolerance, causing a dysfunction of the whole, interconnected system. But, while we have in many instances exceeded the "outer limits" of ecological systems, we have failed at the same time to provide for the needs governed by the "inner limits" of the majority of populations of the world.

Disruptions of ecological systems are, then, connected to the inadequacies of the predominant social systems that govern the developed world and to the international economic order that binds the developed to the developing world.

The Two Characteristics of True Development

Development usually aims at increasing agricultural production, improving public health by attacking diseases characteristic of underdevelopment, intensifying industrialization to create jobs and consumer (as well as capital) goods, and producing power quickly. These goals can be achieved by altering the environment of a given area—that is, by changing the "balance of nature."

The balance of nature is the subject matter of ecology. When it is altered in our favor, we have genuine or true development; when such alteration upsets the balance against our interests, we have environmental deterioration. Therefore, true development and environmental degradation are the opposite ends of the same spectrum.

While it is usually believed that development must cause environmental degradation, or that environmental quality can be achieved only at the cost of slowing down development, we can see that one crucial characteristic of true development is a beneficial alteration of the ecological system (or of the balance of nature). Viewed from the environmental point of view, genuine or true development is to be regarded as synonymous with the improvement of environmental quality.

The other characteristic of true development is the increased access by the common man to vital resources—that is, the equitable redistribution of resources. In this light, the problem of agricultural development or of health-care services, for example, becomes a problem of how to better distribute what is already available in food, fibers, and health-care resources.

To be meaningful, these two characteristics of true development must go hand in hand. True development is, therefore (1) the equitable redistribution of available resources and (2) the beneficial alteration of the ecological "balance" to increase the material well-being of all people (Conference on Problems of the Third World and the Human Environment 1972).

Inner and Outer Limits

The foremost aim of true development should be to meet the basic, minimum needs of people—adequate and nutritious food, access to health care, satisfying employment, reliable shelter, sufficient sources of energy, a healthy environment, and so on—in short, all that human beings need in order to live in a state of biological, physical, social, and psychological fulfillment. Each of these minimum needs has an implication in terms of the environment, since it involves the commitment of a certain quantity of environmental resources per person to satisfy the need. These minimum environmental resources needed to achieve true development are called "inner limits" (Conference on Problems of the Third World and the Human Environment 1972). The concept of "inner limits" was used in the Cocoyoc Declaration (Anonymous 1974) to focus attention on the urgent, unmet needs of most of the world's people. Here, we shall speak, interchangeably, of "reaching the inner limits of development" or of "satisfying the minimum needs of people."

Obstacles to approaching the inner limits are usually of two kinds: (1) a social system characterized by injustice, privilege, and maldistribution; and (2) an ecological system whose productive potential is low or that has been driven beyond its carrying capacity.

If a human ecosystem has surpassed its carrying capacity or if its ecological relationships have been imbalanced, with undesirable effects on its viability and long-term productivity, we shall say that its development has exceeded its "outer limits."

Thus, even after the minimum needs have been achieved, there is a vast margin for effective action between the inner and outer limits of development. We shall return to this concept at the end of the paper.

Violation of Local Outer Limits in the Third World

Something has gone wrong with the balance of nature in practically every part of the Third World. The large body of evidence now available makes this point amply clear.

For example, the valleys on the Pacific coast of Peru are suitable for the intensive, irrigated cultivation of such crops as cotton. One of the valleys, the Cañete Valley, has become well-known in the field of pest control for agricultural production. There, insecticides used to control cotton pests gave rise to highly resistant "superpests" that could not be controlled by any chemical means available: such biological-control mechanisms as parasitic and predatory insects had been destroyed by pesticides. The local "outer limits" in the Cañete Valley had been exceeded.

Then, in 1956, there was a disastrous crop failure. A number of ecologically oriented entomologists, helped by aroused farmers, took immediate steps to outlaw the use of the synthetic organic pesticides that had caused the problem and succeeded in reestablishing the balance of nature (Boza-Barducci 1972).

Farmers in Mexico, Central America, and parts of the Middle East have been less fortunate. In Egypt, a drastic decline in total production of cotton occurred in 1961, the yield dropping from 608 to 402 kilograms per hectare because insects had become resistant to insecticides and because nonchemical control methods had been relaxed on account of the new faith in insecticides (Smith and Reynolds 1972). In 1970, when a similar disaster struck cotton-producing areas of northeastern Mexico (such as the Tampico region), the production of cotton nearly ceased (Adkisson 1970). Large areas of Central America faced a similar failure in the late 1960s and early 1970s.

Cotton is not the only crop that has suffered from a reliance on insecticides. In Malaysia and elsewhere, cocoa, palm oil, rubber, and other crops have undergone heavy attacks by pests unleashed through the use of pesticides (Conway 1972). Such consequences of using pesticides are now commonly expected from this approach to pest control.

Incursion on local outer limits is not the only harm done by pesticides used in agricultural development. In Central America, for example, thousands of highland Indians who annually migrate to the Pacific coast to pick the cotton crop are poisoned by insecticides; hundreds of documented deaths are recorded every year. This, of course, is the ultimate violation of local outer limits.

This repetition of disastrous mistakes is not unrelated to private-market-oriented modes of production. For example, in Central America, as in many other regions, pesticide salesmen who represent foreign manufacturing corporations and who are motivated by profit alone, have a virtual monopoly on the technical information available to farmers and other members of the public. There are no qualifications for dispensing highly toxic poisons other than the ability to sell a product. In most instances, we have been too quick to allow the importation and unchecked sale of modern synthetic chemicals without taking such precautions as setting up laboratories to establish standards and monitor effects on people and ecosystems, or instituting a firm legal system to prevent the exploitation of peoples by profit-motivated corporations.

As a result of intervention by such corporations, every law

proposed in the United States of America for banning or limiting the use of DDT and other pesticides has had a provision that specifically exempts exports of the material. One disaster due to this corporate-oriented policy of dominant, industrialized western countries such as the United States happened in Iraq in 1972, when 80,000 tons of imported wheat and barley, coated with organic-mercury fungicide, caused the mass poisoning of Iraqis. At least 400 individuals died initially, and 5,090 more were admitted to hospitals for treatment. The actual number of persons affected was probably much higher. The fungicide, which often works slowly and enters ecological cycles and food chains, reached people through poultry, meat, river fish, and bread. It had been banned in the United States, Sweden, and a number of other industrialized countries, but this did not bar its export to Iraq by profit-seeking United States corporations (Anonymous 1972). The World Health Organization (WHO) and the Iraqi government have since undertaken an in-depth analysis of the events in this mass-poisoning event. In this and in the Central American situation, we see a clear case of the faulty social system of dominant countries violating the local outer limits of dependent countries.

A further examination of development problems will make this point clearer. In most developing countries, those who own the means of production are forcing a shift from subsistence crops (food, fiber, and other crops raised primarily for use by the local population) to cash crops, often exported to industrialized countries. This shift from using nature for the satisfaction of human needs to exploiting it for profits is inherent in the so-called "green revolution" approach to agricultural development, which has been imposed on many developing countries by market-oriented, industrialized countries and their helpers. The helpers include front organizations such as the Ford and Rockefeller foundations and the Agricultural Development Council in the United States, which are tax-free organizations that frequently further the interests of giant, multinational corporations by sponsoring projects that involve heavy inputs from these corporations.

These "front-runners," who develop and diffuse "miracle" seeds, chemicals, and machinery, in fact act as the charitable edge of a much thicker intrusion into the socioecological setting of the Third World. They have set up CIMMYT (International Centre for Maize and Wheat Improvement) in Mexico, IRRI (International Rice

Research Institute) in the Philippines, and several similar operations. The Food and Agriculture Organization (FAO) of the United Nations and almost all bilateral aid-giving agencies are also influenced by these organizations and, in effect—sometimes in spite of their humanitarian rhetoric—do little more than propagate the interests of the multinational "agribusiness" corporations by driving the wedge deeper into the vulnerable socioecological environment of dependent Third World countries.

On the Indian subcontinent, the green revolution, spearheaded by so-called new "miracle" wheats, has caused a tremendous consolidation of land in the hands of large-scale commercial farmers as thousands of peasants are driven out to give way to the new need for mechanized cultivation and harvesting, uniform irrigation, and pest control, without which the seeds could not flourish.

Far from being the solution to all the problems of the rural areas they were purported to be by their front-runners (Brown 1970, 1971) these "seeds of change" have only succeeded in changing things for the worse. (Brown is one of the main proponents of the green revolution.) By depriving the masses of their subsistence livelihoods, the green revolution—or, more appropriately, the factory-farming approach—has violated not only the outer limits of the ecological system but also the inner limits of the population. (For an alternative point of view on the social effects of the green revolution, see Cleaver 1972.)

Another aspect of this phenomenon occurred in 1968 in Iran, where initial reports of impressive yields in other countries led to the introduction of "miracle" wheats. A new strain of virulent wheat rust, to which the native varieties had been resistant, began to flourish. The new varieties of wheat, which were supposed to be rust-resistant, began to succumb, and many farmers lost their entire crop. There were reports of widespread hunger (Farvar 1970).

In tropical regions like Central America and the Amazon region, a change to vast monocultures has been very destructive of the soil. Just a few years of mechanized agriculture in many areas of the Amazon basin have turned the lush forest into infertile deserts covered with a brick-like, leached soil called laterite (McNeil 1972). This practice of making a quick profit out of humid tropical lands continues as vast jungles in Brazil and elsewhere are opened up to profit-seeking corporations and to large-scale, private plantations (Young

1972). Little attention is paid to more compatible means of meeting human needs from tropical regions, such as "tropical gardening" and shifting agriculture.

The Central American example is instructive in other ways. Two decades of cotton production have eroded more than a millennium's worth of topsoil for the sake of quick cash. Monoculture has also encouraged outbreaks of insect pests that have alarmed the large growers and led to the widespread and intensive use of pesticides, much to the detriment of the environment and human health.

The position of economic dependence that Third World countries are forced into can be seen from the following example. About 2 kilograms of cotton bring $1.00 of foreign currency to Guatemala, of which $0.75, or three-fourths, immediately leaves the country to pay for synthetic insecticides, fertilizers, sprayplanes, and so on. Our calculations show that only 1 or 2 percent of the cotton grown in Guatemala could probably bring in the same amount of net foreign-exchange earnings as raw-cotton export, if finished products (say, Guatemalan shirts) were being exported to the same countries (Farvar 1972). If these shirts and other finished products were made locally as a cottage industry by existing indigenous methods, the household economy would improve and no expensive capital outlays of foreign-made machinery would be required. Even if more technologically advanced manufacturing techniques were employed, many jobs would be created for the local population.

Such an approach to development would also leave most of the area now devoted to cotton (the best and most fertile lands) free for other uses (such as food production), prevent erosion and poisoning of the topsoil, and result in even greater and more meaningful economic growth—provided the people are allowed to participate maximally in the development and benefit fully from their efforts. Here indeed is a case where an exogenous factor—the inherently unjust international economic order—intrudes on the local outer limits of the human ecosystem in a dependent country and prevents even the inner limits from being met by keeping the local population badly fed and housed and in despicable health.

Another instance of the interaction of social and ecological systems can be illustrated by incompletely executed land-reform programs. Over a decade ago, Iran passed its Land Reform Act, which provided for the division and installment sale of large land holdings among peasants in the country. Partly in an effort to encourage modern agricultural production techniques, an exemption was made for those who mechanized their land. One unfortunate result was a "tractor rush" by many large-scale landowners who possessed marginal lands. In rapid succession, pastures, forests, semiarid lands, and steep hillsides succumbed to the mechanical plow. The consequence was erosion and the destruction of many formerly productive pastures and other marginal lands unsuitable for mechanization. The social decision limiting land reform caused the local outer limits to be exceeded in marginal lands; the effects of this were quite widespread. Many nomadic populations witnessed helplessly the shrinking of their summer and winter pasturing grounds: the resources with which they had always met their minimum needs, and which had even supplied many animal products for the country, were fast shrinking.

Public-health schemes constitute another striking example of the interaction of social and ecological systems that affects the inner and outer limits.

Health, says the WHO, is the state of complete physical, mental, and social well-being. In other words, health is a state of balance among the individual's psychological, biological, and social environments. It is therefore closely akin to the state of development in a community. An underdeveloped society is less healthy; development itself, as we have defined it, is a precondition for health. The most effective action that can be taken to eliminate the conditions of unhealthiness is, in fact, action leading to the eradication of underdevelopment—to the redistribution of resources, and the improvement of the environment (Baeza et al. 1975).

Rather than attempting to do away with underdevelopment, the predominant practice has been to adopt the narrow approach of single-disease eradication. In spite of its broad, socially oriented definition of health, the WHO has been the standard-bearer of this concept, developing uniform methodologies for the narrow approach. A case in point is the global malaria-eradication program.

No other disease has marshalled as many public resources for its eradication as malaria. Soon after WHO was formed, a limited number of people, chiefly in the Rockefeller Foundation, were able to persuade the WHO to undertake a global, quasi-military operation, purportedly to wipe out malaria by eradicating the *Anopheles* vectors of malaria. They developed a uniform technology of malaria eradication consisting of four phases: preparatory, attack, consolidation, and maintenance. Even the terminology was taken from military jargon. Encouraged by initial success on a Mediterranean island, the malaria-eradication armies waged chemical warfare in country after country, conquering many villages and intruding on practically every household. Unfortunately, but predictably, the biosocial environment was too complex to allow a victory for this simple-minded holy war.

In Central America, nearly two decades and four insecticides later, we have reached the end of the road. In rapid succession, Dieldrin (sprayed once a year), DDT (sprayed twice a year), Malathion (sprayed three times a year), and Propoxur (sprayed four times a year) have induced physiological or behavioral resistance in *Anopheles albimanus*, the main vector. In Honduras, the incidence of malaria has reached or surpassed pre-eradication levels. In Iraq, southern Iran, Pakistan, India, and Ceylon, the resistance of insects to insecticides or the resurgence of malaria have brought the eradication attempts to a standstill. In Ethiopia, the recrudescence of malaria in the early 1960s caused some 150,000 deaths in an area where malaria had previously been virtually nonlethal: a temporary interruption of transmission had caused the natural immunity of the population to disappear. In South Vietnam, the resistance of parasites to drugs has become a major, insurmountable problem (Malagon, in press; Ahmed, in press; Farvar 1972; Banerji 1974).

In each case, a local outer limit has been reached, with dangerous implications for people in the Third World. A potentially even more serious transgression of local outer limits has occurred as a side effect of reducing malaria eradication to the mere spraying of insecticides: there is increasing evidence that the intradomiciliary spraying of organochloride insecticides has resulted in the unprecedented accumulation in human milk of DDT and its metabolites, benzene hexachloride, heptachlor, and other dangerous compounds in quantities reaching more than 12 parts per million in rural Guatemala, which is over four hundred times greater than the "acceptable daily intake" for human infants. Coupled with protein malnutrition (which is known from experiments with animals to increase drastically the toxicity of insecticides), this new danger threatens to wipe out any benefits to people in areas subjected to the insecticide-based eradication of malaria (Farvar et al., in press).

We now realize that malaria cannot be eradicated by the narrow, quasi-military approach; the conditions favoring transmission have to be changed. These conditions are just as much related to the physical environment as to the socioeconomic and political conditions of people. The boundaries of malaria coincide rather closely with those of the Third World. Where malaria has disappeared, it has been largely due to an improved relationship between man and his environment that is no longer conducive to the transmission of malaria. For example, better housing (including the use of window screens) can successfully interrupt the transmission of parasites by mosquitoes. A national policy leading to well-distributed development at the grassroots level may be far more effective in eliminating the scourge of malaria than any medico-military war on this disease. The masses could be mobilized to participate in habitat management and biological control schemes to wipe out the breeding grounds or reduce populations of the larvae of vector species, and to improve their housing. Some recent experiences by researchers of the Centre for Endogenous Development Studies in western Iran tend to confirm the utility of the habitat management approach.

The control of schistosomiasis is another illustrative case. This snail-borne blood-fluke disease in Africa has become as important a vector-borne public health problem as malaria. It is also a major threat in Asia and Latin America because of ill-conceived water-resources projects, which cause slow warm waters, such as irrigation canals and dams, that create habitat for the vectors of this debilitating disease. [The reader is referred to the paper by Emile A. Malek in this volume for a fuller treatment of the material in this and the next several paragraphs.—Editor]

In the Nile Delta, for example, what was a relatively minor disease affecting no more than a small segment of the population now has very high rates of occurrence, often reaching 75 or even 100 percent. The Aswan High Dam, while yielding some economic gains, such as electricity and permanent irrigation, has at the same time had a tremendous cost in morbidity and mortality due to schistosomiasis (van der Schalie 1972). The approach currently advocated by the WHO is similar to that described for malaria. It usually consists of chemical warfare against the snails or drug treatment of infected persons. The molluscicides (such as copper sulphate and Baylucide), besides having undesirable effects on fish and other aquatic organisms, have failed to control the disease. *Bulinus* snails, for example, have proved to be prolific and effective enemies. The medical approach, too, which usually is based on the use of antimony compounds to kill the worms in human blood, has had severe unexpected effects, including hundreds of deaths per year. A widely hailed drug known as hycanthone was shown recently to be potentially carcinogenic, in addition to producing resistance in schistosomes (Hartman et al. 1971, Rogers and Bueding 1971). Thus, all attempts at eradicating this disease have been unsuccessful.

The local outer limits have been reached in many countries. Headed by physicians with narrow training who followed a simplistic, nondevelopmental approach, the schistosomiasis-control programs aided by the WHO in most countries have unfortunately closely followed the abortive path of malaria eradication, in spite of the best of intentions. Meanwhile, the only country that has been successful in bringing this terrible disease under control is China, where the population was mobilized in vast, integrated programs of health education, habitat management, and snail collection and disposal. Very little use has been made of chemicals in the Chinese program (Cheng 1971); rather, schistosomiasis control has been one integral part of development as a whole.

Water-development schemes have violated local outer limits in other ways as well. The Kariba Dam in Zambia is a case in point. Its completion in 1963 displaced close to 60,000 people who had lived in the river basin. Their agricultural system was destroyed through the loss of land and by fluctuations in the level of runoff from the dam. An epidemic of sleeping sickness occurred, caused by the more favorable habitat for tsetse flies created around the artificial lake. The disease reduced cattle populations in some places to less than one-half their former sizes (Scudder 1972).

The example of Kariba Dam is a revealing one, one where a simultaneous transgression of both local inner and outer limits took place. The inner limits were the population's minimum livelihood, and the outer limits were the resistance of cattle to sleeping sickness. The main purpose of the dam had been the generation of cheap hydroelectric energy for the foreign copper companies in Zambia (the former Northern Rhodesia) and Rhodesia. The welfare of the people directly affected—those who lived in the basin of the Zambezi River—was barely considered. Newly independent, and under great pressure from international copper companies, Zambia could not resist the pressure and the temptation to have a huge, prestigious project at its doorstep.

In fact, reliance on large-scale, panacea-like development projects, such as single-purpose dams or single-disease eradication campaigns, has frequently been the reason why local outer limits are violated.

The Ecosocietal Approach to True Development

The foregoing discussion shows that concentrating on widespread, mass-based solutions would more effectively take into account the local needs (the inner limits) of a population, as well as avoid violating the local ecological imperatives and the natural carrying capacity (the outer limits) of a socioecological region. Orienting national development planning towards a policy of *redistribution* would also bring about a rapid achievement of the inner limits—the basic needs of the populace—a goal without which no development policy can be worthy of its name.

The history of famines and food policies in India amply demonstrates that redistribution can be more important in meeting the inner limits than increased production. During the worst famines in India—a legacy of British colonial rule—there has always been enough food, but it has not been distributed to those who needed it (Bhatia 1967, 1970). As long as a redistributive policy—an indication of social justice—is not implemented, the production of more food and other necessities only make the rich richer and the well fed better fed. The poor remain hungry. The conclusion seems almost self-evident that *an unjust social system is incapable of meeting the inner limits of all the people.* Yet many efforts at development in countries where social injustice, class privilege, and the inequitable distribution of wealth are dominant ignore this fundamental relationship between the social system and the distribution of environmental resources.

Post–World War II development policies in most Third World countries have usually increased the gap between the poor and the rich segments of the population (the "two nations" within a developing country)—which is the opposite of the redistributive characteristic or criterion of true development and which has made it more difficult to meet the inner limits of the population. Secondly, these policies usually have caused an unfavorable balance of nature, environmental degradation, and reduced carrying capacities of local socioecological systems by violating the outer limits. It can be seen, therefore, that development policies have caused a net underdevelopment at the mass level in many Third World countries, as judged on the basis of our two characteristics of true development.

We must therefore distinguish between mere growth and true development: classical development policies have mistaken growth indices (such as per-capita income, gross national product, and capital accumulation) for indices of development; as we have seen, true development must involve the redistribution factor and the environmental-improvement factor. Most of the time, we have had to contend with *growth without development.*

One corollary is that in a socioecological system whose resources are limited or whose outer limits have been violated (the case of most poor countries), it is even more urgent to start with the redistribution of resources (which is an indicator of true development) rather than with increased production through sophisticated technology (which is only a means of achieving growth). Only after sufficient redistribution has brought the population closer to its inner limits is it necessary to take advantage of the margin between the inner and outer limits by intensifying production.

In this connection, it should also be pointed out that the correct approach to development—which we might call the "ecosocietal approach" or the "redistributive-environmental strategy"—would make it irrelevant to discuss vague concepts of "limits to growth" (Meadows et al. 1972), which are usually based on computer models and theoretical calculations of "global outer limits." This pseudo-ecological approach reaches its most dangerous heights in Goldsmith et al. (1972) and in Goldsmith (1972). While it is true that certain global ecological relationships (for example, the nitrogen cycle or the carbon cycle) may ultimately be violated, these are mere abstractions from the Third World's viewpoint. Local outer limits are seen to be far more significant because they have an immediate bearing on the lives and well-being of the masses.

A promising new sign of progress in this direction is the United Nations Environment Programme's interest in what is called "ecodevelopment," which can be considered as equivalent to our definition of an ecosocietal or redistributive environmental approach to development, where a conscious attempt is made to use environmentally appropriate technologies ("ecotechniques") in need-oriented development strategies.

There is an enormous challenge in the redirection of development towards the ecosocietal approach, where the redistribution of resources and the improvement of environmental quality for the masses are the guiding principles. Only when we achieve those two goals will we succeed in meeting the inner limits of true development without violating the outer limits of local socioecological systems.

References

Adkisson, P. L. 1970. A systems approach to cotton insect control. In *Proceedings, Beltwide Cotton Production—Mechanization Conference* (Houston, Texas). Memphis: National Cotton Council of America.

Ahmed, Wasif. In press. Malaria eradication problems in South Asia. In *Environment and Underdevelopment: Third World Perspectives.* Proceedings of the AAAS Conference on Environmental Sciences and International Development.

Anonymous. 1972. *Washington Evening Star.* May 1972.

———. 1974. *The Coyococ Declaration.* Development Dialogue Number 2. Uppsala, Sweden: Dag Hammarskjöld Foundation.

Baeza, F., D. Banerji, V. Djukanovic, H. J. Geiger, O. Gish, M. T. Farvar, M. Rahnema, and G. Sterky. 1975. *Alternatives in Health—An Ecosocietal Approach.* Report of an Expert Panel for the 1975 Dag Hammarskjöld Project on Development and International Cooperation. Uppsala, Sweden: Dag Hammarskjöld Foundation.

Banerji, D. 1974. Social and Cultural Foundations of Health Services Systems. Special Supplement. *Economic and Political Weekly* [New Delhi] (August).

Bhatia, B. M. 1967. *Famines in India: A Study in Some Aspects of the Economic History of India, 1860-1965,* 2nd ed. New York: Asia Publishing House.

———. 1970. *India's Food Problem and Policy since Independence.* Bombay: Somaiya Publications.

Boza-Barducci, T. 1972. Ecological consequences of pesticides used for the control of cotton insects in Cañete Valley, Peru. In *The Careless Technology: Ecology and International Development.* Record of the Conference on the Ecological Aspects of International Development (Warranton, Virginia, 8-11 December 1968), eds. M. Taghi Farvar and John P. Milton, pp. 423-438. Garden City, New York: The Natural History Press.

Brown, Lester R. 1970. *Seeds of Change: The Green Revolution and Development in the 1970's.* New York: Praeger Publishers.

———. 1971. *The Social Impact of the Green Revolution.* International Conciliation Number 581. New York: Carnegie Endowment for International Peace.

Cheng, Tien-Hsi. 1971. Schistosomiasis in mainland China: A review of research and control programs since 1949. *American Journal of Tropical Medicine and Hygiene* 20(1):26-53.

Cleaver, Harry M., Jr. 1972. The contradictions of the Green Revolution. *Monthly Review* 24(2):80-111.

Conference on Problems of the Third World and the Human Environment. 1972. Declaration on the Third World and the Human Environment. [Issued by the independent 41-nation Conference on Problems of the Third World and the Human Environment, sponsored by the OI Committee International in parallel with the United Nations Conference on the Human Environment (Stockholm, 4-16 June 1972).]

Conway, G. R. 1972. Ecological aspects of pest control in Malaysia. In *The Careless Technology: Ecology and International Development.* Record of the Conference on the Ecological Aspects of International Development (Warranton, Virginia, 8-11 December 1968), eds. M. Taghi Farvar, pp. 467-488. Garden City, New York: The Natural History Press.

Farvar, M. Taghi. 1970. Green Revolutions: Ecological determinants of ecological development. Paper presented at the Williams College Conference on Economic Development and Environmental Problems (Williamstown, Massachusetts, September 1970). Unpublished manuscript.

———. 1972. *Ecological Implications of Insect Control in Central America: Agriculture, Public Health and Development.* Ph.D. Dissertation, Washington University, Saint Louis, Missouri. Ann Arbor, Michigan: University Microfilms. (Abstract published in *Dissertation Abstracts International* B-33(5):1939. Order Number 72-24219.)

———. 1973. Is there an environmental crisis in the Third World? *University Quarterly—Higher Education and Society* [London] (Summer).

Farvar, M. Taghi, M. L. Thomas, A. E. Olszyna Marzys, and M. de Campos. In press. Resídos de plaguicidas clorados en la leche humana en Guatemala. [Washington, D.C.] Boletín de la Oficina Sanitaria Pan-americana.

Goldsmith, Edward R. D., ed. 1972. A blueprint for survival. *The Ecologist* 2(1):1-44.

Goldsmith, Edward, Robert Allen, Michael Allaby, John Davoll, and Sam Lawrence. 1972. *A Blueprint for Survival,* by the editors of *The Ecologist.* Boston: Houghton Mifflin Co.

Hartman, Philip E., Kathryn Levine, Zlata Hartman, and Hillard Berger. 1971. Hycanthone: A frameshift mutagen. *Science* 172(3987):1058-1060.

Malagon, Filiberto. In press. Mexico's malaria problem. In *Environment and Underdevelopment: Third World Perspectives*. Proceedings of the AAAS Conference on Environmental Sciences and International Development.

Meadows, Donella H., Dennis L. Meadows, Jørgen Randers, and William W. Behrens III. 1972. *The Limits to Growth: A Report for the Club of Rome's Project on the Predicament of Mankind*. A Potomac Associates Book. New York: Universe Books.

McNeil, Mary. 1972. Lateritic soils in distinct tropical environments: Southern Sudan and Brazil. In *The Careless Technology: Ecology and International Development*. Record of the Conference on the Ecological Aspects of International Development (Warranton, Virginia, 8-11 December 1968), eds. M. Taghi Farvar and John P. Milton, pp. 591-608. Garden City, New York: The Natural History Press.

Rogers, Steffen H., and Ernest Bueding. 1971. Hycanthone resistance: Development in *Schistosoma mansoni*. *Science* 172(3987):1057-1058.

Scudder, Thayer. 1972. Ecological bottlenecks and the development of the Kariba Lake basin. In *The Careless Technology: Ecology and International Development* (Warranton, Virginia, 8-11 December 1968), eds. M. Taghi Farvar and John P. Milton, pp. 206-235. Garden City, New York: The Natural History Press.

Smith, Ray F., and Harold T. Reynolds. 1972. Effects of manipulation of cotton agro-ecosystems on insect populations. In *The Careless Technology: Ecology and International Development*. Record of the Conference on the Ecological Aspects of International Development (Warranton, Virginia, 8-11 December 1968), eds. M. Taghi Farvar and John P. Milton, pp. 373-406. Garden City, New York: The Natural History Press.

van der Schalie, Henry. 1972. World Health Organization Project Egypt 10: A case study of a schistosomiasis control project. In *The Careless Technology: Ecology and International Development*. Record of the Conference on the Ecological Aspects of International Development (Warranton, Virginia, 8-11 December 1968), eds. M. Taghi Farvar and John P. Milton. Garden City, New York: The Natural History Press.

Young, Allen. 1972. Ecology & development in Brazil. *Win* 8(13):10-13.

47. Interdependence: The Technological and Economic Aspects

Robert Muller
United Nations, U.S.A.

Introduction

The number, complexity, and significance of scientific, technological, economic, social, and ecological interdependencies is so great that the subject merits a new discipline: a Science of Interdependencies or Total Earth Science. Separate notions such as technology, economics, social science, and ecology are losing much of their distinctness. The separation of man from "natural" ecosystems would also be an error. Man is a part and product of evolution. A comprehensive scientific and ethical approach is needed to help maximize the physical, mental, and affective capacities of human beings within the limits of inherited conditions and the further evolution of our planet, which henceforth will be dominated by man. We are still far from such a concept of planetary science and ethics, but lately, several major elements have been pushing us in this direction.

Evolution of the Need for a World View

It took three billion years for our star or sun to reach the current stability of light hydrogen explosions which regulate most living conditions on our planet. Life appeared on Earth a few hundred million years ago. Since then, several geological upheavals have changed profoundly the physical conditions, flora, and fauna of the globe. Man evolved some three million years ago. His progress was extremely slow for most of this time and did not significantly affect the environment. Populations grew only in the favorable climatic conditions of Asia, the Middle East, and the Mediterranean, along great rivers. From this phenomenon we have inherited some major traits of the present-day population structure.

Two hundred years ago, the industrial revolution and man's mastery over energy, produced dramatic changes on our planet.

Populations increased rapidly in the northern, colder parts of the globe. The industrial revolution spread from England to continental Europe, Russia, North America, and Japan. Human growth, interventions, and physical transformations began to be a factor of change in the planet's environment.

After World War II, a new acceleration of phenomena with profound repercussions took place. Several waves of scientific and technological developments succeeded each other: atomic energy, aviation and jet propulsion, satellites and space technology, electronics, computers and cybernetics, automation, plant genetics and engineering, industrial agronomy, antibiotics, microbiology, laser technology, and so on.

The nonindustrialized two-thirds of the world were opened to the modern age, its agronomy and industrialization. Its people wanted better lives and longer lives too. The industrial revolution spread relentlessly to an increasing number of countries around the world, via international trade (machinery and technology obtained in exchange for raw materials) and assistance. In the 1950s, the western world taught these countries hygiene and ways to save the lives of their infants. The death rate fell dramatically from an estimated 29-to-34 per thousand in 1939 to 24 per thousand in 1950, and 16 per thousand in 1970, while fertility declined only from 45 per thousand before the war to 41 per thousand in 1970 in these countries. This "baby survival" of the fifties accounts for the fact that today close to 50 percent of the population of the developing countries is less than eighteen years old. By contrast, population growth in the affluent world slowed down during the same period.

The aspirations of the developing countries and the world population problem were the primary global issues brought before humanity's first global institution, the United Nations. Worldwide statistics were assembled on a large number of subjects. Censuses were taken in countries which had never taken a population count. U.N. demographers came up with a world population figure of 2.5 billion people in 1950. Forecasts were made in many fields, especially population, food, health, education, housing, energy, and natural resources. The first U.N. Conferences on the Utilization and Conservation of Resources (UNSCCUR) and on New Sources of Energy were convened in 1949 and 1961 respectively, but they had no impact. The atmosphere was not yet ripe.

In the 1960s, the more numerous independent developing countries in the United Nations stepped up their pressure for economic development in world trade and development conferences and all U.N. and international bodies. In the early 1970s, unprecedented global problems came to the fore in a series of resounding world conferences and crises which opened humanity's eyes to the new conditions prevailing on our planet:

1. 1972—The U.N. Stockholm conference on the environment, after the U.N. Educational, Scientific and Cultural Organization's (UNESCO) earlier conference on the biosphere, lays bare a new view of the world and an unprecedented planetary danger: the potential destruction by man of vital environmental elements. The conference has a deep echo around the world. Ministries of the environment are created in most countries and a U.N. Environment Programme (UNEP) is established.
2. 1972—A book entitled *The Limits of Growth* appears, which has a profound impact in many quarters. In it, daring and oversimplified assumptions show that a world of seven billion people with the average standard of life in America today is practically impossible from the environmental and resources point of view.
3. 1972—Adverse climatic conditions bring the world close to a food shortage and reveal that the human species (now approaching 4 billion people) may be living on a dangerous margin of food reserves. A world food conference is convened in 1974. It establishes a ministerial world food council, calls for the maintenance of food reserves, and adopts a plan for accelerated world agricultural development.
4. 1973—The first world energy crisis occurs. A special session of the U.N. General Assembly is convened in the spring of 1974 to deal with the problem of raw materials and their remuneration in relation to industrial products. The developing countries call for a new international economic order.
5. 1974—The first world population conference is held at the government level. The beginnings of a deceleration in population growth are in sight. The issues of population,

development, resources, and density take a position of paramount importance.

New global problems appear in close succession in the 1970s and become major topics of further U.N. world conferences:

1. A U.N. conference on the law of the sea tries to establish a new order and regime for the seas and oceans.
2. A conference on human settlements and a world employment conference is scheduled for 1976.
3. A world water conference and a world conference on the deserts are planned for 1977.
4. New conferences on science and technology and on outer space are envisaged thereafter.

In the meantime, multinational corporations are becoming a substantial factor of change in a large part of the world. From trade they move increasingly into international production and distribution. Control of these companies is requested by several developing countries. In early 1975, the U.N. Economic and Social Council established a commission and center to deal with this new problem.

At the present time, the continuing upsurge of the industrial and technological revolution, while having not yet reached the entire world, is encountering serious constraints. Economic growth has abated. Many countries, after two hundred years of acceleration in many directions, seem to have reached a moment of pause, evaluation, and interrogation.

Some Major Technological and Economic Interdependencies

The subject is immense. Only a few major interdependencies will be highlighted.

Outer-Space Technology

No other technology has revealed more dramatically the natural interdependencies of our planet. Instruments probing other planets of our solar system have helped develop a planetary science which gives us a better understanding of the physical structure, life chemistry,

The Interaction of Societies and Ecosystems

and equilibria of the Earth. Considerable progress has also been made in solar science and observation. Renewed interest in the central role of the sun for the production of food and other energy, and in human settlement has also developed. Two years ago, a UNESCO conference on the relationship of man and the sun reviewed our knowledge in this field. Perhaps a special science on Earth-solar relations would now be called for. Since the sun is the principal energy source of food production, agriculture should be conceived as a part of a solar science and industry. Carbohydrates industrially produced from petroleum, e.g., rubber, are naturally produced by the sun in the hevea tree. The distinctions between agriculture and industry are diminishing. If some day solar satellites are placed in orbit or stationed on Earth, a new energy interdependence between various regions of the world will be created.

Space technology is also our main instrument for a total environmental and climatic observation of the globe. Thanks to this technology, long-term climatic changes will become detectable; a better knowledge of atmospheric conditions and laws is now possible; and natural disasters are henceforth easier to forecast. A fuller understanding of ocean currents, ocean temperature, earth temperature, desertification, world water balance, ice caps, and atmospheric pollution has been rendered possible through satellite observation. Over time, such observations will contribute immeasurably to man's life and survival on Earth.

Satellites are also accelerating world interdependence through telecommunications and education (Indian education satellite, African and Latin American UNESCO projects). However, the question of direct broadcasting of TV images into sovereign territories has not yet been resolved, and thus far, it has not been possible to devise a common world satellite telecommunications system (INTELSAT and INTERSPUTNIK are two distinct international systems).

There are now more than a dozen countries operating their own satellites. Over one thousand space objects circle around the Earth. Except for the moon and other celestial bodies, for which draft treaties are still under negotiation at the United Nations, a legal regime has been adopted for outer space. This area has been declared a common heritage of humanity. Matters such as the prohibition of weapons, the responsibility for damage caused by space objects, the contamination of space, cooperation in space, and the return of astronauts are legally regulated. The U.N. outer-space agencies, the International Telecommunications Union, the World Meteorological Organization, and the U.N. Environment Programme, provide extensive forums and facilities for cooperation in this new form of world interdependence.

The Atmosphere

The interdependence of the atmosphere, be it among its various layers or its relations with the Earth's water and land masses, finds its expression in the Global Atmospheric Research Program of the World Meteorological Organization (WMO) and in other extensive meteorological and atmospheric programs of the WMO. The U.N. Environment Programme (UNEP) has also been given important responsibilities for the monitoring and environmental protection of the atmosphere. The potential effects of supersonic transport on the ionosphere has not been discussed before international institutions. Some nations have unilaterally prohibited supersonic flights over their territories, for example Austria, Sweden, Switzerland, Netherlands, and the United States.

Aviation, one of the fastest growing technologies and economic developments the world has known in recent years, has had a wide array of repercussions and ramifications on tourism, human settlements, trade, and the environment. World-passenger-kilometers have increased from 60 million in 1955 to 646 million in 1974. Freight kilo-kilometers have increased from 907 million in 1955 to 1.75 trillion in 1974. This phenomenal increase abated somewhat in 1974 because of economic conditions, but aviation will certainly continue to be one of the technologies with the deepest, most far-reaching repercussions on world interdependence. The question of its long-term trends and effects for humanity is particularly complex. The International Civil Aviation Organization, created after World War II, provides a forum within the United Nations for intergovernmental policies on safety, standardization, common facilities, uniform legislation, pollution, and noise abatement.

The Biosphere

The concept of the biosphere and its interdependence has received wide recognition on the international level through UNESCO's Conference on the Biosphere, its ongoing worldwide program on Man and the Biosphere, and the comprehensive biospheric approach of UNEP. The collection of data on the biosphere's conditions have been launched by governments under UNEP, a development that very few persons would have forecast only a few years ago. Observations and data, such as those of the International Hydrological Decade, are being fed into the total program. Humanity is rapidly organizing itself to keep total biospheric conditions under review, to receive warning signals, and to consider proper corrective action. UNEP publishes a yearly report on the state of the environment. Its Project Earthwatch, composed of a Global Environment Monitoring System (GEMS) and an International Referral System (IRS), provides early warning of significant environmental risks and opportunities and ensures that governments have access to the best possible knowledge, opinion, and experience. The task is no less than to evaluate the total effects of man's activities on the biosphere. The recognition of the biosphere's incredibly numerous and complex interdependencies has opened a new age of evolution and has probably saved the human race from serious disasters.

The Seas and Oceans

Rapid technological development of seabed exploitation has led the U.N. to declare the seabed beyond the limits of national jurisdiction as a common human heritage, and to attempt to adopt an international order for the seas and oceans, which cover two-thirds of the planet's surface. The developing countries possess most of the seashores and wish to extend their economic, if not political, sovereignty to 200 miles (about 370 kilometers). The developed countries possess the technology, and they are the ones pressing for the establishment of an international authority to regulate exploitation of the seabed, whose fantastic mineral resources may endanger their exports of raw materials from land that recently commanded better prices. The whole gamut of problems relating to the world ocean, from its numerous uses and scientific investigation to environmental conservation, are before the U.N. Law of the Sea Conference, one of the greatest and most complex international undertakings ever attempted.

Industry, Trade, and Agriculture

The developing world is making rapid strides in industrialization, often under the impact of multinational corporations, which find favorable labor markets and outlets there. (Nearly 25 percent of the gross national product of the nonsocialist world passes through international business.) In view of increasing populations and local consumption, developing countries have now a greater tendency to retain their raw materials and to process them locally. For environmental reasons too, processing at the source is gaining ground. Recent increases in energy prices may also affect the international transport of unprocessed raw materials. For example, the international maritime trade of iron ore increased from 27.2 billion kilos in 1950 to 24 trillion kilos in 1970. But such ore contained a high proportion of useless rocks, an economic system which will become less and less feasible as the pressure on resources intensifies. Thus, Japan, which for environmental reasons has great difficulty in expanding its blast-furnace capacity, has decided to build an iron and steel plant in Brazil. This plant will produce from local iron ore semifinished products for shipment to Japan, where they will be transformed into more refined products. Several similar plants are envisaged. The pattern of trade is likely to undergo significant changes in the future. Its growth has been phenomenal—a 500 percent increase in twenty years. The economics of world production, consumption, and trade are likely to lead to a wider distribution of industries. Transportation costs will have a determining influence. This is being accompanied by a struggle by the developing countries for higher raw materials prices, cartel formations between producers, a link between raw material prices and industrial goods, and a request for an accelerated transfer of technology.

On the agricultural side, the world depends in large degree on the exceptional climatic position and high productivity of a few countries: Canada, the United States, Argentina, and Australia. Thus, the 1972 food crisis has led to a jump of exports of principal agricultural products from $40 billion to $60 billion. But new interdependencies appear here too in the wake of the energy crisis: high productivity of agriculture in the main food-producing countries is largely due to its high energy content (machinery, petro-derived fertilizers). Thus, while surface productivity of corn has trebled from 1945 to 1970 in the United States, in terms of energy, one kilocalorie of energy input yielded 3.8 kilocalories of corn in 1945, but only 2.7 kilocalories in 1970. Coupled with population increases and higher transport costs, this speaks for an urgent increase of agricultural production in all countries, especially in the developing ones. Major policy changes have been initiated as a result of the food shortage and energy crises. International trade requires serious reconsideration in the light of environmental problems, the pattern of economic growth, and transportation costs. Greater self-sufficiency will be emphasized as a trend.

Raw Materials and Energy

The recent crises were foreseeable and were basically the result of a combination of rapid economic growth in the developed countries and the beginning of a serious take-off in several developing countries. The total share of the developing countries' consumption of raw materials and energy has been increasing more rapidly in recent years. The interdependence between population, economic growth, energy needs, and consumption of raw materials is one of the most troublesome interdependencies which has appeared lately on the scene of world evolution and history, and no global policy or vision is yet in sight beyond the grouping of the various competing parties around their respective interests. New world views, concepts, commitments, and mechanisms are urgently needed to better explore, exploit, distribute, manage, and conserve our planet's resources for the benefit of the whole human population, present and future.

Transport and Tourism

The automobile has created dramatic new conditions on our planet and is largely responsible for the opening of new land for settlement and the fantastic development of tourism, especially in the planet's coasts, mountains, and warmer regions. The ecology of these regions is being profoundly transformed. In 1973, foreign tourist arrivals alone, which are generally but a fraction of total tourist movements (around 25 percent), were estimated at 215 million. This was 8.5 times the 1950 arrivals of 25.3 million. Tourism has been a substantial source of income for many countries: aggregate foreign tourist receipts in 1973 were $27.6 billion, excluding expenditures on transport which represented another $7 billion or $8 billion. On the other hand, the adverse effects of tourism seem to be on the increase. In Europe, there is a growing concern that tourism often kills tourism by destroying its original attraction, namely unspoiled natural conditions. At the present rate of growth, French coasts will be packed with 20 million tourists in the year 2000. There were 20,000 pleasure crafts in France in 1950, 220,000 in 1970, and 400,000 in 1974. In Italy, along the mere 50 kilometers of coastline between Cervia and Cesenatico on the Adriatic, there are 3,300 hotels.

The automobile civilization is also spreading rapidly to the developing countries. In 1973, out of a total of 29 million new passenger cars, 17 developing countries produced and assembled 1.2 million units. If a world of 6 billion people were to use the same number of passenger vehicles per population as the United States, the number of cars would reach 2.5 billion.

Water Resources

In some areas of the world, water shortage has been and continues to be the main impediment to settlement and development. In others, water use and misuse is increasing rapidly. Water consumption has increased tenfold in the United States from 1900 to 1970. World water consumption is estimated to increase three to four times before the end of the century: ten times in municipal use, two times in irrigation, fifteen times in industry, twenty times in energy. Even though total water use will represent only one-eighth of the total water precipitation on continents, water is likely to create increasing problems in highly industrialized, populated, and urbanized areas. The problem of pollution is particularly serious with respect to underground water resources, which represent 98 percent of all fresh liquid water on our planet. A world water conference will be convened by the U.N. in Argentina in 1977 in order to provide a global diagnosis of the problem and recommend world and national action. The U.N. Conference on Human Settlements will draw urgent attention to the better management and use of water in our rapidly developing urban civilization. Water will no longer be a cheap and unlimited commodity in an increasing number of courtries. Here, too, the earlier a planetary resources policy is devised the better it will be for us and for our children.

Human Settlements

One of the most disturbing tendencies on planet Earth is the rapid and unprecedented movement of populations from rural to urban areas. By the end of this century, more than half of the world population will live in urban areas. This will mean an increase of 2 billion city dwellers. There will be over 300 cities with more than 1 million inhabitants in the year 2000 as opposed to 11 in 1923. The trend towards urbanization will be particularly active in the developing countries, while in some developed countries decentralization policies or trends are observable. U.N. experts project that of the twelve largest cities in the world in 1985, eight will be in developing countries. Of the twelve fastest growing cities in the world in the next decade, all will be in developing countries—four in Latin America, seven in Asia and the Pacific, and one in Africa. These twelve fastest growing cities will increase in population from 46.5 million to 106.9 million between 1970 and 1985. Sixteen cities will be "megacities" of more than 10 million people in the year 2000. Ten of them will be in developing countries.

A concentration of population and activities on coasts is also observable: two-thirds of humanity live near the coasts. The competing claims of industry (steel plants, refineries, ports), human settlements, fisheries, tourism, and the environment are causing increasing difficulties in these areas, and the United Nations has been holding meetings on this subject.

Other Interdependencies

Innumerable other technological and economic interdependencies are mushrooming on planet Earth, some of them of extreme importance, such as: the interdependence between the level of economic activity in the highly industrialized centers and economic conditions in the developing countries; the interdependence of economic and social aspirations throughout the world; the birth of an interdependent world public opinion, fashions, advertising, imitation, modes of living; the birth of computers and information networks; the growth and development of world organizations; and the emergence of transnational activities, associations, trade unions, and corporations.

Many of these interdependencies involve adverse consequences such as the spread of military techniques and atomic armaments; chain reactions of new social evils, such as alcohol, car accidents, and the increasing number of handicapped; the adverse effects of many so-called economic advances on the quality of life and health of the individual (medicine is more and more on the defensive and is repairing conditions over which it no longer has any control); environmental diseases instead of bacteriological diseases (accidents, cancers—85 percent of which are of man-induced or of environmental origin—etc.); the emergence of world moods and psychoses; and so on.

The subject of interdependence is immense and is one of the most striking characteristics of the present stage of planetary evolution. Even countries which make determined efforts to isolate themselves from detrimental interdependencies are sometimes unable to do so—the USSR and China, for example, which were hit by adverse climatic conditions and forced to purchase huge quantities of wheat from the United States. No society on Earth exists any longer in splendid isolation. Biological, physical, and natural interdependencies inherited from past evolution and inherent in the basic chemistry of our planet are now accompanied by an extremely dense network of man-made interdependencies which are changing the fabric and mechanics of our planet.

Man adapts himself progressively to the new conditions. As labor productivity resulted from the demands of labor, so will an energy productivity arise from the shortages of energy. Similarly, waste and built-in obsolescence are under attack. A no-waste technology and production is being born. Crises can be great blessings in the long run by changing wrong courses or habits. The technological and economic response to the environmental and various other recent crises will change and adapt man's behavior within the constraints of the biosphere and local ecosystems. Any delays or resistances will only be punished by a new crisis and a change in power structure. This applies equally to nations and corporations. There is no way out except a new civilization that will reduce waste, maximize the quality of life for all, and find new technologies tailored to the newly discovered planetary constraints. There will of course continue to be a lot of mischief, going it alone, and attitudes of *"après moi le déluge,"* but even those illusions of freedom and power will sooner or later come to an end. Public authorities will be under increasing pressure to induce and initiate proper corrections and controls.

Interdependencies with the Past and the Future

During this time of unprecedented global problems, we are forced to reach deep into the past and the future. Never has the time range of the human species' total brain power been stretched so far. The species is acquiring a sense of responsibility for the changes it has brought to evolution and to the inherited conditions from our planetary past. This explains the widespread concern of conservationist groups and associations, such as the International Union for Conservation of Nature and Natural Resources (IUCN), the National Audubon Society, and the Sierra Club. It is a deep, real, and well-taken concern. The disappearance of any one living species on Earth through man's action is a grave event in the total, as yet not understood pattern of evolution. If man had found one animal on the moon, millions of dollars would be spent on its conservation! But when it comes to planet Earth, everything is taken for granted.

Fortunately, in addition to the outer-space and ocean heritages, the concepts of natural and cultural heritages of mankind have also been gaining considerable support. Man is beginning to realize that his equilibrium and happiness require a sense of understanding of and participation not only in nature but also in the total stream of history. Many national and international efforts are flowering in order to preserve a past which will increasingly be precious to mankind as time passes. At Abu Simbel, in Egypt, the dilemma "preservation *or* development" has been met by the answer: "preservation *and* development." The same ideal is spreading around the world, seeking for man greater fulfillment and a better equilibrium with his past and total evolution. U Thant's Buddhist ideal of unity in diversity should really serve as a target for the human society on Earth—biologically, culturally, and politically—for it seems to be one of the basic laws of our planetary evolution.

Even greater strides are being made with regard to the future. After World War II, planning in government or private enterprise seldom exceeded a ten-year period. Even in the United Nations and its specialized agencies, forecasts were seldom made for a period of more than a decade or the rest of the present century. All this has changed, to witness:

1. Futurology was born in academic and private circles. It mushroomed and produced a series of "future shocks" such as *The Limits to Growth*. Individuals thus showed greater vision and dimension of mind than institutions. Their efforts soon influenced governments and institutions, for correct ideas know no barrier. Today, there are

few statesmen who are not to some extent futurologists. The increase of the time dimension has simply become part of the new normal intellectual perspective of the planet and of the intensity of developments and the repercussions of present actions or inactions on future generations. Planning for short periods of time is becoming ever more self-defeating and unsatisfactory for the modern human mind, enrobed in its totally new global and long-term planetary conditions.

2. In world institutions too, the time range of forecasts has substantially increased. Rather than allow the spread of private doomsday population extrapolations predicting a world of hundreds of billions of human beings who would destroy the planet, U.N. demographers have decided to provide long-term projections under various assumptions. The latest U.N. projections for the year 2075, issued in February 1975, give as a low, a population of 9.4 billion in that year, as a medium assumption, 12.2 billion, and as a high assumption, 15.8 billion, with a stabilization of the world population at 16 billion at most. The problem is sizeable enough when we think that we are now 4 billion and 29 million people. But at least we know the order of its magnitude and that a geometric catastrophic progression of the world population is already out of question.

3. The time dimension of governmental preoccupations increases year after year. Who would have thought a few years ago that the question of long-term climatic changes would be brought to the United Nations? At the request of the General Assembly, the World Meteorological Organization has consulted the scientific community on this problem. The scientists were divided on the question of whether the Earth was in a cyclical period of cooling which would end around 1980, or whether it was heading towards a new ice age. For, inconclusive as the consultations may have been, at least the international community has decided to look systematically into the problem and to join scientific efforts. The WMO is developing plans for an integrated international effort for the study of climatic changes and their consequences on man's natural environment and on world food production—attention being given to phenomena on time scales from a few months to 100 years.

4. A similar international effort to better understand, study, and observe the movements of our Earth's crust—the international geodynamic program—will be helpful to foresee natural disasters which cost so much to humanity. In the last ten years, there have

been 430 natural disasters in the world, resulting in about 3.5 million deaths, 400 million victims, and damage estimated at $11 billion. The physical effects of dams and of petroleum exploration in coastal areas will be better assessed. The establishment of an Office of United Nations Disaster Relief Coordinator, with responsibilities to study and prevent natural disasters, is another hopeful step in the right direction. Preoccupations about our planet as a whole and our future are no longer merely matters for scientists; they have entered the realm of governmental concerns. The establishment of the United Nations University, with affiliates in various countries, concerned with the world's global problems and having direct access to the United Nations' political assemblies, is another important bridge between the scientific and the political world (similar to what has happened already in the field of the environment through UNEP).

5. The lessons and findings of all sciences begin to trickle through the whole human and institutional fabric. The tremendous advances in astrophysics during the last two decades are shedding a new light on our planet's conditions and evolution. Even if *The Limits to Growth* is wrong in the short term, its conclusions leave a deep impact when one thinks that our little planet will still continue to circle for another six billion years or more around its sun. There is no rush to do everything overnight, nor is there any reason to despair. People and governments from north, south, east, and west; rich and poor; black and white are learning to work together in a new planetary assessment and understanding and will surely resolve the seemingly colossal problems that confront us today. During the last few years the immense strides in science and technology and a series of emerging crises have brought humanity nearer to such an approach. A completely new era of history and evolution accompanied by profound changes in many of our beliefs and basic concepts is foreseeable.

A phenomenon of acceleration has seized man's three main abilities: knowledge, imagination, and organization. Man has made greatest progress in knowledge. In imagination, i.e., in discovery, production, and transformation, through the combination of knowledge and "natural" factors, he is still on the crest of an industrial and technological revolution, the ultimate consequences of which he has not thought out and which have not even reached a large segment of the world. Mankind's managerial ability leaves most to be desired: political progress toward peace, world order, resources management, and optimum individual human lives everywhere are still far out of sight. An immense amount remains to be done. To start with, mankind's global instruments of diagnosis, knowledge, forecasting, warning, and concerted action must be urgently and considerably strengthened.

To this speaker who has lived for almost thirty years in the midst of the transformation of international institutions and world affairs, the United Nations and its specialized agencies are the best available hope and building block toward a global and concerted management of planet Earth. The U.N. system is more and more becoming a central data bank, an incipient brain, a warning system, a nervous system, a heart, and a meeting ground for the definition of a common ethical behavior and fate of the human species. If political leaders understand the new age we have entered, if they set aside their anachronistic differences and guide together the world's huge population on its evolutionary path, an unprecedented era of human fulfillment and happiness will be within our reach.

We stand on the threshold of a new effort to better understand the workings and reasons of human life and to seek forms of happiness which may not be quite synonymous with today's prevailing concepts. The human species will not perish. A collective brain and warning system are being born to it. A new ethics of what is good or bad for humans on this planet is in the making. The world's interdependence, at long last obvious and understood, inspires hope for greater world peace, progress, justice, and order.

48. The Exporting and Importing of Nature

Roderick Nash
University of California, Santa Barbara,
U.S.A.

Introduction

It was a hot July in the Michigan Territory of the United States in 1831, and the settlers of this underdeveloped nation could not believe their eyes or ears: an elegant Frenchman, possibly a nobleman, who called himself Alexis de Tocqueville, wanted to be taken into the woods for fun! Indeed, he had traveled nearly a thousand miles from the eastern seaboard specifically to reach the frontier and experience wilderness. At first, the pioneers naturally assumed that his purpose was lumbering or farming or land speculation. They gathered survey equipment and prepared to take Tocqueville to choice locales. But the Frenchman said "No, thank you." He just wanted horses, food, and a guide who could show him wild nature. He just wanted to look!

Shaking their heads in disbelief, but willingly accepting Tocqueville's money, the Michigan settlers entered the nature-tourism business. They organized a pleasure trip into the wilderness. In retrospect, their attitudes interested Tocqueville as much as did the Michigan landscape. The American pioneer, he reasoned, lived too close to wilderness to appreciate it. On the other hand, people from developed nations, like himself, had built a nearly omnipresent civilization. For them, wilderness was a novelty whose attractiveness increased in proportion to its scarcity (Tocqueville 1959, p. 335; Pierson 1959, pp. 144-199).

Tocqueville's Michigan experience illustrates an axiom of environmental history: nature appreciation, and particularly nature

---

This paper, a preliminary statement in a comparative study of the cultural significance of national parks and nature reserves around the world, was facilitated by a grant from the Natural and Environmental Sciences Division of the Rockefeller Foundation.

protection, are characteristic of highly civilized (some would say "over-civilized") societies. They are full-stomach phenomena. A culture must become civilized before nature preservation becomes economically and intellectually viable. Development is the key to appreciation of the undeveloped.

From this follows the irony of global nature protection. It might be expressed by stating that the development that imperils nature is precisely that which creates a need for nature. In other words, the cultures that have wild nature don't want it, and those that want it don't have it. There are no shortcuts here: the road to the appreciation and protection of nature leads inevitably to and through a highly sophisticated, technological, urbanized civilization. (A pause for definitions: "uncivilized" or "undeveloped" in this discussion should always be taken to have an economic sense; "nontechnological" or "nonurban" are equivalent forms. I also have reference to nations that are "partly" civilized—that is, nations with frontiers and outbacks, such as Canada and Australia. There is no implication that such societies are lacking in other ideas and institutions that define human civilization.)

The corollary to this thesis is that on a world level the primitive is an actively traded commodity; there exists what might be termed an export-import relationship between the wilderness "haves" and the wilderness "have nots." Before development, or during development, a culture or nation is a wilderness exporter. It "sells" wildness to visitors from the developed nations. Nature does not, of course, physically leave the country except in the case of animal trophies. The more common form of export today is through the minds, spirits, and cameras of tourists, but there are also "armchair" tourists who derive pleasure simply from the knowledge that unspoiled nature exists. Their concern about places they may never see has been an important source of support for the protection of nature throughout the world.

Conversely, the developed nation finds its wildness depleted, and therefore "imports" it from less developed countries. The payment is in the currency that tourists like Tocqueville eagerly spend. The philanthropy that funds world nature-protection organizations also constitutes the importation of nature, as does the purchase of books, films, and television specials on foreign wilderness. National parks, wilderness systems, and even the personnel to manage them might be thought of as the institutional "containers" that developed nations send to underdeveloped ones for the purpose of "packaging" the exportable resource.

The Nature Market

It is certainly true that underdeveloped nations may eventually evolve to the economic and intellectual position in which nature protection becomes important. But in the meantime, the preservation of wild places and wild things in the developing nations depends on the existence of developed nations and of the world nature "market."

To be frank, nature protection is the game of the rich, the urban, and the sophisticated: they are the clientele of wildness wherever it exists in the world. They subsidize the decision of the underdeveloped world to protect nature. Without such subsidies, the chances of wildness would be poor in the face of the aspirations of the developing world.

The existence of this export-import relationship is frequently recognized in the discussion of international concepts of nature protection. It is the basis of the idea of a "world heritage trust," expressed most frequently in the United States by Judge Russell E. Train, the present administrator of the U.S. Environmental Protection Agency, and the former chairman of the Council on Environmental Quality in the Executive Office of the President. The principle here is that the developed nations should take steps to ensure the preservation and proper management of extraordinary natural areas in the rest of the world. Of course, this means financial underwriting as well as technical assistance and even political jurisdiction. It is the last point that has kept the concept from getting off the ground: even if an internationally controlled "world park" comes with money attached, few nations welcome it as an enclave in their midst. The developed world, on the other hand, cannot be expected to invest in projects that are vulnerable to political change or economic need. Still, the point that the whole world has in interest in, say, East Africa's wildlife--that the wildlife does not "belong" exclusively to East Africa—is gaining acceptance. Late in 1975, a Convention Concerning the Protection of the World Cultural and Natural Heritage came into force under the auspices of the United Nations. While evading the main point, the enforcement of nature protection, and avoiding the issue of international ownership, the convention marks the first institutional involvement in the exporting and importing of nature (Train 1974, UNESCO 1972).

While Princeton-educated, safari-addicted Russell Train is clearly at the importing end of the spectrum, it is possible to find wilderness exporters with the same ideas. Perez Olindo, the black, American-trained director of Kenya's national parks, also believes that the developed world should exercise its desire for nature protection in the underdeveloped world. Specifically, Olindo proposes that wealthy Americans, frustrated at their inability to establish a tall-grass prairie national park in the Middle West, create the park of their dreams on the grasslands of Kenya (Olindo 1975).

Finally, it is interesting in this regard to hear Reginald Hookway, director of the Countryside Commission for England and Wales, admit that his own country has no more wilderness (Hookway 1975). But, he quickly adds, the English like wild nature, and those who want to experience it simply travel to Norway, Africa, or New Zealand. When challenged that this really is not so simple, Hookway explains that such a trip doesn't amount to much more than a New Yorker journeying to California's High Sierra or Arizona's Grand Canyon. So it is that England's conception of "its" parks and reserves is international. The English are eager importers of nature and glad to support its preservation around the world with time and money. Indeed, many of the international nature-protection movements—particularly those concerning Africa—began in Great Britain.

Wilderness Protection and the United States

For almost the entire nineteenth century, the United States was a classic developing nation characterized by large amounts of undeveloped (wild) land and, in general, exploitative attitudes toward it. There were, to be sure, a handful of Americans sufficiently urbanized to care about nature, but foreigners like Tocqueville led the way. Before him, for example, there was François Auguste René de Chateaubriand, who reveled in the wilds of northern New York and eastern Kentucky in the winter of 1791–1792 (Chateaubriand 1816, pp. 138 ff.). Lord Byron did not visit the New World but celebrated Daniel Boone as a wild man in a wild environment (Beach 1936, Rutherford 1961, and Lovell 1949).

In the middle decades of the nineteenth century, a number of foreign visitors to the United States and its western territories had big-game hunting in mind. With elaborate equipment and large retinues, they shot their way through a West that they clearly regarded as a kind of enormous personal hunting reserve. The presence of Indian guides and white scouts completed the quasi-feudal atmosphere of these wilderness pleasure trips. The best known examples are the expeditions in the 1830s of William Drummond Stewart and Alexander Philipp Maximilian, the Prince of Wied-Neuwied (Porter and Davenport 1963, Wied-Neuwied 1843; for western tourism in general, see Pomeroy 1957).

By the end of the nineteenth century, the proportion of wilderness to civilization in the United States had shifted dramatically. Frederick Jackson Turner made this not only clear but a subject of widespread public concern with his essay on the significance of the 1890 census statement that the frontier no longer existed. It was not the case, of course, that all America was ploughed and paved. Pockets of wildness existed, and some were already protected in national parks, but the spectacle of truly vast wildlands populated with free-ranging big game and Indian hunters had been reduced to memories. The upstart American nation had made it to "developed" status. It had also invented national parks (Nash 1970), but for those who knew the continent's former wildness, Yellowstone (established 1872) and the later parks were only vestiges. Still, it is instructive for the present purposes to note the eastern, urban, artistic, college-educated roots of the national-park idea in men such as Cornelius Hedges, Nathaniel P. Langford, and Thomas Moran. Also significant is the support given the Yellowstone Act of 1872 by railroad entrepreneur Jay Cooke. His eye, clearly, was on the tourist revenue a national park would generate, and the same argument was used to win support among the local population in Wyoming and Montana. In a harbinger of what would occur between nations, Americans from the developed East were already "importing" wildness from the still undeveloped West. For the Yellowstone story, see Bartlett (1974) and Haines (1974).

Theodore Roosevelt and the Wilderness

A change in attitude, favoring wildness, quickly followed the change in America's environmental condition. (Nash [1973] discusses the emerging "cult of the primitive" and cites relevant supporting literature.) So did the start of wilderness "importing." With their own wilderness shrinking rapidly, a stream of wealthy American individuals—the counterparts of Tocqueville—began to flow toward the world's remaining wild places for adventure and pleasure. Theodore Roosevelt is, of course, the classic case in point. His early personal history was thoroughly upper class, urban, and civilized: New York City and Harvard. In 1883, Roosevelt went West and watched the last remnants of the frontier die in the Dakotas (Cutright 1956, Hagedorn 1921, Putnam 1958, and Pringle 1956). Later, working his way up the political ladder in the East, Roosevelt suffered wilderness starvation. In 1888, he joined with other members of America's social and economic elite to form the Boone and Crockett Club, where gentlemen hunters could at least talk about wild land and wild animals (Trefethen 1961, pp. 24 ff.). Roosevelt continued to travel to the wildest places he could find. In fact, when President William McKinley was dying from an assassin's bullet on September 13, 1901, frantic couriers had to race to New York's Adirondack Mountains, where Vice-President Roosevelt was climbing in the wilderness. A panting guide handed him the telegram on a mountain trail!

But the American environment of the early twentieth century could not satisfy Theodore Roosevelt's appetite for wildness. In 1909, he began a year-long safari to Africa, where the concentrations of big wild animals and primitive peoples offered what the American West had lost a generation before. He made other international trips, including one of five months in 1913 on tributaries of the Amazon River that almost claimed his life.

Edgar Rice Burroughs and Tarzan

Americans of the early twentieth century with lesser means and connections than Theodore Roosevelt did their wilderness "importing" vicariously. One of the primary purveyors was a chronically unsuccessful hack writer named Edgar Rice Burroughs who in 1912 struck it rich with the invention of one of America's best-known literary characters: Tarzan of the Apes. Burroughs' contemporaries, and civilized peoples around the world to this day, relished the idea of a white baby abandoned on the west African coast and reared in the jungle by apes. That Tarzan grew into a superman was ample evidence of the esteem with which readers regarded contact with the wild. The choice of Africa for Burroughs' primitivistic fable showed that conditions sufficiently wild to nurture a Tarzan-like sage could no longer be found in the United States, France, Japan, or the dozens of other nations that avidly consumed the tales. The American West, for instance, once produced a Davy Crockett and a Paul Bunyan, but by the twentieth century it had lost its ability to spawn wilderness heroes.

Defining Wilderness

Africans, if they knew about Tarzan at all, undoubtedly dismissed the stories as so much insanity—additional evidence of the irony of global attitudes toward nature. In 1912, Africa was too close to the primitive to appreciate the primitive: it could only be an exporter of wildness.

A good place to begin exploring how the degree of a nation's development forms its view of protecting nature is to look at how different nations or cultures define "wild" places and people. Lacking both extensive and intensive wild country, the citizens of developed nations invariably call places "wilderness" that are only slightly removed from the influence of technological civilization. Underdeveloped peoples have much higher criteria for wildness, but in point of fact many of these societies do not recognize the concept "wilderness" at all. Indicative is the fact that most underdeveloped peoples have no word equivalent to "wilderness" in their vocabularies.

In recent trips to Malaysia, outback Australia, and east Africa, I asked, through interpreters when necessary, for the synonym of "wilderness" in aboriginal languages. The query invariably met with laughter or stares of incomprehension. At length, a word equivalent to "forest" or "nature" was offered: east African Masai, for example, suggested "serenget," as in Serengeti Plains, which signifies an extended place. The idea that it might be wild simply could not be communicated in Masai. It was, after all, their home.

The explanation is that the concept of wilderness was created by civilization. About 15,000 years ago, herding and agriculture introduced the idea and the practice of controlling their environment to a hunting-gathering species that for eons had been controlled by it. The advent of technological civilization accelerated this intellectual revolution and the related assumption of man's difference from and superiority to the rest of life. The controller, in a word, was better than the controlled. Dualism began: man was of a different order from nature. Attendant too was the idea of man's right to order nature in his own interest.

One result was the creation of the concept "wilderness" to define that part of the Earth alien to civilization and its controlling abilities. In other words, it required a town, a cultivated field, or a corral to define the "wild" country beyond its boundaries; it required domesticated animals to conceptualize "wild" ones. The beginning of civilization, in sum, marked the end of a million-year-old hunting and gathering lifestyle in which man simply existed as part of the totality of the natural world (Nash 1975).

To appreciate this difference between civilized and uncivilized or precivilized perspectives, it is only necessary to compare Christian ideas about creation and human superiority to other forms of life with the creation myths of primitive peoples. Invariably, the first man in noncivilized cultures is believed to have come *up* from the Earth—Mother Earth—of which he remains a part both literally and figuratively.

What this means is that civilized people regard as "wilderness" what uncivilized man calls "home." As an example of the resulting perplexities, consider the nineteenth century comments of Chief Luther Standing Bear of the Oglala Sioux: "We did not think of the great open plains, the beautiful rolling hills, and the winding streams with their tangled growth as 'wild'. Only to the white man was nature a 'wilderness' and only to him was the land 'infested' with 'wild' animals and 'savage' people. To us it was tame" (quoted in McLuhan 1971, p. 6).

As Standing Bear saw so well, the problem was that the white man—committed since civilization began to controlling the Earth—feared and hated uncontrolled, "wild" nature. Its presence rebuked him and mocked his efforts, and he was not at home in unordered natural environments. His comfort and security—and his pride—came from dominating, rather than from fitting into, the natural scheme of things: but hunter-gatherers like Standing Bear did not seek to control the environment or bend nature to their will. Consequently, they had no reason to fear or even to think about the uncontrolled. There was no city-country distinction, no "frontier," no forest primeval: every place was home. Without civilization, there was no "wilderness."

The Economics of Wilderness Protection in Developing Countries

These differences in attitude and definition complicate global nature protection. For one thing, the citizens of undeveloped or developing nations have great difficulty understanding what it is that developed peoples wish to "import" or, indeed, why they are in the wildness-importing business at all. Consider the national parks and game reserves in the east African nations of Kenya, Tanzania, and Uganda. For Africans who have been living with wild animals and in a natural environment for as long as they can remember, nature protection makes little sense. An analogous situation would be the astonishment of a resident of New York City that anyone would want to maintain an "urban reserve" between Thirty-second and Forty-second Streets! "Why do it at all?" he would wonder, and, "Why do it for these ten streets rather than for ten others in the vast metropolis?" Just as you can't excite a New Yorker about preserving a taxicab, it is difficult to interest an East African Masai in protecting a lion.

The African is also perplexed at restrictions of his use of the parks for living, farming, and grazing. Not sharing the concept of "wildness" held by the citizen of the developed nation, he sees no reason not to continue to live with animals under natural conditions as he has always done. Thus, when Masai cattle—repeatedly grazing in a national park—are finally shot by rangers, as happened recently in Kenya's Nairobi National Park, the protest is fierce. To continue the analogy with the New Yorker, it would be as if private automobiles entering the urban reserve south of Forty-second Street were confiscated and destroyed.

The Masai and his cattle raise an important issue in world nature preservation: are people—more exactly, lifestyles—part of what is being saved? Developing nations, as a rule, see no issue here and frequently make a place in parks and reserves for aborigines who have traditionally used them without the aid of refined technology. In Malaysia's vast Taman Negara (National Park), for instance, *oran ulu* (literally, men from the headwaters) are permitted to live as they always have, even to the extent of killing wild animals that are otherwise protected in the reserve. The rationale is that the *oran ulu* are part of the ecosystem, just as tigers are. Their presence is regarded as enhancing rather than as detracting from the visitor's experience.

But developed nations have more difficulties accepting the idea of men in wilderness. The aversion even extends to aborigines or at least very ancient immigrants such as the American Indian. The Seminoles were once considered a "feature" of Everglades National Park, and native Americans live in or close to Canyon de Chelly National Monument and Grand Canyon National Park: but as a rule the United States has elected to separate nature preserves and Indian reservations. Still, if the 1832 initial call for a national park in the United States by the artist George Catlin is examined, it clearly includes the idea of wild country *and* wild people in one composite preserve (Catlin 1880, pp. 288 ff.). Although not followed in America, this is an idea readily understood in less developed contexts.

But by and large, east African leaders—if not all the *wananchi* (country people)—have supported national parks and game reserves. The reason is quite simple: money. Parks are profitable. In fact, nature exporting is now Kenya's foremost industry. In Tanzania, a poster in Swahili makes the point clear: "OUR NATIONAL PARKS BRING GOOD MONEY INTO TANZANIA—PRESERVE THEM." Higher motives for protecting nature certainly exist, but from the practical standpoint of maintaining Africa's wild heritage, it is fortunate that the economic argument exists and that it makes sense to Africans. It is fortunate that nature in Africa has an international clientele.

Kenya's Example

The situation has been assessed most perceptively by Professor R. J. Olembo, a member of Kenya's national parks board since the mid-1960s (Olembo 1975). As a high-school student in the 1950s, Professor Olembo recalls having had not the slightest interest in Kenya's parks and reserves. They were, after all, the creation of a colonial government—that of the United Kingdom—already well into the business of nature importing. Admirable as the work of men like Mervyn Cowie (see Cowie 1961, pp. 211 ff., and Cowie undated) and John Owen in Tanzania was, Africans like Professor Olembo did not support it at all. They regarded the parks, quite rightly, as the result of Europeans' determination to pick out and reserve the best parts of the country for themselves. Parks were not for Africans; it was as simple as that. In fact, Professor Olembo clearly recalls his high school teachers using national parks as examples of colonial repression and exploitation. If freedom ever comes, he was told, the feudal relics will go. (Interestingly for the present purposes, Professor Olembo—like many park leaders in developing nations—became inspired to work for nature protection during a trip to the United States.)

Thus, it was with considerable astonishment that Professor Olembo saw *uhuru* (freedom, independence) come to Kenya in 1963 and the parks remain and expand under African administration. His explanation rests on the economic dividends the parks began to pay. Kenyans, Professor Olembo recalls, looked with delighted surprise at the phenomenon of growing numbers of visitors, credit cards in hand, pouring off the jets in Nairobi. Their purpose was even more startling: in sharp contrast to previous whites, this new breed was not interested in slaves, land, mines, government, or religion; they simply wanted to see wild animals in wild settings. And, wonder of wonders to the Africans, they were willing to pay good money for the privilege. At once the parks and reserves took on new interest for Africans. Here was a way to make badly needed money for an ambitious, developing nation without giving up resources, political power, or a part of its independence. The proceeds from the display of wildness were entirely "gravy."

Quickly, Professor Olembo recollects, the movement to abolish the parks and divide their land among the people for economic purposes stopped. President Jomo Kenyatta signed a statement in the fall of 1963 announcing his nation's intention to maintain its nature reserves and inviting the world's help in this effort. In so doing, President Kenyatta drew upon a long history of interest in African nature protection by the developed world. In fact, one root of the international nature-protection movement was a 1900 agreement drawn in London respecting African animals and the African environment. It was never ratified by the participating European nations, but the better-known London Convention for the Protection of African Fauna and Flora, drawn in 1933, was instrumental in launching preservation efforts (American Committee for International Wild Life Protection 1935).

President Kenyatta was also aware that as recently as 1961 the International Union for the Conservation of Nature and Natural

Resources had held a widely publicized symposium in Arusha, Tanzania (Watterson 1963), its purpose—only slightly covert—to smooth the way for transferring national parks and game reserves from colonial to African administration. Here, again, the nature-importing nations were protecting their interests. But their money, not their sentiment, was the most effective argument: the status of African reserves was not secure in the new nations until the flood of tourism began in the mid-1960s.

Africa also received direct technical assistance from the developed world in establishing and operating its parks and reserves. An American team, for example, helped Tanzania plan a national park featuring Mount Kilimanjaro (United States National Park Service 1970), and most Tanzanian park officers have been trained in the College of African Wildlife Management—founded by the United Nations and largely supported by the United States—at the foot of the great mountain. Norwegian foreign-aid money is currently implementing the plans. Other vitally important assistance is coming to Kenya from the Food and Agriculture Organization of the United Nations (FAO). Fully accepting the need for a practical "payoff" from nature reserves in underdeveloped countries, FAO technicians are demonstrating how the cropping of excess animals from the great African herds can help to justify their protection. But the United Nations team is quick to point out that, because of tourism, an animal in the viewfinder of a camera may be far more valuable in the long run than one in the pot! Indeed, FAO calculations show that, on the basis of the revenue they generate, wild animals in an African national park may be the most valuable in the world—race horses included.

Few of those who use economics as an argument for nature protection regard this rationale as more than a temporary stopgap. In time, they reason, the development of the societies in question will produce other motives for nature appreciation; meanwhile, the objective is to carry over a resource of wildness until such time as it is needed for more than its "breadwinning" ability. That this time may be not long in coming in some societies is apparent from their changing profile. Increasing numbers of Kenyans, for example, live in large metropolitan centers—chiefly Nairobi—and increasing numbers have never seen a lion or an elephant. For them, wildness is a novelty, and nature preserves are more likely to appeal as places to visit themselves rather than as nature-exporting institutions.

The Irony of Global Nature Protection

Presently, however, the parks and reserves of developing nations are used almost exclusively by foreigners. This is a natural consequence growing out of the ironical situation of global nature protection. It produces some interesting patterns of park management. One is the tailoring of accommodations and travel in the parks to the tastes of the rich and civilized. This means luxurious lodges, guided Land Rover transportation, and, in general, keeping the wild at arm's length. As Norman Myers (1972) puts it, "The modern traveler likes raw Africa without becoming raw himself." More precisely, the traveler is not allowed to approach wild Africa. Safety considerations figure in this decision, as does respect for the animals' habitat. But the presence of the economic argument cannot be denied: people in hotels spend more than people in tents.

From the standpoint of building an economic case for parks and reserves, luxury tourism is, of course, a decided asset. By deliberately courting persons able to pay "top dollar" for tourist services, park managers in developing nations maximize their land's yield. To be sure, there are liabilities; one is the often deliberate discouragement of campers, backpackers, and climbers in many parklands. A large sign at The Hermitage in Mount Cook National Park, New Zealand, warns those wearing climbing boots to use the rear entrance and to patronize a separate bar. It is illegal to walk in most of East Africa's parks and reserves. Of course, the threat of big wild animals to visitors (and vice versa) is a valid reason for limiting free access, but park personnel and politicians are not so naive as to ignore the fact that visitors in lodges and on organized tours spend more money than "do-it-yourselfers." The same could be said of local people, and the lack of less-than-luxury facilities in some parks is a major factor in discouraging the average citizen from using them. "I want to take my children to a national park," one Nairobi secretary says, "but I don't own a car or camping equipment, and there's no way I can afford the trip" (Mudoka 1975, p. 34).

Since national parks in developing nations are not used by local people, few steps are taken to reduce the monopoly by foreign visitors; the greatest economic returns from protected nature come from concentrating on that clientele. The irony carries other implications for management. One of potentially vital importance concerns the carrying capacities of parklands and the visitor quotas needed to keep use below the critical levels. If, as in the developing nations, parks are justified economically, there are few good arguments against increased visitation. The more visitors the better. But if, on the other hand, aesthetics, science, national-heritage, environmental-ethics, psychological-release, or religion are used to justify protected nature, then there are good reasons for determining carrying capacities. Of course, the ultimate safety valve, even in the developing nations, is public favor. If conditions in the parks become so run-down through overuse that rich foreigners stay away, some type of quota limitations can be expected.

It would be nice—from the standpoint of preserving the resource—to be able to justify and operate all parks according to the nonanthropocentric philosophy now obtaining in the Swiss National Park, where science, not recreation, is the principal rationale for the reserve's existence. But such philosophies carry little weight in the undeveloped world. Switzerland can afford the luxury of low usage and no revenue-producing visitor facilities; its park has other cultural values, principally science and scenery. Tanzania lacks this cultural orientation toward nature; moreover, Tanzania is too poor to afford national parks that do not pay their way. But, significantly (again, the irony of nature protection), the Swiss National Park is land "reclaimed" from intensive economic use (Schloeth 1974). To find wilder landscapes, the Swiss travel to Tanzania!

Conclusion

Addressing the Conference on Conservation of Nature and Natural Resources in Tropical South East Asia, held in Bangkok in 1965, Dr. Gerardo Budowski summarized the basic reason for the irony of global nature protection succinctly. "A feeling for wildlife and recreation is usually strongest among people from urban areas who also happen to be best educated" (Budowski 1968, p. 396). Dr. Budowski recognized that such a spectrum of social conditions, and the resulting attitudes toward nature and its protection, exists within, as well as among, nations.

It may be helpful to think in terms of a world *class* of nature appreciators and protectors; very thin in the developing world and increasingly thick in the developed nations, this social stratum, with its money and political influence, insulates nature from man's exploitative tendencies. Utilizing this class by understanding and encouraging the exporting and importing of nature is—for the time being—the best hope of the world's remaining wildness.

References

American Committee for International Wild Life Protection. 1935. *The London Convention for the Protection of African Fauna and Flora.* Special Publication no. 6. Cambridge, Mass.: American Committee for International Wild Life Protection.

Bartlett, Richard. 1974. *Nature's Yellowstone.* Albuquerque: University of New Mexico Press.

Beach, Joseph Warren. 1936. *The Concept of Nature in Nineteenth-Century English Poetry.* New York: The Macmillan Company.

Budowski, Gerardo. 1968. Protection and management of natural areas in Latin America—Implications for South East Asia. In *Conservation in Tropical South East Asia.* Proceedings of the Conference on Conservation of Nature and Natural Resources in Tropical South East Asia (Bangkok, 29 November to 4 December 1965), eds. Lee M. Talbot and Martha H. Talbot, pp. 395–400. IUCN Publications, New Series, no. 10. Morges, Switzerland: International Union for Conservation of Nature and Natural Resources.

Catlin, George. 1880. *North American Indians; Being Letters and Notes on Their Manners, Customs, and Conditions, Written During Eight Years' Travel Amongst the Wildest Tribes of Indians in North America, 1832-1839,* vol. 1. London: George Catlin.

Chateaubriand, François Auguste René de. 1816. *Recollections of Italy, England, and America, with Essays on Various Subjects, in Morals and Literature.* Philadelphia: M. Carey.

Cowie, Mervyn. 1961. *Fly, Vulture.* London: George C. Harrap and Company, Ltd.

———. Undated. History of the Royal National Parks of Kenya. Unpublished manuscript made available to the author in the course of an interview with Mr. Cowie in Nairobi, Kenya, 10 April 1975.

Cutright, Paul Russell. 1956. *Theodore Roosevelt, the Naturalist.* New York: Harper & Brothers.

Hagedorn, Hermann. 1921. *Roosevelt in the Bad Lands.* Publications of the Roosevelt Memorial Associations 1. Boston: Houghton Mifflin Company.

Haines, Aubrey L. 1974. *Yellowstone National Park: Its Exploration and Establishment.* Washington, D.C.: U.S. National Park Service.

Hookway, Reginald. 1975. Interview at Gloucester, England, 17 April 1975.
Lovell, Ernest J., Jr. 1949 (1950). *Byron, the Record of a Quest: Studies in a Poet's Concept and Treatment of Nature.* Austin: University of Texas Press.
McLuhan, T. C., ed. 1971. *Touch the Earth: A Self-Portrait of Indian Existence.* New York: Outerbridge & Dienstfrey.
Mudoka, Elizabeth. 1975. Interview at Nairobi, 11 March 1975.
Myers, Norman. 1972. *The Long African Day.* New York: Macmillan Publishing Company, Inc.
Nash, Roderick. 1970. The American invention of national parks. *American Quarterly* 22:726-735.
———. 1973. *Wilderness ard the American Mind.* Revised ed. New Haven: Yale University Press.
———. 1975. The Creation of "Wilderness" by Herding and Agriculture. In *EARTHCARE Program/Journal,* ed. Vivien E. Fauerbach, pp. 51-55. New York: National Audubon Society. San Francisco: Sierra Club.
Olembo, R. J. 1975. Interview at Nairobi, Kenya, 18 March 1975.
Olindo, Perez. 1975. Savannahs and Grasslands. Unpublished paper read at the EARTHCARE Conference, New York, 7 June 1975.
Pierson, George Wilson, ed. 1959. *Tocqueville in America: Abridged by Dudley C. Lunt from "Tocqueville and Beaumont in America."* New York: Doubleday.
Pomeroy, Earl S. 1957. *In Search of the Golden West: The Tourist in Western America.* New York: Alfred A. Knopf, Inc.
Porter, Mae Reed, and Odessa Davenport. 1963. *Scotsman in Buckskin: Sir William Drummond Stewart and the Rocky Mountain Fur Trade.* New York: Hastings House Publishers, Inc.
Pringle, Henry F. 1956. *Theodore Roosevelt: A Biography.* Revised and abridged ed. New York: Harcourt, Brace & Company.
Putnam, Carleton. 1959. *The Formative Years, 1858-1886.* Theodore Roosevelt; A Biography, vol. 1. New York: Charles Scribner's Sons.
Rutherford, Andrew. 1961. *Byron: A Critical Study.* Edinburgh: Oliver and Boyd, Ltd.
Schloeth, Robert F. 1974. Problems of wildlife and tourist management in the Swiss National Park. *Biological Conservation* 6(4):313-314.
Tocqueville, Alexis de. 1959. *Journey to America,* trans. George Lawrence, ed. J. P. Mayer. New Haven: Yale University Press.
Train, Russell E. 1974. An idea whose time has come: The World Heritage Trust, a world need and a world opportunity. In *Second World Conference on*

National Parks, Yellowstone and Grand Teton National Parks, U.S.A., September 18-27, 1972, ed. Hugh Elliott, pp. 377-381. Morges, Switzerland: International Union for Conservation of Nature and Natural Resources.

Trefethen, James B. 1961. *Crusade for Wildlife: Highlights in Conservation Progress.* Harrisburg, Pa.: Stackpole Company.

UNESCO. 1972. *Convention Concerning the Protection of the World Cultural and Natural Heritage,* adopted by the General Conference at its 17th Session, Paris, 16 November 1972. New York: United Nations.

U.S. National Park Service. 1970. *Kilimanjaro.* New York.

Watterson, Gerald G., compiler. 1963. *Conservation of Nature and Natural Resources in Modern African States.* Report of a Symposium at Arusha, Tanganyika, September 1961. IUCN Publications, New Series, no. 1. Morges, Switzerland: International Union for Conservation of Nature and Natural Resources.

Wied-Neuwied, Maximilian Alexander Philipp zu. 1843. *Travels in the Interior of North America,* trans. Hannibal Evans Lloyd. London: A. Ackermann and Company.

將欲取天下而爲之吾見其不得已天下神器不可
爲也爲者敗之執者失之故物或行或隨或呴或吹
或强或羸或載或隳是以聖人去甚去奢去泰

*There are those who would
control and then alter
the universe. I see that it
cannot be done.*

*The universe is sacred.
It cannot be changed or held:
changing it ruins it,
grasping it loses it.*

*Because things may be
causal or consequent,
cold or warm,
strong or weak,
secure or endangered,
wise people shun
extremes.*

—Lao Tzu
Tao Teh Ching

Economic Development and Environmental Protection

49. The Urgency of Preserving Natural Ecosystems

Roger Tory Peterson
U.S.A.

I propose to speak about the urgency of preserving ecosystems and the dilemma it poses—specifically about the urgency of preserving natural ecosystems, not the man-dominated ecosystems that are taking over. I shall present the case for wildlife—the vast galaxy of species that lack a means of communicating with us. They have no voice.

Lately, environmental conferences, even those sponsored by organizations dedicated to wildlife and wild places, seem invariably to focus on man's survival. There are nearly 4 billion of us. We will survive as a species for a long time. On the other hand, there are hundreds of species of animals and plants, as well as entire ecosystems, that may disappear before the end of this century. Some ecosystems vanished before they were properly catalogued, let alone understood.

I speak as a concerned observer. During the last few years I have travelled extensively and repeatedly on every continent, and I am deeply disturbed. Antarctica, it would appear, remains the one nearly pristine continent—unviolated because of the ring of ice that has protected it like a chastity belt. After twelve expeditions to the Antarctic and several to the Arctic, however, I have come to realize that the polar regions are, indeed, vulnerable. So far, the wide belt of sea ice has preserved the Antarctic fauna almost intact. Antarctica is the only continent where conservation guidelines were adopted at an early stage. These are embodied in the Antarctic Treaty, ratified in 1961 by twelve governments. There were no indigenous people, no traditions, to offer complications.

By contrast, the history of the subantarctic islands is a conservation "shocker." They were raped, most of them, within a few years of their discovery. South Georgia, the Falklands, the Auklands, Macquarie, and others will never make a complete comeback. It is reassuring, however, to see the increasing numbers of elephant seals bearing their pups among the rusting machinery along the beaches at the old whaling station of Grytviken, which is now a ghost town. Nature, given a chance, reclaims its own.

Recently, the Arctic has come under great pressure, and not only because of the much publicized trans-Alaska pipeline: dozens of projects are under way. The Canadian and the American governments are fully aware of the ecological hazards of undisciplined exploitation where ecosystems are so fragile; wherever one travels in the far north, one now meets teams making environmental impact studies. Some are qualified biologists; others, I suspect, are instant experts or inexperienced neophytes. The ultimate decisions will have to be made on the governmental level.

Some countries which we speak of as "underdeveloped" are no such thing in terms of renewable resources: they have already pushed things to the limit—forest lands denuded, marshes drained, every bit of arable land under pressure. They are actually overdeveloped in this sense. If they possess oil or other mineral resources, they are lucky; if not, they may ultimately face poverty and starvation. But even minerals are no guarantee for the future. They are not renewable; when they are used up, there is only the land.

Recently at a seminar, I was asked to define man's role in nature. It seemed to me that I was given a rather arrogant premise when I was asked to do this: it is exceedingly broad and profound, and therefore, intellectually unmanageable.

If each individual could define his own role in nature (and I think the important word is "in"), we wouldn't have as many problems. It is when we see ourselves out of nature that we court trouble.

First of all, we cannot decide for other people. We might educate or even coerce them, but we cannot decide for them. We must understand this: all of us gathered here have an opportunity to see much more of nature than someone who is trying to exist in a ghetto or a slum. The tundra in Alaska, a mountainside in California, a lake in Africa, has no meaning for someone who is struggling for survival every minute of the day—someone to whom starvation, unemployment, and disease are immediate problems. Our good fortune gives us neither the right nor the ability, nor the wherewithal, to determine the ethics, the morality, or the role in nature of other people. These things, like religion, are properly defined by the individual.

Man has always had a double interest in wildlife, especially birds—on the one hand, aesthetic, personal, impractical; on the other, utilitarian.

Actually, birds are far more than cardinals and orioles to brighten the garden, ducks and quail to fill the sportsman's bag, or warblers and waders to be ticked off by the birdwatcher. They are indicators of the environment—a sort of "ecological litmus paper." They reflect changes in the ecosystem rather quickly: they are an early-warning system, sending out signals.

An intelligent person who watches birds (or mammals, or fish, or butterflies) inevitably becomes an environmentalist. The Honorable Russell E. Train, administrator of the U.S. Environmental Protection Agency, emphasized this point recently:

> The Environmentalist is often portrayed by his critics as a romantic, distracted by fantasies of bluebirds and daisies, a birdwatcher oblivious to the practical needs of making a living. In short, he is not a realist. But I submit to you that he is far more realistic than those who would exploit this Earth for short term profit. He is worried about protecting the birds because Man is linked to them, sharing the same air and water, the same pollutants, the same hazards.
>
> By contrast, it is the wanton polluter, the thoughtless and quick-profit land developer, the promoter of urban sprawl, who are out of touch with the realities.

We invent systems: socialism, fascism, communism, capitalism. Each despises the other; yet they all espouse one creed: "Salvation by Machinery."

Some cultures and religions have fostered the idea of the separateness of man to the extent that even an endangered species is of no consequence. Animals are deemed to have no souls; therefore, they are not worthy of our concern. On the other hand, certain societies embrace the idea of man's oneness with nature.

Two months ago, I was in Nepal. As I travelled through the towns, I was fascinated to see cattle, chickens, and goats roaming the streets as though they were part of the big family. I had expected this, but I was not prepared to see the abundance of kites, house crows, jungle crows, and mynahs scavenging the streets. I thought that, here indeed, people have learned to live with nature in a most amicable way. Then, it occurred to me that those particular birds were indicators of urban decay—rubbish, litter, and filth—a bad sign, really.

I visited a forest, the last of its kind in Nepal—a subtropical ecosystem of limited extent. Serpent Eagles screamed overhead and babblers of several sorts trooped through the undergrowth. There were rare barbets and woodpeckers. In a few hours, I saw more than seventy species of birds, fifty of them new to me. I have never seen a greater variety of butterflies, some very rare. As I walked through the forest, however, I became aware that it was not quite the Eden it had appeared. People were hacking away at the undergrowth and the trees. Women trooped past with great sheaths of leaves for mulch to enrich the fields. At one place, blasting was reducing the mountainside to crushed rock. I wondered how long that unique ecosystem would survive such attrition.

It then occurred to me that here "living with nature" was on man's own terms. Whereas the kites, house crows, mynahs, and bulbuls had come to terms with man, a hundred other species will disappear when these forests die; their death sentence is a certainty.

I have seen similar examples of vulnerable ecosystems in every country I have visited. In upstate New York, for example, there is a special bit of pine barren that is the only home of a rare butterfly, the Karner Blue. There are plans to build a condominium complex that would eliminate the species and its environment. When the developer was told this he commented, "If we don't build *we* will become extinct." Condominiums, of course, can be built in less critical places. There is a precedent in the well known case of the noisy scrub-bird of western Australia, wherein plans for a new town were actually changed in order to spare this endangered species.

In much of tropical America, there is a deep tradition of wildlife exploitation and habitat destruction, making it extremely difficult to foster a new ethic. You can almost measure the attrition of natural habitat year by year in parts of the Andes; the climax forest of tropical mountains is one of the most fragile ecosystems on our planet.

A few years ago, I visited a new national park in the northern Andes with my friend, the late Carlos Lehmann. Brush fires were burning, trees were being felled, livestock roamed freely. The hungry, landless hill people had moved in. They reasoned: "Now that this is a national park, isn't it the property of all to do with as we will?" How can we talk of ethics? Whose ethics, theirs or ours?

I saw a similar thing in the magnificent Simien National Park in Ethiopia last year, where the last of the Walia ibex live. The local tribesmen were setting fire to the endemic vegetation on slopes much too steep to plant their barley. Ethiopia has lost 85 percent of its forest growth within the last fifty years.

I speak of these specific cases, not in criticism, but to illustrate a universal dilemma, a mutual tragedy involving man and nature—a tragedy from which no country is exempt. To insist that population control or technology will solve the dilemma is simplistic. The problems are, in fact, complex and legion; they go far beyond laws and their enforcement into the sphere of applied morality, which is only beginning to dawn on us.

Most people still fail to appreciate the real meaning of extinction: it is not simply the vanishing of the whooping crane, the vicuña, or the Walia ibex; it is the abrupt termination of a long line of evolution.

Every animal on Earth is in the process of becoming something else. If you had wandered around the shores of Lake Rudolph several million years ago and saw the grunting, scratching primate that was in the process of becoming man, you could not have said, "Look at that ape. From those loins someday will come Shakespeare, Leonardo, Ghandi." You could not have known. I am not suggesting that the whooping crane may write a sonnet someday, if we let it stay around long enough. However, it is a line through time. Every single species is a line in evolution.

The whooping crane has become a symbol. If we can afford to lose it, then Saint Peter's in Rome, Jerusalem, and all of the other symbols might also be abandoned. Most of the wildlife conservation effort right now focuses on symbolic animals—the wolf, the peregrine, the condor, the tiger. Yet a thousand lesser creatures—rare butterflies, reptiles, small mammals—are going by the board. Are they any less important?

They are no less important in the final analysis, but some are more important than others only because they are capable of becoming symbols—"public-relations" symbols that capture the imagination and create awareness; furthermore, they represent whole ecosystems. We are dealing with masses of people who may be vague about ecosystems; but one important goal they seem to comprehend and to act on is preserving species.

Of course, the ultimate observation is that man is the endangered species. A cliché, but true: the extinction of another species is a road marker of our own decline. Each extinct species is like a "mine canary," the canaries that miners used to warn them of lethal gas leakage. We are all down the mine shaft together. If we create the physical, political, chemical, and moral environment in which other forms of life cannot survive, by what insane arrogance do we think we could?

It is difficult not to feel negative about the prospects for the survival of wildlife. Rather than living within the natural system, we have maintained the widespread and persistent use of poisons—certainly the most atrocious example of assaulting nature. Now that there are restrictions on their use in America, some manufacturers are unloading their wares on so-called "developing nations." Few of the world's ecosystems have been entirely spared from these biocides.

It has been suggested that we may be eventually destined to have only a single, almost universal ecosystem—a man-dominated one where, aside from his crops, only dandelions, thistles, and other highly tolerant weeds of the roadside will survive, and where cockroaches and houseflies will be the dominant insects, rats the most successful mammals, and where only sparrows, pidgeons, starlings, and a handful of other birds will enliven the air.

It is a drab prospect. I like to think that we can live with nature on a more liberal plane than that. Uniformity of any sort is stifling to the human spirit and to ultimate survival. Variety must be preserved. To preserve variety in the natural world, we must hold on to every ecosystem if only to refurbish the Earth when man himself has reached a dead-end and has gone to his extinction—or, more desirably, has evolved into a "super-hominid" who is enlightened enough to live with nature.

Our world may seem badly afflicted, but it is not mortally ill. It is a world in which the great old philosophical concept of nature has been for a century strengthened and promoted by Darwin's discovery of organic evolution, Haeckel's discovery of ecology, and Mendel's discovery of genetics. Now, we are beginning to learn what nature is, discovering its rules and laws. We can dissect it to a certain extent, analyze its mechanisms, measure its parts, and even repair it.

The civilized man readily accepts, not only the humane ethic, but also the conservationist's philosophy—as well as the environmentalist's point of view. All are essential to a better and more civilized world in which there is a reverence for life. To this end, it is most urgent that we safeguard the environment to save the ecosystems that support life—all life.

50. The Developed Nations in an Interdependent World

Victor Ferkiss
Georgetown University, U.S.A.

The economic, social, and political interdependence of the nations of the contemporary world is so well established in the minds of all of us as to have achieved the status of a cliché. Wheat from North America feeds millions in the Eastern Hemisphere, oil from the Middle East fuels the economies of Western Europe, Japan, and, increasingly, North America. Fashions in dress and music, educational techniques, and even family-rearing practices quickly spread from nation to nation and continent to continent. Local wars, whether in Southeast Asia or the Middle East, excite our legitimate fears that great powers may become involved, resulting in a global conflagration.

But most people are much less conscious of the extent to which changes in the physical environment or conditions in various nations affect all other nations. This is true because most of us still fail to recognize the extent to which our lives are affected by—in fact, depend upon—the air and water, soil and seas, that constitute the ecosphere. Our infatuation with—indeed, worship of—technology seduces us into believing that we can make the Earth conform to our wishes rather than having to conform our desires and aspirations to what the objective conditions in the physical universe will permit. Yet, increasingly, we have been forced to take cognizance of the way in which pollution of the air and the sea by industrial or agricultural practices in some nations affects conditions in other nations and the atmosphere and weather system of the whole globe—of how the Soviet Union's plans for altering the flow of its rivers, for instance, might affect climate in Western Europe, of how the deforestation of the Amazon basin could affect the process of photosynthesis on a world scale.

Before we can fully comprehend the way in which the interdependent global system affects our lives, however, we need to understand the way in which the various economic, social, and political interdependencies we readily admit, and the physical interdependencies we increasingly discern, are in turn interrelated. We need to know how political and economic decisions in one nation affect physical conditions in another, and how physical conditions in some nations affect social conditions in nations both neighboring and distant. With a moment's reflection, it is easy to understand how drought in one country will affect the price level in another that consumes the first country's agricultural produce, or how a high demand for a particular product by consumers in one country can lead to lumbering or mining activities that can devastate the countrysides of distant lands. But while particular incidents of this kind come readily to mind, it is much more difficult to create an intellectual portrait of our global system in which such interactions are included as all-important features.

But creation of such a picture of the world is absolutely necessary if we are ever going to understand the problems involved in protecting natural areas on a global scale and in a global context. It is not necessary now, nor is it possible here, to create a detailed master model of world environmental problems analogous to the input-output matrices with which economists attempt to map the economies of nations. But we must, at least as a starting point, achieve a clear sense of the major dynamics inherent in the international systems of human and physical relationships that have an effect on natural areas. Some forces that menace natural areas in developing nations are largely or purely indigenous, and thus fall outside the scope of our present discussion. Despite the ahistorical views of some who speak and write as if no one had ever disturbed the "balance of nature" by shortsighted or rapacious action until modern, Western industrial civilization triumphed throughout the world, primitive peoples, and even advanced pre- or non-industrial civilizations, have wreaked havoc on their own natural habitats and surroundings. Such activities are being repeated in less developed nations throughout the world today. They can, perhaps, be mitigated by the actions of international bodies or the examples and precept and influence of groups within individual developed nations, but that is not our direct concern here. The problem we are dealing with here is more limited, although, perhaps in terms of total impact, it is a more significant one: to what extent is the integrity of natural areas in less developed nations (or, for that matter, in such regions as the Arctic, the Antarctic, or the open seas) menaced by pressures generated by the desires and activities of the developed nations; what are these pressures; how can they be reduced; and what are the preconditions for their reduction?

Some of these pressures are of a directly politico-military nature. Modern war is inevitably a destroyer of nature. The story is told of a Japanese officer captured in the Pacific campaign during World War II who was asked which of the opponents he had fought against, the Americans or the Australians, were the best jungle fighters. The Australians, he readily answered: the Americans first tear down the jungle, and then they fight. We have seen this judgment dramatized by the "ecocide" that accompanied the U.S. intervention in Vietnam. It goes without saying that any future wars fought with modern weapons, even at a subnuclear level, would be utterly destructive of the terrain in which it was fought, while, of course, a major nuclear war would be an "ecocatastrophe" of almost unimaginable magnitude. Preventing wars must be of special concern to those interested in the preservation of wilderness, but the prevention of major wars is necessarily also an overriding goal for anyone concerned with the future of humanity. It is true that we must never forget the importance of peace, even though our personal professional or activist concerns are concentrated elsewhere; but it is also true that everyone cannot do everything, and the major thrust of the attention of those of us here must be upon those forces that menace natural areas even in peacetime.

But while war may be a remote or tangential problem as far as the menace to natural areas is concerned, we should not forget the role that military expenditures and preparation for war play in the economic life of the developed nations. Ironically, the economic waste involved in military spending is a major underpinning of the general worship of economic growth. Economies where the normal workings of the market system—however problematic such a concept may in fact be—cannot keep people employed or industries profitable, often turn to military spending as a way of maintaining so-called prosperity. Where effective demand fails as an economic spur, often because of maldistribution of wealth, the military procurement officer stands ready to take up the slack. An even more important factor, perhaps, is the influence that fear of defeat in war by a technologically superior, or simply a more industrially powerful, foe has in spurring programs of industrial expansion generally. Because our potential enemy may have more or better weapons, we must produce more steel, dig more coal, and till more fields in order to earn money to buy arms, and so on, *ad nauseam.*

More subtle, perhaps, is the influence that the military metaphor of strength and conflict has in determining national images and aspirations. In many nations, natural areas are directly menaced by population growth that stems, not only from normal unchecked procreative pressures, but also from the desires of governments to promote population growth as a means to "national power." Yet how long would sheer numbers be valued as an index of national greatness in a world from which armed conflict had been eliminated as an option of national policy? Preparation for war in a world in which war is assumed to be a normal social force can be a far greater and more immediate danger to the future of natural areas than the possibility of war itself.

Nevertheless, the single greatest source of pressure against natural areas operating within developed nations and menacing their natural areas and the natural areas in developing nations as well is the general pressure for economic growth. This is not to say that if lower growth rates prevailed within developed nations or in the world economy generally, natural areas would be safe. Some experts argue that the primitive slash and burn method of agriculture destroyed virtually all virgin stands of vegetation in Africa, save on the highest mountains or in the most dense swamps, before the era of colonization, during a period when what was in effect a no-growth economy was in existence. A handful of squatters and/or hunters can have a scarring and disfiguring impact on wilderness without their activities having any relation to a growing "gross national product" (GNP), even today. A handful of vandal hikers can write their names with paint on cliffs without generating any but the most infinitesimal economic demand for more paint. Activities such as these, though they cannot be ignored, are essentially minor and gratuitous menaces to the integrity of natural areas. But in a society whose highest priority is economic growth, the menace to natural areas is necessary and ubiquitous, and any triumphs over it will be both difficult and precarious. The greatest single danger to the future of natural areas, within nations or globally, comes from pressures to exploit them, directly or indirectly, in order to promote the growth of the economy generally.

To lay initial stress on the importance of economic factors is not to assert that they are either the sole cause of danger to natural areas nor that they act independently of other factors. Far from it. As I shall point out later, how and to what extent economic factors are able to affect natural areas is conditioned by political systems, and these in turn are conditioned by basic beliefs about the nature of man and the universe. But economic factors are the means through which most of the pressures on natural areas operate, and it is necessary to understand the economic aspects of the problem before we can ask how politics and philosophy can modify their operation.

First of all, we should be clear as to what we mean when we talk about the problems of growth in an economy. This is a difficult task because there is no generally accepted and readily accessible terminology for discussing these issues. Economics deals with scarce goods (and services) that are exchanged in a marketplace or allocated in some other fashion. In our contemporary economic thinking, the amount of economic activity that is measurable through market-type transactions is equated with the physical or even the general well-being of a population: we assume that money can buy happiness. This is not hyperbole. It is an inevitable consequence of the fact that when older philosophies of human life that assumed the existence of objective needs were replaced by the philosophy of utilitarianism, it had to be assumed that the concept of "need" was meaningless and all that could be spoken of were "desires." Desires were subjective: the only way in which the desires of human beings could be understood was in terms of how much individuals valued their satisfaction. Money existed as a measure of how much different people valued different things—never mind that people did not possess money with which to register their desires to the same extent; that is another problem. The more money that was available, the more people could satisfy desires, i.e., buy happiness. Therefore the formula: higher GNP equals more money available, therefore more opportunities for happiness. To ask whether GNP measured any kind of substantive economic well-being, even in the physical sense, was impossible given the abandonment of any objective standard of human needs. Growth in GNP necessarily meant growth in well-being (happiness) under such a system.

Increasingly today we have come to recognize that the relationship between increased GNP and increased happiness is not all that simple. But not only are growth in GNP and growth in well-being not directly related, neither of them is directly related to growth in such environmental problems as pollution, resource depletion, energy consumption, and so forth. They may, and indeed usually are, positively correlated with these environmental problems, but not in any necessary or symmetrical fashion. Therefore, growth in GNP may or may not mean increased environmental hazards, including threats to the preservation of natural areas, just as it need not mean increased well-being. But (and herein lies our difficulty), while it is easy to measure—and therefore to speak of—growth or non-growth in terms of GNP, we have no terminology or easy means of measurement for growth in physical stress on the environment, any more than we do for the satisfaction of human needs. Yet what concerns environmentalists is not GNP growth per se, but growth in stress on the environment. It is within this frame of reference that we must be concerned with the creation of a no-growth society. Perhaps the best term to emerge so far as a substitute for the psychologically and politically unfortunate term "no-growth society" is the "sustainable society," a society in which production and consumption patterns are balanced within the context of ecological balance, so that they can continue indefinitely without menacing the Earth's life-support systems.

What changes in the economic systems of developed nations would be necessary to create a sustainable society? First and all important would be the substitution of some form of overall economic planning, informed by ecologically viable norms, for the present so-called market system as the means of deciding what will be produced, in what quantities, and in what manner. Obviously, in no modern nation—certainly not the United States—is the market today autonomous in the classical economic sense. Oligopolistic and monopolistic competition combine with government intervention to make a mockery of traditional concepts. But, while economic freedom is a myth, the controls and influences that presently exist do not add up to coherent planning, least of all to planning for sound goals.

Such planning would have to include among its elements at least the following. There would be comprehensive land-use planning to assure that land was neither wasted nor used in a destructive manner. Industrial, agricultural, residential, recreational, and wilderness uses (and we need to remember that the latter two are not necessarily synonymous) would be segregated and/or related in such a manner as to preserve their integrity. Through such planning, natural areas could be preserved on a sounder basis than is possible under the present *ad hoc* system of creating occasional national parks and wilderness areas. Concurrently, planning would aim at reducing the use of resources through regulations and economic incentives designed to encourage better insulation of homes, more efficient automobiles, recycling of waste materials, and so forth.

Not only physical materials, but space, must be "recycled." In many urban areas, there are empty—often virtually abandoned—tracts of land that could be turned into attractive housing sites as an alternative to new "developments" farther out in the countryside. Many industrial sites could be "recycled," so as to avoid the pressure to despoil natural areas, to provide locations for new manufacturing or transportation facilities. Planning would also be alert to the need to provide alternative employment for workers and industrial establishments that would be adversely affected by conservation measures.

Economic planning, in short, would not be merely fiscal or monetary—aiming at creating or sustaining demand—but structural in nature: the way to deal with the real problem of unemployment in the automobile industry is not to create a renewed demand for more cars, but to find alternative roles for redundant employees or suppliers. Comprehensive economic planning would integrate environmental assessment and technology assessment within its general processes and would concern itself with the basic interrelationships among new, or possible, technologies and their economic and environmental impacts.

Many will immediately shy away from the idea that such long-range overall planning is desirable, even if possible, or that it is necessary for the protection of natural areas. But realism dictates that, even if it were possible in an era of unlimited growth to create new national parks, for instance, or to maintain the integrity of old ones, this is not enough. Parks such as the Everglades can be destroyed by the effect of growth on the watertable throughout the whole state of Florida, just as hunting and agricultural practices outside their borders menace African parks. An economy that encourages the proliferation of off-track vehicles or simply cries out for more timber and coal constantly, exerts inexorable pressures on forests and wilderness. In addition, the preservation of wilderness areas can be achieved only by decreasing the rising demand for their use, which in part reflects the fact that outdoor recreation areas within existing urban areas are scarce, often difficult of access, and frequently unsafe. Comprehensive urban planning on a megalopolitan level is necessary as part of an integrated plan for the preservation of natural areas.

Nor can the argument stop here. Despite the fact that GNP does not measure happiness, the general fact remains that the standard of living of most of humanity has in fact risen during the recent generations of industrial expansion throughout the world, and that, even within developed nations, there are a great many people who are still poor—relatively or absolutely. To forget this fact is to validate the charges of elitism often leveled against environmentalists. It will be absolutely impossible to protect natural areas through the creation of a sustainable society without confronting head-on the problem of economic redistribution within developed societies. How far this redistribution will go in different societies will depend on local possibilities and problems. Swedish society continues to operate successfully in a situation of relative economic equality that would be regarded as revolutionary in North America. But major redistribution must take place if the pressures for unchecked growth generated by the desire of the less advantaged to increase their absolute share is not to continue and, in doing so, make protection of natural areas impossible. Whatever the political antecedents of such a movement for redistribution may be, over the long run this too will have to be part of the process and agenda of comprehensive economic planning.

But just as the pressures for growth within developed nations necessarily menace natural areas in the developed nations themselves, they also necessarily menace natural areas in less developed nations as well. This is a major consequence of global interdependence. Increasingly, developed nations draw upon the resources of the less developed for the necessary elements for their own economic processes. Developing nations are willing to destroy their own wilderness areas or allow the import of polluting industries to their shores in order to raise their own, often desperately poor, standards of living. In order to reduce these pressures upon their less developed global neighbors, the developed nations will have to change the patterns of their own economic processes and goals; otherwise, they are in effect paying the developing nations to destroy their environments and natural areas. But to reduce these demands on developing nations will involve not only a change in the living standards in developed nations, once cheap raw materials from abroad are no longer available: it will also have a negative impact on rising GNP in developing countries. Already suspicious and resentful at being told by developed nations not to follow their often sad example, the developing countries would see their own aspirations for prosperity thwarted by the actions of the developed countries. To compensate for this, the developed nations would have to give the less developed nations greater economic assistance than they have in the past, and to do so, it must be remembered, within the context of economies of their own that were no longer growing according to past patterns.

What all this amounts to is that the halting of environmentally pernicious growth in the developed nations and in the world generally would require a dual redistribution of wealth—within the developed nations and between the developed and the as yet undeveloped. Such a redistribution would require the most careful and comprehensive economic planning on a global scale. Such global planning would, as the 1974 Cocoyoc Declaration recommends, have to employ a redefinition of development that would provide the basis for curbing and reversing "overdevelopment" in some nations, as well as seeking to end the anomalies of "underdevelopment" in others; and it would seek to combine increased national "self-reliance" with international control over the global "commons."

Such a change in economic goals and policies, however logical, will seem utopian. We cannot speculate here in detail as to the political scenario within which it might take place. But we can and must ask ourselves what, logically, its political and philosophical prerequisites might be.

The first requisite for the global protection of natural areas is the creation of governments that can govern: economic reorientation and redistribution on the scale described above will require strong and effective political leadership. Executives must be able to dramatize to their citizens the need for measures to protect natural areas and must be able to mobilize public opinion behind them. They must also be able to keep their own houses in order. One of the pervasive problems of governments throughout the world—democratic and nondemocratic alike—is the tendency of entrenched bureaucracies to go their own ways, regardless of the wishes of the general population or the legally established political leadership. When such bureaucratic kingdoms become a threat to measures designed to protect natural areas, they must be brought to heel. The United States Army Corps of Engineers is a well known and spectacular example of such a bureaucracy, but all governments tend to have ministries or agencies with similar tendencies. Yet the Corps of Engineers could not behave as it traditionally has without strong legislative support for its activities.

The problem of divided government and the consequent ambivalence of policy is more acute in presidential than in parliamentary systems, at least in theory. But all legislative bodies, however they relate to executive branches of government, will have to devise internal mechanisms that enable them to take holistic views of ecological and environmental problems and to not let special interests, often working through powerful committees, have their own way with natural areas. Insofar as committees have power vis-à-vis legislatures as a whole, their memberships must be so constituted as to enable the committees to take an overall view of the problems with which they must deal and must be broadly based so as to represent all citizens, not just those special interests most directly affected.

In nations where the judiciary plays an important role in the making of political and economic policy, as it does in common-law systems, above all in the United States, its processes must be available to the public to the greatest possible extent so as to enable them to seek injunctive relief, obtain declaratory judgments, or initiate class-action suits.

Indeed, it can be argued that political processes in all branches and at all levels of government must be opened to popular participation if natural areas are to be protected, because the greatest menace to our environment comes from the concentrated power of the alliance between dominant economic interests—committed to unlimited growth in traditional terms—and dominant political elites who accept the need for and personally benefit from such growth. This is as true in the developing nations, where growth and foreign aid have tended to benefit primarily a small minority at the top—as it is in developed nations run by corporate or ministerial bureaucracies.

But the issue is a complex and delicate one. It could be argued that education and a certain degree of affluence contribute to an interest in such matters as ecology or conservation and to the ability to appreciate the aesthetic and moral values of natural areas, while the short-run benefits of growth gained through the destruction of a particular natural area seem to be the most pressing concern of local workers or farmers. The implication of this argument—perhaps especially in some developing nations—is that only an educated elite will have the vision necessary to undertake measures to protect natural areas. If this is the case, then opening up the processes of government to populistic initiatives and inputs will make the far-sighted planning necessary to sustain ecological balances difficult if not impossible.

This line of argument is not to be dismissed lightly, and certainly it would be fanatic as well as foolish to insist that in every nation and at all times popular understanding and overt approval are the necessary underpinnings of sound policies. But taken on balance it can be argued that over the long run it is the people of the Earth as a whole who have the most to lose from the degradation of their habitats and that, unless they realize this and support measures for their protection, these measures cannot succeed anyway. Greater openness to popular pressures—political and legal—may make the path of special interests easier in some ways, but if insulating planning procedures and measures for environmental protection from popular pressures often may make sense tactically in particular instances, strategically it is a blind alley.

One difficult problem will lie in the realm of relating the actions of individual governments to international programs for environmental protection, whether these involve the protection of natural areas directly or their protection indirectly through a measure of economic restructuring or redistribution. Even when the executives of various governments agree to or feel bound by international agreements, implementing these agreements can be difficult domestically. Democratic governments, even when they can claim to be bound by international agreements, will have to struggle with legislatures in order to be able to implement the agreements. Nondemocratic governments will have to fight resistance by economic or bureaucratic cliques. One of the main tasks of political leadership in an interdependent world will be to mesh domestic governance with international measures and to overcome nationalistic pressures that may militate against agreements designed to protect natural areas.

Up to now, we have looked primarily at changes in economic and political structures and institutions that are prerequisite to the protection of natural areas in an interdependent world. We have postulated the need, in the economic realm, for a sustainable society in which the traditional theoretical role of the market in promoting growth is subordinated to a planned economy aiming at ecological equilibrium and the redistribution of wealth within and among nations. It has been argued that such economic changes will require strong governments able to make and implement plans to protect natural areas and the ecological balance they require to exist—governments in practice stronger than usually has been the norm in democratic societies. At the same time, we have suggested, somewhat paradoxically, that such governments must be open to popular influence and participation.

As I noted earlier, such proposals sound utopian. They are. But they are utopian only in the absence of the philosophical reorientation that is their necessary foundation. Current ideologies throughout the world—whether socialist or liberal capitalist—accept the basic premise that only through growth in the production of physical goods and services and through the technological overcoming of natural limitations can the happiness of mankind be secured. Despite this, the traditional belief systems of various cultures have always recognized the dangers to societies as well as to individuals of greed and pride, and the value of contact with nature to the human psyche. Recent scientific developments have supported those traditional beliefs by showing us the necessity of limiting growth and preserving natural areas for the physical survival of the human race.

If—and only if—the revolutionary implications of this view of the world are understood generally can, not only natural areas, but human civilization itself, survive. If such a revolution in cultural outlook does take place (it may be hopelessly optimistic to believe that it can or will) the changes in the economic and political systems of the developed nations and of the world, suggested above, can be judged as means to reformulated ends. If the projected revolution in human consciousness does not take place, then these changes will not be seriously considered. Thus, we have a situation in which two interdependent utopian visions add up to realism.

Make no mistake about it: the preservation of natural areas is not an isolated or technical problem. It is one aspect of the whole problem of environmental and ecological preservation, which in turn involves all facets of economics and politics. For problems of such magnitude, nothing short of revolutionary and holistic solutions are possible.

51. Ecologically Sound Development in a Developed Country: Canada

Jeanne Sauvé
Department of the Environment, Canada

It is with considerable temerity that I venture into this difficult and largely uncharted realm of what we have called "ecologically sound development." It is particularly difficult because we in Canada are at present essentially a traditional growth society, having achieved a high level of economic and technological development, urbanization, and cultural adaptation to consumption. The implications of the profound social changes we are considering are staggering; with the time available, I will only be able to provide a mere sketch of a preliminary conceptual framework. The task of signposting the path of transition from our present state of exponential growth to some future equilibrium society seems currently beyond our capability, particularly when we contemplate the social and economic disruptions that such changes may exact. Yet it seems to me that we must do this by choice rather than by necessity or accident, either of which could prove more costly. It was André Maurois who suggested the wisdom of giving up with good grace things that eventually would be taken from you anyway.

Wherever we look, we see the symptoms of crisis and malaise—environmental and psychosocial. What the Club of Rome called the human predicament pervades our world. In the words of André Malraux, *"Le monde contemporain domine la Nature sans dominer le progrès—il a donné naissance à la première civilisation qui n'est pas plus d'accord avec elle-même."*

More and more of our attention is being focussed on the nature and consequences of what we in North America call progress. We realize, of course, that progress involves more than the single factor of economic growth. But, in our culture, economic growth has come to be seen as the indispensable bridge over which we must pass to gain access to all other roads to social progress. The problem, however, has been that the ideology of progress has become one-sided and is dominated by economic imperatives. And through this domination we have violated a universal principle of nature, namely that continued and unlimited growth is incompatible with stability. We are the only species whose population violates this principle.

Thus, I would like to suggest to you that this still-dominant world view of development equated with economic growth (most often uncontrolled and exponential economic growth) is being seriously challenged by an emerging world view that, for want of a better term, we are calling "organic growth" or ecologically sound development. Regardless of how we perceive the limits to our world, there is an almost universal consensus on the need to protect the integrity of the biosphere.

I would further propose that the possibility of realizing a

radically new form of development must be closely related to the social, political, and economic conditions that exist in each country at a given point in time. It is this concept that truly excites me as a Canadian, for I believe that Canada possesses a unique set of physical, political, social, and cultural conditions that, at least theoretically, should give us an excellent opportunity for embracing the principle of ecologically sound development. Let me amplify this point before I return to the concept of ecological soundness.

Canada is a country of some 22 million people, more than 75 percent of whom live in urban communities. They are mainly located in a relatively narrow strip of land a couple of hundred miles north of the Canada-United States border. We have one of the world's highest gross national products and energy consumption per capita. Thus our lifestyle closely parallels that of the United States.

Yet, within our country there are regions and political jurisdictions of great diversity. While we are one of the few countries in the world that is more than self-sufficient in natural resources, both renewable and nonrenewable, these resources are not uniformly distributed, nor is population or industrial and agricultural activity. Moreover, despite the richness of our resource base, we rely heavily on the import of manufactured and processed products. The combination of our sheer size—a land mass that is the second largest in the world, bordered by three oceans and with climates and terrains ranging from polar to temperate—coupled with our relatively small population, is unique. Moreover, a large part of our land mass is hinterland, resource-rich though ecologically fragile, inhabited by native peoples with many of the social characteristics of the peoples who constitute the "Third World." We are a bilingual country, with a parliamentary, federal form of government that works within the framework of a constitution written over 100 years ago, long before there was concern for the natural environment. There is, therefore, a complicated sharing of legislative powers and regulatory responsibilities between the federal and provincial governments.

It is this very diversity, in conjunction with our unique mix of physical and social conditions, that is the basis for both challenge and hope. In some ways, Canada is a microcosm of the world, and our problems are the same as those that beset the world at large. In our search to find solutions, we hope to contribute to the creation of viable social models that many other countries may find valuable.

However, to be realistic, I must caution you that I speak as an elected representative and officer of a government that represents all the people of Canada. I do not wish to misinform you: there is as yet no large constituency for the social changes we are discussing, although the climate of concern is shifting in that direction. Ecologically sound development in all its implications represents a radical change; an essential ingredient for its achievement is an informed, concerned, committed, and responsible public. The latter goal cannot be achieved without individual responsibility. Each and every one of us must be increasingly conscious of what he (or she) is doing and how that affects the environment and, indeed, all ecosystems.

Consider the value of our understanding the full implications of our lifestyles—of understanding, for example, what is involved when we buy a new car: the raw materials that go into making the car; the manufacturing process; the allocation of land to build highways; the energy and resources used, both in the manufacture and the operation of the car; the emissions generated; and, ultimately, the waste it constitutes when it is finally disposed of.

This is not to say that we shouldn't have such goods, but simply that we should not be frivolous or wasteful. Perhaps we don't need to replace our goods so often; perhaps they could be better designed to create less pollution and use less energy; or perhaps we could invent superior alternatives.

But such profound changes in personal responsibility, lifestyles, and values would be greatly assisted if we could also achieve a much higher level of national consensus on goals, an enhanced sense of mutual sharing and mutual interdependence, and a common perception of the problems and the necessary solutions. The real situation is dynamic; it is to be hoped that social goals and individual responsibility are converging.

Given our diversity, the problems are not just environmental or economic. They are also social and cultural. It is through the achievement of national social goals based on distributive justice and integrated with our environmental and resource programs that we can create a new system of development that is ecologically sound.

The concept of ecologically sound development rests on the application of ecological principles, not just to industrial activity alone, but to the restructuring of all of our institutions and our culture. Ecology is the science of systems, of indivisibility, of connectedness. Ecology imposes an interdisciplinary mode on thinking and mutuality on acting. This illuminates the potential for symbiotic relationships between people, institutions, regions, and states. We are awakened to the fact that, in our ecosystem, everything affects everything else. We can now sense the full import of the term "ecologically sound development."

We might add to this concept one modification, namely, appropriateness. We would therefore speak of development which is ecologically sound and appropriate. Appropriateness allows for the marked distinctions in the starting conditions of countries. Among these starting conditions are the level of economic development, the resource base, population, population distribution, social and cultural institutions, and so on. Even among developed nations such starting conditions will differ, perhaps radically.

We are attempting to develop a conceptual scheme for appropriate and ecologically sound development in Canada. The overriding concepts are that the principles of ecology should permeate the nature of social institutions, the culture and lifestyle of people, and the industrial activities designed to secure the goods necessary to survive and thrive. But if these concepts are to develop and flourish and pass from the arena of ideas to reality, we must resolve the major conflicting visions of global and national priorities—the achievement of social equity and the ecological imperative of living within the constraints of the world that is ours. This resolution of the apparent conflicts between development for the achievement of distributive justice and development that is ecologically sound is a fundamental task. Resolving this conflict into forms of ecologically sound development is neither sacrificing the present for the future nor sacrificing the future for the present. On the contrary, it is applying the wisdom of ecological concepts to the design of social institutions, life-support systems, and lifestyles.

We must begin by affirming as undeniable that a certain threshold of economic development is required by all societies to provide for their basic needs of survival and fulfillment. We must equally affirm that the excessive production and consumption patterns of the economically developed countries is wasteful and, in itself, a source of major environmental, psychological, and resource disruption. To achieve the desired condition is to pass through a stage of basically essential economic development. To eschew essential economic development is to accelerate the "ecocide" of the planet. We must therefore affirm the need for each country to achieve a certain level of economic development and to redefine the form of this development in harmony with the principle of ecologically sound development. What therefore becomes a critical priority is to define an appropriate form of development applicable to countries at radically different stages of economic development. While the guiding perspective is ecological, the details of appropriateness relate to unique and distinctive social, political, cultural, and technological characteristics of a particular nation or region. Moreover, appropriate development will involve both internal and external policies of mutual interdependence.

Canada is beginning to develop such policies and programs. At the present time, we have not issued an overall policy statement or guidelines on economic development, but we have taken a number of steps designed to protect the environment. These steps started out being mostly responsive in nature: they focussed on bringing existing problems under control. They are gradually becoming more anticipatory, by trying to ensure that problems don't arise in the future.

Let me give you some specific examples. Many of you will recall the widespread concern a few years ago about mercury contamination in fish and the consequent decision to ban certain fish for commercial sale. An investigation showed that mercury was present in particularly high concentrations in the discharges from certain industries. In consultation with the industries involved, we drew up specific control measures and agreed upon a timetable so that, as of today, the discharge of mercury compounds into the environment by these industries has been substantially reduced.

We have also developed emission standards for a number of other industry sectors, such as the pulp and paper, chlor-alkali, and petroleum-refining sectors. At the same time, because specific control responsibilities often reside with provincial governments, we have been developing a number of air and water-quality objectives for ambient conditions that form the basic framework within which pollution-control programs are developed.

Important pieces of federal legislation with respect to these matters include the Canada Water Act, the Clean Air Act, the Fisheries Act, the Northern Inland Waters Act, the Arctic Waters Pollution Prevention Act, and the Motor Vehicle Safety Act. Of course, there are many other pieces of federal and provincial legislation that contribute to the total Canadian effort to come to grips with environmental and ecological questions.

Today, however, we are looking ahead; we have a new bill in Parliament to which we refer as the Contaminants Bill. When passed, this bill will give us the power to require prospective users of selected substances, judged to be harmful to humans or the environment, to submit detailed information about the proposed use, the quantities to be used, and the possible consequences of their use—all before the substances are put into routine use. If the government is not satisfied it will be able either to prohibit their use or to impose certain restrictions.

We also have introduced procedures specifically designed to encourage ecologically sound development proposals for major projects built with federal funds or on federal lands. We refer to our "Environmental Assessment and Review Process," which is basically an administrative rather than a legal procedure. A proponent of a major construction project is required to prepare, for consideration by government, a comprehensive evaluation of the environmental impact of the project; if harmful effects appear likely, modifications are recommended. This process is still being developed and refined on the basis of experience gained during the past year.

The formulation of policies, however, requires certain basic knowledge in the areas of conserving the resources of the biosphere. I should add that Canada is participating in UNESCO's Man and the Biosphere Program, an international, interdisciplinary research program designed to produce basic knowledge and to identify viable policies for dealing with the practical problems of conserving the resources of the biosphere. In Canada, we have identified four areas of special interest: urbanization, forestry and agricultural management practices, coastal ecosystems, and Arctic and Subarctic development. We are in the midst of developing selection criteria and soliciting proposals for research in these four areas from universities, research institutes, government, and industry. We are giving particular emphasis to the development of proposals that will incorporate both natural and social approaches.

This is not the only area in which we are active internationally. We have, of course, a long common boundary with the United States and we are constantly discussing environmental problems produced or anticipated in one country as a result of actions taken or proposed by the other. We have found particularly useful and effective for dealing with these matters a joint institution we refer to as the International Joint Commission. One of the many encouraging developments has been the Great Lakes Water Quality Agreement, and we are looking critically, on both sides of the border, to measuring up to the expectations of that agreement.

On the broader international front, Canada has been in the forefront of nations at the Law of the Sea Conference, arguing for managerial responsibility of fisheries by coastal states. We participate actively in the United Nations Environment Programme, the Organization for Economic Cooperation and Development's Environment Committee, the NATO Committee on the Challenges of Modern Society, and the ECE Senior Advisers on Environmental Problems.

Speaking of social factors, and remembering the emphasis I have placed on the sense of individual responsibility, I should like to mention our experience with public participation. I am convinced that this will become increasingly important, but it is not without its difficulties. We experimented with public participation as long ago as the preparatory work for Canada's participation in the United Nations Conference on the Human Environment, held in Stockholm in 1972. At that time, we held a series of one-day public debates based on the United Nations Conference Secretariat's background papers. We received a large number of briefs that ranged widely in size and quality. We felt they were helpful to us in getting a feel for public concern at that time, and even in framing our formal positions for the conference itself.

We sought similar public inputs in connection with Canadian participation at the Law of the Sea Conference, and again for the World Population Conference. There is also provision for some form of public input to the Environmental Assessment and Review Process. The problem remains, however, to establish the proper form for this extension of participation that neither sacrifices the public's right to participate nor the government's obligation to act.

In this respect, I would like to mention another effort that we have been promoting. We refer to it as the Conserver Society Project, a term that should be thought of in relation to today's "consumer society." We are really trying to explore the basis of public attitudes and to examine ways and means of changing from a consuming to a conserving attitude.

As you can see, in Canada, we have begun in a very modest way to plan and implement our evolution towards a more appropriate development. Canada's commitment to environmental protection—to EARTHCARE—is not transitory. Indeed, it is embodied in the following goals taken from Environment Canada's ten-year planning guide:

- The Canadian water environment will support fish, marine mammals, and recreation at desirable levels;
- Fish, forest, land, and wildlife will be managed on an optimum-sustainable-yield basis;
- Programs for Canada's contribution to the preservation of genetic resources will be implemented; and
- "Clean" energy will be a normal Canadian requirement.

Ambitious goals, and costly. We are asking the people of Canada to spend more than $6 billion in the next ten years. That is around $1,200 for every family in Canada, a lot of money in anybody's language. But is it costly at all when you consider the alternatives? Surely all of us in this room think not.

52. Environmentally Sound Development: The Search for Principles and Guidelines

Noel J. Brown
United Nations Environment Programme

Introduction

I would like to congratulate the authors of the EARTHCARE Petition for their bold and timely initiative, and to express my hope for its widest possible circulation, endorsement, and success. We in the environmental community have a special interest in the promotion of environmental quality as a human right that is protected by international norms and supported by effective national measures.

Rights to EARTHCARE could very well hold the key to our long-term survival. More importantly, however, I believe that the concept ties in very closely with the theme of "Interdependence for Survival," which was kept sharply in focus by the preceding papers. I fully appreciate the authors' remarks: they provide an appropriate context for the comments I would like to make on the question of environmentally sound development. Let me quickly emphasize that my approach to the subject is not to pretend any special insight or keys for solution, but to appeal for an intensification in the search for principles and guidelines for development, and perhaps to add the perspective of the "Third World" to our deliberations.

Let me add also, that perhaps one of the most significant aspects of our deliberations is that they come at a time when there is considerable ferment in thinking on the question of development. This is due in part, I suppose, to the growing crises in the world's economic affairs and to the belated awareness that traditional concepts of development, with their narrow sectoral perspectives, are inadequate for meeting the accelerating global challenges and the insistent claims of the Third World for a larger share of the world's production capacity.

International Development Strategy

The ferment is registered most tellingly at the United Nations, which is currently in the process of reviewing and appraising the international development strategy, which is some fifteen years old and which had aimed at achieving measurable progress in the lot of the developing countries by 1980. The United Nations' review now fully acknowledges the fact that the development strategy was launched at a time when the international community as a whole was not fully aware of environmental issues and before relevant principles and measures had been formulated and agreed to.

It is interesting to note that the only specific reference to environment in the strategy was a provision expressing the hope that governments would intensify international efforts to arrest the deterioration of the human environment, take measures towards its improvement, and promote activities that would help maintain the ecological balance on which human survival depends. At its inception, however, the strategy did not propose any specific actions or policy measures, nor did it recommend ways to handle the environmental consequences of the targeted growth. This remains one of its shortcomings, and I hope it will be remedied in the current mid-term review and appraisal.

The New Economic Order

Coincident with the review is the Third World's demand for the fashioning of a new international economic order based on principles of equity, trade, and price stability, and for the acceptance by all states of the sovereignty of the nations concerned over natural resources and their appropriate management. A Charter on the Economic Rights and Duties of States provided the legal and judicial framework that would govern relations among states in the new international economic order.

These developments appear to be extensions of a similar reappraisal at the national level, where deep inflation and recession—particularly in the industrialized world—are forcing governments to question many of the assumptions on which economic development was premised and on which the dynamic growth of the last two decades was based—particularly since the present crisis does not appear to be another cyclical downturn, but could represent a break with past economic patterns. Moreover, these developments take place against the background of the nearly universal realization that the environment is continuously affected by economic growth and social change, and that unrestrained growth could very well propel economic societies to transgress those "outer limits" on which the stability of the biosphere depends. In effect, when placed together, the trends could force important shifts in international economic orientations that are likely to have far-reaching consequences for strategies of national and international development.

Added to these trends is the general mood of dissatisfaction with traditional concepts of development, with their narrow, sectoral orientation, their obsession with growth quantitatively defined, and their nearly total disregard for the natural order or the environmental context within which all human activity takes place. Such dissatisfaction certainly underscores the need for both a theory and a philosophy of development that is environmentally sound and of general applicability.

Theories of Development

What is most ironic about the present situation is that, despite nearly 500 years of economic progress in the West, an effective theory of development applicable to the needs of the overwhelming majority of the world's people still remains to be formulated. For the Third World, this is a matter of great significance, because Third World nations have benefited least from traditional prescriptions, as even the most cursory review of the more popular formulas will attest.

For example, the so-called "takeoff theory," with its rather mechanical point of takeoff, supported and propelled by the industrialized countries, has proved totally erroneous. Nor did the "trickle-down theory" prove any better—a theory which focused on the development of the industrial sector, with its extremely small industrial modernizing elite who would prosper at the expense of the majority of the population. The "closing-the-gap theory," on which the development decades were based, has proved hardly more reliable: the gap continues to widen and the real population explosion occurs among the poor.

What remains to be defined, therefore, is a comprehensive and integrated concept of development, sufficient to meet basic human needs without endangering the carrying capacity of the biosphere. Needed also is a redefinition of the goals of development and the formulation of social norms to guide economic decision-making at all levels. In this connection, the addition to the development debate of the environmental dimension, with its emphasis on the "quality of life," is most timely indeed.

Development and the Quality of Life

That debate, which achieved global prominence at the Stockholm conference, provided what was perhaps the most creative opportunity ever for reexamining the premises on which the development process is based and for clarifying the central elements in the relationship between development and environment. On the other hand, the debate gave a measure of legitimacy to the Third World's claim that the central environmental priority of the Third World was development and that solutions to environmental problems in developing countries would be reached through development.

The Declaration of the United Nations Conference on the Human Environment (see Appendix A) specifically states:

> In the developing countries most of the environmental problems are caused by under-development. Millions continue to live far below the minimum levels required for a decent human existence, deprived of adequate food and clothing, shelter and education, health and sanitation. Therefore, the developing countries must direct their efforts to development, bearing in mind their priorities and the need to safeguard and improve the environment.

But perhaps the most significant result of the discussion of this question at Stockholm was recognition of the organic relationship and complementarity between the quality of man's physical environment, the growth of the population, and the pattern and nature of the use of resources and space on the one hand, and such factors as social goals and value systems, socioeconomic systems and institutions, production systems, life and consumption styles, international patterns of trade, and the distribution of wealth and income on the other. The conference provided an integrated concept of development; this, in turn, provided the conceptual framework within which the newly established United Nations Environment Programme (UNEP) hopes to contribute to the formulation of principles and guidelines for environmentally sound development.

"Ecodevelopment"

In this connection, it might be useful to identify a few of the ideas currently under discussion within the UNEP community. The first idea is what Maurice Strong calls "ecodevelopment," which seems more directly applicable to the rural sectors of the world and to human-scale economics oriented toward human well-being. The aim of ecodevelopment is to identify alternative patterns and styles of development that are specially adapted to those resources of particular "ecoregions" necessary for providing food, housing, health care, and education for the human population.

The concept of ecodevelopment considers man the most valuable resource of an ecoregion and seeks to contribute to his fulfillment through the quality of human relationships and through respect for diversity: natural resources are exploited and managed both from the standpoint of the current generation's well-being and in terms of its obligations to future generations.

This process would strictly forbid the depredation and exhaustion of some nonrenewable resources, as well as waste, and would thereby significantly extend the life of known resources. Applied to tropical and subtropical regions, ecodevelopment would rely on the natural capacity of the region for photosynthesis in all its forms—a point made most emphatically by Professor René Dubos in his "Introduction to EARTHCARE." In effect, ecodevelopment would ensure the use of local energy sources and technologies and so reduce reliance on energy from commercial sources.

For its implementation, ecodevelopment would call for the establishment of a "horizontal" authority capable of integrating, within a policy framework, the various sectors of the development process and facilitating the complementarity of the different activities involved. To be effective, such an authority would require the meaningful participation of the populations concerned and affected, and, consequently, the satisfaction of the real needs of the participants.

Sound Management

Complementing the concept of ecodevelopment is the establishment of machinery for sound environmental, societal management, which again, according to Maurice Strong, involves the management of the whole system of relationships of individual activities that affects human development and well-being. Envisaged is the development of better techniques for allocating the real costs of activities to those who benefit from them, and for assigning real value to such traditionally free goods as water and air. Envisaged also is the development of better methods of evaluating beforehand the full consequences of decisions that significantly affect both the physical and social environment. This would mean, in effect, the integration and incorporation of environmental considerations at all stages of the decision-making process in order to minimize undesirable side effects, which, for the most part, are responsible for prevailing environmental problems.

In short, what is required is a new ecological approach to the management of those activities and processes by which we shape our future. Such an approach fully acknowledges the fact that many crucial social and economic decisions are made on the basis of inadequate consideration of their implications and consequences. To be successful, however, effective environmental management will require significant changes in the values, attitudes, and behavior of those concerned with the management of society and a drastic revision of our concepts of evaluating the future.

The Governing Council's Third Session

At its third session, held in Nairobi, Kenya, from 17 April to 2 May 1975, the Governing Council of UNEP reviewed both the concept of ecodevelopment and environmental management, and encouraged the executive director to undertake their further elaboration and development. The governing council also made the following general recommendations and observations:

1. The systematic examination of how the environmental dimension was reflected in development activities within the United Nations system, paying particular attention to environmental issues of the new international economic order;
2. Encouragement of the development of human resources and technologies;
3. Support of governmental and regional bodies in identifying and integrating environmental considerations in development programs;
4. The promotion of self-reliance with regard to technical capability, and the development of environmentally sound technology—a matter closely related to the debate on the transfer of technology;
5. The promotion and support of development activities that explicitly preserve and enhance the environment; and
6. The increase of knowledge on the state of the environment and the environmental issues that must be considered in the development process.

In this connection, the council recognized, however, that all countries needed more and better information on tolerances and on techniques for achieving environmentally sound development, and that UNEP should collect and disseminate such information so that developing countries, in particular, could make sound choices and set standards. (In any expansion of knowledge on the state of the environment, the recently established Earthwatch Program can be of vital importance, since it can be designed to provide a data base on critical environmental issues; to keep this up to date is vital.)

Finally, the council felt that mechanisms and methods should be developed to transmit information to planners and policymakers at various stages of their deliberation and programming process.

Such considerations, I believe, accurately reflect the premises on which UNEP approaches this question—namely, that environmentally sound development is development without destruction.

53. Opportunities for Environmental Protection in the Developing World

William L. Finger
International Executive Service Corps, U.S.A.

If we had known a mere twenty years ago what we know now about the vital importance of environmental protection, what a difference it would have made in the nature and the location of the plants and other installations that constitute the economic structure of this country! It is estimated that in this country alone total capital expenditure for environmental protection between 1972 and 1981 will amount to about $275 billion, or approximately 2.5 percent of our gross national product (GNP) during those years. About 40 percent of the expenditures will be made by government agencies and 60 percent by private enterprises. These expenditures will be made to correct past mistakes and to comply with present standards of environmental protection.

How many countries of the "developing world," especially the oil-rich countries around the Persian (or Arabian) Gulf, are twenty-five years or more behind the United States in their industrial development? Some are just on the threshold. What an opportunity they have to take advantage of all the knowledge about environment protection that has been developed in recent years—and so avoid many of the mistakes we made—in their drive to improve their industries and diversify their economies.

The International Executive Service Corps (IESC) helps the developing countries in this respect and finds them receptive, although occasionally one hears from a developing country the comment that the United States and other industrialized countries became rich by exporting pollution, and that now it is the turn of the developing countries to benefit by exporting pollution. Of course, appropriate measures to protect the environment do add appreciably to the cost of any business (as indicated, our capital expenditures for this purpose amount to about 2.5 percent of our GNP), and any country that fails to set adequate standards in protecting the environment could obtain a short-term cost advantage in international competition; but it is certain that, in the long run, it would be causing serious damage to all countries, including itself.

It is to be hoped that developing countries will not allow national pride or short-term self-interest to prevent them from taking advantage of the lessons which have been learned the hard and costly way by the United States and other industrialized countries.

IESC, a nonprofit corporation, was formed about ten years ago by a group of distinguished business and professional leaders and was based on a few very simple and logical premises: first, the most serious and dangerous economic problem in the world is the wide gap between the industrialized and the developing countries; second, if nothing is done about this gap, the whole world, including the United States and other industrialized nations, will become even more chaotic and difficult to live in than it is now; third, the answer to closing the gap is not money but knowledge—managerial and technological knowledge; fourth, the United States has literally tens of thousands of businessmen who have successfully completed their careers and who are still mentally and physically able to work and who want something challenging to do. Most thoughtful people would agree with these premises, and they form the basis of IESC, which was inaugurated in the Rose Garden of the White House in 1964. Our sole mission is to help the economies of the developing countries through the assignment of advisors, mainly retired executives, to solve specific problems. The executives receive reasonable expenses for themselves and their spouses, but no other compensation except the very great satisfaction of rendering a much needed service. We have some 8,000 such volunteer executives in our files, of whom about 50 are specialists in pollution control. It is not a "giveaway" program because we do not believe that anything given away will be appreciated or used effectively. The clients pay a very meaningful contribution toward the cost of the assignment, more than they pay most of their salaried employees. IESC receives its funds from three sources: the client, just mentioned; the U.S. Agency for International Development; and some 200 prominent U.S. sponsor corporations and a similar number of foreign sponsor corporations that we have assisted. The first chairman of our board was Mr. David Rockefeller; the second chairman was Mr. George Woods, former head of the World Bank; and the third and present chairman is Mr. Peter McColough, chairman of Xerox. The president and chief executive officer is Mr. Frank Pace, Jr., former U.S. director of the budget, secretary of the army, and head of General Dynamics Corporation. Our board of directors consists of some sixty-five distinguished business and professional leaders.

IESC does not ask anything for itself or for the United States. We do not seek jobs or promote takeovers. Our only mission is to help the client, thereby helping the economy of the country, and thereby helping ourselves indirectly by improving the world business climate.

In the last ten years, we have completed over 4,500 assignments in some fifty developing countries, ranging literally from airplanes to zippers, and involving every type of skill—general management, production, marketing, finance, and so on. We complete about 700 projects a year. Our main thrust is toward business, but we also help governments where such help will have a favorable effect on the economy. For example, we have assisted in the establishment of bureaus of standards; we have advised ministries of transport of agriculture, and of finance; and we have assisted in the administration of educational institutions.

Practically all of our volunteer executives find an assignment in a developing country one of the most rewarding experiences of their lives, and we have been told many times by our friends in the developing countries that an IESC executive is the best ambassador the United States can send, because they are there to help, not to ask anything for themselves.

How does all this tie in with the environment? It is very simple. Because our executives are all successful and recently retired from companies that are very conscious of the importance of environment protection, they naturally have that consideration very much in mind in advising their clients, whether it be in the construction of a steel mill or aluminum plant, the development of a shopping center, the most effective use of agricultural lands, the erection of a fertilizer factory, or the planning of a paper and pulp mill. They have the environment in mind, not only because of their own recent experiences in this country, but also because they know that it is very much in the long-term interest of their clients to do so.

Not only do IESC executives consider the effect on the environment as a matter of course on any assignment, but they have also carried out a number of projects dealing specifically with environment protection. For example, a paper company in the Philippines asked for advice in establishing a wildlife preserve to be used for studying the feasibility of developing a tropical-rain-forest center. The assignment was successfully carried out by Dr. Joseph Shomon, Director of the Nature Centers Planning Division of the National Audubon Society.

Right now we are advising the Ministry of Economic Affairs in Taiwan on the control of air, water, and noise pollution on occupational disease, and on protective equipment. This project is being handled by Dr. Willis G. Hazard, a graduate of Harvard, professor of industrial hygiene at Harvard, an official of the U.S. Public Health Service during World War II, and, most recently, Director of Industrial Hygiene of Owens-Illinois, a company with 70,000 employees. A somewhat similar project was completed in Taiwan last year when Mr. Hartselle D. Kinsey, former vice-president of Union Carbide, advised the Industrial Development Bureau on the control of solid-waste pollution from the petrochemical and caustic soda-chlorine industries.

A company in Bogota, Colombia, making cement and concrete was given help in the control of air pollution caused by dust, fumes, and smoke, and our man for Bogota was Mr. Clifford N. Stutz, sanitary engineer for the Monsanto Company. The state of Sabah in East Malaysia requested assistance in combating pollution of all kinds in Kota Kinabalu, which is being developed as a tourist center, and this help was given by Mr. Howard Almgren, a mid-career executive from San Diego, California, in charge of waste-treatment processes of the Santee County Water District. In Singapore, two projects were carried out for the Economic Development Board on the control of water and air pollution through the treatment and disposal of industrial wastes.

Members of the Audubon Society in particular will be interested in the request of a Philippine foundation for help in establishing an aviary to house birds typical of the Philippine Islands under optimum conditions, using the latest scientific principles for creating a natural and harmonious environment for the birds. This assignment was carried out by Dr. Kenton C. Lint, Curator of Birds of the San Diego Zoological Society.

Views differ as to what is a desirable rate of growth for any country, but we can be sure that the developing world will proceed with plans for economic expansion. As it does so, let us hope that it will take full advantage of all the knowledge that has been acquired in recent years about protection of the environment. IESC is proud to make some contribution toward that objective through the 700 executives whom it sends out each year to give the developing countries the benefit of their practical experiences, including an awareness of the vital importance of practicing EARTHCARE every day.

Institutional Measures for the Preservation of Ecosystems

54. Institutional Approaches to Global Protection of the Environment

J. Michael McCloskey
Sierra Club, U.S.A.

In a recent book on international environmental law, Ludwig Teclaff (1974) concludes by looking far ahead and wonders whether

> ... technology will make economic expansion compatible with preservation of the life-sustaining environment. If not, then eventually, by agreement or otherwise, the expansion of production will have to be reduced or stopped altogether, and, predictably, the emphasis would shift in the international community in general, and in international law in particular, from development to distribution. It is also possible to foresee, in such a society, the emergence, after perhaps a period of unrest, of a central authority which would assure the distribution of limited resources on an equal or an unequal basis, depending on how the period of unrest were resolved.
>
> Finally, far beyond the horizon, lies the dreamland of the environmentalists—the goal to which the environmentalist movement may eventually lead—a rationally ordered community in which the rights to existence of all, or almost all, species are recognized, limited only by the similar rights of others. In such a community, man would cease to be a destroyer and become a benevolent steward with the responsibility of assuring the survival of other species within the limitations imposed by the life-sustaining

capacity of the environment and within a legal system enlarged to encompass interests other than those purely human.

In looking at these same problems, another set of authors sees the period of unrest that Teclaff speaks of as being close at hand. In writing on "Environmental Decay and International Politics," Linda Shields and Marvin Ott (1974) observe:

> The risk of conflict stemming from competition for vital natural resources and energy supplies is particularly critical because it concerns elements which, combined with technological capabilities, now largely define a nation's power potential. One response, manifested in the current round of U.N. conferences regarding the law of the sea, will be attempts to strengthen the fabric of international law and organization. But because of the wide-ranging implications of energy and resources for state interests, the principal trend may be towards internalizing national sources and supplies where possible—a process which has apparently already begun, as evidenced by "Operation Independence." It is conceivable that in the future sovereignty will be substantially defined in terms of the autonomy of a nation's energy supply.
>
> The limitations of this approach are obvious.
>
> . . . If the constructive, rather than the destructive, possibilities of this situation are to be realized, it will be because both producer and consumer nations will perceive their physical security and economic well-being, i.e. their vital national interests, to be imperiled by conflict and safeguarded by cooperation.
>
> There are no defenses which individual states can establish against widespread environmental deterioration except through coordinated international action, no aggressor except that each state is a polluter and depletor, and no victims except that every state is dependent on the biosphere. But any initiative for cooperation will come primarily from the realization on the part of individual states of the extent to which their security is threatened in the absence of cooperation.

Thus, Shields and Ott suggest the immediate alternatives are increased conflict, increased tendencies toward self-sufficiency, or increased efforts to develop cooperative international institutions. Since no one would welcome the prospect of increased conflict, hope must lie with the other two alternatives.

But is increased self-sufficiency the answer? It would depend, I suppose, among other things, on whether it is premised on controlling demand or simply expanding exploitation. "Project Independence" in this country is premised on expanding exploitation, and, accordingly, it appears that it can be won only at a high environmental price. It means increased mining and drilling in fragile and hazardous environments, with high environmental costs, high economic costs, declining recovery of net energy, and depleted reserves.

Might not Adam Smith have had a point—at least in one respect—when he urged free trade on the supposition that everyone benefits most when each nation concentrates on doing what it can do more economically than its neighbors? The United States may never be able to produce much new oil at as low an economic or environmental price as the Middle East. One could envision an international regime in which trade would develop along channels that reflect minimal amounts of environmental impact as well as relative economic efficiency. Environmental impact analyses could be done to determine which countries can produce and ship a raw material with the least adverse environmental impact. Of course, no regime exists to bring order of this type into international trade, but it is intriguing to speculate whether the need for greater economic order internationally might usher in new international institutions that could also help to protect the environment.

Raymond Dasmann (1975) views the prevailing web of international trade as the pattern peculiar to what he calls "biosphere people," whom he contrasts to "ecosystem people." He says that ecosystem people are "members of indigenous traditional cultures and those who have seceded from, or been pushed out, of technological society." Biosphere people are tied in with the global technological civilization.

> Ecosystem people live within a single ecosystem, or at most two or three adjacent and closely related ecosystems. They are dependent upon that ecosystem for their survival. If they persistently violate its ecological rules, they must necessarily perish.
>
> Biosphere people draw their support, not from the resources of any one ecosystem, but from the entire biosphere.
>
> Local catastrophes that would wipe out people dependent on a single ecosystem may create only minor perturbations among the bio-

sphere people, since they can simply draw more heavily on a different ecosystem.

Today, with the increase in human numbers and the enormous pressure being exerted on all ecosystems, one of the distinctions between ecosystems and total biosphere is being broken down. No longer can biosphere people remain buffered against the breakdown of particular ecosystems. A drought in India or North America now has global repercussions. The entire biosphere is now becoming as closely interconnected through human endeavor as the most delicately balanced ecosystem within it (pp. 6 and 9).

Dasmann thus suggests that the stability of the economic system on which biosphere peoples depend may be faltering, and perhaps we are already seeing signs of that. Diminishing returns may thus set in if we continue to create greater economic interdependencies. If we persist in spreading this pattern, more international institutions may be needed to restrain biosphere people and to regulate the equilibrium of this increasingly unstable mechanism.

Recently, writing in the *IUCN Bulletin*, Robert Allen (1975) made the point that heavy manipulation and exploitation of nature carries implications for political institutions. In discussing spreading development, he writes:

The more functions we try to take over from other species, the more dependent we become on our own technologies and management systems. And the more we are dependent on them, the more disciplined we ourselves must be—the more we must subordinate ourselves to the demands of efficiency and expediency. Discipline on such a scale means loss of freedom, loss of individuality, and the constant anxiety, nervous tension and small-mindedness that comes when more and more ordinary men and women are in a position to make bigger and more disastrous mistakes. In fact, discipline on such a scale would probably be impossible to sustain.

Thus, we face a series of paradoxes. As Roderick Nash pointed out, an environmental movement can come into existence only as heavy development produces environmental problems. But institutions to solve these problems may come into existence internationally only as development programs and trade relations deepen. But as that happens, this system of trade may move into a precarious equilibrium, which may need more discipline and authoritarian control to survive—and that may come at the high cost of political freedom.

There has to be a better way—a way that will lead the overdeveloped countries back toward the ways of ecosystem peoples and that will encourage the underdeveloped countries to avoid the mistakes of the biosphere peoples.

The authors of the following papers will fill in some of this middle ground of what may be done to evolve limited institutions for controlling development in order to protect the environment.

References

Allen, Robert. 1975. Interdependence: The trend we cannot buck. *IUCN Bulletin*, New Series 6(3):9-10.

Dasmann, Raymond F. 1975. National Parks, Nature Conservation, and "Future Primitive." Paper prepared for the South Pacific Conference on National Parks, Wellington, New Zealand, 24-27 January 1975. 14 pages (mimeographed). (Published in *The Ecologist* 6(5):164-167 and in Kenneth Brower, [1975?] *Micronesia: Island Wilderness.* San Francisco: Friends of the Earth.)

Shields, Linda, and Marvin Ott. 1974. Environmental decay and international politics: The uses of sovereignty. *Environmental Affairs* 3(4):743-767.

Teclaff, Ludwik. 1974. The impact of environmental concern on the development of international law. In *International Environmental Law*, eds. Ludwik A. Teclaff and Albert E. Utton, pp. 229-262. Praeger Special Studies in International Politics and Government. New York: Praeger Publishers.

55. Environmental Impact Analysis: The First Five Years of the National Environmental Policy Act in the U.S.A.

Oliver Thorold
London, England

Introduction

In an important sense, the United States' National Environmental Policy Act (NEPA) of 1969 (Public Law 91-190, signed into law on 1 January 1970), marked the arrival of a new kind of law. For while earlier environmental laws had addressed particular risks or impacts, or had single-factor objectives, NEPA calls for the examination of the total environmental consequence of a decision and attempts to bring about a process of environmental thought reform in all institutions of the federal government. It is therefore of relevance to all kinds of environmental concern. The novelty of the approach lies not so much in the ambitious policy declared in the act, but in the "action-forcing" provision designed to implement it.

During the past five years, all agencies and departments of the federal government, before taking major actions significantly affecting the environment, have been obliged to prepare environmental impact statements (EIS's) describing not only the expected impact of the proposed action, but also that of the alternatives. If the concept of environmental impact assessment is not wholly new, it has never before been introduced into the machinery of government in such a general and systematic form. No doubt that is why so much interest has been aroused. Some twenty-one states have already adopted laws patterned on the federal model, and the influence has spread abroad. Canada, France, Australia, and Israel have either announced or enacted comparable programs, while in Britain and Germany the concept is being studied.

After five years and nearly 6,000 EIS's, opinions both public and private seem to confirm that the exercise is well worthwhile. Though most commentators still stress the need for improvement

The Preservation of Ecosystems 661

both in the quality of statements and in their influence on decisions, few now dispute that federal decision-making is more sensitive to environmental considerations.

The lessons to be learned from NEPA are far too valuable to be neglected overseas. For although NEPA shows all the hallmarks of U.S. legislation, the problems with which it deals are more general.

This paper has been compiled partly from the voluminous literature on NEPA, and partly from the author's inquiries and observations made during October 1973 and April 1974 in the United States. In it, the more important provisions of the act are described, together with the judicial and administrative gloss it has acquired. Following this description, some of the "teething" problems are examined, and a preliminary judgment on the first five years offered.

The Act and Its Overseers

Environmental Impact Statements

NEPA is best known for its requirement (NEPA, Section 102[2][C]; see Healy 1973) that all agencies of the federal government shall

> ... include in every recommendation or report on proposals for legislation and other major Federal actions significantly affecting the quality of the human environment, a detailed statement by the responsible official on—
>
> 1. the environmental impact of the proposed action,
> 2. any adverse environmental affects which cannot be avoided should the proposal be implemented,
> 3. alternatives to the proposed action,
> 4. the relationship between local short-term uses of man's environment and the maintenance and enhancement of long-term productivity, and
> 5. any irreversible and irretrievable commitments of resources which would be involved in the proposed action should it be implemented.

Today it requires a conscious effort of memory to realize that this provision was a comparatively late addition to a bill whose principal aim was to declare a national environmental policy.

The National Environmental Policy

The need for the policy had been debated for many years. In full form, and as embodied in Section 101 (a) of the act, it reads:

> The Congress, recognizing the profound impact of man's impact on the interrelations of all components of the natural environment, particularly the profound influences of population growth, high-density urbanization, industrial expansion, resource exploitation, and new and expanding technological advances and recognizing further the critical importance of restoring and maintaining environmental quality to the overall welfare and development of man, declares that it is the continuing policy of the Federal Government, in co-operation with State and local governments, and other concerned public and private organizations, to use all practicable means and measures, including financial and technical assistance, in a manner calculated to foster and promote the general welfare, to create and maintain conditions under which man and nature can exist in productive harmony, and fulfill the social, economic and other requirements of present and future generations of Americans.

This declaration of policy served to extend the legal mandates of all federal agencies, whose principal authorizations had, until NEPA, not usually made clear their power or duty to give heed to unquantifiable environmental factors. Section 102(1) of the act now puts that point beyond doubt:

> The Congress authorizes and directs that, to the fullest extent possible ... the policies, regulations, and public laws of the United States shall be interpreted and administered in accordance with the policies set forth in this Act.

Whether or not the statement of policy in Section 101(a) contains "law to apply" (on which point the cases are not entirely clear), it provides the basic direction for implementation of the act.

Soon after NEPA passed into law it became clear that the courts would enforce the obligation in Section 102(C) requiring the preparation of impact statements. Moreover, the doctrine of "standing to sue" was at the same time changing, with the courts expressing themselves increasingly ready to take a liberal view of standing requirements.

An important part of NEPA's history therefore concerns the lawsuits (numbering over 300 to date) brought by citizen groups seeking injunction against agency actions until adequate EIS's had been prepared. Given the uncertainties and ambiguities in the act, in particular the question of deciding what constitutes a "major action," or what amounts to a significant effect on the environment, the court rulings resulting from citizen litigation have done much to clarify the act's meaning.

The readiness of the courts to enforce NEPA's provisions may be due in part to the absence of any other purposely built scheme for enforcing the act. The Council on Environmental Quality (CEQ) and the Environmental Protection Agency (EPA) both have certain responsibilities, although their joint powers and duties fall short of real supervision or regulation.

The Council on Environmental Quality

CEQ was established under Title II of the act (Section 202) as a nonexecutive body in the Executive Office of the President. The council consists of three members, who have a supporting staff of under a hundred. It is charged with the duty of advising the president in the preparation of his annual "Environmental Quality Report," and, in addition, its functions include gathering timely and authoritative information on enviornmental trends, reviewing existing federal-agency policies and recommending changes where necessary, conducting surveys and studies, issuing reports, and carrying out environmental monitoring.

Executive Order No. 11514. By executive order (Executive Order No. 11514, dated 15 March 1970), CEQ was further charged with issuing guidelines to federal agencies on the preparation of EIS's. Although the guidelines are not themselves binding law, they have been extensively adopted and applied by the courts, and in their most recent version (August 1973) they have in turn incorporated and codified the more important court decisions handed down up to that date. Thus, both the courts and CEQ have been involved in the process of elucidating and clarifying the act's meaning, each adopting principles from the other.

The Environmental Protection Agency

The role of EPA is not set out in NEPA itself, but received attention shortly after in Section 309 of the Clean Air Act of 1970, which requires the administrator of EPA to review and comment in writing on the environmental impact of any matter relating to the Clean Air Act, proposed regulations or legislation, newly authorized federal projects for construction, and any major federal action to which Section 102(2)(C) applies. In the event that the administrator finds any of the above "unsatisfactory from the standpoint of public health or welfare or environmental quality," he must publish his "determination" and refer the matter to CEQ.

Although this provision would appear to provide some machinery through which unsatisfactory projects would come, via EPA and CEQ, to the attention of the president, it appears to have been neglected. However, EPA does undertake two kinds of review. EIS's in draft are classified under three headings—Category I: Adequate; Category II: Insufficient information; and Category III: Inadequate. The proposal described in the statement is assessed as either "LO" (lack of objection), "ER" (environmental reservations), or "EU" (environmentally unsatisfactory). This system of review, it seems to be widely agreed, leaves much to be desired. Moreover, it has been suggested that EPA, for political reasons, has endeavored to avoid the use of "EU" ratings when possible, preferring to rate "ER" in order to avoid direct confrontations with other agencies.

The Long Reach of NEPA

In Section 102(2)(C), Congress enacted that EIS's were to be prepared on legislative proposals and "major" federal "actions" "significantly" affecting the quality of the human environment. From this terminology, the courts and CEQ have evolved principles that assist agencies to identify the kinds of actions covered, and to identify "major" actions with "significant" effects. Thus, the latest CEQ guidelines suggest that "actions" include but are not limited to:

   1. Recommendations or favorable reports relating to legislation including requests for appropriations. The requirement for following the Section 102(2)(C) procedure

The Preservation of Ecosystems

applies to (i) agency recommendations on their own proposals for legislation, and (ii) agency reports on legislation initiated elsewhere. In the latter case only the agency with primary responsibility need prepare.
2. New and continuing projects and program activities: directly undertaken by federal agencies; or supported in whole or in part through federal contracts, grants, subsidies, loans, or other forms of funding assistance, or involving a Federal lease, permit, license certificate, or other entitlement for use.
3. The making, modification, or establishment of regulations, rules, programs, and policy.

The breadth of meaning given to "actions," in part through the courts upon the challenge of an environmental group, shows that NEPA has a long reach. Virtually every possible form of federal involvement is covered, however slight.

Upon the problem of determining what is a "major" action with "significant" impact, Frederick Anderson has written:

> NEPA's legislative history shows that Congress intended to interrupt business-as-usual and affect decision-making at the lowest agency levels. ... Judicial failure to implement this purpose could have meant that the Act would hardly have deflected the wheels of government from their accustomed ruts. For this reason the courts have policed the lower boundaries of NEPA's application with greater than ordinary vigilance and have worked in concert with the CEQ guideline-setting process to keep the threshold low. (Anderson 1973)

Some idea of the extent of the impact-statement program and its impact upon different agencies is gained from Table 1, in which the numbers and distribution of statements prepared to 1 July 1974 are presented. It shows that some 5,430 EIS's have so far been prepared, with the Department of Transportation and the U.S. Army Corps of Engineers best represented.

Statistics give certain misleading impressions, however, for while many of the statements filed by the Department of Transportation are brief and straightforward, dealing largely with road projects partly funded by the department, the EIS on the liquid-metal fast-

Table 1. Environmental Impact Statements Prepared by Agencies of the Federal Government Through 1 July 1974

Agency	1970	1971	1972	1973	1974 (6 mo.)
Department of Transportation	61	1,293	674	432	192
Corps of Engineers	119	316	211	243	174
Department of Agriculture	62	79	124	166	94
Department of the Interior	18	65	107	119	72
Atomic Energy Commission	32	22	65	28	28
Federal Power Commission	5	15	65	15	7
Department of Housing and Urban Development	3	23	26	22	10
Department of Defense	5	27	24	19	12
General Services Administration	3	34	6	24	11
Environmental Protection Agency	0	16	13	26	7
National Aeronautics and Space Administration	0	22	3	3	1
All Others	11	60	70	51	36
Total	315	1,946	1,371	1,148	644

Source: Council on Environmental Quality 1974.

breeder reactor program from the Atomic Energy Commission, for example, constitutes a major review of a new technology, running to many volumes.

The CEQ Guidelines

Numerous court cases provide guidance on the "threshold," while guidelines adopted by the agencies themselves have become more reliable as they incorporate the lessons of past court cases, and as they survive further legal challenge. The CEQ guidelines suggest some general formulary principles:

> The Statutory clause ... is to be interpreted by agencies with a view to the overall, cumulative impact of the action proposed, related Federal actions and projects in the area, and further actions contemplated. ... Proposed major actions, the environmental impact of which is likely to be

highly controversial, should be covered in all cases. In considering what constitutes major action significantly affecting the environment, agencies should bear in mind that the effect of many Federal decisions about a project or complex of projects can be individually limited but cumulatively considerable. This can occur when one or more agencies over a period of years puts into a project individually minor but collectively major resources, when one decision involving a limited amount of money is a precedent for action in much larger cases or represents a decision in principle about a future major course of action. . . .

The significance of a proposed action may also vary with the setting, with the result that an action that would have little impact in an urban area may be significant in a rural setting, and vice versa. . . .

In many cases, broad program statements will be required in order to assess the environmental effects of a number of individual actions on a given geographical area (e.g., coal leases), or environmental impacts that are generic or common to a series of agency actions (e.g., maintenance or waste handling practices), or the overall impact of a large-scale program or chain of contemplated projects (e.g., major lengths of highway as opposed to small segments). . . .

Agencies engaging in major technology research and development programs should develop procedures for periodic evaluation to determine when a program statement is required for such programs. Factors to be considered in making this determination include the magnitude of Federal investment in the program, the likelihood of widespread application of the technology, the degree of environmental impact which would occur if the technology were widely applied, and the extent to which continued investment in the new technology is likely to restrict future alternatives. Statements must be written late enough in the development process to contain meaningful information, but early enough so that this information can serve as an input in the decision-making process. . . . In any case a statement must be prepared before research activities have reached a stage of investment or commitment to implementation likely to determine subsequent development or restrict . . . later alternatives. Statements on technology research and development programs should include an analysis not only of alternative forms of the same technology that might reduce any adverse environmental impacts, but also of alternative technologies.

From the cases come specific illustrations. They show that NEPA has been applied to a loan from the Farmers Home Administration for the construction of a golf course and park in Texas

(*Texas Committee on Natural Resources* v. *United States,* 430 F. 2d 1315 [5th Circuit]); to a contemporary suspension by the Interstate Commerce Commission of rail freight rates to permit a 2.5 percent surcharge on the shipment of recyclable metal scrap (*City of New York* v. *United States,* 337 F Supp. 150); to the repair and expansion of a canal towpath (*Berkson* v. *Morton,* 2 ELR 20659 [D. Md. 1971]); to permits issued by the U.S. Army Corps of Engineers under the Refuse Act permit program (*Kalur* v. *Resor,* 335 F. Supp. 1. [D.D.C. 1971]); to a Corps of Engineers project to clear vegetation from 88 kilometers of river (*Sierra Club* v. *Laird,* 1 ELR 20085 [D. Ariz. 1970]); to a grant for construction of a 42-kilometer highway in a national forest (*Upper Pecos Association* v. *Stans,* 328 F. Supp. 332 [D.N.M. 1971]); to an urban-renewal project in Washington, D.C., covering five blocks (*Zlotnick* v. *Redevelopment Land Agency,* 339 F. Supp. 793 [D.D.C. 1972]); and to the construction of an incinerator at a medical center (*Montgomery County* v. *Richardson,* 2 ELR 20140 [D.D.C. 1972]).

And NEPA has been held not to apply to a Department of the Interior plan to fence 2,750 hectares of land acquired to protect a reservoir (*Maddox* v. *Bradley,* 345 F. Supp. 1255 [N.D. Texas 1972]); to the same department's plan to lease 50 hectares of land next to land already leased for coal extraction (*Jicarilla Apache Tribe of Indians* v. *Morton,* 2 ELR 20287 [D. Ariz. 1972]); to an EPA grant for construction of a regional sewage plant (*Howard* v. *E.P.A.,* 2 ELR 20745 [W.D. Va. 1972]); or to the construction of a Postal Service bulk-mail facility covering 25 hectares in a 135-hectare industrial park (*Maryland-National Capital Park and Planning Commission* v. *United States Post Office,* C.A. 349 F. Supp. 1212 [D.D.C. 1972]).

It needs to be pointed out that some of the decisions referred to above took into account the long-term cumulative impact to be expected, and in some cases the "triggering" effect of the particular project concerned. But in any event, since the first set of CEQ guidelines called upon agencies to develop their own guidelines defining the actions to which NEPA applies, the threshold lines have become clearer (although a court, unlike the agency itself, is not bound by such a guideline).

In the fields of energy, water resources and transport, nearly all projects or licensing are now covered. The Federal Power Commission's first set of regulations placed the cut-off for hydro projects at 2,000 horsepower. The policy of the Federal Highway Administration is to prepare statements on virtually every project funded by it, whether wholly or partly, and on all authorizations. The Department of Housing and Urban Development has a harder line to draw, but in its own guidelines has adopted a principle that any housing project involving over 100 units merits a statement. Certain agencies, such as the Atomic Energy Commission, hardly deal with matters falling outside NEPA.

николи

NEPA and the Environmental Protection Agency

While in principle NEPA applies to actions with beneficial consequences as well as those that are adverse, its applicability to EPA has been a matter of some dispute. It has for some time accepted that it should prepare statements on its grant-giving actions (principally for sewage-treatment plants), but only as from October 1974 has it accepted (voluntarily) that statements should be prepared on its other activities, such as the setting of national ambient air-pollution standards, designating sites for ocean dumping, or cancelling pesticide registrations. Since these involve some careful "balancing," it is perhaps as well that the costs and benefits be set out explicitly.

Environmental Impact Statements

Timing

If EIS's are to have the desired effect on policies, they must be prepared at a time when options are genuinely open. Inevitably, during NEPA's first few years EIS's were found to be of poor quality, a reflection no doubt of the fact that NEPA did not exempt projects or actions uncompleted yet at an advanced state. With the passage of time, it has become possible to prepare the analysis at an earlier point in a project's life.

So far, most of the matters falling within NEPA's scope have been analyzed in a single EIS. But the logic of the concept often requires several EIS's, corresponding to the various "option-foreclosing" decisions taken along the way. There is now developing a trend towards "programmatic" EIS's such as that by the Atomic Energy Commission on the Liquid-Metal Fast-Breeder Reactor Program.

If EIS's are prepared at an early stage, it follows that the information available may not be complete, and perhaps in some cases seriously deficient. Between the need to prepare early enough to look at the alternatives realistically and late enough to have adequate information available, a balance must be struck. In its fifth annual report (Council on Environmental Quality 1974) CEQ suggested that

> ... an environmental analysis needs to be prepared as a rough approximation during the initial planning of a project and then gradually refined as the planning of the project proceeds and as alternatives are identified, analyzed, and perhaps discarded. ... This procedure is analogous to ways in which the economic analysis of a project is currently made. ... (Council on Environmental Quality 1974, pp. 411-412)

Content

Section 102(2)(C) sets out the required content of an EIS, though without specificity. As a result of court decisions and agency guidelines, it is now possible to describe the requirements in more detail. Though the EIS must be "detailed," there are limits to the amount of detail which is desirable. The CEQ guidelines suggest that the information, data, maps, and diagrams should be adequate to permit an assessment of potential environmental impact by commenting agencies and the public. The content requirements not explicitly set out in Section 102(2)(C) include:

1. A description of the area likely to be affected;
2. Population and growth characteristics of the area;
3. Interrelationships and cumulative impacts of the proposed action and related projects;
4. Relationship of the proposed action to other land-use plans and policies, and controls for the affected area;
5. Both positive and negative aspects of the action;
6. Secondary or indirect, as well as direct primary consequences;
7. Alternatives including those not within the authority of the responsible agency;
8. A rigorous exploration and objective evaluation of the impacts of all reasonable alternatives, including the alternatives, including the alternative of taking no action;

The Preservation of Ecosystems 671

9. A description of the unavoidable impacts, including a clear statement of how other avoidable impacts will be mitigated;
10. Where they have been prepared, cost-benefit analyses; and
11. The consequences of accidents which might occur.

In the least six cases, courts have endorsed the fundamental principle that NEPA is above all an environmental "full-disclosure" law, and calls for discussion of all known impacts. Court decisions also make clear that EIS's are to be "nonconclusory." The Calvert Cliffs' decision (*Calvert Cliffs' Coordinating Committee v. Atomic Energy Commission*, 449 F. 2d. 1109 [D.C. Cir. 1971]) further established the important principle that agencies are not entitled to rely upon the certification of other agencies that their environmental requirements would be satisfied and thus called for independent evaluation.

Commenting Procedures

In order to achieve the coordination of policies that had previously been lacking, NEPA Section 102(2)(C) laid down, in outline, a procedure for obtaining comments on impact statements:

> Prior to making any detailed statement, the responsible Federal official shall consult with and obtain the comments of any Federal agency which has jurisdiction by law or special expertise with respect to the environmental impact involved. Copies of such statements and the comments and views of the appropriate Federal, State and local agencies, which are authorized to develop and enforce environmental standards, shall be made available to the President, the CEQ and to the public . . . and shall accompany the proposal through the existing agency review processes.

Accordingly, there has developed a system under which all EIS's are first prepared in draft, then circulated to other agencies for comment as well as being made available for public comment. With the aid of the comments received, the agency then proceeds to a final EIS, which then accompanies the proposal through the agency review processes.

NEPA was wholly silent on the timescale of the procedure, but the CEQ guidelines lay down that "no administrative action subject to Section 102(2)(C) is to be taken sooner than 90 days after a draft EIS has been circulated for comment," and "neither should such administrative action be taken sooner than 30 days after the final text of an environmental statement (together with comments) has been made available." Those required or wishing to comment on draft EIS's have levelled the complaint that this 30 to 45 day period is too short for the preparation of worthwhile comments, while agencies who by virtue of their mandates must comment on a large number of EIS's from other agencies have faced a major addition to their workload.

Some projects, no doubt, would be delayed by a longer commenting period, but many have been planned over so many years that even a doubling of these time periods would be unlikely to introduce serious delay. The CEQ guidelines suggest that the draft EIS should be prepared for comment prior to the "first significant point of decision" in the agency review process. The comments from other agencies and from the public can then become a genuine input into the decision.

Depending on the magnitude of an action, the degree of interest in it, and its complexity, agencies should consider holding public hearings, according to the guidelines. But this may be less necessary if public involvement has already been secured through other means.

The Five Years of Experience

After five years, it remains a matter of great difficulty to assess the real changes brought about by NEPA. Obviously, nobody is in a position to say what might have been had the act not been passed. Moreover, the transitional difficulties were considerable: NEPA passed into law, as has been noted, without any form of "grandfather clause." Most agencies were therefore confronted by severe problems at the start, for they had not only to apply the act to new projects but also to many existing ones. Before the act passed, no serious attempt seems to have been made to forecast the manpower required to implement it. Indeed, it would have been impossible to do so reliably, for it proved to be partly as a result of lawsuits and court rulings that the scope of the requirements became clear.

For example, the Calvert Cliffs' decision (which required the Atomic Energy Commission to consider a broader range of issues and also to report on many ongoing licensing operations) was said to have caused the agency to increase its impact statement-writing staff from 20 to 200. Similarly, the Greene County case (*Greene County Planning Board v. Federal Power Commission*, 455 F. 2d. 412 [2d. Cir. 1972]) (which established that the Federal Power Commission could not rely on an EIS prepared by a prospective licensee) caused the FPC's staff to rise from 22 to 75. The early period must be judged with this in mind. And, given that it is of the greatest importance that EIS's be prepared at an early stage of an action or project, those concerning projects already well advanced were unlikely to be of the best.

Certainly there is a body of evidence confirming that many EIS's prepared in the first three years were well below the standard of quality necessary to offer much prospect to improving decision making. The litigation tended to highlight the worst cases, and therefore does not give a balanced picture. But several other studies based on fair samples bear the point out.

A Report by the Comptroller General of the United States to a House of Representatives Subcommittee on Fisheries and Wildlife Conservation of November 1972, for example, found that the usefulness of a group of EIS's studies had been impaired by:

1. Inadequate discussion of, and support for, identified environmental impacts;
2. Inadequate treatment of reviewing agencies' comments; and
3. Inadequate consideration of alternatives and their environmental impacts.

Serious defects and omission were pointed out in each EIS covered.

A study was published in 1973 by the Center for Science in the Public Interest. [This is a group of public-interest scientists, based in Washington, D.C., whose staff investigates issues involving energy, the environment, consumers, and food administration; publishes reports; and initiates legal action.—Editor] The study examined statements produced by the Federal Highway Administration through June 1972 for urban highway projects in cities of larger than 50,000 population. It showed that, inter alia, 86 percent did not consider mass-transit alternatives, 18 percent failed to mention noise, 54 percent failed to mention impact on nearby property values, and 34 percent failed to discuss community disruption. Of the 795 highway statements reviewed by EPA between November 1971 and April 1973, 45 percent were found to contain inadequate information (Sullivan et al. 1973).

Alternatives to the Proposed Action

The complaint that alternatives were inadequately considered runs through almost all of the early comments on NEPA's implementation. One of the lessons to be learned is that alternatives need to be considered at the proper time. Little purpose seems to be served by requiring each EIS on a power station to rehearse the economic and other arguments on wind, tidal, and solar power. Equally, it may be doubted whether EIS's on urban highways should look at mass-transit alternatives in each and every case.

Indeed, there soon emerged a more or less standard treatment of such alternatives (often verbatim copies of other statements), which was slotted in as "boiler plate" for each EIS. Many kinds of alternatives can be realistically discussed only at a general level—or, perhaps more accurately, in some kind of hierarchy that reflects the stage by stage "option-foreclosing" inherent in project planning and program planning—and a trend is developing in this direction.

The Standard of Analysis

The standard of analysis also seems to have come under fire, a problem closely related to the quality and quantity of the staff hired to analyze. It was put to the author that those unable to competently analyze a project would have recourse to prolix description, with obvious effects on readability, value, and impact. Similarly, it was suggested that while certain agencies were skillful at quantifying benefits, they were less successful at assessing costs. In consequence, the very problem that NEPA was designed to correct endures as long as EIS's are inadequate. But once again, one is bound to reflect that this may be related to the way in which NEPA was introduced. As late as April 1974 it was suggested to me that at least three-quarters of the projects emanating from the Department of the Interior had been "in the pipeline" before 1970, and it may be too much to expect that model EIS's can be prepared under those circumstances.

Almost all commentators seem to agree that the quality of EIS's is rising. The relentless litigation here played a part, for agencies soon came to realize that failure to prepare an EIS that would survive judicial review on grounds of adequacy could involve even greater delay. Provoked by both lawsuits and aggressive guidelines from CEQ, agencies have increasingly stiffened their own internal guidelines, and have stipulated procedures for their own staffs to follow. Those agencies, like the Army Corps of Engineers, which made strenuous efforts to assemble qualified staffs to carry out the analyses, were able to show notable improvements in quality.

CEQ's fifth annual report (Council on Environmental Quality 1974) records that universities, consulting firms, agencies, industry, and environmental groups have channeled substantial effort into developing an understanding of environmental forecasting. All federal agencies involved in the energy field are cooperating on a study of the environmental effects of different types of energy systems.

Secondary Impacts

Another area calling for methodological research is that of secondary impacts. At its simplest level, the difficulty arises when sewage works or water mains are constructed. Inevitably, they may create conditions in which residential development is encouraged, even if current plans do not so state. Likewise, the construction of a deepwater port will probably trigger considerable onshore developments. Early EIS's were particularly weak on analysis of secondary effects, a failing that has attracted criticism. In conjunction with other agencies, CEQ is now sponsoring research that might help to cure this defect. Secondary impacts of the type mentioned will need to be reconciled with local land-use plans, a fact that highlights the important if obvious point that systems of environmental analysis and land-use planning should be closely interrelated.

Mere preparation of EIS's by backroom analysts will not ensure that projects are environmentally sound. The importance of full communication between program planners and those who prepare and write the statements can hardly be overstated. In the early years it was apparent that this communication was lacking, which is not surprising, but as the personnel concerned develop working relationships things should improve. Internal agency guidelines can help by establishing stages in project planning-procedures at which environmental evaluation must be carried out.

Notwithstanding the initial difficulties the act has plainly had an effect, more it is thought on some agencies than others. The Army Corps of Engineers has been widely mentioned as an agency which took the spirit of the act to heart after a reluctant start. Indeed the corps itself has published statistics showing just how many of their studies and projects have been affected. Up to June 1973 the corps together with the Board of Engineers for Rivers and Harbors reported 197 projects or studies modified, 23 abandoned, deferred, or restudied, 39 temporarily or permanently delayed, and a further 6 subject to internal negative reports. A further seven had been stopped by court action (Anonymous 1973). The corps is so continually involved in projects with environmental impact that it may not be wondered at that they were affected in such a dramatic way. Yet other agencies have also been affected:

> The application of the NEPA review procedures [to the AEC] has resulted over the last 3 years in many modifications and changes in nuclear plant design, including redesign of intake structures and major cooling systems, modifications of the thermal plume and the radiological and the chemical waste systems, rerouting of transmission lines, installation of fish screens, redesign of causeways, revision of environmental monitoring plans, and new studies of alternative cooling systems. (Council on Environmental Quality 1974, p. 377)

In personal interviews most agency staff involved with environmental evaluation would mention specific cases in which they felt that a project had been abandoned on account of NEPA review, and others which had been changed. Equally, a timely adjustment to a project may save the need for NEPA review—as where a highway planned to cross a public park is rerouted around the park thus ceasing to have "significant" environmental impact.

The CEQ has been active in sponsoring studies on NEPA's impact on different agencies. The Forest Service, they found, complied "substantially with both the letter and the spirit of NEPA and with the Council's guidelines." The Bureau of Land Management was found to have supplemented its planning system at critical points to fulfill the requirements but had failed to prepare statements on their management framework plans. The navy, largely as a result of the attitude of senior officials, had made substantial efforts to prevent environmental problems before they arose.

In short, it seems fair to say that substantial benefits have already been reaped, and that the system is still passing through the stage of improvement. The activity of preparing an impact statement, looked at alone, of course costs money. In its third annual report (Council on Environmental Quality 1972, p. 258) the CEQ suggested that the budgetary costs of the personnel involved could be as much as $65 million per annum, but then point out that the Cross-Florida barge canal, which was stopped by the president after an impact statement had been prepared, had already cost $50 million and would have cost a further $130 million to complete. Viewed against the costs of projects abandoned as a result of NEPA review, the investment should please fiscal conservatives.

While NEPA has yet to reveal its full potential, it can be commended as an ambitious but realistic approach to improving the quality of government decision making. From U.S. experience in the field much can be learned about both the developing science of impact assessment and the ways in which an assessment system should be set up. Impact assessment is no substitute for other kinds of environmental control, but it provides the institutions of government and the public with a perspective on decisions which has for too long been lacking.

References

Anderson, Frederick R. 1973. *NEPA in the Courts: A Legal Analysis of the National Environmental Policy Act.* Washington, D.C.: Resources for the Future.

Anonymous. 1973. *Effect of NEPA on Corps Studies and Projects.* Washington, D.C.: U.S. Army Corps of Engineers.

Council on Environmental Quality. 1972. *Environmental Quality: The Third Annual Report of the Council on Environmental Quality.* Washington, D.C.: U.S. Government Printing Office.

———. 1974. *Environmental Quality: The Fifth Annual Report of the Council on Environmental Quality.* Washington, D.C.: U.S. Government Printing Office.

Healy, Martin. 1973. The Environmental Protection Agency's duty to oversee

NEPA's implementation: Section 309 of the Clean Air Act. 3 *Environmental Law Reporter* 50071.

Sullivan, James B., Paul A. Montgomery, and Andrew Farber. 1973. *Evaluating Highway Environmental Impacts.* Washington, D.C.: Center for Science in the Public Interest.

56. Habitat Management: Problems and Prospects

Margaret M. Stewart
State University of New York at Albany,
U.S.A.

Introduction

The big question we all face is how to maintain diversity of species and the integrity of natural communities in the face of human increase topped with human ignorance and greed. We now have behind us more than 100 years of experience and research in habitat and species management, from which we have gleaned a basic understanding of community ecology. Although we have much to learn about most species and can predict only generalities on a large scale, we know in general what sorts of things will happen with various kinds of disturbances.

When we talk about habitat management, we usually think about one of three levels: the single species, interaction between two coexisting species, or interactions at the community level. Operationally, we cannot deal with large-scale units, such as biomes, continents, or oceans, Ideally, we like the idea of no management—just letting nature manage—but practically, we have gone too far for that except in some wilderness or selected areas. We should keep in mind that wilderness areas exist today only because in the past they have been inconvenient for man to exploit. They have been too cold, too steep, too hot, too dry, too rocky, or too wet. Aldo Leopold said, "The first principle of intelligent tinkering is to save all the parts." That is not always done in management.

What are the implications of habitat management? In most situations we are talking about man's exploitation of some part of an ecosystem, or competition between man and the species he wishes to eliminate or utilize in some way. If we really "manage" a habitat, we may be arresting succession and holding community development at a "preclimax" stage that is suited to our needs, or favoring one species at the expense of others. In the past, habitat management has involved primarily aspects of (1) agriculture; (2) resource use, such as timber production; (3) prevention of over-exploitation with mining or fishing; and (4) maximum production of game species.

Attitudes toward habitat management for any specific part of the world reflect very much the history and cultural patterns of that area. Attitudes change with time and place. In the United States, we still have a frontier attitude toward nature. Changing that attitude so that financial resources can be expended to preserve and improve species and habitats is a slow process.

I will present some of the major problem areas in habitat management as I see them, concentrating on subjects of interest to this group. My remarks are aimed primarily toward patterns in the United States, but I will draw a few examples from other countries with which I have had personal experience. In each of these major aspects of habitat management, attitudes and actions of sensitive citizens who are concerned about their environment can play a major role in influencing action. Finally, I will suggest "avenues for action," toward habitat improvement.

Urban Wildlife

The research division of the New York State Department of Environmental Conservation has four subdivisions of its habitat-management section: (1) farm lands, (2) wetlands, (3) forest lands, and (4) urban wildlife. The most recent of these is the consideration of urban wildlife. This is not a new concept, for Olmstead established Central Park in New York City, planting it mostly by hand, over 100 years ago. Since 73.5 percent of our population is urban (Hauser 1971), management has a strong responsibility to consider urban wildlife. By wildlife, I mean plants as well as animals. The great rise in sales of house plants and pets reflects people's needs for having a bit of nature close to them. We should look to those needs and bring natural settings into the urban scene. More and more environmental and wildlife literature is speaking to the problems of nature in the urban setting (George and McKinley 1974, Husek 1975).

A surprising number of animals, as well as plants, can survive in cities. I am constantly amazed at seeing skunks, opossums, chipmunks, and raccoons, as well as the usual squirrels and rabbits, in residential areas of cities. Nighthawks, swifts, swallows, and kestrels are frequent city dwellers. Mérida, one of Mexico's most beautiful cities, has its colony of vultures, which adorns the walls of the city and helps to make the city one of the cleanest in Mexico. What a neat thing to have an unpaid and natural garbage disposal unit!

We are not doomed to seeing only pigeons, starlings, and house sparrows in cities. What a pleasure it has been, on my concrete campus, to see the house finches, with their melodious songs, give competition to the house sparrow. Recently, mockingbirds have moved onto the campus. When I visited Berkeley, it was a delight to find California quail scooting about under the shrubs of residential gardens. With a little encouragement from bird feeders (Geis 1974), songbirds have increased greatly in cities, a change brought about by the people themselves, with the encouragement of only a few general articles and the enthusiasm of many "birders." With small changes, we could further encourage wildlife diversity even in the central city. The heavy use of zoos and parks would be relieved by more "pocket parks" which could provide readily accessible enjoyment to many!

The domination of cities by pigeons (Woldow 1972) and dogs has gone on long enough, and the problems they create are becoming unmanageable. We have a responsibility to manage domesticated species as well as wild ones. This is an area where management falls squarely in the hands of the citizenry. Citizens cannot blame "someone else."

The nuisance of and damage by dogs in New York City is substantial. Installing protective shields around the base of trees has increased the cost of planting a tree by $15 (Andresen 1974). I have often wondered how many persons have had serious falls or broken bones resulting from slipping on dog droppings. It is easy to criticize elephant management, but are we dealing with our own habitat management?

Funding

A major problem with management in the past has been its emphasis on game species and lack of attention to other organisms. This is not the desire of most managers, but, rather, the result of funding. Most funds have come from the sale of hunting and fishing licenses and from taxes on arms and ammunition. Since sportsmen have provided so many funds, they have had a right to demand emphasis on their interests. Until more state or federal funds are appropriated for general habitat and nongame-species use, the sportsmen's interests will continue to be served.

The state of Washington has done something about this problem and, I hope, other states will follow with solutions. In 1973, the citizens of Washington passed Referendum 33, which provided the State Department of Game with funds to protect, manage, and enhance nonhunted wildlife. Funds come from the purchases of personalized auto license plates; an income of $450,000 was predicted for the first year (Rieck 1975). Emphases are on marine biology, urban wildlife, wildlife ecology, protection of nongame species, education, and interpretation. A comprehensive inventory of nongame species is being conducted, and the information it yields will lead to a comprehensive management program for all species within the state.

I have just received a copy of a beautiful book, the *Arkansas Natural Area Plan,* prepared by the Arkansas Department of Planning (1974). Its aim is to provide information for the general public and the legislature to assist them in understanding the importance of preserving natural areas of the state. This book is a fine example of cooperative efforts of professionals and enlightened nonprofessionals working toward a common goal.

During 1975, the New York State Legislature raised hunting-license fees as much as $1 to $3.50 for residents and as much as $34.50 for nonresidents. This fee hike was supported by the sportsmen's lobby in an effort to dislodge a similar amount for wildlife research from the state's general fund. If nonconsumptive users of our wildlife would give similar financial backing, then perhaps we could make substantial headway in research on nongame species and general habitat problems. As long as major funding comes from special-interest groups, we cannot complain that their interests are being served.

Durwood Allen (1962) said that "deer are being managed by the legislature," a complaint that applies to many environmental problems. Until we can somehow gain enough public support to influence legislation, we are helpless. Every concerned citizen, through letters to legislators, can influence decisions. Sensitive environmentalists have not often enough used their full strength to influence policy decisions by legislative bodies; when they do, the results are often astounding.

Single-Species Management

Management practices too often focus upon a single species—either a game species or an agricultural or otherwise economically valuable one. Although this is understandable, we must take into account how such management will affect the total community.

Forest production provides an excellent example. We know that the most productive stages of forest succession are the earlier ones (Horn 1975, Odum 1969). Where maximum harvest is desirable, softwoods such as fast-growing pines are maintained. But a pine woods is a far less diverse community than the later mixed-deciduous forests, and we must realize that we will eliminate many species by producing an even-aged pine stand.

In the early 1960s, I studied wildlife on the magnificent high plateau, the Nyika, in Malawi. I was surprised to find large stands of Mexican pine *(Pinus patula)* growing on the plateau, which supports a fire-controlled grassland. Some years ago, the Forestry Department decided that the plateau could best be used for timber production and the pine scheme was undertaken. The scheme was soon abandoned because of the problems with hauling out the timber.

I decided to check the numbers and diversity of small mammals that could invade the forest patches from adjoining grasslands. Species diversity dropped in the forest, although a few species could adjust to the new habitat. Numbers decreased markedly with increasing distance from the forest margin. About the only species that benefited from the pine patches was the lion, which used them for cover between periodic forays to catch a zebra or one of the compound cattle.

Forest-management practices often involve the controversy over clear-cutting. Scientific research (Likens et al. 1969) has shown us the dangers of clear-cutting. Yet the cost of harvesting becomes the management tool used by paper companies.

Usually, one of the best signs of good management in natural habitats is the diversity of species. In general, stability is a consequence of diversity (Odum 1969): it seems to be easier for a community with many species to recover from natural disasters and man's ravages than those with few species.

An excellent study showing how the presence of insecticides reduced the diversity of a seemingly simple community was made by Cantlon (1969). The application of insecticides resulted in drastic repercussions throughout the community, and demonstrated the interrelatedness of the organisms within it. All management practices that reduce community diversity should be examined carefully before they are applied on a widespread basis.

Economic Interests

About two years ago, at a luncheon meeting of the Northeastern Wildlife Conference, I happened to sit next to the ecological advisor to one of the large Florida development corporations. In the course of our conversation, I questioned him about the status of bald eagles in Florida. When he sensed that I was chiding land developers for destroying habitat, he retaliated with emphatic statements and strong questions concerning "preservationists." "How many dollars is one eagle nest worth? Well, *how* many?" When I hedged, he jumped in again: "You see, you can't prove it is worth any dollars at all! That is the trouble with you conservationists—you don't understand that the important things are those which people pay money for." He refused to listen to my side of the argument. And surely, this is a major problem of habitat management. Management practices are indeed set, to a great degree, by the political voices backed by the most money. We have known this for a long time.

Economic interests are strongly linked to the quest for energy sources, a major problem that conservationists, and habitats, will face in the next few years. It will be a difficult task to save valuable habitats from destructive changes due to the construction of pipelines, off-shore drillings (see "Perils of the Petroleum Imperative" in *Audubon,* May 1975), and the Bureau of Land Management (see the *Wilderness Report* for April 1975). Land development of any sort means that constant attention must be given to preservation of natural habitats within developed areas. Too often, developers give only lip service to proper management; public outcries—if they are loud enough—will provide the pressure they need. Mandated environmental impact statements, even though they may be inadequately performed, often give needed attention to possible habitat degradation.

Protection and Enforcement

Protected areas are worthless without enforcement of protection laws. This is a problem in almost every reserve and national park in the world. Let us look at a few examples. Currently, I am studying in Jamaica the effects of introduced species and man-altered habitats on native frog populations. Jamaica has established all forests over 5,000 feet (1,525 meters) as reserves for the protection of invaluable watersheds. Yet vertical mountain slopes below that elevation are under virtually complete cultivation, and the local forest manager is reluctant to throw out local citizens who have their gardens above the 5,000-foot (1,525-meter) contour. In fact, no one seems to know just where the line is. Current government policy in Jamaica is to use all private lands. No longer can anyone hold large tracts without developing them. This policy, which may be laudable as a means of solving immediate human problems, will surely create havoc with native plants and animals, and could eventually impoverish the island. Fortunately, the tops of the limestone karst "haystacks" are inaccessible to the people and will, I hope, endure as natural refugia for their unique communities, with species such as the giant snoring tree frog and the Jamaican parrot.

Dr. Anne LaBastille recently surveyed the New Saona National Park in southeastern Dominican Republic. She told me of a situation there in which a mere handful of squatters are gradually destroying the treasures of the area. Rare palms are being defoliated for thatch. Extremely rare conchs are collected and sold for $25 each. (What a fortune that must be to a local peasant!) The few large forest trees are being burned for charcoal. In the face of human poverty and the persistence of human endeavor, it doesn't take long to produce a biotic desert. Special protection must be given quickly to those communities that are the most fragile. But pressures on governments are the only way to effect changes, and governments could hardly care less for a few rare species when their people face starvation.

Protection means protection from overuse and "overloving" too. Overuse is a problem in many of our national parks and will destroy them unless policies change soon (Darling and Eichhorn 1967). We should raise entrance fees for national parks. Since parks are used by the low-income populace much less than they are by those able to pay higher fees, no one group would be eliminated. Persons who use the parks only because they provide cheap vacations would be discouraged.

Trails in the High Peak country of the Adirondacks are so overused that they have become deep gullies or mudholes. Should we resort to paving trails as has been done on parts of the Appalachian Trail? Should not the hiker pay a small trail tax and tax on camping gear, just as the hunter pays for a hunting license and tax on arms and ammunition? These fees could go toward maintenance and toward study of habitats and nongame species. New policies must be developed to adjust to overpopulation in our parks as well as in our cities. Can we change our policies before it is too late?

Fire as a Management Tool

One of the oldest habitat managers is natural fire. Our Great Plains, the pine barrens of Long Island and New Jersey, our southeastern coastal plains, and the African savannah have been molded by fire over evolutionary (biological) time. Only species that can tolerate or adjust to fire grow in these communities. Some species are so adapted to fire that they cannot reproduce without it. Our attitudes toward fire in this country were set by the devastating forest fires that followed the huge lumbering operations around the turn of the century. Only in the last few years has the general press carried articles treating the benefits of fire (Johnson 1973, Moser 1974). Even the Forest Service and the Park Service are admitting that occasional light fires are necessary to prevent less frequent but devastating fires (Dodge 1972, Sterba 1974), and many results of research concerning the use of fire have been presented in Komerek (1974).

[See also Dr. Jan van Wagtendonk's paper, "Wilderness Fire Management in Yosemite National Park," this volume.—Editor]

Although I watched my father use fire in the management of our farm in Piedmont, North Carolina, during my childhood, I learned to fear it and think of it only as a destructive force. Not until I went to Africa did I realize just how necessary it is in maintaining many communities. The Forestry Department of Malawi has an excellent staff. They maintain a series of experimental burning plots on the Nyika Plateau in order to study the effects of various burning regimes. This gave me an excellent opportunity to prove—or so I thought—that burning was detrimental to the small mammal, lizard, and snake populations of the grassland. I carefully observed areas immediately after a burn and never found a dead animal. I set out traplines on burned and unburned plots and found, much to my surprise, more small mammals on *burned* plots than on the unburned grasslands (Stewart 1972).

We know that when fire-controlled communities are protected from burning, the character of the community changes all too quickly and that succession proceeds toward the climax forest of the area. This is happening in the New Jersey Pine Barrens as civilization encroaches; it is happening in the Albany Pine Bush (Breen 1973). Fire plugs have even been installed in one area that is a city park in the Albany Pine Bush: fire just isn't safe once buildings invade an area. Maintaining a fire-controlled community once civilization has encroached is extremely difficult, but if we don't manage such communities we will lose them and many of the specialized species that have evolved through long adaptation to fire (Breen 1973, Lemon 1968, and Wright 1974).

Enlightenment

There is only one form of management that can hope to alter policies and result in effective programs: education for attitudes of conservation. Although we have made great strides in recent years toward creating a concerned, educated public, naiveté is rampant. Public schools are not able to cover all fronts, and natural history and environmental concerns frequently give way to more traditional subjects. Brandwein (1966) defined education "as the conservation of man." It has "its prime concern with the fitness of man to the environment."

Each year, students with lots of biochemical "know-how," but with an almost total ignorance of basic natural history and ecological principles, enter my classes, having studied biology in high school. Many of my colleagues have pressured me to stop teaching a course in field biology, telling me that it has no place in a university curriculum. This situation is not uncommon in our colleges and universities, in spite of the help of the "environmental revolution" of the late 1960s. Even now, we seem to be forgetting environmental concerns in curricula. A major policy of every institution related to the subject of the environment must be to educate, to interpret knowledge of natural resources to the public. This is the sort of endeavor on which a single individual or organization can make a great impact. Only with an enlightened public can we hope to influence legislation wisely. It is absolutely essential to obtain accurate information about habitats and species before launching "causes." Misguided persons with loud voices and religious zeal can cause damage that may never be undone.

Introduced species or feral domesticated species are major dangers to native species. Damage to habitats from such species may be severe (Courtenay and Robins 1975, Elton 1958, Laycock 1966). A prime example of the damage done by a feral species is that of the "wild horse" of the southwestern United States, which is now protected by law. However, it is eating food and destroying fragile plant communities needed by the native ungulates of the area. The superior competitive strength of the horses may well eliminate the large, native hoofed mammals; in time, the horses will either starve or move to other areas, repeating the destruction. In this situation, pressure from "preservationists" elicited legislation that was more damaging than helpful.

During 1974, a similar situation developed on Round Island, Mauritius, in the Indian Ocean. Round Island, like the Galapagos Islands, has several species of animals found nowhere else in the world. Their fragile island habitat was being rapidly destroyed by introduced goats and rabbits. The goats could be shot, and studies showed that the devastating rabbits could be killed with strychnine-soaked hay, which would not be eaten by other species. So-called "animal lovers" mounted a huge campaign to prevent the use of strychnine on the rabbits, and the governor cancelled the corrective measures. So, the unique species may be lost forever, and soon the rabbits will die of starvation, for they will have destroyed their habitat!

Feral Dogs

A problem that is much closer to all of us, and one that has received little attention, is the problem of dogs as pest animals (Andresen 1974, Beck 1973, Sears 1974 and 1975). Since 46 percent of all American households have at least one dog (Feldman 1974), this is a responsibility for nearly half of us in the United States. Free-roaming packs of dogs—feral and tame—have become a major destructive force to sheep especially, to say nothing of ponies, steers, pigs, poultry, and people. In 1970, 6,809 dog bites were reported (Beck 1973). Controlling pest species has long been a responsibility of habitat management. We must not dodge that responsibility lest "man's best friend" become man's closest enemy!

Dangerous Species

A most delicate and difficult problem concerns preserving and managing those species that come into direct competition—real or imagined—with man. These may be the elephant of Tsayo (Corfield 1973), the timber rattlesnake of the northeastern United States, the timber wolf (*New York Times,* February 23, 1975; Engelhart and Hazard 1975), the coyote (Laycock 1974, McMahan 1975), or the mountain lion. We are still a long way from the adequate management of such species.

Yellowstone National Park presents an example of improper management, albeit a very difficult situation. The numbers of grizzly bears are low in the park because too many are being shot, to "protect" the people. But bad management by park personnel may be responsible for the nuisance bears, because garbage dumps have lured the bears away from their natural foods to the neighborhood of populated areas where they have too often been shot (Craighead and Craighead 1967). It is time that the people were kept away from the bears, for Yellowstone is one of the last spots in the contiguous states of the United States where grizzlies still exist. The political aspects of the grizzly-bear problem were outlined by Regenstein (1975). Recent studies indicate that with changes in management policies, the situation may be improving (Cole 1974).

Predator-Prey Relationships

There is no more controversial subject in management than predator control (Howard 1974). Although the scientific literature is filled with documented research establishing the necessary role that predators play in maintaining the stability of natural communities (for general discussions see Allen 1962, Russo 1964, and Flader 1974), we are still shooting and poisoning predators or putting bounties on their heads. I was distressed to read (*New York Times*, February 25, 1975) that Alaska is on the verge of killing off the wolves in an area south of Fairbanks, in order to replenish the moose herd there. I was distressed because the moose population is lower on account of excessive hunting and a hard winter, not because of wolf predation.

The absence of predators has had a significant role in the death of many elk in Yellowstone National Park during the spring. Without wolves to check them, elk populations have exceeded the carrying capacity of the vegetation. They are not strong enough to withstand a hard winter. The same sort of situation exists with Adirondack deer (Bromley 1964; Hesselton et al. 1965; Severinghaus 1962, 1973, 1974). The herd is too large for the habitat to support, and deer die during hard winters. They did not have enough food the previous summer! Still hunters lobby for more deer. Perhaps what we need instead is the reintroduction of the native mountain lion to the area, to help keep the deer population in check.

Reintroduction of native species to areas they once inhabited is under discussion in the New York State Department of Environmental Conservation. Reintroduction of wolves to the Adirondacks has been suggested (Engelhart and Hazard 1975). A wide-ranging pack species might not live close to man and his stock, but the mountain lion—a solitary, gentle beast that is wary of man—might succeed in the large, uninhabited woodlands of the Adirondacks. Deer populations would be improved by the presence of a natural predator.

Obviously, cultural patterns and historical attitudes are too strong to be broken overnight. Education and interpretation on this front have fallen far short of the ideal.

Successes in Habitat Management

There have been many failures, as well as some rewarding successes, in the history of habitat management. Both the failures and the successes have taught us much about the ways in which natural communities function. Somehow, it is always easier to point to failures rather than successes. The protection of endangered species has worked with many species, such as the trumpeter swan, the bison, the beaver, and the sea otter.

Another surprising success has been in the effectiveness of nest boxes for wood ducks. Wood duck populations were alarmingly low following extensive logging early in the century. Now, the species is doing very well. During the 1930s, farmers were encouraged to build farm ponds. No one realized what an asset these ponds would be to migrating waterfowl. Many people put up wood duck boxes and built ponds, and the returns in terms of waterfowl were tremendous. With appropriate information and encouragement, many citizens can contribute to fine management practices. Controlling the level of water in marshes along the East Coast of the United States is another management effort that has increased the populations of water birds and reduced mosquito populations as well. Good management benefits the entire biotic community, not just one species.

International Cooperation

More and more, as the world "shrinks," cooperation among nations is becoming an essential factor in global preservation. Management of the marine environment is the most obvious area where new policies must be established. Another area is the understanding of the biology of migrating animals, or those using waters shared by two states or nations, since habitats in one country can be destroyed by pollution from another country. International understanding and cooperation is a necessary factor in habitat-management policies.

Role of the Individual

Our first responsibility is to inform ourselves through accurate information. The Sierra Club and the National Audubon Society are excellent sources of information. Supporting such organizations through membership is a first step. Using a balanced point of view to influence our representatives in government is one of our best "avenues for action." Participation of any kind in the governmental process can influence the course of action. One must find out the attitudes of candidates and support those in favor of population controls and the conservation of resources. One must encourage knowledgeable and competent persons to run for office and assist them where possible. Participation by informed, responsible citizens in local-improvement associations is desperately needed. Such participation is time consuming and rarely yields financial remuneration, but without informed statements, habitats will be lost to pollution or the bulldozer.

The support of scientific research by concerned citizens is desperately needed. Funds for research are limited as defense funds drain our national coffers. The cost of one bomber could finance an unbelievable amount of ecological research. In terms of cost, ecological research is the least expensive kind of scientific research and is surely as important as cancer research. In fact, the results of ecological research might lead to environmental improvement that could very well reduce the amount of cancer. Legislators are constantly restricting funds for research and often deride the subject matter of important studies. The House of Representatives has already approved a bill (Bauman Amendment, passed April 9, 1975) that gives Congress veto power on all National Science Foundation grants (Shapley 1975).

We lack basic knowledge about most of our common species, to say nothing of the rarer ones. More and more, specific knowledge is required before any action concerning critical habitats can be taken. We rarely have the knowledge we need. Citizens can assist by sending letters to congressmen, asking them to support environmental research. The key words in the Canada Wildlife Act are: "Wildlife research, conservation and interpretation." We, too, can use these words as guidelines for our efforts.

Ultimately, our concerns are for the management of man and for his survival. When we concern ourselves with habitat and other species, our concerns are, indeed, for man. If the world is unfit for other organisms, it is unfit for man, and he will soon join the list of extinct species. Overpopulation will destroy us faster than will any other form of mismanagement. If we "do not understand and appreciate the absolute and final dependence of Man on his organic environment, we can expect innocent, but uninformed and genocidal political action under our democratic system" (Benson 1975). Every one of us can play a major role in informing ourselves, in informing others, especially our children and our legislators, of man's dependence on nature and of the destructiveness of overpopulation and the overuse of our natural resources.

Acknowledgments

C. W. Severinghaus and Eric Fried of The New York State Department of Environmental Conservation provided information on the department's management policies.

References

Allen, Durwood L. 1962. *Our Wildlife Legacy.* Revised ed. New York: Funk and Wagnalls.
Andresen, J. W. 1974. Dog control. [Letter to the Editor.] *Science* 186(4162): 394.
Arkansas Department of Planning. 1974. *Arkansas Natural Area Plan.* Little Rock: Arkansas Department of Planning.
Beck, A. M. 1973. *The Ecology of Stray Dogs: A Study of Free-ranging Animals.* Baltimore: York Press.
Benson, D. A. 1975. A long-term plan for socio-economic studies related to Canadian wildlife. In *Transactions, Thirty-eighth Federal-Provincial Wildlife Conference,* pp. 125–129. Ottawa: Canadian Wildlife Service.
Brandwein, P. F. 1966. Origins of public policy and practice in conservation: early education and the conservation of sanative environments. In *Future Environments of North America,* eds. F. Fraser Darling and J. P. Milton, pp. 628–647. Garden City, New York: Natural History Press.
Breen, Peg. 1973. The Albany Pine Barrens. *The Conservationist* 28(2):3–5.
Bromley, A. W. 1964. Moose River deer. *The Conservationist* 18(6):13–15, 40–41.
Cantlon, J. 1969. The stability of natural populations and their sensitivity to technology. In *Diversity and Stability in Ecological Systems.* Brookhaven Symposia in Biology No. 22. Upton, New York: Brookhaven National Laboratory.
Cole, G. F. 1974. Management involving grizzly bears and humans in Yellowstone National Park, 1970–73. *BioScience* 24(6):335–338.
Corfield, T. F. 1973. Elephant mortality in Tsavo National Park, Kenya. *East African Wildlife Journal* 11:339–368.
Courtenay, W. R., Jr., and C. R. Robins. 1975. Exotic organisms: an unsolved, complex problem. *BioScience* 25(5):306–313.

Craighead, J. J., and F. C. Craighead, Jr. 1967. Management of bears in Yellowstone National Park. Unpublished research report.

Darling, F. Fraser, and N. Eichhorn. 1967. *Man and Nature in the National Parks.* Washington, D.C.: The Conservation Foundation.

Dodge, M. 1972. Forest fuel accumulation—A growing problem. *Science* 177 (4044):139-142.

Elton, C. S. 1958. *The Ecology of Invasions by Animals and Plants.* London: Methuen and Co.

Engelhart, S., and K. Hazard. 1975. Wolves in the Adirondacks. *The Conservationist* 30(2):9-11.

Feldman, B. M. 1974. The problem of urban dogs. [Editorial.] *Science* 185 (4155):903.

Flader, Susan L. 1974. *Thinking Like a Mountain: Aldo Leopold and the Evaluation of an Ecological Attitude toward Deer, Wolves, and Forests.* Columbia: University of Missouri Press.

George, C. J., and D. McKinley. 1974. *Urban Ecology: In Search of an Asphalt Rose.* New York: McGraw Hill.

Geis, A. D. 1974. The new town bird quadrille. *Natural History* 83(6):54-61.

Hauser, P. M. 1971. The Census of 1970. *Scientific American* 225(1):17-25.

Hesselton, W. T., C. W. Severinghaus, and J. E. Tanck. 1965. Deer facts from Seneca Deport. *The Conservationist* 20(2):28-32.

Horn, Henry S. 1975. Forest succession. *Scientific American* 232(5):90-98.

Howard, W. E. 1974. Predator control: whose responsibility? *BioScience* 24(6): 359-363.

Husek, V. P. 1975. New directions in wildlife management. *The Conservationist* 3(2):3-5.

Johnson, J. W. 1973. Fire—destroyer and creator. *The Conservationist* 28(1): 34-37, 48.

Komarek, E. V., ed. 1974. *Proceedings, Annual Tall Timbers Fire Ecology Conference.* Tallahassee, Florida: Tall Timbers Research Station.

Laycock, George. 1966. *The Alien Animals: The Story of Imported Wildlife.* Garden City, New York: Natural History Press.

———. 1974. Travels and travails of the song-dog. *Audubon* 76(5):16-31.

Lemon, P. C. 1968. Fire and wildlife grazing on an African plateau. Proceedings Annual Tall Timbers Fire Ecology Conference. 1968: 71-88.

Likens, G. E., F. H. Bormann, and N. M. Johnson. 1969. Nitrification: Importance to nutrient losses from a cutover forested ecosystem. *Science* 163 (3872):1205-1206.

McMahan, Pamela. 1975. The victorious coyote. *Natural History* 84(1):42-50.

Moser, D. 1974. New fire on the mountain. *Audubon* 76(5):72-86.

Mostert, N., J. G. Mitchell, and B. Gilbert. 1975. Perils of the petroleum imperative. *Audubon* 77(3):18-80.

Odum, E. P. 1969. The strategy of ecosystem development. *Science* 164(3877): 262-270.

Regenstein, Lewis. 1975. *The Politics of Extinction.* New York: Macmillan Publishing Co.

Rieck, C. 1975. Washington's non-game program, 1973. In *Transactions, Thirty-eighth Federal-Provincial Wildlife Conference,* pp. 97-101. Ottawa: Canadian Wildlife Service.

Russo, John P. 1964. The Kaibab North deer herd—its history, problems, and management. *Wildlife Bulletin* 7. Phoenix: Arizona Game and Fish Department.

Sears, Paul B. 1974. Dog control. [Letter to the Editor.] *Science* 186(4162):394.

———. 1975. Reply to a critical dog. [Letter to the editor.] *Science* 188(4190): 780-782.

Severinghaus, C. W. 1962. The future for deer in New York. *The Conservationist* 17(2):2-4.

———. 1973. A modest proposal to improve deer habitat. *The Conservationist* 27:37.

———. 1974. Deer population—a wildlife roller coaster. *The Conservationist* 28:36-38.

Severinghaus, C. W., and R. Gottlieb. 1959. Big deer vs. little deer. *The Conservationist* 14(2):30-31.

Shapley, D. 1975. News and comment. *Science* 188(4186):338-341.

Sterba, J. P. 1974. Rangers refute Smokey Bear and let forest fire spread. *New York Times,* October 15, 1974, p. 41.

Stewart, Margaret M. 1972. Fire related to habitat preference of small mammals on the Nyika Plateau. *The Society of Malawi Journal* 25(1):33-42.

Woldow, N. 1972. Pigeon and Man: A spotty friendship. *Natural History* 81(1): 26-37.

Wright, H. E., Jr. 1974. Landscape development, forest fires, and wilderness management. *Science* 186(4163):487-495.

57. National Parks: Atonement for Environmental Sins

Nathaniel P. Reed
Department of the Interior, U.S.A.

Introduction

National parks have become an international movement. This is a cause for rejoicing: no one would challenge the assertion that parks are an excellent tonic for the human spirit. Yet today I have a caveat to give on the subject of national prks: every nation that establishes national parks should beware of the tendency to use them as an excuse to be self-satisfied and negligent about the rest of its environment.

To judge the growing international status of the national-park movement, we need only take a brief look at the recommendations and follow-up of the Second World Conference on National Parks, held in Yellowstone and Grand Teton National Parks in 1972. The conference generated twenty solid recommendations, nearly all of which have been seized upon for substantial follow-up by the International Union for Conservation of Nature and Natural Resources (IUCN), often with the assistance of the World Wildlife Fund (WWF). Let me outline a few of the recommendations and the actions that have been taken to implement them.

Selected Recommendations of the Second World Conference on National Parks

Conservation of Tropical Rain-Forest Ecosystems

The conservation of tropical rain-forest ecosystems (Recommendation 2) is a major project for both IUCN and WWF. During 1974, regional conferences were held in Caracas and Bandung. Specific efforts to project rain forests have been made in Puerto Rico, Sri Lanka, Dominica, and the Ivory Coast.

Marine National Parks. The establishment of marine national parks (Recommendation 4) is a very active area that has given rise to regional conferences in Tehran for the promotion of marine parks in the northern Indian Ocean and in Wellington for the South Pacific. Specific programs include the Mediterranean and coral reefs in the Sudanese Sea.

Regional Systems of National Parks and Other Protected Areas. Another active area is the establishment of regional systems of national parks and other protected areas (Recommendation 7), with efforts underway in Central America, east Africa, the South Pacific, and eastern Europe.

The Integrity of National Parks and Equivalent Reserves. IUCN continually intercedes with governments in efforts to protect parks from unsuitable development or other deleterious uses, in accordance with Recommendation 11.

Exchange of Information. To further the exchange of information (Recommendation 16), IUCN is putting together a handbook on the planning and management of national parks and is planning an international magazine on the subject.

I think this short list of recommendations and specific actions will suffice to make the point that the Second World Conference on National Parks has been most fruitful.

The World Wildlife Fund

I want to pause and pay special tribute to the World Wildlife Fund (WWF), which, through its international headquarters in Switzerland and affiliates in twenty-six countries, has been a driving force for worldwide conservation. Its actions for the protection of endangered species are widely known. Equally important is its work to save endangered habitats and valuable ecosystems.

WWF has played a major role in the protection and establishment of scores of national parks. This work has grown significantly since the Second World Conference on National Parks. As only one example, I would mention the WWF's significant support of the recent effort by six Central American countries to establish a regional system of national parks and other natural and cultural resources. This is in direct response to Recommendation 7 of the conference.

Cooperation between the United States and Canada

Close cooperation in national-park management has long characterized United States–Canadian relations: the two nations have exchanged personnel for training purposes and devised ways of protecting ecosystems that span international boundaries; in 1932, the United States and Canada established the Glacier-Waterton International Peace Park in Alberta and Montana.

There is an outstanding opportunity for further cooperation in the far northwest, in Alaska: the proposed American Wrangell–St. Elias National Park in Alaska would be adjacent to Kluane National Park in the Yukon Territory. I look forward to productive joint action with the Canadians to protect this major mountain ecosystem.

Proposals for Parks in Alaska

While we are on the subject of Alaska, I want to give you a brief introduction to the proposals for new parks that are now pending in the Congress. The Alaska Native Claims Settlement Act not only settled the aboriginal claims of Alaskan Indians, Eskimos, and Aleuts to land in Alaska, but it also directed the administration to formulate proposals for creating new national parks, refuges, forests, and wild and scenic rivers out of federal land in Alaska.

We emerged from this task with recommendations for over 30 million acres (12 million hectares) of new park lands—a doubling of the existing nation-park system. The intense beauty and stupendous proportions of these proposed parks are factors of Alaska's untouched and fragile vastness.

I hope the Congress will respond favorably to our proposals. If it does, we will have been able to preserve—not just bits and pieces of the natural environment, as in the lower forty-eight states—but several ecosystems in their entirety.

Darker Thoughts on National Parks

Let me turn now to some of my darker thoughts. In the U.S. and perhaps elsewhere, the establishment of national parks has tended to deflect attention from other environmental problems. Even more, the creation of parks has served to lessen our guilt over the environmental havoc we have wreaked just about everywhere else in the country.

Often, parks in the U.S. have come to play the role of a lovely symbolic rug, under which the nation hides deep scars in the floor and sweeps all manner of trash. I have seen distinct traces of this attitude in other nations: placing all of a nation's environmental stock in national parks threatens to become an international phenomenon.

The National Malaise

I hope it may be helpful in checking that trend if we take a detailed look at its causes and manifestations in the U.S. Anyone who has observed this country at all during the last few years must have perceived that it is a nation very much ill at ease with itself. Our doubts and anxieties become more and more striking as we outwardly celebrate a noisy Bicentennial but inwardly brood over our values and priorities.

Of course, our malaise can be attributed to a number of sources—the gloomy state of the economy, the energy crisis, the sad end of our Vietnam ordeal. But there is more to it than these immediate causes, something much deeper and more pervasive.

I believe that our national uneasiness is prompted by a mind-set that has permeated the thinking and policies of our national leaders for generations, but which is outmoded and must be abandoned. This mind-set might be labeled "the high-standard-of-living ethic," but that is an awkward phrase. There is a great deal of what I call the U.S. No. 1 mentality in this mind-set. Perhaps the handiest description of this way of thinking is simply "American materialism." At any rate, the philosophy I am speaking of can be expressed in a statement such as the following: Happiness flows directly from the high standard of living that technology and a free-enterprise economy will inevitably produce if they are allowed free rein.

The Course and Consequences of American Materialism

This mentality took strong hold in the U.S. after the Civil War, during what Mark Twain called the "Gilded Age." Checked somewhat by the two world wars and the great depression, the American tendency to materialistic grossness enjoyed its most blooming success during the post–world-war booms. Only in the past few years—as a consequence of the nation's energy and economic doldrums—has the drive for unlimited production and quick wealth relented somewhat.

I need not dwell on the unfortunate direct results of this mentality—the unchecked depletion of our natural resources, the dizzying conglomeration of business into ever larger and more top-heavy super-corporations, the nurture of artificial needs for frivolous consumer goods, the rush of rural people to the cities, and the corresponding economic and spiritual impoverishment of our countryside.

Even more important are some of the indirect, intangible effects of the U.S. drive for success and bigness: the demeaning of work from an integrated task that one saw through to the end, to the meaningless insertion of one item into an endless process; the loss of a sense of connection with the land; the loss, in short, of a sense of perspective.

Americans no longer seem to know who they are, how they fit into the universe, what their role on the Earth is. The Greek philosopher Protagoras said, "Man is the measure of all things." Americans today seem unable to measure themselves, let alone anything else.

The Origins of U.S. Materialism

One may ask at this point how this state of affairs developed in a nation to which men came because it was fresh and wild and unexploited. How did this happen in the nation that originated the national-park system and whose environmental movement in the 1960s and 1970s has inspired the other industrial nations? What happened to the "fresh, green breast of the new world," as F. Scott Fitzgerald put it in *The Great Gatsby*? The answer, I believe, lies in the U.S. passion for the big and the spectacular.

Because our natural resources were of a staggering magnitude, we convinced ourselves we could exploit them voraciously and endlessly. Because we set aside gigantic, spectacular areas as national parks, we felt we could march through the rest of our resources like soldier ants. We drew a cordon around the parks and left almost everywhere else a free-fire zone.

And so, for a century (Yellowstone, the first and certainly one of the most spectacular national parks, was established in 1872) we have lived a "Jekyll-and-Hyde" existence: kindly Dr. Jekyll has established a few oases in the form of national parks, while villainous Mr. Hyde has plundered the rest of our corner of the planet; Dr. Jekyll has complacently saved a few grandiose spots, while Mr. Hyde has run amok over millions of acres of the American landscape.

A Curious Proposal

The April 1975 issue of *Development Forum,* a United Nations publication, contains a curious proposal for national economic parks, wherein "nothing would be allowed to interfere with pure economic growth" and where the National Environmental Degradation Act would protect industry from all environmental laws! Any pro-environmental activity in the park would require an economic impact statement. The article does not specify what would happen to people living and working in the economic parks, but undoubtedly they would collect hazardous-duty pay and have a life expectancy of about twenty-seven. This obviously would require us to reinstitute child labor so that twenty-one-year-old retirees would have time to build up a pension. The factory owners and their stockholders would, of course, never set foot in such an unhealthy environment. Or, if they did, an economic impact statement would be required to justify turning the toxic operations off.

For much of their history, Americans have neglected their land outside national parks to such an extent that our situation today faintly resembles these make-believe national economic parks; and the national parks have made up a system unto itself with relatively little influence outside its boundaries. The millions of Americans who have visited them and then gone back to their wasteful urban lives have been like people who go to church on Sunday but practice no charity during the rest of the week. Like church for these people, the national-park system has become a way of atoning for the country's environmental sins, the ideal salve for the conscience of the U.S. materialist.

In addition, the very size and grandeur of the parks have rendered them a national symbol that seems to indicate that the environment is "okay." The cities might be smog-choked, but we could always get away to the pure air of Rocky Mountain or Sequoia-

Kings Canyon national parks. Lake Erie might be moribund, but Old Faithful is still spouting. And so, in an emotional and ethical sense, Americans have exploited their national parks.

"Not Parks Apart"

Friends of the Earth calls its monthly newsletter *Not Man Apart*. If I may borrow the phrase, I would entitle one theme of the rest of my talk "Not Parks Apart." The title of a book by the British economist, E. F. Schumacher, *Small Is Beautiful*, characterizes another theme.

By "not parks apart," I mean two things. First, parks should not be set apart as mere token fulfillments of a much more demanding environmental ethic; parks should be considered only a first step—however graceful and attractive a step it may be—on the path toward man's organic relationship with the Earth and its resources. Parks should not merely shine, they should radiate.

National parks are one piece of evidence in the inquiry as to whether man can enter into a trust relationship with the natural resources he has inherited. By themselves, parks certainly do not make the case, but if the lessons parks teach and the values they exemplify can extend beyond—if the park experience can inculcate a sense of everyday environmental stewardship in visitors—then parks can become the central evidence in a positive verdict for mankind and his environment.

Second, parks should not always be placed far from concentrations of population. They should not exist only in the distance, separate from ordinary life. Instead of continually replacing abandoned and useless buildings in the central city with high-rise, glass-and-brick "quickies," I would like to see the worldwide revitalization of downtown areas with open space and greenery.

Small Is Beautiful

I would like to see small parks that nestle and weave among business and residential districts, rather than lying isolated at the city's edge. I would like to see many of the activities conducted in our national parks—guided nature tours, plant workshops, talks on man's proper relationship with the environment, and so on—conducted in such city parks on a smaller scale.

In very few major cities of the world can a citizen escape the effects of the materialist ethic—smog, traffic, dirt, noise, and ugliness of all kinds. My vision is of small patches of green sprinkled throughout every city, providing recreational opportunities for people of as many ages and lifestyles as possible.

I believe that "small" can be beautiful. For this to happen, however, the U.S.—and other nations as well—must lower its sights and recover its perspective. We need to stop gauging success in materialistic terms alone. We need to pare down our needs for creature comforts and consumer items. We have to regain a sense of modest proportions in the demands we make on our resources. We need to refine our jaded thirst for colossal pleasures and opulent displays, so that we can again experience the joy in small things done well. We need to appreciate the deep-running currents of life as well as the soaring peaks.

The article entitled "Thinking Small," which was published in the June 2, 1975, issue of *Newsweek*, gives several examples of the attitude I would inculcate. Multiple-industry companies are beginning to shed their peripheral concerns and to specialize in the one or two things they do best. Firms formerly bent on expansion at all costs are now content to make and sell the amount of goods actually demanded.

The U.S. anxiety for more and bigger is being transformed into a call for smaller and better. The automobiles projected to come out of Detroit during the next few years are a highly visible embodiment of this trend. Perhaps, as we redirect our vision to necessities as opposed to luxuries and to bettering the quality of our lives as opposed to dashing madly onward and upward, we will be able to reacquire the measure of ourselves and then of all things.

Conclusion

Let me conclude by emphasizing that national parks are needed around the world more than ever. But we cannot afford to be lulled by our successful international park program into thinking we have done enough. The world's peoples are going to continue using our natural resources, no doubt about that. But we must begin to use them in accord with insights and ethics we have gained from our park experiences.

The worldwide inquiry as to whether man can live in equilibrium with the Earth is still going on. The jury will consist of future generations, and the verdict is not yet in. When it ultimately comes down, the verdict can go one of two ways—for mankind and his environment or for an eventual breakdown of the planet. The shaping of this verdict is the greatest challenge of the final twenty-five years of this century.

58. Individual Initiatives

F. Wayne King
New York Zoological Society, U.S.A.

Introduction

There are some 1,400 national parks and reserves scattered throughout some 144 countries of the world. There are many times that number of state, provincial, local, and private wildlife and botanical refuges, reserves, and sanctuaries. They vary in size from a few hectares to millions of hectares. Each preserves some bit of habitat, some group of plants or animals, some species, some ecosystem or fragment of an ecosystem.

Surely one of the most effective methods of protecting natural ecosystems or inhabitants of the ecosystems, the wild species, is to include them in sanctuaries and national parks. Of course, not all of the parks, reserves, or sanctuaries provide total protection from exploitation. Many are merely dots on a map in some governmental office. Many are poorly staffed or administered. Some lack the legal base necessary to prevent violation of the area. Some have already been lost to land, mineral, timber, or wildlife exploiters, or converted to agricultural use by invading pastoralists and farmers who simply moved in and stayed; others are being lost to pollution. With all these recognized shortcomings, national parks, wildlife and botanical refuges, and protected natural scenic areas still seem to be the best method yet found to conserve for future generations many of the wild species and ecosystems on which they depend.

The vast majority of national parks and reserves are administered by national or local governments. Because of this, we sometimes overlook the active role played by individuals in establishing and maintaining these sanctuaries. Individuals can preserve ecosystems in ways other than just including them in parks, but exploring how some reserves came into being reveals a lot about the types of action that are effective in other conservation efforts. A few examples will illustrate this.

Early Initiatives

Yellowstone National Park

Yellowstone National Park, the world's first national park, was established through the efforts of a few individuals. During the first half of the 1800s, explorers of the Yellowstone area brought back glowing stories of the scenic wonders of geysers and hotsprings, and abundant wildlife. Such tales focused scientific attention on the area; and in 1870, an expedition that included Nathaniel Langford, General H. D. Washburn, Lieutenant G. C. Doane, and Judge Cornelius Hedges headed for Yellowstone—an expedition that led to the formation of history's first national park.

After seeing the Yellowstone wilderness, these men were concerned that exploiters would destroy the area before it could be enjoyed by other Americans. They discussed the possibility of taking out land claims in order to protect the area. Judge Hedges suggested that that the area should belong to all the people of the United States, to the nation.

On returning from the expedition, Nathaniel Langford wrote a series of articles that were published in *Scribner's Monthly* (May and June 1871), and these won wide public support for preserving the Yellowstone area for all citizens. At the same time, Judge Hedges petitioned Congress to act to protect the area. Their combined efforts won congressional approval, and in 1872, President Grant signed the bill that established Yellowstone National Park. Two individuals, Nathaniel Langford and Judge Hedges, were the driving force behind saving Yellowstone.

Sequoia National Park

The story of Yellowstone is not unique. Other individuals took the initiative in other places. In the late 1880s, Colonel George W. Steward led the fight to save the giant sequoia trees of the California Sierras. His efforts persuaded Congress to support the creation of the Sequoia National Park in 1890.

The rapid influx of visitors to these new parks caused governments everywhere to take note, and national parks were soon established in Canada, Australia, New Zealand, Africa, Asia, and Europe. Many of these first parks were the result of individual initiatives.

Kruger National Park—South Africa

In the 1880s and 1890s, President Paul Kruger of South Africa became concerned about the rapid disappearance of wildlife caused by hunters and farmers in that nation. The blaubok, quagga, and cape lion were already extinct, and the mountain zebra, bontebok, blesbok, white rhino, and white-tailed wildebeest were nearly extinct. He urged establishment of game reserves to protect the animals but encountered considerable opposition. After several years of political combat over the issue, Kruger managed in 1894 to proclaim the first federal game reserve in South Africa, the Pongola Game Reserve. This was followed in 1897 by the formation of the Hluhluwe Game Reserve and, in 1898, by creation of the Sabie Game Reserve. Over the years, the Sabie reserve has grown through the annexation of adjacent land, and in 1926, its name was changed to Kruger National Park in commemoration of President Kruger's foresight and personal initiative.

Swiss National Park

In 1909, internationally known Swiss zoologists, Paul and Fritz Sarasin, urged the creation of an alpine park to preserve some of the Swiss mountain ranges. Their campaign was joined by Dr. Coaz of the Federal Forestry Service, and five years later, their combined efforts were rewarded by the formation of the Swiss National Park.

American Bison Society

At the turn of this century, the American bison, the familiar buffalo of the American frontier, had all but disappeared from the face of the Earth. The species which, in the United States, had ranged as far east as Pennsylvania and Tennessee, and may have numbered as many as 60 million, had been gunned down to the last twenty-nine wild specimens in Yellowstone National Park, and only a few hundred remained in Canada. The majority of bison were in privately owned captive herds and in the New York Zoological Park and the U.S. National Zoological Park. A 1903 census revealed a total of 1,010 bison remaining on the North American continent. Then, in 1905, anxious to save the species from extinction, William T. Hornaday, the first director of the New York Zoological Society, offered the U.S. government a nucleus herd of bison from its Zoological Park, if the government would establish a fenced range for them in Oklahoma, where the species had once been abundant.

That same year, writer Ernest Harold Baynes proposed that a society be formed for the sole purpose of saving the bison from extinction. William Hornaday joined Baynes, and together they organized the American Bison Society with Hornaday as its first president.

Through the efforts of these two individuals, the Bison Society, and the Zoological Society, Congress, and the president were persuaded to establish not only the Wichita National Bison Range in Oklahoma in 1907 (known today as the Wichita National Wildlife Refuge), but also the Montana National Bison Range in 1907 and the Wind Cave National Bison Range of South Dakota in 1913. Each of these empty reserves was stocked with bison from the New York Zoological Park, and the bison, which today can be seen in many other parts of the country, owe their existence to the prodding and forceful threats of Hornaday and Baynes.

Volcanoes and Virunga National Parks—Africa

In 1921, famed naturalist, artist, and taxidermist Carl Akeley visited the Virunga Volcano area of central Africa to collect the group of five mountain gorillas on exhibit today in the cloud forest diorama in the African Hall of the American Museum of Natural History in New York. Akeley was fascinated by the mountain gorillas and worried that they would be destroyed and the Virunga area despoiled by later visitors. He drew up plans for the creation of a gorilla sanctuary and presented them to King Albert of Belgium. In 1925, King Albert designated the area of the Mikeno, Karisimbi, and Visoke mountains as a gorilla national park. It was enlarged in 1935 and renamed the Albert National Park. Today part of it is known as the Virunga National Park in Zaire, and part is the Volcanoes National Park in Rwanda.

Taman Negara National Park—Malaya

In the early 1930s Theodore Rathbone Hubback, a former hunter-turned-photographer, campaigned to save Malaya's animals when he saw that its wildlife was disappearing as a result of overexploitation. He was particularly concerned about the decreasing populations of Seladang, Javan, and Sumatran rhinoceroses. His persistent efforts to stop the slaughter resulted not only in the announcement of the first national park in Malaya in 1938—the King George V National Park (known today as the Taman Negara National Park)—but also the creation of the Malayan government's wildlife department.

Recent Initiatives

The creation of national parks and wildlife sanctuaries through the interest and action of individuals is not a thing of the past. It is still occurring today.

Arctic National Wildlife Range—Alaska

In the 1950s, under a grant from the New York Zoological Society, Olaus Murie visited the Brooks Range of northern Alaska to survey that area's wildlife. His findings, published in 1958, urged the creation of a wildlife sanctuary for the Arctic north slope of Alaska. That sanctuary was established by Congress as the 3,460,000-hectare Arctic National Wildlife Range.

Dudwa Wildlife Sanctuary—India

In 1959, Arjan Singh, a hunter turned conservationist, began a campaign to protect the wildlife of northern Uttar Pradesh, India. Through a persistent personal effort of leasing hunting rights on state lands, and then not hunting, of physically driving out herders and their cattle, and actually purchasing the land, he was able to establish the small private wildlife preserve called Tiger Haven. His campaign, however, did not stop there. Singh continued his fight on behalf of the chital, swamp deer, tiger, leopard, monkeys, crocodiles, and birds of the area until he found an ally in Charan Singh, an Uttar Pradesh forest minister. These two men were able to persuade the state government to legally establish the Dudwa Wildlife Sanctuary in 1967. Since then, Arjan Singh has pursued efforts to assure the adequate financing, protection, and administration of the sanctuary.

Tortuguero National Park—Costa Rica

In the 1950s, Archie Carr began his well-known studies of the green sea turtles that nest on the rookery beach at Tortuguero, Costa Rica. His interest in protecting the green turtles and the rookery led to the formation of the Caribbean Conservation Corporation (CCC), a nonpolitical organization that promotes sea turtle conservation in the Caribbean. Books and popular and scientific publications that have originated from Dr. Carr and the CCC have prompted several governments, including Costa Rica, to protect their sea turtle resources. Furthermore, they have encouraged Costa Rica to set aside the rookery beach and the surrounding forest as the Tortuguero National Park, an action accomplished by presidential decree in the early 1970s.

Marine and Wildlife Sanctuaries in Argentina

In 1965, William G. Conway, director of the New York Zoological Society, visited Patagonia, Argentina, where he surveyed the sea lion, elephant seal, penguin, and sea bird colonies scattered along that coast. On returning to the United States, he published several articles urging the protection of a number of these wildlife areas and describing their potential as a tourist attraction. Several of the articles were translated into Spanish and republished in newspapers in Argentina. The interest generated by these publications led the government of Chubut Province to create, in 1969 and 1970, six provincial wildlife sanctuaries, one each at Punta Delgada, Punta Loma, Punta Norte, Punta Piramides, Punta Tumbo, and Camerones.

About that same time, Roger Payne, of the Zoological Society staff, began his studies of the southern right whale at Golfo San José on the Patagonian Coast. The interest his research created in the people of Argentina encouraged the governor of Chubut Province to protect the right whales. In 1974, Governor Don Benito Fernández established the world's first right whale sanctuary, the Golfo San José Provincial Marine Reserve. With more than a little luck, the action taken by the governor will preserve the right whales of Argentina while the majority of the other great whales is being blown to extinction by the exploding harpoons of the whaling nations.

Monte Verde Cloud Forest Reserve—Costa Rica

In the early 1970s, a group of biologists at the Tropical Science Center in San José, Costa Rica, moved to protect an area of unique montane cloud forest in that Central American nation. Led by Joseph Tosi and L. R. Holdridge, the science center staff sought to acquire the land which was the sole habitat of the diminutive Monte Verde toad *(Bufo periglenes)*. Through wide public solicitation of funds they were able to raise money to purchase the land and provide for a resident warden. Their efforts today seek to increase the size of the Monte Verde Cloud Forest Reserve to include areas of critical habitat for a variety of mountain forest plants and animals.

The Success and Continuing Need of Individual Initiatives

Over the past thirty years, many wilderness areas and wild species have been saved by the mere presence of a field botanist or zoologist studying a particular species, or archeologist or paleontologist seeking to unravel the history of man or the Earth in a critical area. The presence of these researchers and the publication of their findings have been enough to focus world attention on the need to protect a unique resource. World news media, particularly television, have played an important role in bringing these scientists into the public eye. This attention persuades governments to take action to protect the resources under study.

The names of a few of these individuals command instant recognition for the ecosystem research or wild species conservation they have performed: Jane Goodall and the Gombe chimpanzees; George Schaller and the gorillas of central Africa, the tigers of India, and the ungulates and snow leopards of the Himalayas; Stanley Temple and the Indian Ocean birds of prey; Diane Fossey and the mountain gorillas of Rwanda and Zaire; Sir Peter Scott and numerous waterfowl species; Adelmar Coimbra-Filho and the golden lion marmosets; Bernhard and Michael Grzimek and the Serengeti; L.S.B. Leakey and Olduvai; George S. DeSilva and the orangutans and sea turtles of Sabah; Federico Medem and the crocodilians of South America; Kim Hon Kyu and the cranes of the Han River and Demilitarized Zone in Korea; and Roger Payne and the humpback whales of Bermuda and right whales of Argentina. In fact, today it seems that one of the most effective means of protecting a wild species or ecosystem is to put a scientist in the field to study it. Unfortunately, one of our most pressing shortcomings in the field of resource conservation is not having enough qualified field scientists, especially scientists from developing countries.

The actions of some private citizens who have worked hard to preserve wildlife areas have not resulted in wilderness parks, but nonetheless have been significant in preserving several ecosystems and a few wild species. For example, although he raises cattle on his large Venezuelan ranch, Tomas Blohm protects many of the wild species that are found there. He has steadfastly resisted efforts to eliminate the caimans from his ranch, efforts originating from commercial hide-hunters and from individuals who mistakenly think the reptiles threaten the cattle. Today, the ranch is the site of several studies on the ecology and behavior of these interesting crocodilians.

Literally hundreds of other examples could be cited of the actions and interests of individuals which led to the establishment or continued maintenance of national parks and wild sanctuaries, but these few are sufficient to illustrate how individuals can take initiative.

After discovering the need to protect a particular area, ecosystem, or species, they have lobbied (aggressively in some cases) legislators or government agencies charged with resource protection in order to gain protection through legislation or administrative promulgation. They have focused public attention, both local and global, on the need for protecting the resource. They have studied the resource and gained enough knowledge about it to suggest how it can best be conserved. When no other protection is available, they have purchased the land on which the resource occurs in order to provide personal protection; and they have done these things as often as necessary to maintain the protection.

Recognition of the role played by individuals is not intended to disparage the role of governments and governmental agencies in protecting ecosystems and wild organisms. Obviously the two must work together if the most benefit is to accrue to the resource. Where conservation is most effective, governmental agencies are most effective. Governmental agencies are themselves guided by the actions and interests of their individual officers, as are nongovernmental conservation organizations. Where individual citizens and individual government officials unite in their effort to conserve a resource, they can, to rephrase a well-known expression, "Prevent the movement of mountains." They can also preserve the view from the mountain, and the animals and plants that live on it.

Few parks will be created, fewer still will be preserved, and few ecosystems will remain intact as man's population expands unless individuals become actively concerned about their loss. Thank goodness that a few individuals are willing to set themselves apart from the "silent majority" in order to lead the drive for conservation of wilderness and wild creatures.

59. Biotic Impoverishment

Thomas E. Lovejoy
World Wildlife Fund, U.S.A.

The Energy of Living Systems

Unheralded except by a very few, a far more fundamental energy crisis is taking place on our planet than that represented by our concerns of 1974 and 1975 over fossil fuels. It concerns not the energy that drives our technology but rather the energy that drives living systems.

That energy is ultimately derived from the sun, and it may seem absurd to talk of an impending crisis, for will not the sun continue to shine—at least on a time scale relevant to human existence? The sun certainly will continue to bathe our planet with its energy, and, providing we do not foul the atmosphere too badly or relatively permanently, it will continue to be available for use by green plants. And, provided green plants are not too greatly disturbed, they can continue to convert that energy, through photosynthesis, into a form consumable by those animals and plants not blessed with the possession of chlorophyll.

The rapidly developing problem is a consequence of the effects of technology and its products on the biological energy systems of our planet. The effect is one of biotic impoverishment, simultaneously a reduction in the Earth's inventory of species and a reduction in biological productivity of the Earth's surface. The two effects are linked but not entirely.

Accelerating Rates of Extinction

The reduction in the number of species on the planet is the result of soaring extinction rates, rates continuously accelerating to levels way above normal. It is extraordinarily difficult to measure these rates—to a considerable extent because the inventory of the Earth's biota is woefully incomplete: somewhere between 70 and 85 percent of the species on our planet have yet to be described by science. We do know that conspicuous species, larger and higher life forms, such as predators at the top of food chains, have been exterminated at accelerating rates over historical time. At the moment, it would appear as if extinction rates of these kinds of species have temporarily diminished, but this is an artifact of our increased ability to monitor populations of species; we know that, actually, extinction rates are higher than ever and that the rates are accelerating. If we look forward to the results of current trends in population and technological impact, it is reasonable to envision extinction rates reaching 100 to 1,000 times normal by the end of this century.

We tend to look at extinctions on a species-by-species basis, searching for individual effects of this irreversible process. We are usually far too ignorant to assess the specific biological (as opposed to aesthetic) effects of the loss of a species, although there are very few who would not be given pause at the contemplation of the loss of somewhere between 10 and 50 percent of the planet's stock of species.

Species are the biological units that transform solar energy into forms consumable by man. The reduction in species number therefore automatically reduces the number of ways in which this energy is available to us. Also, to the extent that the level of biological productivity is a function of biological diversity, these extinctions are reducing the productivity of the planet, essentially reducing its capacity to support man and other forms of life.

Other Forms of Biotic Impoverishment

Extinction of species is not, however, the only form of biotic impoverishment taking place nor the only way in which biological productivity is being reduced. In area after area, through various stresses placed on ecosystems, man is causing local extinction, reducing the number of species in a particular ecosystem although the species removed may be surviving elsewhere.

The net effect of these stresses—whether it be phosphate pollution of the Great Lakes, the dumping of wastes from our coastal cities into the oceans, the dumping of pesticides on a variety of ecosystems, or the paving over of large portions of the Earth's surface for highways and parking lots—is to reduce the world biological product, which might be termed WBP, in contrast to the economic term GWP or Gross World Product. That portion of WBP which represents solar energy trapped by green plants and fixed as carbon is termed net primary productivity; it is estimated to be roughly 160 billion tons of carbon fixed or, in energetic terms, 840 trillion kilowatt hours. The factors limiting this form of productivity include the amount of sunlight falling per unit area on the planet and the maximum possible efficiency of photosynthesis. In addition, there are often—varying from place to place—other limiting factors such as water, which can in part be overcome by irrigation, or trace elements, which can in part be overcome by fertilizer. It must be pointed out, however, that efforts to remove the effects of limiting factors can often reduce productivity elsewhere.

Reduction of Productivity

The effects of the reduction of ecosystem productivity can be far reaching. Copper smelting activities in Sudbury, Ontario, have brought the biological productivity of a 1,000-square-mile area close to zero through the emission of oxides of sulphur; the plume from the smokestacks can be detected all the way to the Atlantic. Over much of the northeastern United States, sulphur oxides from fuel combustion are increasing the acidity of rain—at times, rain as acid as lemon juice. These acid rains will certainly have an impact on biological productivity in the Northeast—including agricultural productivity. What the effect will be is hard to assess, and to this end, in mid-May the first International Symposium on Acid Precipitation and the Forest Ecosystem will be held at the Ohio State University. [The proceedings have been published by the U.S. Forest Service as USDA Forest Service General Technical Reports NE-23 and NE-26 (1976), both available from Forest Service, U.S. Department of Agriculture, Upper Darby, Pennsylvania, 19082. See also: *Ambio* 5(5 and 6) (1976).—Editor] While the exact effects of acid rain are yet to be determined, we should keep in mind that the pH factor is as fundamental a biological parameter as temperature—and nobody would be complacent about a proposed alteration of world temperature; neither should we be complacent about the alteration of pH.

Biotic impoverishment is undoubtedly the greatest problem we face today; truly it dwarfs most other concerns. If the World Biological Product were monitored in terms of net primary productivity, it would be seen that we are close to the time when a measurable reduction could be detected. If we have not passed that point, we are close to it. This means the planet's capacity to support man is decreasing, just as, at conservative estimates, the human population will be increased by 2 billion before the end of the century.

In addressing this problem most conservation organizations, the World Wildlife Fund included, tend to deal with pieces of it—with endangered species and/or endangered ecosystems. Dealing with endangered species mostly means ecosystem conservation: Project Tiger is really Project Tiger Ecosystem. Our projects tend to be in underdeveloped countries, which usually need help the most, and as much as possible we seek out local involvement.

The odds in this struggle are high and stacked against the conservationists and the World Biological Product. The problems often seem impossible to solve. We can applaud Argentina's province of Chubut for establishing a sanctuary for the right whales of Peninsula

The Preservation of Ecosystems

Valdez, but we still must be concerned about whether the whales will be hunted during the portion of the year they are absent, as about the consequences of the possible development of an antarctic krill fishery, or the general treatment of the oceans as a world disposal unit. The best planned and managed of ecosystem reserves whether in the northeastern United States or elsewhere is susceptible to the effects of acid rain.

New Projects

Consequently, we are ever vigilant for innovation and projects with high leverage. Our activities have therefore included a project to redefine the principles for management of living resources which has yielded four proposed principles for "Optimum Ecological Resource Management." Another project involves computer simulation models for the management of endangered species. These models developed by Professors Miller and Botkin at Yale use the population dynamics of a species to predict the effects of various management decisions or nondecisions. An early version of the model demonstrated that despite a population of close to 193,000 sandhill cranes, a hunting rate of 12,500 would bring the species to extinction in nineteen years. This model when perfected will be made available widely and promises to be a powerful and exciting new weapon in the conservationists' armory.

Even these projects fall far short of halting the process of biotic impoverishment, which in terms of species extinction has the chilling characteristic of irreversibility. Too large a segment of the human population is either wrapped up in cocoons of concrete, steel, and glass or too involved in struggles to exist in already degraded environments to be aware of the accelerating erosion of the Earth's life-support systems. Values of both those who govern and are governed must change rapidly. It must be realized that World Biological Product is not to be ignored or considered an obstacle, but rather that Gross World Product can only be sustained in any long-term fashion provided it is compatible with World Biological Product. WBP can exist without GWP, but the reverse is not true. The extent to which we and our descendants are secure depends on the extent to which EARTHCARE as a term is taken to heart and the extent to which biotic impoverishment, the funeral march of the Earth's biota, is halted.

60. Private Action for the Global Protection of Natural Areas

R. Michael Wright
The Nature Conservancy, U.S.A.

Introduction

Worldwide private action for the protection of natural areas encompasses a broad and disparate range of endeavors. In this paper, I will present an overview—a synopsis of the problem—and several examples of private activity, and proposals for the future; also, I will briefly demonstrate the significance of biotic diversity—the reason why natural areas must be protected. The several types of activity to be examined are: treaties—early proposals and new initiatives such as the World Heritage Convention and Man and the Biosphere (MAB) programs; national efforts—The Natural Trust and The Nature Conservancy; and international cooperation—Lake Nakuru, Coto Doñana, Dominica, and Monteverde. The proposal for the future is an approach to the protection of natural areas that is consistent with the increased, but as yet not universal, recognition of the need to preserve global diversity.

There are two kinds of constraint on global action. The first is the fact that natural areas are established inside national borders, thereby limiting international jurisdiction; and the second constraint is due to the limited perception of the need to preserve natural areas.

Natural Areas and Global Action

The global protection of natural areas has yet to be generally accepted as an international problem, although some attention has been given to it (e.g., Russell and Landsberg 1971). Hargrove (1972) defines the term "international problem." Decisions about the future of natural areas are made at least in theory—at the national level. The results arouse global concern, but they are not subject to international jurisdiction. There simply is no consensus among nations that the problem so affects the reasonable interests of other states as to overcome national sovereignty and to come within the competence of the international public order.

Although the destruction of natural areas is one of the few kinds of irreversible environmental disruptions there are, the duration, degree of certainty, and magnitude of the effects of such destruction, are not well understood. Detrimental results appear to be long-deferred, cumulative, and even somewhat speculative. Lack of appreciation of the impact of destruction of natural areas can, in large part, be attributed to a historic failure to recognize the significance of ecological diversity. Ecological diversity is the broadest manifestation of the rationale for preserving natural areas (UNESCO 1974) and is also a potential conceptual framework within which to evaluate other international environmental problems.

Privately undertaken action for the global protection of natural areas has in the past, and will in the future, struggle with these limitations.

Natural Areas and Global Diversity

The movement to protect natural areas has been unable to contend with the incremental destruction of ecosystems or with the loss of their unique and qualitatively most important components. The challenge of private action, then, is twofold: first, to create a global consensus that ecological diversity is the uniting interest of species and biotic communities—mankind and nations are fundamentally dependent upon this diversity which we have heretofore ignored, disdained, and even sought to destroy; and second, to accept the moral and ethical obligations which this fundamental dependence implies.

Man simplifies nature, thereby reducing natural diversity and preventing his own use of the land for his greatest benefit. When infertility results, he has traditionally moved, relying on natural regeneration to restore the land to health or (more recently) has turned to costly artificial interventions. Despite the ill-defined role of various natural organisms in ecological succession, their elimination continues; the landscape is altered at an ever-increasing scale and pace. The genetic pool to be drawn upon for natural or man-assisted regeneration shrinks, and what remains is widely, perhaps too widely, dispersed. History has demonstrated the eventual human costs when the Earth, unable to heal itself, bears indefinitely the scars inflicted by man, or recovers but with reduced complexity and carrying capacity.

As Rolf Edberg (1969) has so felicitously said:

> In every part of the world, man has worn down the film of life. . . . The good earth is life, life more pirmary and in its way more essential than that of individual vegetable and animal species, which are, after all, only secondary expressions of a living earth. Earth is a dynamic process. . . . Its topsoil, produced so very slowly, can be consumed—diluted, impoverished, squandered in dust clouds over the oceans—all too quickly. Just as individual plant and animal species can be decimated or exterminated, so the life of the earth itself can wither and vanish—with ominous consequences for secondary life, the higher forms of life that have arisen out of the juices and salts of soil.
>
> This applies also to man himself. "My friend, who is not an offspring of the earth?"—a perspicacious question posed in the Babylonians' Gilgamesh epic. When we pillage the earth, we are simply indulging in an indirect version because it affects generations still unborn.

Each species is unique and capable of producing an almost infinite abundance under the proper circumstances. Each species might someday be used in man's service: the search for beneficial uses is continual, and the results are often unexpected. In medicine, penicillin is but the most dramatic example. Recently, the blood of the ignoble horsecrab, the subject of (fortunately) unsuccessful extermination campaigns, has been found to be an extraordinarily sensitive diagnostic tool for detecting spinal meningitis in children (Thomson 1975).

Genetic diversity is the primary resource of the agricultural scientist for creating new, high-yield strains and preventing subsequent threats to sophisticated hybrids. Natural predators, identified in their natural systems, may also provide the escape from the pesticide treadmill. The essential issue is not whether each species has potential adaptability, but that, with a broad genetic pool, there is a high probability that some species—we cannot anticipate which—contain critical, irreplaceable, and therefore invaluable resources for meeting unforeseen human needs. This natural legacy is being squandered, often to no alternative purpose.

Man's "monocultural" agricultural system yields high, but often unreliable, benefits from the land. We must look to the diversity and the resulting stability contained in natural systems if "monocultures" are to be more stable—if catastrophic fluctuations in food supply are to be avoided, if man-created grasslands are to hold out against the encroaching desert with the efficiency of their natural cousins.

Man's continued manipulation of the environment necessitates an ability to assess the stress he imposes on natural systems to determine whether a new natural stability has been reached or whether technology must give way. Such monitoring will rely on the rare, specialized species apt to have characteristics highly specifi-sensitive to certain substances (indicator species), which may accurately reflect environmental influences and impacts in a way that technology cannot duplicate.

Mankind has undertaken to experiment on a massive scale without benefit of control or ecological baseline for comparison. (According to Leopold [1966], ecological baseline is "a base datum of normality, a picture of how healthy land maintains itself as an organism . . ." [p. 274] .) An artificial laboratory cannot create the complex, often unexpected, chains of dependency. Only the direct observation of nature will reveal whether change is trivial or serious, and—if the change is serious—the tools for correction.

Natural areas, unaltered by the hand of man, can serve his needs in a multitude of other ways (Dasmann et al. 1973): they can attract tourists who seek renewal from wildlife and undisturbed beauty or serve as recreational escapes for local people from fast-growing urban "deserts." In the form of undisturbed natural watersheds, they can secure pure water for cities, prolong the life of hydroelectric dams that supply power to the cities, and prevent the erosion that makes soil the world's most common pollutant. In the form of natural marshes, natural areas provide the food essential for fisheries upon which whole populations depend; or they may support the native species, such as antelope, that can provide human sustenance in a more ecologically wise manner than introduced species.

Finally, mankind has, from prehistory, constructed monuments to his nations and his progress—most recently, capitals or dams, some of them magnificent and necessary, others frivolous. Only now has the preservation, for our children and our children's children, of the irreplaceable things that make a nation truly unique been recognized as a patriotic value. The time must come, as for example it recently has come in Venezuela, for each country to expand its definition of "national interest" to include the preservation of its natural heritage and the future this assures. President Carlos Andrés Pérez of Venezuela, in a speech given at Mirafloris Palace, Caracas, 29 May 1974, states: "The natural resource . . . constitutes part of an image of values that not only corresponds to the inventory of the physical environment, but also to the political and spiritual order of the nation. . . . natural resources are the integral wealth of society. . . . It must also be known that nature is included within the proceedings of culture and associated with its fate."

For a more detailed treatment of these issues, see The Nature Conservancy (1975, Part I) and Programme on Man and the Biosphere (UNESCO 1973).

Natural Areas and International Treaties

Despite the increased awareness among people of their value, natural areas and the biotic diversity they sustain will continue to be threatened. Choices will be made between immediate national development and future global options; between personal economic values and often still unknown probabilities that may not be directly beneficial. Decisions will be made that, in the aggregate, will have incremental and irreversible, but largely recognized, effects. The worth at some future time of a species that has been eliminated can never be known.

These problems are being met by agreements among governments—agreements that often are due to private initiatives. International agreements continue to be the focus of much private conservation effort. The result of this effort is considerable: all major international agreements to protect the natural environment have been negotiated in this century; however, only recently have natural areas been their subject.

Traditionally, treaties, with only a few exceptions, have focused on the protection of migratory species, particularly their protection from hunting. The threats posed by pollution were not recognized, nor was the loss of habitat deemed the proper subject of international jurisdiction. Species-specific agreements, the most common of which deal with various fisheries depleted by uncontrolled exploitation, generally have been based upon financial considerations. The solution to depletion was not preservation per se, but research into the prerequisites for sustained yield.

There are exceptions to these generalizations, such as the Fur

Seal Convention (1911), which for its own particular reasons was effective. The Convention on Nature Protection and Wildlife Preservation in the Western Hemisphere (1941) lacked provisions for its enforcement but it nevertheless was concerned with habitat and sanctuaries, and it constituted the embryo of many recent conservation efforts. The African Convention Relative to the Preservation of Flora and Fauna in Their Natural State (1933) and the more recent African Convention for Conservation of Nature and Natural Resources (1968) are also notable exceptions. And the Convention on International Trade in Endangered Species of Wild Fauna and Flora, while having its antecedents in the migratory-species treaties, establishes an important alternative basis for international jurisdiction.

More recently, The Convention for the Protection of the World Cultural and Natural Heritage, as well as The Man and the Biosphere Program (MAB), have brought the subject of natural areas directly into the international arena. Neither proposes international obligations that could be the subject of coercive sanctions, but both represent an important recognition of national responsibility for natural areas. Properly established and funded, in cooperation with the International Union for Conservation of Nature and Natural Resources (IUCN), these could serve as the essential international clearinghouses for technical and financial conservation assistance in the global protection of natural areas. The implementation of the World Heritage Convention and MAB, with the awareness they create and the support they provide for nations that opt for conservation, should be a prominent—but not the principal—goal of private international conservationists. By themselves, these programs will be insufficient.

The global protection of natural areas can be "internationalized" only when the interests and welfare of nations are better protected (or protected to an extent unobtainable by themselves) by multinational cooperative efforts to establish a system of global identification and support. However, with the present level of appreciation, neither is likely to achieve parity with the pressures for growth and alteration that are rapidly destroying the natural world, nation by nation. It is essential, therefore, that private action focus on support of individual countries which have, and will retain into the foreseeable future, sovereignty over natural areas and the ecological diversity they contain.

Natural Areas and National Preservation

Since the last century, the most effective conservation efforts have been on the national scale—a movement which often emanated from individual or private organization efforts and, more often than not, involved the establishment of parks with spectacular geographic or wildlife features (the "last-biggest-most-rarest-least") on existing government-owned land. The success in the United States of private groups such as The Boone and Crockett Club and of individual philanthropists, such as John D. Rockefeller, Jr., and many of lesser financial resources, is impressive (Udall 1963, Dorst 1970). While governments have become involved in the establishment of parks, the acquisition and maintenance of lands for scientific reserves or natural areas have usually remained the province of private organizations. An exception to both of these generalizations is the Soviet system of *zapovedniks*, which in both orientation and extent may be the closest approach to a systematic attempt to preserve biotic diversity. [See P. R. Pryde and V. A. Bonssoff's papers on *zapovedniks* in this volume.—Editor] The creation or support of such indigenous conservation groups within other nations should be a major priority of private international conservationists. Two such organizations can be briefly described.

The National Trust for Places of Historic Interest and Natural Beauty (U.K.)

Private organizations in England have been influential as the original leaders of the conservation movement and now carry the principal burden in creating nature preserves, sometimes through acquisition, but more often through agreements with landowners. The County Naturalists' and Conservation Trusts, the Scottish Wildlife Trust, the Field Studies Council, the Royal Society for the Protection of Birds, the Society for the Promotion of Nature Reserves and, above all, the two National Trusts—for England and Wales, and for Scotland—have created such preserves and now manage them.

The National Trust for Places of Historic Interest and Natural Beauty, registered under the Company's Act of 12 January 1895, was described recently as "unarguably the world's greatest conservation society" (Green 1975). It has a share of about 1 percent of the total area of England and Wales (160,000 hectares) and is the largest single landowner after the Crown and the state. Much of its ownership is in the heart of national parks or on the coast, giving it influence far beyond what even its impressive statistics would indicate.

The trust considers itself totally private; however, by U.S. standards, it has at least a quasi-public character, and the trust's effectiveness has been enhanced, in part, by governmental actions. An Act of Parliament in 1907 enabled the trust to declare its land inalienable without the special will of Parliament. Ninety-five percent of trust land is so declared, and only once in seventy years has land been taken from the trust against its will. The same act gave the trust's regulations for protecting its preserves the force of law. After succeeding acts, the trust could hold lands purely for income purposes and receive funds and lands from local governments, personal property from individuals, and conservation easements (protective covenants), which were in gross and in perpetuity. Of great importance was the exemption from death duties, which could reach 75 percent of an estate, for gifts given to the trust and declared inalienable. The exemption was later extended to lands subject to life estates, endowment funds, and gifts of personal property. Finally, properties could be offered in lieu of death duties, and a government fund was established to compensate for lost revenues from this and earlier exemptions. In fact, the government has provided substantial amounts to the trust's fundraising campaigns. Despite the considerable involvement of the government, the trust must be considered a private effort to preserve Britain's natural as well as historic heritage. It has not allowed its special relationship to inhibit its confrontations with government in defense of trust properties or of the countryside in general.

The Nature Conservancy (U.S.A.)

In 1917, the Ecological Society of America established the Committee for the Preservation of Natural Conditions, thus recognizing the fact that the nation's natural areas were endangered. The committee undertook the first comprehensive inventory of natural areas in the United States including those known to be protected. In 1946, the committee became a separate entity, the Ecologists Union, but subsequently decided that only a direct acquisition role in natural areas preservation could make the group effective. Borrowing the name of an earlier established British group, the 342-member Ecologists Union became "The Nature Conservancy" in 1950.

The new organization spent several years experimenting with various methods of land preservation. Then, in 1953, the first independent project was undertaken at Mianus River Gorge in Westchester County, New York, when a group of Connecticut residents asked for affiliation with the conservancy in order to raise money to preserve and protect the gorge. Today, The Nature Conservancy is the only national, nonprofit, publically supported conservation organization in the United States whose resources are solely devoted to acquisition and preservation of ecologically and environmentally significant natural land. Nearly 300,000 hectares of forests, swamps, marshes, prairies, mountains, and beaches, in over 1,350 separate projects across the continent and in the Caribbean, have been protected through the efforts of the conservancy since the acquisition of its first preserve twenty years ago. Together, The Nature Conservancy and the Trust for Historic Preservation represent the United States to the British National Trust and British Conservancy.

Since its first project, the conservancy's acquisitions have developed into five distinct categories relying on financial resources as well as staff, tax, legal, scientific, and business expertise:

1. Purchasing key ecological areas with funds available from a national project "revolving fund" that is repaid through public fundraising, both nationally and locally (39 percent).
2. Receiving gifts of land from concerned donors (both individual and corporate) for use in research, education, and other environmentally relevant purposes (40 percent).
3. Acquiring areas for government conservation agencies, at their request, before the agencies are able to do so, for future conveyance to the requesting agency (13 percent). Such cooperative projects with government agencies have been possible through the flexibility of established lines of credit with several banks and loans on specific projects from banks, insurance companies, and foundations.
4. Assisting other private or public conservation bodies to acquire and protect natural lands, including several major joint projects with the National Audubon Society (7 percent).

5. Acquiring conservation easements or other less-than-fee interests (1 percent).

Recently, returning somewhat to its roots, the conservancy has undertaken to systematically inventory ecologically significant areas, communities, species, or features as part of its State Natural Heritage Program. Working closely with state governments, the goal of this program is to create an integrated approach to identifying areas, establishing priorities, and, ultimately, protecting the full array of each state's natural lands.

The conservancy has also recently initiated an international program (referred to below in the Dominica case) to assist in the preservation of areas, particularly in other countries of the Western Hemisphere.

The number of private conservation groups in the United States alone is awe-inspiring. This brief description of The Nature Conservancy does not diminish the role of organizations such as the Sierra Club, the Natural Resources Defense Council nor of a myriad of others that, through litigation or lobbying, seek to protect the natural environment, nor of the numerous effective groups that deal directly with natural areas—the Western Pennsylvania Conservancy, the Society for the Protection of New Hampshire Forests, the National Audubon Society, and the Trustees of Reservations, to name but a few.

It would be fatuous to suggest that the National Trust or The Nature Conservancy could or should be transferred, *in toto*, to other nations. They can offer lessons and inspiration, but each nation must evolve organizations compatible with its cultural and historic roots. However, international conservationists should strongly encourage and support such broad-based private conservation constituencies as indigenous counterbalances in favor of irreplaceable national assets in their own nation's debate over development.

Natural Areas and International Cooperation

One of the most intriguing aspects of global protection combines national action with international support. The projects I describe below are by no means an exhaustive sampling of local conservation efforts assisted by private international groups. They do not include, for example, numerous national efforts in which individuals, governmental or private, or local private organizations were critical, nor do they represent the more usual, but no less important, method of private international assistance in which, through "good offices," governments have established parks by decree. The examples demonstrate, however, that with a will to act, nations and international private conservationists can in fact cooperate to save land whose loss would make the world poorer. From such action can come a broader recognition of the importance of Earth's natural areas.

Lake Nakuru

Lake Nakuru, situated in the Great Rift Valley in Kenya, has been described as "the greatest ornithological spectacle on Earth." Set amidst mountains, the lake is rimmed in pink by some 500,000 to 2 million lesser flamingos. The sight is enhanced by thirty other species of birds that depend on the lake, including cormorants, darters, grebes, spoonbills, seven or eight species of herons and egrets, and between 3,000 and 30,000 great white pelicans. To preserve this unique concentration, the area was officially declared a park in 1961; however, by focusing on the spectacular rather than the ecological (and in the face of considerable local opposition), the park covered only the lake itself and a narrow strip of surrounding land. This sowed the seeds that led to the need for private and international involvement.

By 1971, conflicts arose over existing and proposed uses of land around the lake. Urban growth, as well as a proposed road bypass, encroached on the lake; sewage and pesticide runoff increased; and the large, passively ranched holdings surrounding the lake began converting to intense, small-holding agriculture. These threats to the integrity of the lake led to a meeting between concerned African officials and representatives of UNESCO and the World Wildlife Fund (WWF). It was determined that a quick ecological survey could establish priorities and clarify issues.

The survey proved to be of critical importance to the subsequent process. (The report, by Kai Curry-Lindahl and John Hoperaff, was published in *Africana* 5[5] [1974].) While recognizing the spectacle, the survey noted that the lake's international significance is really the extreme biological import of its unique ecosystem, which encompasses all of the basin and the interrelationships of organisms and habitats. In fact, the diverseness of the terrestrial habitat with its seventy species of mammals create the optimal conditions in the lake upon which most of the world's lesser flamingos depend. The survey identified eleven distinct major habitat types in the basin, only two of which, the lake and its immediate shoreline, were in the park. It was recommended that the park be expanded from its original 5,600 hectares to 20,600 hectares, noting this national and global asset could not be moved—for the lake there was no alternative. There was an alternative, however, for the pattern of urban and agricultural growth. With proper information and direction, growth and preservation were not incompatible.

This report, with its boundary-extension plans, was adopted by the Parks Department. WWF made the raising of funds for the acquisition of the privately held land (including additional lands to relocate small landowners), its top priority. The process was long and often frustrating because of considerable intransigence from the town and all the complexity of multiple-party negotiations. The underlying structure was a legal agreement between WWF and the Kenyan government, whereby the former agreed to provide funds, while the latter would negotiate and acquire—compulsorily if necessary—and subsequently protect the expansion areas, in accord with international standards. The perseverance of a few individuals who refused to abandon the project when good sense, perhaps, dictated, and the commitment of top levels of government to preservation when jurisdictional disputes arose, helped sustain the project. In addition, without private support and the financial commitment it represented, the area would probably have been lost. The fact that it was not lost has been economically as well as ecologically justified. Lake Nakuru is now Kenya's second most visited national park, and its revenues are constantly increasing.

The Coto Doñana

The Coto Doñana or Marismas in far southwestern Spain, at the delta of the Guadalquivir River, is considered the most important wildlife sanctuary in Western Europe. Bounded by the Atlantic Ocean and impenetrable marshes, it has been insulated from disturbances for 350 years. Here is found the last stronghold of two of Europe's rarest creatures, the Spanish imperial eagle and the Spanish lynx, as well as, at various times, half the total number of bird species inhabiting Europe. A diverse habitat of marsh, savannah, scrub, woodlands, and dunes provides habitat for hundreds of thousands of migrating songbirds, ducks, geese, and waders. Other birdlife includes a flock of 1,000 flamingos (one of the last two in Europe), three species of vultures, and up to 15,000 herons, spoonbills, egrets, and storks. Red deer, fallow deer, and wild boar roam the area amid African plant and insect life invading from the south.

The marshes *(marismas)* were increasingly being subjected to drainage for eucalyptus forests and rice farming under a government program. In 1960, the other protective barrier was breached by international speculators developing a huge coastal holiday resort. Luckily, three expeditions—in 1952, 1956, and 1957—to the Coto resulted in a popular book (Mountford 1958) and movie, whose appearance coincided with the threat. These works aroused public opinion and were bolstered by a scientific report from a Spanish zoologist.

In retrospect, it is difficult to establish whether the move to save the Coto led to the establishment in May 1961 of WWF International, or vice versa; however, lengthy negotiations began between the fledgling WWF and three owners of the critical 67-kilometer-square Coto Doñana portion of the delta. An option was obtained at well below commercial value, primarily as a result of the involvement of one owner's son in the earlier expeditions, and it was exercised in December 1963. Most of the more than $500,000 was secured by loans when the new organization began fundraising throughout Europe. The prestige of the Spanish government was aroused, and it agreed to contribute one-third the amount, establish a biological study center, and declare the area under permanent protection as a place of national importance. A contract to such effect (with a reverter) was signed, and in June 1965, WWF turned over title to the land to Spain.

The drive to preserve the *marismas* did not end with the Coto Doñana. During the initial fundraising and thereafter, the La Nuevas and Guadiamor areas were acquired through WWF and the Spanish government, thereby securing the essential ecosystem elements (as at Lake Nakuru) to provide a viable preserve, as well as adding alternative types of habitat.

Management of the Coto Doñano sanctuary has been the subject of considerable controversy since its establishment, especially with competing adjacent land use. Clearly, international involvement and concern do not cease with the fundraising—yet patiently, over time and with the trust and goodwill established during the initial acquisition, these problems are apparently being resolved one by one.

Dominica

The island of Dominica lies midway on the chain of Caribbean islands that stretches from the Virgin Islands south to Trinidad, and has the only large expanse of undisturbed flora remaining in the Lesser Antilles. The rugged interior mountains, which often receive more than 760 millimeters of rain per year, provide habitat for some 135 species of birds, a small opossum, the agouti (a rabbit-size rodent), and boa constrictors. The *crapaud* (a large woodfrog), the blue-headed hummingbird, and two species each of parrots (the imperial and the red-necked, both endangered), snakes, lizards, and bats, are known only on this island. However, it is the lush tropical rain forest, including approximately 5,000 species of vascular plants, many of which are endemic, that assures Dominica's international significance and has resulted in an extensive, long-term scientific study of the island, the Bredin-Archbold-Smithsonian Biological Survey of Dominica.

The pristine forest, dominated by trees rising over 35 meters and festooned with lianas and covered with epiphytes (air plants), also brought Dominica to the attention of a Canadian lumber company in 1967. The question became how to utilize a needed resource without jeopardizing an irreplaceable national asset.

The island's uniqueness had been recognized by a number of scientists, who had made various suggestions for establishing preserves over the years, but it took the lumber proposals to catalyze action. In 1969, the Conservation Foundation, working with a new regional organization, the Caribbean Conservation Association (CCA), obtained a grant and undertook an investigation of the park proposals. The resulting report was delivered to the government in June 1970. It recommended the setting aside of Crown Lands, promoting the gift to The Nature Conservancy of a large, private tract containing the best rain forest in the park area, and the establishment of a national park service (Eddy et al. 1970). The report was readable, concise, and, while an exhortation to establish the park, clearly delineated the economic, resource, and ecological values protected, as well as its scientific significance. These benefits included tourist and educational potential as well as protecting water supply, hydroelectric potential and avoiding erosion problems.

Efforts continued to build and maintain local interest in the park during the succeeding years after the Conservation Foundation involvement terminated and the lumber company went bankrupt. The CCA, who along with interested Dominican officials and the private landowner kept the idea alive, was instrumental in obtaining a Canadian aid grant. This was to be administered by a private conservation group, the Canadian Nature Federation (*née* Canadian Audubon), to assist the government with the specific delineation of boundaries so as to avoid title disputes, drafting appropriate legislation and regulations, and working with park development. This phase is now underway, and the results are expected to be presented to the government in the immediate future.

Also during this intervening period, The Nature Conservancy initiated a new international program to assist with the conservation of natural areas outside the United States, and Dominica became the first priority. The gift of the private 380-hectare tropical rain forest was of great significance in itself, but it also displayed an international commitment in support of the government's decision to set aside over 9 percent (approximately 6,800 hectares) of Dominica's land area as a park. A number of visits to resolve complex legal issues and procedures and to establish cooperative relations with the government culminated in the gift to the conservancy in January 1975. The conservancy is working with the government to manage the area as part of the park through a lease arrangement and expects, ultimately, to donate the area to the people of Dominica for inclusion in the park.

The project's success is the result of numerous international individuals and groups—the gift offer along with the timbering proposals that crystallized action—but primarily local people in Dominica and CCA who persisted over the long term. With the establishment of the park, properly protected, the future of Dominica's heritage and its unique place in the Caribbean seems well on its way to being assured.

The Monteverde Cloud Forest Preserve

The Monteverde Cloud Forest Preserve straddles the continental divide in the Tilarán Mountains of northwestern Costa Rica. The astonishingly high biotic diversity of the area is seen in its six distinct ecological communities, composed of over 2,000 species of plants, 100 species of mammals, and 222 species of birds, including the black guan, the great green macaw, the resplendent quetzel, the three-wattled bellbird, the bare-necked umbrella bird, Baird's tapir, three species of monkeys, the jaguar, the ocelot, and the most brilliant of toads, the golden toad, whose range is confined to a small part of this cloud forest. As in the case of Dominica, however, it apparently is the flora (particularly the epiphytic growth which may be among the world's richest—a single tree may host up to 100 species), which is of greatest significance and provides the essential habitat to be protected.

Earlier attempts to exploit economically the Monteverde forests capitulated to the ecological realities of rain-forest soil, wind, and rain. Then, throughout Costa Rica, speculative land deals, particularly the sale of small tracts to unwary Americans, became lucrative and threatened to destroy forests nationwide with no productive economic benefit. An additional concern was rapid clearing of the forest, particularly in dry areas, in large part to supply beef to U.S. markets.

The preservation of Monteverde was initiated in 1972 by a married scientific team that had earlier undertaken research in the area. With perseverance and the use of limited personal funds, they were able first to acquire squatters' rights within a 326-hectare tract and then convince a private company to donate their presumed title to the Tropical Science Center, a nonprofit Costa Rican scientific research association. Some 400 adjacent hectares were added, although not without difficulty, through acquisition of squatters' rights on the government land. The Monteverde Quaker Community agreed (by means of a cooperative landholding agreement) to retain their 800 hectares of communal forest north of the preserve in undisturbed condition, thus simultaneously assuring their own water supply and contributing to conservation.

Partly as a result of the attention and activity, and partly out of personal conviction, the government has protected the surrounding land from squatting and timbering according to the Forest Reserve

Act. The decision not only retains the future option of officially establishing a park, but protects the watershed, assuring irrigation of the truly productive Padivid plains in dry season, providing flood protection, and preventing sedimentation streams during the rainy season on the Atlantic coast.

With the initial purchases, the Tropical Science Center appealed for financial support from the international community. The first responses came from the Explorer's Club of New York and the Philadelphia Conservationists. Subsequent small grants came from The Nature Conservancy, the National Audubon Society, the International Council of Bird Preservation, a number of other groups, and, particularly, various individuals. By December 1974, the preserve had reached an effective size of 1,600 hectares.

Legal fees and management costs remain, along with the need for an additional 6,000 hectares to provide self-sustaining habitat for the larger mammals and birds. Pressure continues from extensive forest clearing, and land prices are escalating (although the costs for an area of such biotic diversity remain extraordinarily reasonable by our standards). The task of obtaining the necessary funds is now being undertaken as part of the WWF/U.S. tropical rain-forest program in cooperation with other conservation groups.

General Observations

Nakuru, the Coto, Dominica, and Monteverde are in many respects singular; however, some general observations are possible:

1. The individual does matter. In each case, there were a few concerned persons, often a scientist or nature lover, some local and some not, whose affection and determination usually identified the area and sustained the preservation idea over the years. As Aldous Huxley has said:

   When a piece of work gets done in the world, who actually does it? Whose eyes and ears do the perceiving, whose cortex does the thinking, who has the feelings that motivate, the will that overcomes obstacles? Certainly not the social environment, for a group is not an organism but only a blind unconscious organization. Everything that gets done within a society is done by individuals.

2. The scientific value of each project was initially evaluated, often by the individuals referred to above. The surveys resulted in a document that created interest, delineated the limits of the proposal, and explained benefits and potential. While explaining scientific value, some also referred to the economic implications of the alternatives. These documents, with their recommendations for action, were a form understandable and useful to decision makers and to the broader public.
3. The preservation of each area was of unquestionable significance, but the identification was fortuitous and usually in reaction to an imminent threat. It will be increasingly necessary in the future to establish, wherever possible, priorities and to institute preservation earlier—while retaining the flexibility to react defensively if necessary. In this manner, international projects will be clearly compatible with local preservation needs and capabilities.
4. The time, effort, and costs of multinational projects indicate a need to concentrate on a limited number of large projects. Although there were numerous small holdings in Nakuru and Monteverde, the cases are also generally characterized by large holdings, either governmental or in non-native (individual or corporate) ownership.
5. The four ostensibly private projects had definite government involvement in different degrees, ranging from acquiescence to active negotiation. While private international action can provide a catalyst for particular projects, it is essential, if conservation is to have any long-term impact, that the ethic be integrated into the everyday actions of governments and people. Likewise, the projects were not completely international initiatives, but involved local individuals from their inception. All are essentially self-liquidating in relation to international involvement in ownership or management, if not concern.
6. At some point in each case, it was necessary to explain the private international group's intentions to local officials who were justifiably suspicious of a wave of new conservation "missionaries." It was possible to overcome their suspicion for two reasons: (a) a true respect was displayed

for the needs of local people, as well as those of the natural environment; and (b) a rational reason was given for the international involvement. In each case, there was something to offer on specific aspects of the problem—whether it was funds; tax, legal and acquisition expertise; scientific advice for the initial evaluation or subsequent management planning; or negotiations with foreign owners—multinational corporations or individuals—with possible tax benefits that were not locally applicable.

7. The four projects encountered considerable legal technicalities of different degrees and costs resulting from noncompatible procedures and approaches. As the legal details of various private-government agreements and the tax and legal requirements of foreign acquisition are to be the subject of a separate investigation, suffice it to say that expert advice is essential. The management of each preserve has faced, or will face, its difficulties from conflicting land use and new demands, particularly as local and international attention turns to new challenges. These two factors require the broadest possible base of local support and patience and perseverance on the part of international groups for actions that may extend over a period of years.

8. Ultimately, the decision will be national, yet international recognition is important to balance the debate on how one might grow. For example, in the United States, groups and individuals in the field find that to increase their effectiveness and create legitimacy for conservation concepts, the value of their efforts must be recognized.

9. Finally, each instance of internationally assisted acquisition, while extraordinarily complex and limited to only one project, has a great advantage. In a world of verbal exhortation, something tangible results: ecologically significant land has been saved. From action on specific projects can come a broader understanding and constituency for natural areas generally, and ultimately for the creation within each country of its own "national environmental interest."

Natural Areas: The Future Imperative

International efforts traditionally seek to have nations abstain from or oppose certain actions. In this, there has been some limited success. It will be more difficult to overcome institutional inertia, short of perceived crisis, when affirmative national action is required, and the preservation of natural areas will require positive action to generate a sufficiently strong countervailing force to the powers that are altering the face of the globe.

Private conservationists must recognize that inequality of interest and capability—differences in immediate versus long-term self-interest that exist among nations—will defeat attempts to institute "internationalized" control over the future of natural areas. Instead, the goal must be to provide tools, information, and guidance to national efforts in order to save a local impact and to establish the model for action that is currently lacking.

What is needed is a systematic program on a national scale. Ultimately, the implications of growth and equity within the global system must be faced; however, even on the assumption that alteration will continue unmitigated for the foreseeable future, it can be channeled into less destructive patterns. A rural nation that pursues development and economic growth does not thereby express a desire for a degraded environment or depleted wildlife. What is inevitable is that unbalanced, ill-planned development (often to feed the consumption of others) jeopardizes human needs as well as the ecosystems and the future options they contain. Unique areas are a necessity; they improve the quality of life.

The future of private action in the international regime must be to exhort nations to take conservation action; to strongly support nations which have made the commitment to preserve their heritages; and, where exploitation continues, to work with governments and indigenous private groups to affect (1) the rate, (2) the direction, and (3) the quality of such alteration.

The goal of such an effort should be to provide data that will help determine whether the remaining fragments of unspoiled environments are to be meaningfully distributed, in the recognition that ecological components are neither universally distributed nor perfectly interchangeable. The rational channeling of continued exploitation can have a profound impact on whether or not what remains is a systematic array of lands containing adequate, viable samples of the range of major ecological formations, communities, and habitats, and illustrating the degree of variation within each. The choice of the pattern of exploitation or growth can be accomplished at a minimum of economic cost to the nation if we dare to assume that everything will not be altered beyond a semblance of natural condition and if we can choose the pattern of distribution of what survives.

Such an approach must seek to establish a pattern of growth that can yield realistic economic gains without the needless and irreversible destruction of precious natural resources. The approach has three principal stages.

1. *Inventory.* An inventory of the prime natural features and areas throughout the country must be undertaken. The identification process can include economic evaluation as well as historic, archaelogic, and recreation potential, but must focus on basic diversity and its distribution. It should be on a countrywide basis and should be in a form capable of use by political decision makers, and should be suitable for continual updating.
2. *Setting of Priorities.* Priorities among sites must be established to present clear choices. Policy will emphasize the protection of rare and systematic samples rather than random pieces left over as unprofitable for exploitation—which is, at present, often the case. Other areas are open for exploitation although efforts should be made to see that this is accomplished in an ecologically sound manner. As Aldo Leopold expressed it:

> ... A system of conservation based solely on economic self-interest ... tends to ignore, and thus eventually eliminate, many elements in the land community that lack commercial value, but that are (as far as we know) essential to its healthy functioning. It assumes, falsely ... that economic parts of the biotic clock will function without the uneconomic parts. (Leopold 1966, pp. 229–230)

3. *A Plan for Protection and Action.* Implementation will require a plan for protection and action to ensure compliance. While the will to action is the most difficult step, it is made easier because an objective basis for identifying the best areas provides support, and the mere existence of a specific number of identified priority areas facilitates national planning for protection through regulation, legislation, or acquisition.

Having gained a conception of the problem and adequate information and priorities, it will be necessary to identify or create an organization that is able to coordinate related land efforts, apply and update information, and be adaptable to change.

Although such an approach may seem obvious, it exists in no nation of the developing world and only incompletely in industrialized nations. It is in the rural nations where development and growth must occur to satisfy basic human needs and, therefore, where the environment will continue to be manipulated. If conservationists position themselves against such legitimate aspirations, they will quickly become irrelevant in the face of complex national needs. As the Oriental fable puts it:

> Once upon a time a monkey and a fish were caught up in a great flood. The monkey, agile and experienced, had the good fortune to scramble up a tree to safety. As he looked down into the raging waters, he saw a fish struggling against the swift current. Filled with humanitarian desire to help his less fortunate fellow, he reached down and scooped the fish from the water. To the monkey's surprise, the fish was not very grateful for this aid. (Foster 1962)

Therefore, the "natural-heritage approach" just described assists nations to identify and clarify the alternatives to scooping the fish from the water—alternatives on how such growth can occur at little economic cost when it is channeled to meet the needs of both man and nature. It is a positive approach, consistent with the nations to control their own destinies and resources. It recognizes the overriding need for resource policies within which thoughtful development can take place, unnecessary mistakes minimized, damaging paths—well-travelled by others—avoided; and what is nationally and internationally important preserved for the future.

Nowhere is the relationship between man's survival and the elimination of future options through the destruction of natural areas more acute than it is in the developing nations of the Earth. It is there that the most magnificent and least understood ecosystems exist and where the threat is most persistent and persuasive.

Some natural areas are lost in response to the legitimate aspirations for human survival and growth, but the vast majority are sacrificed needlessly through inadvertence or the culturally destructive and ecologically ignorant exploitation often undertaken for transitory economic gains.

The great tragedy is that many nations, lacking proper resource inventories, cannot even determine what they are losing. The effort to preserve the vital remaining natural areas of the globe must ultimately occur; to delay will only escalate the cost and further jeopardize human and natural values. At present, there is still the chance to prevent the loss of these unique and fragile lands of unparalleled natural abundance and diversity.

Acknowledgments

Dr. Robert Jenkins and Stephen Steinhour, Esq., gave me invaluable information, ideas, and insight for writing this paper, but I alone bear responsibility for the presentation, and no statements made herein should be imputed to them. I obtained the information on Lake Nakuru and the Coto Doñana through the kind assistance of the World Wildlife Fund.

References

Caldwell, Lynton K. 1972. *In Defense of Earth: International Protection of the Biosphere.* Bloomington: Indiana University Press.

Dasmann, Raymond F., John P. Milton, and Peter H. Freeman. 1973. *Ecological Principles for Economic Development.* New York: John Wiley & Sons.

Dorst, Jean. 1970. *Before Nature Dies.* Trans. Constance D. Sherman. Boston: Houghton Mifflin Co.

Edberg, Rolf. 1969. *On the Shred of a Cloud: Notes in a Travel Book.* Trans. Sven Ahman. University, Alabama: University of Alabama Press.

Eddy, William, Robert Milne, and Leonard Godfrey. 1970. *Dominica: A Chance for a Choice. Some Considerations and Recommendations on Conservation of the Island's Natural Resources.* Washington, D.C.: The Conservation Foundation.

Foster, George M. 1962. *Traditional Cultures, and the Impact of Technological Change.* New York: Harper & Row, Publishers.

Green, Timothy. 1975. National Trust guards everything from park to pub. *Smithsonian* 6(1):72–81.

Hargrove, John Lawrence, ed. 1972. *Law, Institutions, and Global Environment: Papers and Analyses of the Proceedings of the Conference on Legal and Institutional Responses to Problems of the Global Environment* (Arden House, 1971). Dobbs Ferry, New York: Oceana Publications.

Leopold, Aldo. 1966. *A Sand County Almanac: With Other Essays on Conservation from "Round River."* New York: Oxford University Press.

Mountfort, Guy. 1954. *Portrait of a Wilderness: The Story of the Coto Doñana Expeditions.* London: Hutchinson and Co. (Publishers).

The Nature Conservancy. 1975. *The Preservation of Natural Diversity: A Survey and Recommendation Prepared for U.S. Department of the Interior.* Final Report. Contract no. CX0001-5-0110. Arlington, Virginia: The Nature Conservancy.

Russell, Clifford S., and Hans H. Landsberg. 1971. International environmental problems: A taxonomy. *Science* 172(3990):1307–1314.

Thomson, Peggy. 1975. Value is extracted from a nuisance. *Smithsonian* 6(1): 40–45.

Udall, Stewart L. 1963. *The Quiet Crisis.* New York: Holt, Rinehart and Winston.

UNESCO. 1973. *Conservation of Natural Areas and of the Genetic Material They Contain.* Final Report of the Expert Panel on Project 8. MAB Report Series no. 12. Programme on Man and the Biosphere (MAB). SC.73/Conf. 619/2.

———. 1974. *Final Report of the Task Force on Criteria and Guidelines for the Choice and Establishment of Biosphere Reserves.* MAB Report Series no. 22. Programme on Man and the Biosphere (MAB). SC.74/Conf. 203/2.

61. United Nations Environment Programme

Kai Curry-Lindahl

United Nations Environment Programme, Kenya

Introduction

The United Nations Environment Programme (UNEP) has its origin in the first United Nations Conference on the Human Environment held in Stockholm, Sweden, June 1972. The conference was convened to bring to the attention of governments and the people of the world evidence that man's activities are damaging the natural environment and are giving rise to risks for his own survival and well-being. The conference provided the basis for cooperative action to deal with this new challenge. At Stockholm, representatives of 112 nations agreed on:

1. A Declaration on the Human Environment constituting the first acknowledgment by the community of nations of the new principles of behavior and responsibility which must govern their relations in the environmental age;
2. An action plan consisting of 109 recommendations calling on governments, U.S. agencies, and international organizations, governmental and nongovernmental, to cooperate in taking specific measures to deal with the wide variety of environmental problems; and
3. Proposing institutional and financial arrangements for carrying out the action plan and providing a continuing mechanism to facilitate international cooperation in the environmental field within the U.N. system.

Principles 2, 3, and 4 of the Declaration on the Human Environment were of particular interest to this conference. Therefore, it is worth recapitulating them:

Principle 2. The natural resources of the earth, including the air, water, land, flora and fauna, and especially, representative samples of natural ecosystems must be safeguarded for the benefit of present and future generations through careful planning or management, as appropriate.

Principle 3. The capacity of the earth to produce vital renewable resources must be maintained and, wherever practicable, restored or improved.

Principle 4. Man has a special responsibility to safeguard and wisely manage the heritage of wildlife and its habitat which are now gravely imperilled by a combination of adverse factors. Nature conservation, including wildlife must therefore receive importance in planning for economic development.

In December 1972, the U.N. General Assembly met in New York and accepted this declaration and the 109 recommendations of the action plan. Furthermore, the assembly stated in a resolution that it was "aware of the urgent need for a permanent institutional arrangement within the United Nations system for the protection and improvement of the human environment," and proceeded to create:

1. A Governing Council for the Environment Program composed of fifty-eight member countries elected by the General Assembly;
2. A small secretariat to serve as a focal point for environmental action and coordination within the United Nations system to be headed by an executive director elected by the General Assembly on the nomination of the secretary general; and
3. An Environment Coordination Board under the chairmanship of the executive director.

The General Assembly voted unanimously to: (1) locate the headquarters of the new secretariat in Nairobi, Kenya; and (2) elect fifty-eight member states of the Governing Council on a basis of sixteen seats for African states, thirteen seats for Asian states, ten seats for Latin-American states, thirteen seats for western European and other states, and six seats for eastern European states. Hence, UNEP was conceived in Stockholm in June 1972 and born six months later in New York; it is still the baby in the U.N. family.

The 1972 U.N. Conference on the Human Environment recognized the great importance of ecological principles in the planning and utilization of the environment and, consequently, for human society. Indeed, in establishing UNEP in response to the recommendations of this conference, governments specifically sought to create the kind of new international machinery needed to assist them in dealing with this new complex of interrelated issues.

The Internal Organization of UNEP

The prime task of UNEP is to act as an international coordinator for the environment, especially within the U.N. system. As an organ of the U.N., it reports to the General Assembly through the U.N. Economic and Social Council. Administratively, the executive director reports to the U.N. secretary-general. This means that, although UNEP has its own governing council, it is a part of the United Nations. This is somewhat different from the status of U.N. agencies such as the U.N. Educational, Scientific, and Cultural Organization (UNESCO), the Food and Agriculture Organization (FAO), the World Health Organization (WHO), and others which are sovereign within the U.N. system. Nevertheless, it is up to UNEP to coordinate the activities of all U.N. organizations in the field of environment.

Under its executive director, UNEP's Secretariat is organized in two main bureaus: (1) the Bureau of the Program, and (2) the Bureau of the Environment and Fund Management. Each bureau is headed by an assistant executive director.

The Bureau of the Program is organized as follows:

1. Division of Geophysics, Global Pollution and Health
2. Division of Ecosystems and Natural Resources
3. Division of Economic and Social Programs
4. Division of Technical Assistance and Training
5. International Referal Systems (IRS)
6. Global Environmental Monitoring System (GEMS)
7. Library

The Bureau of the Environment Fund and Management is organized as follows:

1. Environment Fund
2. Division of Program Management
3. Division of Communication
4. Division of Administration

In the executive director's office there are two offices responsible for policy planning and external relations. In addition, there are three senior advisers on the following subjects: (1) ecology and conservation; (2) economy; and (3) energy and environment.

The Environment Fund was established to provide voluntary additional finance for new environmental initiatives, including those envisaged in the Stockholm Conference Action Plan. The total of the resources expected to be available to the fund in its initial five-year period (1973-1977), based on amounts already received and announced, slightly exceeds the initial $100 million target. So far, more than fifty nations, among which are developing countries, have contributed to the fund, bringing the total pledged over this five-year period to some $100 million.

By mid-1974, some 180 projects had been financed by the fund. Areas covered in the projects include: marine pollution (giving priority to the Mediterranean area); human settlements; arid zones (particularly the drought-affected areas of the Sahelian region); training, education, and information activities; patterns of resource use; environment and development strategies; the development of GEMS and IRS, etc. Most of these projects have been joint ventures involving the U.N. system as well as scientific and professional institutions.

UNEP has liaison offices in New York and Geneva and regional offices for Africa in Nairobi, for Latin America in Mexico City, and for western Asia in Beirut. Another regional office for Asia and the Far East is being established.

Cooperation with Other Organizations

With the rise of the conservation movement after World War II, ecological principles began to be stressed as important factors for human society in planning its environment, and particularly with respect to the utilization of renewable resources. Many at the EARTHCARE Conference contributed significantly to these views and paved the way for an environmental understanding which at present is of such a tremendous importance for the future of mankind.

The United States, through the federal government and numerous powerful nongovernmental conservation organizations, has played a particularly significant role in this process. Conservation has attained recognition as a major issue in the life of this nation, reflecting the remarkable progress of the conservation movement during the 1960s, when the general attitude shifted from indifference or ignorance to keen interest and the passage of significant legislation governing the use of natural resources. Yet for half a century the

United States has been several decades ahead of all other nations in the field of conservation; that is not to say, however, that there are no serious conservation problems in the United States.

On the international scene, the International Union for the Conservation of Nature (IUCN) has pioneered ecological principles as a basis for conservation, management, and utilization of renewable natural resources. The International Biological Program (IBP) and UNESCO's Man and the Biosphere (MAB) are other expressions of international efforts to investigate, through long-term ecological research, the productivity of various biomes and ecosystems and the way in which these resources can be managed and utilized without causing adverse environmental effects from which man will ultimately suffer.

Individual and collective citizen action throughout the world helped create UNEP. Nongovernmental organizations (NGO's) took an active part in preparing for the Stockholm conference. Many hundreds of their observers contributed to an "NGO Declaration" to the conference plenary, stating that "our planet's resources are limited . . . its life support systems are vulnerable . . . the combined effect of modern technology, consumption, and population growth can place our whole planetary life at risk." The results of the Stockholm conference were communicated to millions of members of NGO's, and global environmental issues became a priority in many NGO programs.

During the first session of the UNEP Governing Council in Geneva, in 1973, the first World Assembly of Nongovernmental Organizations Concerned with the Global Environment, set up an Environment NGO Liaison Board to maintain communications and effective working relationships among NGO's and with UNEP in achieving common environmental goals. At the second session of the UNEP Governing Council, an NGO Service Center was established in Nairobi to coordinate NGO activities with UNEP.

Within UNEP we appreciate these working relations with NGO's as being of great importance and stimulation. I am pleased to say that we have especially good relations with the United States NGO's. Therefore, we very much appreciate the invitation to participate in this important EARTHCARE Conference. Obviously, all activities of UNEP are planned and implemented in a way which gives full consideration to ecological factors as far as these are known. But, as you are all aware, there are still many facets of ecology that man does not yet understand. June 5, World Environment Day, the day on which this conference was initiated, is a particularly fitting date, because it commemorates the opening of the Stockholm conference on 5 June 1972, evoking the spirit and the hopes which characterized it.

Role and Priorities

The three general aims of UNEP are to provide improved knowledge for the rational management of resources in the biosphere, to encourage an integrated approach to planning and development, and to assist countries in dealing with their environmental problems. UNEP has also been given the task of coordinating and catalyzing environmental activities within the U.N. system. In formulating programs and projects, UNEP has to ensure their compatibility with the U.N. International Development Strategy and with the World Plan of Action for the Application of Science and Technology to Development. As you see, these commitments put human interests in the center, but it is up to UNEP to see that they are not detrimental to long-term productivity and other values represented by renewable natural resources. This approach has led UNEP to the concept of "ecodevelopment," which is designed to help the people of a given ecoregion realize the full development potential of its resource endowment and environmental conditions; in this way, maximizing the use of human resources and skills can produce the kind and quality of life to which the people of an ecoregion aspire without destroying the resource base on which sustained development depends.

UNEP has been given several subject areas as priorities: human health and well-being; human settlements and habitat; ecosystems, with particular attention to climatic changes; arid lands; desertification; grasslands; woodlands; and tropical forests, including its wild animals. Resources such as water, soil, endangered species, and genetic diversity play an important part in the various programs. Other major subject areas are natural disasters, pest management systems, environment and development, and international environmental law. The "earthwatch" system of global environmental monitoring (evaluation and information dissemination), is now in the process of being established by UNEP. As well as the International Referral System, obviously, education, training, and information are important forms of assistance to which UNEP will contribute.

These activities for the long-term benefit of human society cannot be successfully performed if full attention is not paid to ecological principles. UNEP is fully aware of this. Therefore, this conference with its approach to wilderness areas and with its wealth of ecological data is very useful to us in our attempts to understand and solve the environmental problems of today.

The primary aims of UNEP are: to promote coordination and proper balance in environmental planning and development, and to initiate, where necessary, international activities in environment, especially within the U.N. system, with special reference to those problems that are transnational, regional, and global.

At the Third Governing Council, meeting in Nairobi, it was emphasized that UNEP, as the "environmental conscience" of the U.N. system, should be extending its global programs, concentrating on a few priority problem areas, and intensifying its coordinating, catalytic role in an integrated development strategy for the fulfillment of man's basic needs.

The Third Governing Council also decided:

1. To provide $1.5 million to support the program and preparations for the U.N. Conference on Desertification 1977, in which, by decision of the General Assembly, UNEP will be directly involved;
2. That UNEP should be actively involved in the environmental and ecological aspects of water development programs and in the preparations for the U.N. Water Conference to be held in Argentina in 1977;
3. To endorse the three-level programmatic approach as the basic process for the development of the program and for the management of environmental issues within the U.N. system. In this approach:

   a. Level One, through its presentation of the "State of the Environment," seeks to identify emerging problems requiring the attention of governments;
   b. Level Two (the program) is based on the objectives and priorities established by governments and defines a program of action to meet these objectives;

c. Level Three (the fund program) consists of those actions and projects identified within the program framework at Level Two, which are selected for support by the UNEP fund;

4. To give total support to the work on arid lands and grazing ecosystems. It was suggested that savannahs should be included in the program area;
5. That woodlands and forest ecosystems in tropical regions be given priority consideration by UNEP;
6. To give greater attention to the maintenance of ecosystems.

There was wide agreement on the program for ecosystems—sites and samples (national parks and reserves), endangered species, and wildlife—and its importance among UNEP activities was highlighted. Several delegations suggested that the area deserved greater emphasis and more financial support. The delegations appreciated the program to establish a network of national parks and reserves, and supported the secretariat's approach to organizing the related activities regionally, which would provide a more harmonious basis for cooperation between countries. It was pointed out that UNEP could help in the development of special guidelines for the selection and creation of such reserves. Furthermore, delegates

7. Expressed concern for the disappearance of plant and animal species, and particularly supported the activities aimed at protecting endangered species. One delegate requested that the secretariat pay special attention to whales, and another to those migratory species for which the protection of certain habitats did not provide sufficient safeguard;
8. Emphasized the close relationship between the program activities concerned with ecosystems, sites and samples (national parks and reserves), endangered species and wildlife, and genetic and other biological resources, and expressed the hope that work on these areas would be closely integrated;
9. Expressed concern about soil degradation, erosion, and overuse, as well as eutrophication, and welcomed the related activities proposed by the Secretariat; and

10. Noted that the Stockholm Conference on Human Environment called on the International Whaling Commission to adopt a ten-year moratorium on commercial whaling and asked governments to strengthen the commission and and increase international research efforts; and that the governing council at its first meeting endorsed these recommendations, requesting that support be given to research on marine mammal populations and on whales and small cetaceans in particular.

Global Environmental Problems

Seldom has man used a system of ecologically sound long-term land use planning. Independent of the pressures he places on the lands and waters, the utilization of our most valuable elements usually results in environmental failure. This situation is quite clear from the testimony of the lands and waters themselves in both developing and developed countries. Whatever region or climatic areas one focuses on in tropical, temperate, or subarctic parts of the world, the evidence of unwise land use is abundant.

Therefore, in the interest of national prosperity and human progress in the various countries of the world, it is imperative that problems of ecology, conservation, management, and utilization of a nation's renewable natural resources be accorded priority with respect to economic, social, and scientific planning. In subtropical and tropical countries with fast-growing populations, the nutritional need alone justifies national efforts to improve the efficiency and increase the productivity of various forms of land use without causing long-term deterioration. Yet over the last century in general, and the last two decades in particular, exploitation of renewable natural resources in most countries has been ecologically unwise, resulting in the destruction, on an increasingly larger scale, of soil, vegetation, and wild animal populations, and the contamination of water. Should this environmental degradation continue for another decade without energetic and efficient measures to stop it, there is little hope for the future of quite a number of nations.

Obviously, progress—agricultural, industrial, scientific, and educational—is the only road to prosperity. However, progress does not mean to continue using methods and applying policies which disregard the ecological realities. Elementary ecological knowledge and conservation principles must be considered at the same level as social and economic issues in all development planning in order to avoid the process of gradual environmental decline leading to a point of no return.

The situation just described has global implications, but it is particularly true for the tropics and subtropics. Can the answer to water and land use problems be found in management and legislation? It certainly can, provided it is ecologically sound, effective, and long-term in its perspective. It also takes social courage and a firm political will to implement what is ecologically necessary.

In my opinion, however, all attempts to plan and manage the renewable natural resources would not lead to man's progress as long as the human population growth continues at the present rate. This is the basic conservation problem on which all other serious conservation problems depend. The key to the global population issue is the acceptance by each nation of the responsibility for sound national population policies by which the level, growth, and distribution of its population are related to its available resources, its capacity to manage these resources according to ecological principles, and to the kind and quality of life to which its people aspire. Equally important is the need for each nation to accept the corollary responsibility to assure that the demands of its population on the natural resources and environment beyond its national boundaries do not impair the rights and interests of other nations.

Since all the renewable natural resources are interacting, solutions to environmental problems connected with these resources require an integrated approach. Likewise, land-use planning and a land-use policy must necessarily be based on the totality of renewable natural resources—on the whole unit consisting of interrelated resources. In the past, planning, policies, management, and utilization of renewable natural resources have been almost exclusively organized along sectoral lines, because ecological considerations were mostly absent.

An ecologically based, integrated management approach to the planning of renewable natural resources would bridge the potential of these associated resources and constitute an insurance of their maximum productivity on a sustained basis.

The last governing council of UNEP recognized that the problems of population, resources, environment, and development are interrelated. Solutions to these problems must form part of an integrated strategy directed towards coordinated objectives, to which UNEP will contribute within the framework of its specific environmental tasks. Fundamental to these objectives is the importance of meeting man's aspirations for the fulfillment of his basic needs.

Increasing evidence now exists that consumption of fossil fuels, the increase in nuclear reactors, and the introduction of one thousand new man-made chemical compounds per year into the environment is giving rise to serious risks to human health and well-being.

Short-term economic interests are threatening the future of the world's oceans and marine resources. Neither the meeting at Caracas in 1974 nor the recent Geneva meeting have given adequate concern to the environment of the oceans. UNEP's recommendations to the conference have not received the priority they deserved. Other urgent problems are the hasty development of nuclear power, relentless desertification, and the accelerating destruction of water, soil, wild vegetation, and wild animal resources.

With a world population of more than 3.8 thousand million, which is projected to reach at least 6.5 thousand million by the end of the century, we face problems concerning our potential capacity to support populations while providing acceptable conditions of life. This is particularly true in tropical and subtropical regions where renewable natural resources are quickly disappearing.

Progress in the conservation and wise management of whale populations is still disappointingly absent and almost paralyzed by the attitude within the International Whaling Commission of the few remaining countries actively engaged in commercial whaling. Not only marine mammals but also marine turtles are threatened, another economically very important group of animal resources. Locally also, marine fish populations are exposed to continued overharvesting. Likewise, the threats to endangered species of land animals continue to mount. In all, 359 species and subspecies of mammals, birds, reptiles, amphibians, and fishes have been exterminated during historic time, and 1,032 are today endangered by extinction. For this deplorable state of affairs man alone is responsible.

Achievements

The growing concern among nations with environmental problems and with the necessity of conservation, management, and wise utilization of renewable natural resources—as well as a deeper understanding of the fact that all human exploitation and manipulation of natural resources have implications for an ecosystem—is the result of the preparatory work of the Stockholm conference, the conference itself, and the post-conference activities of UNEP. This is perhaps so far the most important achievement.

Other steps forward are still rather small but it may be worthwhile to mention some examples:

1. The Barcelona 1975 Action Plan, adopted by the sixteen Mediterranean nations at a conference in Spain. This plan aims to save the Mediterranean Sea as a vital marine ecological system. This agreement means that the Mediterranean resources will no longer be treated as individual national assets but as an entire region and part of a single ecosystem.

    The action plan includes a coordinated program of research, monitoring, and exchange of information covering, initially, six projects, which the plan considers to be the most dangerous threats to the Mediterranean:

    a. Pollution of marine waters by oil and petroleum hydrocarbons.
    b. Pollution of marine organisms by metals, particularly mercury.
    c. Pollution of marine organisms by DDT and other chlorinated hydrocarbons.
    d. Effects of pollutants on marine communities and ecosystems.
    e. Coastal transport pollution problems.
    f. Coastal water quality control.

    These will be pilot projects of at least two years' duration. The first three will help to keep a check on marine pollution by performing a monitoring function.

2. Progress achieved in the field of national parks and equivalent reserves. A correlated series of regional conferences in East Africa (Seronera, Tanzania, 1974), Central America (San José, Costa Rica, 1974), the South Pacific (Wellington, New Zealand, 1975), and Teheran (Iran, 1975)—for marine national parks and reserves in the Mediterranean, Red Sea, Gulf of Persia, and the northern Indian Ocean; all yielded encouraging initial results.

As far as marine national parks are concerned, these regional conferences culminated in a world conference on the subject, convened in Tokyo in May 1975 with participants from thirty-two countries. But what has been achieved so far is just a beginning, particularly for UNEP. Much remains to be done before the U.N. system can be said to be fully attuned to the new conservation approach to development required by the ecological realities of the population-environment-resource equation. UNEP's own responsibilities and program are still far from adequately integrated into the overall structure and activities of the U.N. family.

The Future

It is not necessary to spell out to the participants in this conference the conservation problems of the future. They are obvious and manifold. In my opinion, however, there are three elementary undertakings which are absolutely necessary for the future of our planet. They are:

1. To halt the ongoing, accelerating destruction of renewable natural resources in the tropics and subtropics. Such an effort requires a transitional period during which bilateral and multilateral aid schemes must be concentrated in areas where local human populations are, at present, obliged to destroy the environment on which they depend in order to survive. If these populations are not offered alternative food and fuel resources there is no way to halt the present environmental destruction.
2. To plan a global restoration and revitalization scheme for destroyed renewable natural resources in order to regain

the potential productivity of man-made deserts and other sterile lands and waters. This requires a giant financial effort, which only developed countries can afford, and international cooperation of a previously unmatched scale.
3. To promote worldwide conservation education by all media and institutions directed to all age and social groups as a necessary background to understanding why, in the long run, an ecological strategy will restore the environment of our planet and, in the global community, confidence for the future. This is also a giant task but easier to realize than numbers 1 and 2 above; moreover, it should precede them.

I believe that most of you at the EARTHCARE Conference are convinced that we have very little time at our disposal to solve the environmental problems of the world. In this way, the conference has certainly served to indicate, constructively, what these urgent problems are.

62. Zapovedniks: National Preserves of the Soviet Union

Philip R. Pryde
San Diego State University, U.S.A.

Introduction

It is both an honor and a surprise to be filling in today for Mr. Borissoff, who could not be here. It is fitting that he was invited, since the Soviet Union occupies one-sixth of the world's land surface and probably contains at least that percentage of its resources, both biotic and mineral. Also, Mr. Borissoff is one of the principal officials concerned with the management of national preserves systems in the Soviet Union; he is secretary-general of the Central Laboratory on Nature Conservation, which is within the USSR Ministry of Agriculture and which manages most Soviet preserves. It goes without saying that he could give you much greater insight than I can into the nature and management of Soviet preserves, but I will try to give you a brief sketch of their most important characteristics. More exhaustive descriptions are given by Borissoff himself (1971 and this volume) and by Pryde (1972 and 1976). (See also Schoenbaum [1976].)

The Origins and History of *Zapovedniks*

The present system of state preserves—*zapovedniks*—in the Soviet Union is an outgrowth of preserves that existed before the 1917 revolution. There were not many preserves at that time, and most were estates that belonged to the nobility. In many cases, these were hunting preserves, a few of which had been preserved for that use for hundreds of years; thus, a long tradition is associated with

---

Printed here is the edited transcript of Dr. Pryde's extemporaneous remarks on the subject of *zapovedniks*. After the conference, Mr. Borissoff very kindly submitted a paper on the same subject. For the sake of completeness, and in the hope that they will complement each other, both Dr. Pryde's and Mr. Borissoff's papers are published herein.—Editor

these preserves. After the revolution they were set up as a state system of preserves.

The preserve system grew rapidly through the 1920s, 1930s, and 1940s, so that, in 1950, it consisted of about 128 different areas. Unfortunately, there was a reorganization of the network in 1951; it was greatly reduced, from 128 to 40 preserves and from about 12.5 million to about 1.4 million hectares—a decrease in area of almost 90 percent. The reorganization illustrates the fact that while *zapovedniks* can be created easily, without some of the parliamentary procedures that the creation of national parks must go through in the United States, for example, they can also be abolished easily.

Since 1951, the system has increased again. It consists of 107 preserves and 8.96 million hectares at the present time (see Appendix E). Thus, the Soviet system is again one of the largest in the world.

In addition to the state system of *zapovedniks*, or national preserves, the same sort of preserve system exists at the union-republic and at the province level. These are similar to the systems of state and county parks in the United States.

The Salient Characteristics of *Zapovedniks*

The word *zapovednik* comes from another Russian word that means either "restricted" or "forbidden"—a derivation that emphasizes the fact that *zapovedniks* generally are not intended to be open to the public in the sense that national parks encourage visitation by the public in the United States. Only a very few of them—perhaps only half a dozen to a dozen—are widely visited by the public. A few of those near major cities and in the Caucasus are open to tourism; the rest are primarily scientific preserves.

Zapovedniks are not really analogous to what we call national parks in the United States. They are defined by Soviet administrators as areas forever withdrawn from economic utilization, for scientific research and for cultural and educational purposes. Scientific study is the main feature of their management. If one had to suggest a comparison, it would probably be more accurate to compare them to the system of national wildlife refuges in the United States than to its system of national parks.

Two types of scientific research go on in *zapovedniks*. The first is passive research—the observation and study of the habits of their wildlife and plants. The second type is active research, which involves the introduction of exotic species, acclimatization, and so forth. In this respect, *zapovedniks* are different from what are called wilderness areas in the United States.

In his paper, "The Exporting and Importing of Nature," Roderick Nash points out that wilderness is a difficult term to define and is sometimes even difficult to find in the vocabularies of some countries. In a sense, this is true in the Soviet Union, where there is no statutory provision for wilderness similar to the Wilderness Act of 1964 in the United States. In the Soviet Union, *zapovedniks* are spoken of, not as wilderness areas, but as "standards of nature"—the control part of a sometimes uncontrolled experiment in economic growth and the transformation of nature.

Another, perhaps minor, distinction is that many American national parks—McKinley, Yellowstone, Grand Canyon, Yosemite, and so forth—were created around major unique geologic phenomena; generally, this is true of some Soviet preserves, but not of most. For instance, most of the highest mountains in the Soviet Union are not situated within *zapovedniks*.

Problems in the Management of *Zapovedniks*

There is no single management agency in the Soviet Union. While the USSR Ministry of Agriculture administers some *zapovedniks*, and some are administered by the Academy of Sciences, most are operated by union-republic ministries. This is, perhaps, not an important distinction, since wilderness areas in the United States are managed by one of two separate agencies.

Similarly, there is no unified management policy for *zapovedniks*; some are very tightly excluded from tourism, while others are open to wide-scale tourism; some are reserved for passive research, and others involve active research as well. There is considerable variation. Here, too, as long as the management system can be justified and is logical within the context of a given preserve, this may not be an important distinction.

Tourism is a problem of great concern in the Soviet Union because the average Soviet citizen is becoming more mobile, the consequence of a developing large-scale automobile industry—with all that that implies. Visits to *zapovedniks* have increased, just as visits to national parks in the United States increased with the advent of the automobile in the 1910s and 1920s. The Soviets recognize the implications of the increase. One of their approaches is to define the network of *zapovedniks* primarily as off limits to tourists and to talk about the creation of many national parks—although only three have been created to date. (From the way they have been described in the literature, the proposed national parks may be closer to what Americans call national recreation areas than to national parks.)

The accomplishments of *zapovedniks* are many. I would say that the most outstanding of them have been in the area of wildlife preservation. As in the United States, many species have been threatened with extinction in the Soviet Union. In some respects, the problem has been worse in the Soviet Union because of the two world wars—particularly the Second World War, during which much of the country was under occupation, when millions of people were starving and were willing to eat anything available, including wild animals.

The species the Soviets have had the most success in reestablishing is the European bison or wisent *(Bison bonasus)*, which is of particular interest to Americans because of their own problems with reestablishing the North American bison *(B. bison)*. Also, the Soviets have reestablished the beaver *(Castor fiber)*, the kulan *(Equus h. hemionus)*, the saiga *(Saiga tatarica)*, the Siberian tiger *(Panthera tigris longipilis)*, and many other species.

A study of the Soviet preserve system reveals two facts that are of worldwide interest. First, both the problems and the potentials of reserve areas are universal and tend to be independent of national, economic, and political systems: one finds similar problems and, often, similar successes almost everywhere one looks. Second, the problems must be approached on a worldwide basis: no individual country, not even the largest in the world, can successfully preserve many types of species, habitats, and biotypes unless there is worldwide cooperation in this regard.

References

Borissoff, Vladimir A. 1971. Soviet system of protected natural areas. *National Parks & Conservation Magazine* 45(6):8-14.

Pryde, Philip R. 1972. *Conservation in the Soviet Union.* Chapter 4, "Zapovedniki in the Soviet Union," pp. 45-67. London: Cambridge University Press.

———. 1976. Nature preserves and national parks in the Soviet Union. Unpublished paper presented to the Twenty-third International Geographical Congress, Moscow, July and August 1976. (To be published in the proceedings of the congress, ed. Gordon Nelson, Faculty of Environmental Studies, University of Waterloo, Waterloo, Ontario, Canada.)

Schoenbaum, Thomas J. 1976. Natural area preservation in the Soviet Union and the United States: A comparative perspective. *American Journal of Comparative Law* 24:521-539.

63. Zapovedniks

Vladimir A. Borissoff
Central Laboratory on Nature Conservation,
USSR

Introduction

Zapovedniks* are scientific institutions established primarily for carrying out research on tracts of land and bodies of water in the Soviet Union especially set aside for that purpose. Of the twenty or so categories of protected natural areas in the Soviet Union, the *zapovedniks* are considered to be the major ones. They are specially protected natural areas intended primarily to preserve nature for intensive study—not for display—and are investigated by their own resident research units. Therefore, *zapovedniks* were not created for recreation or other forms of public enjoyment; for these purposes, national and natural parks, recreation areas, protected landscapes, and other areas have been set aside.

---

*When speaking about Soviet nature reserves of the highest protection category, many authors, myself included, exclusively use the Russian word *zapovednik* (nature reserve; English plural, *zapovedniks*). This usage is dictated not merely by the absence of any fully corresponding terms in other languages, but also, and more importantly, by the existence of a set of features unique to *zapovedniks*.

Nature is protected in *zapovedniks* for the same basic reason it is in most, if not all, other nature-reserve systems in the world—namely, to perpetuate plants and animals in their natural environment, with a minimum of interference from humans. The aim of the research pursued in *zapovedniks* is much broader than the mere study of flora, fauna, soils, geomorphologies, or specific curiosities; rather, it is the long-term and intensive exploration of relationships among various components of natural ecosystems in *zapovedniks* and their comparison with those of ecosystems that have been modified by man.

Special nature reserves (including botanical preserves, wildlife sanctuaries, protected landscapes, and natural monuments) protect individual elements or sets of elements of ecosystems, whereas *zapovedniks* are intended to protect entire ecosystems.

History

The concept of *zapovedniks* has evolved since they were first established as strictly nature reserves, often with rather narrow purposes. Many of them were species-specific, aimed at preserving, for example, the sable, beaver, or desman. [*Desmana moschata,* the Russian desman, is an aquatic, mole-like mammal of southeastern Europe and central western Asia.—Editor] But their *raison d'être* was not only to preserve and restore stocks of rare and vanishing animals; their primary goal was to perpetuate a series of habitats throughout the country in order to study the protected species in close connection with their ecosystems.

The necessity of simultaneously protecting and studying the ecosystems of which the protected species were integral parts was proven by increases of protected populations, and it is justified both by the need to find the best ways of propagating the species over large territories and by the eminent role these natural islands play in the ocean of man-disturbed nature.

While it may be sufficient to study individual species by means of field expeditions, the thorough study of organism-population-environment systems requires year-round observations and records; expeditions do not yield enough information in this regard. To satisfy the need for year-round data, a permanent scientific staff has been appointed for each *zapovednik*. Thus, the development of the scientific capability of *zapovedniks* has passed through three stages: first, protection and observation alone; second, the establishment of research facilities; and third, the appointment of scientific associates as integral parts of the *zapovedniks'* staffs, special scientific divisions, scientific councils, and so on. Other types of nature reserves may have research facilities that outside researchers may use; *zapovedniks*, however, have their own resident scientists, who work only in a specific *zapovednik*, which may not have facilities for external researchers.

The Voronezh State Zapovednik provides a case history. Created in 1927 to protect and restore the beaver (the total beaver population in the country was estimated at 900), this *zapovednik* was intended also to protect and study other valuable wildlife and forest ecosystems, to elaborate biological methods of forest protection, and to promote conservationist thinking. The primary goal was reached: since the Experimental Beaver Farm, established in 1932, first bred captive specimens in 1934, over 2,200 beavers from this *zapovednik* have been reintroduced in thirty provinces of the USSR as well as in the German Democratic Republic, Poland, and Mongolia; the total beaver population in the Soviet Union is now over 60,000. Besides the beaver farm, the *zapovednik* has five research laboratories, a weather station, a library, and a museum. Natural environmental conditions are monitored at more than 100 permanent test plots. Results of research at this institution are contained in the twenty volumes of *Acta of the Voronezh State Zapovednik,* as well as in various other journals and books. Started as a beaver reserve, this scientific unit now has as its official tasks the interdisciplinary study of the nature of pine and oak forest islands in the forest-steppe zone; the exploration of peculiarities of their ecosystems; the search for ways of restoring primeval forest types and their biocoenoses; and the study of methods of restoration, propagation, and wise use of beaver and red deer in the USSR.

To conduct these studies, the *zapovednik* needs experts in soil science, hydrology, geobotany, botany, forestry, entomology, phytopathology, mammalogy, ornithology, parasitology, animal physiology, plant biochemistry, microbiology, and so on.

Thus, *zapovedniks* are created to discover the laws ruling nature that are hidden from the noninitiated. They offer the permanent research facilities necessary for long-term observation and systematic comparison of ecosystems, for studying natural dynamic processes (e.g., interrelationships of forests and steppe vegetation, coniferous and broad-leaved forests), and so on. Expeditions collect numerous samples and supply information about places and dates of specimen collection and gather data on the distribution of species in certain associations, the interaction of the vegetative cover with its environment, the length of the fruit-bearing period of many uncommon plant species, and the practical and ecological value of certain plants and animals. Scientists working in such institutions are in a good position to enrich our knowledge of plants and animals with observations and perspectives often ignored by field expeditions, whose members face an unavoidable fragmentation of their task. Prof. S. Stankov, the eminent Soviet botanist, has emphasized that such data are especially important and valuable in the study of critical polymorphic plant species.

From another standpoint *zapovedniks* differ from other research institutions in that they have an intact natural environment within which to work and are charged with protecting the natural areas that are the subject of their research. These peculiarities demand a particular approach to research planning which stipulates that only projects that deal with nature-protection measures should be undertaken.

Management

Further, the protection of nature within the boundaries of a *zapovednik* gives rise to many specific problems—just as in any strictly protected nature preserve. To preserve one element of an ecosystem is quite feasible *in vitro*, but to do it *in vivo* is almost unthinkable unless the entire ecosystem is preserved. In specifying an area as a nature reserve, we produce a kind of artificial environment: within nature there is no demarcation between use and nonuse spheres. Ecological disequilibrium is caused mainly by deficiencies in or disturbances of the many linked elements of an ecosystem. But disequilibrium may also be caused by the proliferation of one element. Until we study the natural mechanisms within nature reserves, we cannot effectively manage ecosystems.

Soviet *zapovedniks* vary in size from about 100 to about 1 million hectares. Large ones may be considered more or less self-regulating ecosystems, and the schemes for their protection do not evoke many discussions. The protection of small *zapovedniks*—which often are very important because of certain features marked for preservation—is difficult because adverse forces operate both outside and inside the reserves.

There is another facet to consider. A single *zapovednik*, even a well-functioning one, is a minor achievement. A network of reserves with common or similar objectives operates much more successfully. Better yet is a system of protected natural areas; only such a system can provide the most comprehensive results.

An adequate national system of protected natural areas should include a network of reserves of different kinds, a scientific center providing leadership and coordination of research, and an administrative center governing the system. This is an ideal scheme, one that is far from being achieved in most instances.

Zapovedniks are legally established by the councils of ministers of the fifteen Soviet republics, with the consent of the USSR *Gosplan* (state planning authority), and are administered by a variety of special bodies. Scientific advice is provided by *ad hoc* public commissions or by special research institutions. Because *zapovedniks* are administered by a variety of special bodies, coordination between the republics and the other agencies that administer *zapovedniks* is performed by an appropriate subdivision of the USSR Ministry of Agriculture (Main Board of Nature Conservation, Nature Reserves, and Hunting Economy), usually by means of all-union conferences. The network of state *zapovedniks* now [mid-1976] consists of 107 units covering nearly 9 million hectares (see Appendix E).

Improvement of the network is well underway. The impetus for creating *zapovedniks* came from a desire to save valuable tracts of land in regions of intense economic development. This is why the majority of our first *zapovedniks* were formed in quite industrialized provinces: Ilmen, in the Ural Mountains; Astrakhan, in the Volga delta, where fishing and hunting have developed; the Ukrainian Steppe Zapovednik plots, in densely populated regions; and so on. Parts of Siberia and the European North have more recently seen intensive use, and we are promoting new *zapovedniks* in these regions. The general idea is to have at least one *zapovednik* in each major geographical subdivision of the country.

The natural wealth protected and studied in our *zapovedniks* is an integral part of the world's heritage, and the Soviet system of protected natural areas is ready to cooperate with other national systems. The initiation of cooperation in this field between the Soviet Union and the United States is a good example, and we would like to be sure that biosphere reserves now being discussed have the same organizational features as *zapovedniks*.

... a land of brooks and water, of fountains
and springs, flowing forth in valleys and hills,
a land of wheat and barley, of vines and fig
trees and pomegranates, a land of olive trees
and honey.
—Deuteronomy 8:7–8

And Abimelech fought against the city all of that
day; he took the city, and killed the people that
were in it; and he razed the city and sowed
it with salt.
—Judges 9:45

... the whole land brimstone and salt, and a
burnt-out waste, unsown, and growing nothing,
where no grass can sprout ...
—Deuteronomy 29:23

POSTSCRIPT

64. The Invasion of Cyprus and the Huge Environmental Price

Renos Solomides
*Association for the Protection
of Cyprus Environment, Cyprus*

The following description of the environmental consequences of the recent war on Cyprus was submitted by Mr. Renos Solomides, chairman of the Association for the Protection of Cyprus Environment, one of the cooperating sponsors of the EARTHCARE Conference. While Mr. Solomides' description was not presented at the conference itself, its publication in this volume is particularly appropriate because Dr. Kurt Waldheim, secretary-general of the United Nations, who had been scheduled to present a World Environment Day address during the opening ceremony of the EARTHCARE Conference, was called to Cyprus at the last moment to participate in sensitive negotiations on the conflict. Dr. Waldheim's address was presented instead by Mr. Ismat T. Kittani, his executive assistant. Mr. Solomides' remarks are appropriate as well because of the growing threat that military proposals and actions may pose to wilderness and other natural areas.

The editor has provided a short, select bibliography on this important subject.

Until recently, Cyprus was a happy island, perched in the eastern basin of the Mediterranean Sea. It had enjoyed a very wide reputation, not only for the traditional hospitality of its people, but also for its healthy climate, plentiful sunshine and sea blessings, rich flora and bird life, and wonderful and rare butterflies and wildflowers. Cyprus had been hoping that the rest of the world would respect its serenity and unspoiled environment and would allow it to remain an oasis amidst a troubled and polluted world.

The motorcar, the increased demand for minerals and building materials, and the increased pollution from mechanization were slow to invade our country; nonetheless, the "writing was on the wall," and some progressive-minded people set up this society, the Association for Protection of Cyprus Environment. The association started to pressure the government of Cyprus to legislate measures to save the environment from the encroachments of man and his technology, to protect the environment for coming generations. The government, as well as the people, has responded in many and effective ways.

Little did we suspect that these noble endeavors would be so cruelly jeopardized by man's own blind mania for conquest, destruction, and militarist expansion. Little did we suspect that the Greek junta would envy this small country's freedom and prosperity, or that Turkey, our big and mighty neighbor, a member of the United Nations and the Council of Europe, would use the Greek junta's *coup d'état* in Cyprus as a pretext to use lethal hardware against unarmed people and mute creatures of nature.

We mourn the loss of lives—about 5 percent of the active population, plus innocent lives of noncombatant people—and we shall go on licking our wounds for years; over 200,000 people have been rendered homeless and jobless and 3,000 young men are still missing, most of them feared killed in cold blood. But the destruction of inanimate things—of our unique physical environment—is no less painful. It is a catastrophe that cannot be expressed in any pecuniary terms.

One-third of this country's population fled, leaving behind a flourishing environment in and around 180 villages, 10 small towns, and 3 large cities—households, farms, groves, stores, industrial and other business estates, ports, and so on, worth almost $6 billion. Undoubtedly, the disruption of continuous protection and care is causing extensive damage to and pollution of the environment of Cyprus.

Thus, more than 100 square kilometers of citrus orchards and three other crops—producing not only an income of $60 million a year, but also oxygen and natural beauty—were left unattended and unirrigated so long that most of them are drying up. Even if they should manage to survive, it will take at least ten years to restore the damage—corresponding to an income loss of almost $600 million.

In the occupied areas, considerable numbers of livestock and poultry in industrial farms, barns, and the backyards of houses have been deprived of food and water. Many have died and decayed, giving rise to serious pollution and constituting a grave health hazard. The loss was estimated at 48,000 pigs, 280,000 sheep and goats, 12,000 cattle, and 1,400,000 poultry.

Refugee life itself is another hazard to health. Because of insufficient sanitary facilities, there is an ever-present threat of epidemics, especially in view of the lack of medical facilities in isolated areas.

A heavy toll has been paid by the famous Cyprus forests. Indiscriminate napalm bombing has started various fires of vast dimensions, causing a catastrophe that is by far the worst in the country since the recording of forest fires was initiated in 1885. About 340 square kilometers—one-fifth of the island's main forests—have been completely devastated. The disaster can only be compared to the fire that devastated the maritime pine forest of France in 1948.

Superb forests of pine, cedar, cypress, and numerous other species—including many rare endemic species—have become black graveyards. Experts have stated that these excellent biotopes, many of which were selected or declared nature reserves, will require at least a century to return—if they return at all—to their original condition.

The loss in timber is more than $7 million, while the cost of reforestation will be about $6.5 million. Irrespective of the waste of money, what is vitally important is the fact that it will take more than fifty years to restore the forest to its preinvasion state, even if it were possible to achieve the impossible task of reforesting the whole area now.

The worst of the fires occurred in the old Paphos and Troodos forests—woodlands that provide the only natural habitat of the Cyprus moufflon *(Ovis ammon orientalis)*, a race unique in the world and one that has long been in danger of extinction. It took years of great care and special protection to restore the moufflon's depleted populations, but now, with one-third its habitat devastated, the status of this wild animal is more precarious than ever.

The adverse effects of the war on soil and on general ecological relationships can easily be understood, but they cannot be assigned any monetary value.

And what does one say about the looting of entire abandoned towns and villages; the destruction of ancient sites and tombs, Byzantine churches, and historic monuments, or their abandonment to the elements and their subsequent collapse; the desecration of religious places and relics; the pillage of antiquities; and the continuing trafficking in ancient objects, icons, and historical items? Unfortunately, the continuous military occupation of almost half of our country by invading forces, disrespectful of all signs and symbols of civilization and culture, perpetuates and worsens this tragic situation.

The Association for the Protection of Cyprus Environment would be grateful for support from any part of the world in arousing public opinion to the calamity that has befallen Cyprus's unique natural environment and glorious cultural heritage, and in bringing pressure to bear on those responsible for its ruin and unprecedented degradation to let this place live in peace. Then the people of Cyprus shall strive to restore the country's natural environment and rediscover their happiness.

Let all decent human beings, all those dedicated to EARTHCARE, stand solidly together and vow to do away, once and for all, with war, that perennial scourge of man and his environment.

Bibliography

Barnaby, Frank. 1976. Towards environmental warfare. *New Scientist* 69(981): 6-8.

International Commission of Enquiry into U.S. Crimes in Indochina. 1972. *The Effects of Modern Weapons on the Human Environment in Indochina: Documents Presented at a Hearing Organized by the International Commission of Enquiry into U.S. War Crimes in Indochina.* Stockholm.

McClintock, Michael, E. W. Pfeiffer, Robert Williams, Warner L. Wells, and Susan Zolla. 1970. *Environmental Effects of Weapons Technology.* A Scientists' Institute for Public Information Workbook. New York: Scientists' Institute for Public Information.

McClintock, Michael, Ruth B. Russell, Herbert Scovill, Jr., Edith Brown Weiss, Arthur H. Westing, and Susan Zolla. 1974. *Air, Water, Earth, Fire: The*

Impact of the Military on World Environmental Order. International Series no. 2. San Francisco: Sierra Club.

Orians, Gordon H., and E. W. Pfeiffer. 1970. Ecological effects of the war in Vietnam. *Science* 168:544-554 (1 May 1970).

Stockholm International Peace Research Institute. 1976. Geophysical Warfare. *SIPRI Yearbook.* Stockholm.

Tschirley, Fred H. 1969. Defoliation in Vietnam. *Science* 163:779-786.

APPENDIX A

Declaration of the United Nations Conference on the Human Environment

The United Nations Conference on the Human Environment,
Having met at Stockholm from 5 to 16 June 1972,
Having considered the need for a common outlook and for common principles to inspire and guide the peoples of the world in the preservation and enhancement of the human environment,

I

Proclaims that:

1. Man is both creature and moulder of his environment, which gives him physical sustenance and affords him the opportunity for intellectual, moral, social and spiritual growth. In the long and tortuous evolution of the human race on this planet a stage has been reached when, through the rapid acceleration of science and technology, man has acquired the power to transform his environment in countless ways and on an unprecedented scale. Both aspects of man's environment, the natural and the man-made, are essential to his well-being and to the enjoyment of basic human rights—even the right to life itself.

2. The protection and improvement of the human environment is a major issue which affects the well-being of peoples and economic development throughout the world; it is the urgent desire of the peoples of the whole world and the duty of all Governments.

3. Man has constantly to sum up experience and go on discover-

ing, inventing, creating and advancing. In our time, man's capability to transform his surroundings, if used wisely, can bring to all peoples the benefits of development and the opportunity to enhance the quality of life. Wrongly or heedlessly applied, the same power can do incalculable harm to human beings and the human environment. We see around us growing evidence of man-made harm in many regions of the earth: dangerous levels of pollution in water, air, earth and living beings; major and undesirable disturbances to the ecological balance of the biosphere; destruction and depletion of irreplaceable resources; and gross deficiencies, harmful to the physical, mental and social health of man, in the man-made environment, particularly in the living and working environment.

4. In the developing countries most of the environmental problems are caused by under-development. Millions continue to live far below the minimum levels required for a decent human existence, deprived of adequate food and clothing, shelter and education, health and sanitation. Therefore, the developing countries must direct their efforts to development, bearing in mind their priorities and the need to safeguard and improve the environment. For the same purpose, the industrialized countries should make efforts to reduce the gap between themselves and the developing countries. In the industrialized countries, environmental problems are generally related to industrialization and technological development.

5. The natural growth of population continuously presents problems for the preservation of the environment, and adequate policies and measures should be adopted, as appropriate, to face these problems. Of all things in the world, people are the most precious. It is the people that propel social progress, create social wealth, develop science and technology and, through their hard work, continuously transform the human environment. Along with social progress and the advance of production, science and technology, the capability of man to improve the environment increases with each passing day.

6. A point has been reached in history when we must shape our actions throughout the world with a more prudent care for their environmental consequences. Through ignorance or indifference we can do massive and irreversible harm to the earthly environment on which our life and well-being depend. Conversely, through fuller knowledge and wiser action, we can achieve for ourselves and our

posterity a better life in an environment more in keeping with human needs and hopes. There are broad vistas for the enhancement of environmental quality and the creation of a good life. What is needed is an enthusiastic but calm state of mind and intense but orderly work. For the purpose of attaining freedom in the world of nature, man must use knowledge to build, in collaboration with nature, a better environment. To defend and improve the human environment for present and future generations has become an imperative goal for mankind—a goal to be pursued together with, and in harmony with, the established and fundamental goals of peace and of worldwide economic and social development.

7. To achieve this environmental goal will demand the acceptance of responsibility by citizens and communities and by enterprises and institutions at every level, all sharing equitably in common efforts. Individuals in all walks of life as well as organizations in many fields, by their values and the sum of their actions, will shape the world environment of the future. Local and national governments will bear the greatest burden for large-scale environmental policy and action within their jurisdictions. International cooperation is also needed in order to raise resources to support the developing countries in carrying out their responsibilities in this field. A growing class of environmental problems, because they are regional or global in extent or because they affect the common international realm, will require extensive co-operation among nations and action by international organizations in the common interest. The Conference calls upon Governments and peoples to exert common efforts for the preservation and improvement of the human environment for the benefit of all the people and for their posterity.

II
Principles

States the common conviction that:

Principle 1

Man has the fundamental right to freedom, equality and adequate conditions of life, in an environment of a quality that permits a life of dignity and well-being, and he bears a solemn responsibility to protect and improve the environment for present and future generations. In this respect, policies promoting or perpetuating *apartheid*, racial segregation, discrimination, colonial and other forms of oppression and foreign domination stand condemned and must be eliminated.

Principle 2

The natural resources of the earth, including the air, water, land, flora and fauna and especially representative samples of natural ecosystems, must be safeguarded for the benefit of present and future generations through careful planning or management, as appropriate.

Principle 3

The capacity of the earth to produce vital renewable resources must be maintained and, wherever practicable, restored or improved.

Principle 4

Man has a special responsibility to safeguard and wisely manage the heritage of wildlife and its habitat, which are now gravely imperilled by a combination of adverse factors. Nature conservation, including wildlife, must therefore receive importance in planning for economic development.

Principle 5

The non-renewable resources of the earth must be employed in such a way as to guard against the danger of their future exhaustion and to ensure that benefits from such employment are shared by all mankind.

Principle 6

The discharge of toxic substances or of other substances and the release of heat, in such quantities or concentrations as to exceed the capacity of the environment to render them harmless, must be halted in order to ensure that serious or irreversible damage is not inflicted upon ecosystems. The just struggle of the peoples of all countries against pollution should be supported.

Principle 7

States shall take all possible steps to prevent pollution of the seas by substances that are liable to create hazards to human health,

Appendix A

to harm living resources and marine life, to damage amenities or to interfere with other legitimate uses of the sea.

Principle 8

Economic and social development is essential for ensuring a favourable living and working environment for man and for creating conditions on earth that are necessary for the improvement of the quality of life.

Principle 9

Environmental deficiencies generated by the conditions of under-development and natural disasters pose grave problems and can best be remedied by accelerated development through the transfer of substantial quantities of financial and technological assistance as a supplement to the domestic effort of the developing countries and such timely assistance as may be required.

Principle 10

For the developing countries, stability of prices and adequate earnings for primary commodities and raw materials are essential to environmental management since economic factors as well as ecological processes must be taken into account.

Principle 11

The environmental policies of all States should enhance and not adversely affect the present or future development potential of developing countries, nor should they hamper the attainment of better living conditions for all, and appropriate steps should be taken by States and international organizations with a view to reaching agreement on meeting the possible national and international economic consequences resulting from the application of environmental measures.

Principle 12

Resources should be made available to preserve and improve the environment, taking into account the circumstances and particular requirements of developing countries and any costs which may emanate from their incorporating environmental safeguards into their development planning and the need for making available to them, upon their request, additional international technical and financial assistance for this purpose.

Principle 13

In order to achieve a more rational management of resources and thus to improve the environment, States should adopt an integrated and co-ordinated approach to their development planning so as to ensure that development is compatible with the need to protect and improve environment for the benefit of their population.

Principle 14

Rational planning constitutes an essential tool for reconciling any conflict between the needs of development and the need to protect and improve the environment.

Principle 15

Planning must be applied to human settlements and urbanization with a view to avoiding adverse effects on the environment and obtaining maximum social, economic and environmental benefits for all. In this respect, projects which are designed for colonialist and racist domination must be abandoned.

Principle 16

Demographic policies which are without prejudice to basic human rights and which are deemed appropriate by Governments concerned should be applied in those regions where the rate of population growth or excessive population concentrations are likely to have adverse effects on the environment or the human environment and impede development.

Principle 17

Appropriate national institutions must be entrusted with the task of planning, managing or controlling the environmental resources of States with a view to enhancing environmental quality.

Principle 18

Science and technology, as part of their contribution to economic and social development, must be applied to the identification, avoidance and control of environmental risks and the solution of environmental problems and for the common good of mankind.

Principle 19

Education in environmental matters, for the younger generation as well as adults, giving due consideration to the underprivileged, is essential in order to broaden the basis for an enlightened opinion and responsible conduct by individuals, enterprises and communities in protecting and improving the environment in its full human dimension. It is also essential that mass media of communications avoid contributing to the deterioration of the environment, but, on the contrary, disseminate information of an educational nature on the need to protect and improve the environment in order to enable man to develop in every respect.

Principle 20

Scientific research and development in the context of environmental problems, both national and multi-national, must be promoted in all countries, especially the developing countries. In this connexion, the free flow of up-to-date scientific information and transfer of experience must be supported and assisted, to facilitate the solution of environmental problems; environmental technologies should be made available to developing countries on terms which would encourage their wide dissemination without constituting an economic burden on the developing countries.

Principle 21

States have, in accordance with the Charter of the United Nations and the principles of international law, the sovereign right to exploit their own resources pursuant to their own environmental policies, and the responsibility to ensure that activities within their jurisdiction or control do not cause damage to the environment of other States or of areas beyond the limits of national jurisdiction.

Principle 22

States shall co-operate to develop further the international law regarding liability and compensation for the victims of pollution and other environmental damage caused by activities within the jurisdiction or control of such States to areas beyond their jurisdiction.

Principle 23

Without prejudice to such criteria as may be agreed upon by the international community, or to standards which will have to be determined nationally, it will be essential in all cases to consider the systems of values prevailing in each country, and the extent of the applicability of standards which are valid for the most advanced countries but which may be inappropriate and of unwarranted social cost for the developing countries.

Principle 24

International matters concerning the protection and improvement of the environment should be handled in a co-operative spirit by all countries, big and small, on an equal footing. Co-operation through multilateral or bilateral arrangements or other appropriate means is essential to effectively control, prevent, reduce and eliminate adverse environmental effects resulting from activities conducted in all spheres, in such a way that due account is taken of the sovereignty and interests of all States.

Principle 25

States shall ensure that international organizations play a co-ordinated, efficient and dynamic role for the protection and improvement of the environment.

Principle 26

Man and his environment must be spared the effects of nuclear weapons and all other means of mass destruction. States must strive to reach prompt agreement, in the relevant international organs, on the elimination and complete destruction of such weapons.

16 June 1972

APPENDIX B

Convention on Wetlands of International Importance, Especially as Waterfowl Habitat

The Contracting Parties,

Recognizing the interdependence of man and his environment;

Considering the fundamental ecological functions of wetlands as regulators of water regimes and as habitats supporting a characteristic flora and fauna, especially waterfowl;

Being convinced that wetlands constitute a resource of great economic, cultural, scientific and recreational value, the loss of which would be irreparable;

Desiring to stem the progressive encroachment on and loss of wetlands now and in the future;

Recognizing that waterfowl in their seasonal migrations may transcend frontiers and so should be regarded as an international resource;

Being confident that the conservation of wetlands and their flora and fauna can be ensured by combining far-sighted national policies with co-ordinated international action;

Have agreed as follows:

Article 1

1. For the purpose of this Convention wetlands are areas of marsh, fen, peatland or water, whether natural or artificial, permanent

or temporary, with water that is static or flowing, fresh, brackish or salt, including areas of marine water the depth of which at low tide does not exceed six metres.

2. For the purpose of this Convention waterfowl are birds ecologically dependent on wetlands.

Article 2

1. Each Contracting Party shall designate suitable wetlands within its territory for inclusion in a List of Wetlands of International Importance, hereinafter referred to as "the List" which is maintained by the bureau established under Article 8. The boundaries of each wetland shall be precisely described and also delimited on a map and they may incorporate riparian and coastal zones adjacent to the wetlands, and islands or bodies of marine water deeper than six metres at low tide lying within the wetlands, especially where these have importance as waterfowl habitat.

2. Wetlands should be selected for the List on account of their international significance in terms of ecology, botany, zoology, limnology or hydrology. In the first instance wetlands of international importance to waterfowl at any season should be included.

The inclusion of a wetland in the List does not prejudice the exclusive sovereign rights of the Contracting Party in whose territory the wetland is situated.

4. Each Contracting Party shall designate at least one wetland to be included in the List when signing this Convention or when depositing its instrument of ratification or accession, as provided in Article 9.

5. Any Contracting Party shall have the right to add to the List further wetlands situated within its territory, to extend the boundaries of those wetlands already included by it in the List, or, because of its urgent national interests, to delete or restrict the boundaries of wetlands already included by it in the List and shall, at the earliest possible time, inform the organization or government responsible for the continuing bureau duties specified in Article 8 of any such changes.

6. Each Contracting Party shall consider its international responsibilities for the conservation, management and wise use of migratory stocks of waterfowl, both when designating entries for the List and when exercising its right to change entries in the List relating to wetlands within its territory.

Appendix B

Article 3

1. The Contracting Parties shall formulate and implement their planning so as to promote the conservation of the wetlands included in the List, and as far as possible the wise use of wetlands in their territory.

2. Each Contracting Party shall arrange to be informed at the earliest possible time if the ecological character of any wetland in its territory and included in the List has changed, is changing or is likely to change as the result of technological developments, pollution or other human interference. Information on such changes shall be passed without delay to the organization or government responsible for the continuing bureau duties specified in Article 8.

Article 4

1. Each Contracting Party shall promote the conservation of wetlands and waterfowl by establishing nature reserves on wetlands, whether they are included in the List or not, and provide adequately for their wardening.

2. Where a contracting Party in its urgent national interest, deletes or restricts the boundaries of a wetland included in the List, it should as far as possible compensate for any loss of wetland resources, and in particular it should create additional nature reserves for waterfowl and for the protection, either in the same area or elsewhere, of an adequate portion of the original habitat.

3. The Contracting Parties shall encourage research and the exchange of data and publications regarding wetlands and their flora and fauna.

4. The Contracting Parties shall endeavour through management to increase waterfowl populations on appropriate wetlands.

5. The Contracting Parties shall promote the training of personnel competent in the fields of wetland research, management and wardening.

Article 5

The Contracting Parties shall consult with each other about implementing obligations arising from the Convention especially in the case of a wetland extending over the territories of more than one Contracting Party or where a water system is shared by Contracting Parties.

They shall at the same time endeavour to co-ordinate and support present and future policies and regulations concerning the conservation of wetlands and their flora and fauna.

Article 6

1. The Contracting Parties shall, as the necessity arises, convene Conferences on the Conservation of Wetlands and Waterfowl.

2. These Conferences shall have an advisory character and shall be competent inter alia:

- a. to discuss the implementation of this Convention;
- b. to discuss additions to and changes in the List;
- c. to consider information regarding changes in the ecological character of wetlands included in the List provided in accordance with paragraph 2 of Article 3;
- d. to make general or specific recommendations to the Contracting Parties regarding the conservation, management and wise use of wetlands and their flora and fauna;
- e. to request relevant international bodies to prepare reports and statistics on matters which are essentially international in character affecting wetlands.

3. The Contracting Parties shall ensure that those responsible at all levels for wetlands management shall be informed of, and take into consideration, recommendations of such Conferences concerning the conservation, management and wise use of wetlands and their flora and fauna.

Article 7

1. The representatives of the Contracting Parties at such Conferences should include persons who are experts on wetlands or waterfowl by reason of knowledge and experience gained in scientific, administrative or other appropriate capacities.

2. Each of the Contracting Parties represented at a Conference shall have one vote, recommendations being adopted by a simple majority of the votes cast, provided that not less than half the Contracting Parties cast votes.

Appendix B

Article 8

1. The International Union for the Conservation of Nature and Natural Resources shall perform the continuing bureau duties under this Convention until such time as another organization or government is appointed by a majority of two-thirds of all Contracting Parties.

2. The continuing bureau duties shall be, inter alia:

   a. to assist in the convening and organizing of Conferences specified in Article 6;
   b. to maintain the List of Wetlands of International Importance and to be informed by the Contracting Parties of any additions, extentions, deletions or restrictions concerning wetlands included in the List provided in accordance with paragraph 5 of Article 2;
   c. to be informed by the Contracting Parties of any changes in the ecological character of wetlands included in the List provided in accordance with paragraph 2 of Article 3;
   d. to forward notification of any alterations to the List, or changes in character of wetlands included therein, to all Contracting Parties and to arrange for these matters to be discussed at the next Conference;
   e. to make known to the Contracting Party concerned, the recommendations of the Conferences in respect of such alterations to the List or of changes in the character of wetlands included therein.

Article 9

1. This Convention shall remain open for signature indefinitely.

2. Any member of the United Nations or of one of the Specialized Agencies or of the International Atomic Energy Agency or Party to the Statute of the International Court of Justice may become a party to this Convention by:

a. signature without reservation as to ratification;
b. signature subject to ratification followed by ratification;
c. accession.

3. Ratification or accession shall be effected by the deposit of an instrument of ratification or accession with the Director-General of the United Nations Educational, Scientific and Cultural Organization (hereinafter referred to as "the Depository").

Article 10

1. The Convention shall enter into force four months after seven States have become Parties to this Convention in accordance with paragraph 2 of Article 9.

2. Thereafter this Convention shall enter into force for each Contracting Party four months after the day of its signature without reservation as to ratification, or its deposit of an instrument of ratification or accession.

Article 11

1. This Convention shall continue in force for an indefinite period.

2. Any Contracting Party may denounce this Convention after a period of five years from the date on which it entered into force for that Party by giving written notice thereof to the Depository. Denunciation shall take effect four months after the day on which notice thereof is received by the Depository.

Article 12

1. The Depository shall inform all States that have signed and acceded to this Convention as soon as possible of:

a. signatures to the Convention;
b. deposits of instruments of ratification of this Convention;
c. deposits of instruments of accession to this Convention;
d. the date of entry into force of this Convention;
e. notifications of denunciation of this Convention.

2. When this Convention has entered into force, the Depository shall have it registered with the Secretariat of the United Nations in accordance with Article 102 of the Charter.

IN WITNESS WHEREOF, the undersigned, being duly authorized to that effect, have signed this Convention.

DONE at Ramsar this 2nd day of February 1971, in a single original in the English, French, German and Russian languages, in any case of divergency the English text prevailing, which shall be deposited with the Depository which shall send true copies thereof to all Contracting Parties.

APPENDIX C

1. The Antarctic Treaty

On December 1, 1959, a treaty was signed in Washington at the end of a twelve-nation conference on peaceful international scientific cooperation in Antarctica. The twelve original signatories were Argentina, Australia, Belgium, Chile, France, Japan, New Zealand, Norway, South Africa, the Soviet Union, the United Kingdom, and the United States; they have been since joined by Czechoslovakia, Denmark, the Netherlands, Poland, Rumania, the German Democratic Republic, and Brazil. The treaty entered into force on June 23, 1961.

Text of the Antarctic Treaty

The Governments of Argentina, Australia, Belgium, Chile, the French Republic, Japan, New Zealand, Norway, the Union of South Africa, the Union of Soviet Socialist Republics, the United Kingdom of Great Britain and Northern Ireland, and the United States of America,

Recognizing that it is in the interest of all mankind that Antarctica shall continue forever to be used exclusively for peaceful purposes and shall not become the scene or object of international discord;

Acknowledging the substantial contributions to scientific knowledge resulting from international cooperation in scientific investigation in Antarctica;

Convinced that the establishment of a firm foundation for the continuation and development of such cooperation on the basis of freedom of scientific investigation in Antarctica as applied during the International Geophysical Year accords with the interests of science and the progress of all mankind;

Convinced also that a treaty ensuring the use of Antarctica for peaceful purposes only and the continuance of international harmony in Antarctica will further the purposes and principles embodied in the Charter of the United Nations;

Have agreed as follows:

Article I

1. Antarctica shall be used for peaceful purposes only. There shall be prohibited, *inter alia*, any measures of a military nature, such as the establishment of military bases and fortifications, the carrying out of military manoeuvres, as well as the testing of any type of weapons.

2. The present Treaty shall not prevent the use of military personnel or equipment for scientific research or for any other peaceful purpose.

Article II

Freedom of scientific investigation in Antarctica and cooperation toward that end, as applied during the International Geophysical Year, shall continue, subject to the provisions of the present Treaty.

Article III

1. In order to promote international cooperation in scientific investigation in Antarctica, as provided for in Article II of the present Treaty, the Contracting Parties agree that, to the greatest extent feasible and practicable:

   a. information regarding plans for scientific programs in Antarctica shall be exchanged to permit maximum economy and efficiency of operations;
   b. scientific personnel shall be exchanged in Antarctica between expeditions and stations;
   c. scientific observations and results from Antarctica shall be exchanged and made freely available.

2. In implementing this Article, every encouragement shall be given to the establishment of co-operative working relations with those Specialized Agencies of the United Nations and other international organizations having a scientific or technical interest in Antarctica.

Article IV

1. Nothing contained in the present Treaty shall be interpreted as:

   a. a renunciation by any Contracting Party of previously asserted rights of or claims to territorial sovereignty in Antarctica;
   b. a renunciation or diminution by any Contracting Party of any basis of claim to territorial sovereignty in Antarctica which it may have whether as a result of its activities or those of its nationals in Antarctica, or otherwise;
   c. prejudicing the position of any Contracting Party as regards its recognition or non-recognition of any other State's right of or claim or basis of claim to territorial sovereignty in Antarctica.

2. No acts or activities taking place while the present Treaty is in force shall constitute a basis for asserting, supporting or denying a claim to territorial sovereignty in Antarctica or create any rights of sovereignty in Antarctica. No new claim, or enlargement of an existing claim, to territorial sovereignty in Antarctica shall be asserted while the present Treaty is in force.

Article V

1. Any nuclear explosions in Antarctica and the disposal there of radioactive waste material shall be prohibited.

2. In the event of the conclusion of international agreements concerning the use of nuclear energy, including nuclear explosions and the disposal of radioactive waste material, to which all of the Contracting Parties whose representatives are entitled to participate in the meetings provided for under Article IX are parties, the rules established under such agreements shall apply in Antarctica.

Article VI

The provisions of the present Treaty shall apply to the area south of 60° South Latitude, including all ice shelves, but nothing in the present Treaty shall prejudice or in any way affect the rights, or the exercise of the rights, of any State under international law with regard to the high seas within that area.

Article VII

1. In order to promote the objectives and ensure the observance of the provisions of the present Treaty, each Contracting Party whose representatives are entitled to participate in the meetings referred to in Article IX of the Treaty shall have the right to designate observers to carry out any inspection provided for by the present Article. Observers shall be nationals of the Contracting Parties which designate them. The names of observers shall be communicated to every other Contracting Party having the right to designate observers, and like notice shall be given of the termination of their appointment.

2. Each observer designated in accordance with the provisions of paragraph 1 of this Article shall have complete freedom of access at any time to any or all areas of Antarctica.

3. All areas of Antarctica, including all stations, installations and equipment within those areas, and all ships and aircraft at points of discharging or embarking cargoes or personnel in Antarctica, shall be open at all times to inspection by any observers designated in accordance with paragraph 1 of this Article.

4. Aerial observations may be carried out at any time over any or all areas of Antarctica by any of the Contracting Parties having the right to designate observers.

5. Each Contracting Party shall, at the time when the present Treaty enters into force for it, inform the other Contracting Parties, and thereafter shall give them notice in advance, of

   a. all expeditions to and within Antarctica, on the part of its ships or nationals, and all expeditions to Antarctica organized in or proceeding from its territory;
   b. all stations in Antarctica occupied by its nationals; and
   c. any military personnel or equipment intended to be introduced by it into Antarctica subject to the

conditions prescribed in paragraph 2 of Article 1 of the present Treaty.

Article VIII

1. In order to facilitate the exercise of their functions under the present Treaty, and without prejudice to the respective positions of the Contracting Parties relating to jurisdiction over all other persons in Antarctica, observers designated under paragraph 1 of Article VII and scientific personnel exchanged under sub-paragraph 1(b) of Article III of the Treaty, and members of the staffs accompanying any such persons, shall be subject only to the jurisdiction of the Contracting Party of which they are nationals in respect of all acts or omissions occurring while they are in Antarctica for the purpose of exercising their functions.

2. Without prejudice to the provisions of paragraph 1 of this Article, and pending the adoption of measures in pursuance of sub-paragraph 1(e) of Article IX, the Contracting Parties concerned in any case of dispute with regard to the exercise of jurisdiction in Antarctica shall immediately consult together with a view to reaching a mutually acceptable solution.

Article IX

1. Representatives of the Contracting Parties named in the preamble to the present Treaty shall meet at the City of Canberra within two months after the date of entry into force of the Treaty, and thereafter at suitable intervals and places, for the purpose of exchanging information, consulting together on matters of common interest pertaining to Antarctica, and formulating and considering, and recommending to their Governments, measures in furtherance of the principles and objectives of the Treaty, including measures regarding:

    a. use of Antarctica for peaceful purposes only;
    b. facilitation of scientific research in Antarctica;
    c. facilitation of international scientific cooperation in Antarctica;
    d. facilitation of the exercise of the rights of inspection provided for in Article VII of the Treaty;
    e. questions relating to the exercise of jurisdiction in Antarctica;

f. preservation and conservation of living resources in Antarctica.

2. Each Contracting Party which has become a party to the present Treaty by accession under Article XIII shall be entitled to appoint representatives to participate in the meetings referred to in paragraph 1 of the present Article, during such times as that Contracting Party demonstrates its interest in Antarctica by conducting substantial scientific research activity there, such as the establishment of a scientific station or the despatch of a scientific expedition.

3. Reports for the observers referred to in Article VII of the present Treaty shall be transmitted to the representatives of the Contracting Parties participating in the meeting referred to in paragraph 1 of the present Article.

4. The measures referred to in paragraph 1 of this Article shall become effective when approved by all the Contracting Parties whose representatives were entitled to participate in the meetings held to consider those measures.

5. Any or all of the rights established in the present Treaty may be exercised as from the date of entry into force of the Treaty whether or not any measures facilitating the exercise of such rights have been proposed, considered or approved in this Article.

Article X

Each of the Contracting Parties undertakes to exert appropriate efforts, consistent with the Charter of the United Nations, to the end that no one engages in any activity in Antarctica contrary to the principles or purposes of the present Treaty.

Article XI

1. If any dispute arises between two or more of the Contracting Parties concerning the interpretation or application of the present Treaty, those Contracting Parties shall consult among themselves with a view to having the dispute resolved by negotiation, inquiry, mediation, conciliation, arbitration, judicial settlement or other peaceful means of their own choice.

2. Any dispute of this character not so resolved shall, with the consent, in each case, of all parties to the dispute, be referred to the International Court of Justice for settlement; but failure to reach

Appendix C 793

agreement on reference to the International Court shall not absolve parties to the dispute from the responsibility of continuing to seek to resolve it by any of the various peaceful means referred to in paragraph 1 of this Article.

Article XII

1. (a) The present Treaty may be modified or amended at any time by unanimous agreement of the Contracting Parties whose representatives are entitled to participate in the meetings provided for under Article IX. Any such modification or amendment shall enter into force when the depositary Government has received notice from all such Contracting Parties that they have ratified it.

(b) Such modification or amendment shall thereafter enter into force as to any other Contracting Party when notice of ratification by it has been received by the depositary Government. Any such Contracting Party from which no notice or ratification is received within a period of two years from the date of entry into force of the modification or amendment in accordance with the provisions of sub-paragraph 1(a) of this Article shall be deemed to have withdrawn from the present Treaty on the date of the expiration of such period.

2. (a) If after the expiration of thirty years from the date of entry into force of the present Treaty, any of the Contracting Parties whose representatives are entitled to participate in the meetings provided for under Article IX so requests by a communication addressed to the depositary Government, a Conference of all the Contracting Parties shall be held as soon as practicable to review the operation of the Treaty.

(b) Any modification or amendment to the present Treaty which is approved at such a Conference by a majority of the Contracting Parties there represented, including a majority of those whose representatives are entitled to participate in the meetings provided for under Article IX, shall be communicated by the depositary Government to all the Contracting Parties immediately after the termination of the Conference and shall enter into force in accordance with the provisions of paragraph 1 of the present Article.

(c) If any such modification or amendment has not entered into force in accordance with the provisions of sub-paragraph 1(a) of this Article within a period of two years after the date of its communication to all the Contracting Parties, any Contracting Party may at any time after the expiration of that period give notice to the depositary Government of its withdrawal from the present Treaty; and such withdrawal shall take effect two years after the receipt of the notice by the depositary Government.

Article XIII

1. The present Treaty shall be subject to ratification by the signatory States. It shall be open for accession by any State which is a Member of the United Nations, or by any other State which may be invited to accede to the Treaty with the consent of all the Contracting Parties whose representatives are entitled to participate in the meetings provided for under Article IX of the Treaty.

2. Ratification of or accession to the present Treaty shall be effected by each State in accordance with its constitutional processes.

3. Instruments of ratification and instruments of accession shall be deposited with the Government of the United States of America, hereby designated as the depositary Government.

4. The depositary Government shall inform all signatory and acceding States of the date of each deposit of an instrument of ratification or accession, and the date of entry into force of the Treaty and of any modification or amendment thereto.

5. Upon the deposit of instruments of ratification by all the signatory States, the present Treaty shall enter into force for those States and for States which have deposited instruments of accession. Thereafter the Treaty shall enter into force for any acceding State upon the deposit of its instrument of accession.

6. The present Treaty shall be registered by the depositary Government pursuant to Article 102 of the Charter of the United Nations.

Article XIV

The present Treaty, done in the English, French, Russian and Spanish languages, each version being equally authentic, shall be deposited in the archives of the Government of the United States of America, which shall transmit duly certified copies thereof to the Governments of the signatory and acceding States.

Ratifications

The following original signatories to the Antarctic Treaty ratified the Treaty on the dates indicated.

Country	Date of Deposit
Argentina	June 23, 1961
Australia	June 23, 1961
Belgium	June 26, 1960
Chile	June 23, 1961
France	September 16, 1960
Japan	August 4, 1960
New Zealand	November 1, 1960
Norway	August 24, 1960
Republic of South Africa	June 21, 1960
Union of Soviet Socialist Republics	November 2, 1960
United Kingdom	May 31, 1960
United States of America	August 18, 1960

Accessions

The following countries acceded to the Antarctic Treaty on the dates indicated.

Country	Date of Deposit
Czechoslovakia	June 14, 1962
Denmark	May 20, 1965
Netherlands	March 30, 1967
Poland	June 8, 1961
Roumania	September 15, 1971
German Democratic Republic	November 19, 1974
Brazil	May 16, 1975

2. Agreed Measures for the Conservation of Antarctic Fauna and Flora

Preamble

The Governments participating in the Third Consultative Meeting under Article IX of the Antarctic Treaty,

Desiring to implement the principles and purposes of the Antarctic Treaty;

Recognizing the scientific importance of the study of Antarctic fauna and flora, their adaptation to their rigorous environment, and their inter-relationship with that environment;

Considering the unique nature of these fauna and flora, their circumpolar range, and particularly their defencelessness and susceptibility to extermination;

Desiring by further international collaboration within the framework of the Antarctic Treaty to promote and achieve the objectives of protection, scientific study, and rational use of these fauna and flora; and

Having particular regard to the conservation of principles developed by the Scientific Committee on Antarctic Research (SCAR) of the International Council of Scientific Unions;

Hereby consider the Treaty Area as a Special Conservation Area and have agreed on the following measures:

Article I
[Area of Application]

1. These Agreed Measures shall apply to the same area to which the Antarctic Treaty is applicable (hereinafter referred to as the Treaty Area) namely the area south of 60° South Latitude, including all ice shelves.

2. However, nothing in these Agreed Measures shall prejudice or in any way affect the rights, or the exercise of the rights, of any State under international law with regard to the high seas within the

Appendix C

Treaty Area, or restrict the implementation of the provisions of the Antarctic Treaty with respect to inspection.

3. The Annexes to these Agreed Measures shall form an integral part thereof, and all references to the Agreed Measures shall be considered to include the Annexes.

Article II
[Definitions]

For the purposes of these Agreed Measures:

- a. "Native mammal" means any member, at any stage of its life cycle, of any species belonging to the Class Mammalia indigenous to the Antarctic or occurring there through natural agencies of dispersal, excepting whales.
- b. "Native bird" means any member, at any stage of its life cycle (including eggs), of any species of the Class Aves indigenous to the Antarctic or occurring there through natural agencies of dispersal.
- c. "Native plant" means any kind of vegetation at any stage of its life cycle (including seeds), indigenous to the Antarctic or occurring there through natural agencies of dispersal.
- d. "Appropriate authority" means any person authorized by a Participating Government to issue permits under these Agreed Measures.
- e. "Permit" means a formal permission in writing issued by an appropriate authority.
- f. "Participating Government" means any Government for which these Agreed Measures have become effective in accordance with Article XIII of these Agreed Measures.

Article III
[Implementation]

Each Participating Government shall take appropriate action to carry out these Agreed Measures.

Article IV
[Publicity]

The Participating Governments shall prepare and circulate to members of expeditions and stations information to ensure understanding and observance of the provisions of these Agreed Measures, setting forth in particular prohibited activities, and providing lists of specially protected species and specially protected areas.

Article V
[Cases of Extreme Emergency]

The provisions of these Agreed Measures shall not apply in cases of extreme emergency involving possible loss of human life or involving the safety of ships or aircraft.

Article VI
[Protection of Native Fauna]

1. Each Participating Government shall prohibit within the Treaty Area the killing, wounding, capturing or molesting of any native mammal or native bird, or any attempt at any such act, except in accordance with a permit.

2. Such permits shall be drawn in terms as specific as possible and issued only for the following purposes:

  a. to provide indispensable food for men or dogs in the Treaty Area in limited quantities, and in conformity with the purposes and principles of these Agreed Measures;

  b. to provide specimens for scientific study or scientific information;

  c. to provide specimens for museums, zoological gardens, or other educational or cultural institutions or uses.

3. Permits for Specially Protected Areas shall be issued only in accordance with the provisions of Article VIII.

4. Participating Governments shall limit the issue of such permits so as to ensure as far as possible that:

  a. no more native mammals or birds are killed or taken in any year than can normally be replaced by natural reproduction in the following breeding season;

b. the variety of species and the balance of the natural ecological systems existing within the Treaty Area are maintained.

5. The species of native mammals and birds listed in Annex A of these Measures shall be designated "Specially Protected Species," and shall be accorded special protection by Participating Governments.

6. A Participating Government shall not authorize an appropriate authority to issue a permit with respect to a Specially Protected Species except in accordance with paragraph 7 of this Article.

7. A permit may be issued under this Article with respect to a Specially Protected Species, provided that:

a. it is issued for a compelling scientific purpose, and
b. the actions permitted thereunder will not jeopardize the existing natural ecological system or the survival of that species.

Article VII
[Harmful Interference]

1. Each Participating Government shall take appropriate measures to minimize harmful interference within the Treaty Area with the normal living conditions of any native mammal or bird, or any attempt at such harmful interference, except as permitted under Article VI.

2. The following acts and activities shall be considered as harmful interference:

a. allowing dogs to run free,
b. flying helicopters or other aircraft in a manner which would unnecessarily disturb bird and seal concentrations, or landing close to such concentrations (e.g. within 200 m),
c. driving vehicles unnecessarily close to concentrations of birds and seals (e.g. within 200 m),
d. use of explosives close to concentrations of birds and seals,
e. discharge of firearms close to bird and seal concentrations (e.g. within 300 m),

f. any disturbance of bird and seal colonies during the breeding period by persistent attention from persons on foot.

However, the above activities, with the exception of those mentioned in (a) and (e) may be permitted to the minimum extent necessary for the establishment, supply and operation of stations.

3. Each Participating Government shall take all reasonable steps towards the alleviation of pollution of the waters adjacent to the coast and ice shelves.

Article VIII
[Specially Protected Areas]

1. The areas of outstanding scientific interest listed in Annex B shall be designated "Specially Protected Areas" and shall be accorded special protection by the Participating Governments in order to preserve their unique natural ecological system.

2. In addition to the prohibitions and measures of protection dealt with in other Articles of these Agreed Measures, the Participating Governments shall in Specially Protected Areas further prohibit:

a. the collection of any native plant, except in accordance with a permit;
b. the driving of any vehicle.

3. A permit issued under Article VI shall not have effect within a Specially Protected Area except in accordance with paragraph 4 of the present Article.

4. A permit shall have effect within a Specially Protected Area provided that:

a. it was issued for a compelling scientific purpose which cannot be served elsewhere; and
b. the actions permitted thereunder will not jeopardise the natural ecological system existing in that Area.

Article IX
[Introduction of Non-Indigenous Species, Parasites, and Diseases]

1. Each Participating Government shall prohibit the bringing into the Treaty Area of any species of animal or plant not indigenous to that Area, except in accordance with a permit.

Appendix C

2. Permits under paragraph 1 of this Article shall be drawn in terms as specific as possible and shall be issued to allow the importation only of the animals and plants listed in Annex C. When any such animal or plant might cause harmful interference with the natural system if left unsupervised within the Treaty Area, such permits shall require that it be kept under controlled conditions and, after it has served its purpose, it shall be removed from the Treaty Area or destroyed.

3. Nothing in paragraphs 1 and 2 of this Article shall apply to the importation of food into the Treaty Area so long as animals and plants used for this purpose are kept under controlled conditions.

4. Each Participating Government undertakes to ensure that all reasonable precautions shall be taken to prevent the accidental introduction of parasites and diseases into the Treaty Area. In particular, the precautions listed in Annex D shall be taken.

Article X
[Activities Contrary to the Principles and Purposes of These Measures]

Each Participating Government undertakes to exert appropriate efforts, consistent with the Charter of the United Nations, to the end that no one engages in any activity in the Treaty Area contrary to the principles or purposes of these Agreed Measures.

Article XI
[Ships' Crews]

Each Participating Government whose expeditions use ships sailing under flags of nationalities other than its own shall, as far as feasible, arrange with the owners of such ships that the crews of these ships observe these Agreed Measures.

Article XII
[Exchange of Information]

1. The Participating Governments may make such arrangements as may be necessary for the discussion of such matters as:

   a. the collection and exchange of records (including records of permits) and statistics concerning the numbers of each species of native mammal and bird killed or captured annually in the Treaty Area;

- b. the obtaining and exchange of information as to the status of native mammals and birds in the Treaty Area, and the extent to which any species needs protection;
- c. the number of native mammals or birds which should be permitted to be harvested for food, scientific study, or other uses in the various regions;
- d. the establishment of a common form in which this information shall be submitted by Participating Governments in accordance with paragraph 2 of this Article.

2. Each Participating Government shall inform the other Governments in writing before the end of November of each year of the steps taken and information collected in the preceding period of 1st July to 30th June relating to the implementation of these Agreed Measures. Governments exchanging information under paragraph 5 of Article VII of the Antarctic Treaty may at the same time transmit the information relating to the implementation of these Agreed Measures.

Article XIII
[Formal Provisions]

1. After the receipt by the Government designated in Recommendation I-XIV (5) of notification of approval by all Governments whose representatives are entitled to participate in meetings provided for under Article XI of the Antarctic Treaty, these Agreed Measures shall become effective for those Governments.

2. Thereafter any other Contracting Party to the Antarctic Treaty may, in consonance with the purposes of Recommendation III-VII, accept these agreed Measures by notifying the designated Government of its intention to apply the Agreed Measures and to be bound by them. The Agreed Measures shall become effective with regard to such Governments on the date of receipt of such notification.

3. The designated Government shall inform the Governments referred to in paragraph 1 of this Article of each notification of approval, the effective date of these Agreed Measures and of each notification of acceptance. The designated Government shall also inform any Government which has accepted these Agreed Measures of each subsequent notification of acceptance.

Article XIV
[Amendment]

1. These Agreed Measures may be amended at any time by unanimous agreement of the Governments whose Representatives are entitled to participate in meetings under Article IX of the Antarctic Treaty.

2. The Annexes, in particular, may be amended as necessary through diplomatic channels.

3. An amendment proposed through diplomatic channels shall be submitted in writing to the designated Government which shall communicate it to the Governments referred to in paragraph 1 of the present Article for approval; at the same time, it shall be communicated to the other Participating Governments.

4. Any amendment shall become effective on the date on which notifications of approval have been received by the designated Government and from all of the Governments referred to in paragraph 1 of this Article.

5. The designated Government shall notify those same Governments of the date of receipt of each approval communicated to it and the date on which the amendment will become effective for them.

6. Such amendment shall become effective on that same date for all other Participating Governments, except those which before the expiry of two months after that date notify the designated Government that they do not accept it.

Annexes to These Agreed Measures

[Annex A (Specially Protected Species) and Annex B (Specially Protected Areas) are omitted.—Editor]

Annex C: Importation of Animals and Plants

The following animals and plants may be imported into the Treaty Area in accordance with permits issued under Article IX (2) of these Agreed Measures:

    a. sledge dogs;
    b. domestic animals and plants;
    c. laboratory animals and plants.

Annex D: Precautions to Prevent Accidental Introduction of Parasites and Diseases into the Treaty Area

The following precautions shall be taken:

1. *Dogs:* All dogs imported into the Treaty Area shall be inoculated against the following diseases:

    a. distemper;
    b. contagious canine hepatitis;
    c. rabies;
    d. leptospirosis *(L. canicola* and *L. icterohaemorrhagicae).*

Each dog shall be inoculated at least two months before the time of its arrival in the Treaty Area.

2. *Poultry:* Notwithstanding the provisions of Article IX (3) of these Agreed Measures, no living poultry shall be brought into the Treaty Area after 1st July, 1966.

APPENDIX D

Resolution Adopted by the Tropical Forests Panel

Introduction

The Tropical Forests Panel of the EARTHCARE Conference (Dr. Lawrence S. Hamilton, chair) endorsed Resolution 7 of the Eleventh General Assembly of IUCN (Banff, Alberta, Canada, 16 September 1972), and urged that a sixth clause be added to the resolution.

The text of the original resolution follows:

Text of Original IUCN Resolution

7. Conservation and Development of Tropical Rain Forests

Recognizing that present and planned agricultural, grazing and forest exploitation activities, sometimes involving agrarian reform schemes, are resulting in major impacts on tropical rain forests and often lead to their complete disappearance and replacement by secondary communities, thus depriving the countries concerned of a valuable potential resource;

The 11th General Assembly of IUCN meeting in Banff, Canada, in September 1972:

1. *Urges* all development programmes which involve the manipulation of tropical rain forests should be based upon ecological analysis and principles and the application of appropriate technologies that can result in a sustained

yield from the resource with minimum adverse effects upon the environment;
2. that the governments of those countries in which companies extracting timber from tropical forest lands are based, should exercise increased controls over the operations of such companies undertaken abroad to oblige them to take all the precautions that would avoid the degradation of tropical forest ecosystems;
3. that important and unique areas within tropical rain forests should be set aside for management as national parks, sanctuaries and reserves to insure the conservation of representative natural formations and species, as well as genetic resources;
4. that critical areas within tropical rain forests such as upper watersheds, riverine and estuarine zones, slopes and areas subject to accelerated erosion be given special treatment including the restriction of harmful land-use practices, and the use of reforestation or other protective measures; and
5. that countries with large tropical timber resources be given financial assistance by appropriate national and international agencies to help maintain their forest resources.

The text of the sixth clause, recommended by the Tropical Forests Panel, follows:

6. that governments take steps to provide and develop a local research capability and training of personnel in ecological land-use planning.

APPENDIX E

Zapovedniks in the Soviet Union
Philip R. Pryde

The table that follows provides a list of all the *zapovedniks* in the Soviet Union as of mid-1976. It gives detailed up-to-date information, which is often difficult to obtain. The data are based on these sources: A. G. Bannikov, ed. (1969), Zapovedniki Sovetskogo Soyuza, *Kolos,* Moscow; A. G. Bannikov (1974), Po Zapovednikam Sovetskogo Soyuza, *Mysl',* Moscow; and Tsentralnaya Laboratoriya Okhrany Prirody (1976), Zapovedniki, zapovedno-okhotnichi khozyaistva i natsionalnyye parki SSR (typescript).

RSFSR has been used as an abbreviation for Russian Soviet Federated Socialist Republic.

Zapovedniks of the Soviet Union

Name of Zapovednik	Location (republic)	Year(s) Established	Area (hectares)
Adzhameti	Georgia	1958, 1946	4,849
Aksu-Dzhabagly	Kazakh	1926	74,416
Algetskiy	Georgia	1965	5,919
Alma-Ata	Kazakh	1960, 1931	89,531
Altai	RSFSR	1968, 1932	863,828
Aral-Paigambar	Uzbek	1961	3,093
Askaniya-Nova	Ukraine	1921, 1898	11,054
Astrakhan	RSFSR	1919	62,423
Badai-Tugai	Uzbek	1971	6,497

***Zapovedniks* of the Soviet Union (continued)**

Name of Zapovednik	Location (republic)	Year(s) Established	Area (hectares)
Badkhyz	Turkmen	1941	87,680
Baikal	RSFSR	1969	165,724
Barguzin	RSFSR	1916	263,176
Barsa-Kelmes	Kazakh	1939	18,300
Bashkir	RSFSR	1958, 1930	72,140
Basutchai	Azerbaidzhan	1974	117
Batsara-Babanur	Georgia	1935–1960	3,812
Berezina	Belorussia	1925	76,201
Bolshekhekhtsir	RSFSR	1964	44,928
Borzhomi	Georgia	1960, 1935	18,048
Chapkyalyai	Lithuania	1975	8,543
Chatkal	Uzbek	1947	35,256
Chernomore	Ukraine	1927	64,806
Darvin	RSFSR	1945	112,611
Dilizhan	Armenia	1958	28,799
Galichya Gora	RSFSR	1969	238
Gyok-Gyol	Azerbaidzhan	1965, 1925	7,131
Girkan	Azerbaidzhan	1969, 1936	2,900
Grini	Latvia	1957	799
Ilmen	RSFSR	1920	30,380
Issyk-Kul	Kirgiz	1948	733,000
Kabardino-Balkar	RSFSR	1976	53,303
Kandalaksha	RSFSR	1932	35,026
Kanev	Ukraine	1968, 1931	1,030
Karakul	Uzbek	1971	21,021
Karpatskiy (Carpathian)	Ukraine	1968	12,672
Kavkaz (Caucasus)	RSFSR	1924, 1888	263,485
Kedrovaya Pad	RSFSR	1916	17,897
Khingansk	RSFSR	1964	58,903
Khopyor	RSFSR	1935	16,178
Khosrov (Garni)	Armenia	1958	23,000
Kintrish	Georgia	1960	7,166
Kivach	RSFSR	1931	10,460
Kodry	Moldavia	1971	5,117
Kolkhida	Georgia	1959, 1946	500
Komsomolsk	RSFSR	1963	31,958
Krasnovodsk (Gasan-Kuli)	Turkmen	1968, 1932	262,037

Appendix E

Zapovedniks of the Soviet Union (continued)

Name of Zapovednik	Location (republic)	Year(s) Established	Area (hectares)
Kronotskiy	RSFSR	1967, 1934	964,000
Kurgaldzhino	Kazakh	1968, 1959	183,704
Kyzylagach	Azerbaidzhan	1929	88,360
Kyzylkum	Uzbek	1971	8,749
Kyzylsu	Uzbek	1975	16,002
Lagodekhi	Georgia	1912	17,668
Lapland	RSFSR	1930	161,254
Lazov (Sudzukhe)	RSFSR	1935	116,524
Lugansk	Ukraine	1968	992
Malaya Sosva	RSFSR	1976, 1929	92,921
Mariamdzhvar	Georgia	1935	1,040
Matsalu	Estonia	1957	13,500
Mordov	RSFSR	1936	32,220
Moritssala	Latvia	1957, 1912	859
Mys Martyan	Ukraine	1973	240
Naurzum	Kazakh	1965, 1934	89,642
Nigula	Estonia	1957	2,766
Nurata	Uzbek	1975	22,537
Oka	RSFSR	1935	22,896
Pechoro-Ilych	RSFSR	1930	721,322
Pinega	RSFSR	1974	41,244
Pirkulin	Azerbaidzhan	1969	1,521
Pitsunda-Myussera	Georgia	1957–1966, 1934	3,770
Polesskiy	Ukraine	1968	20,097
Prioksko-Terrasnyy	RSFSR	1945	4,945
Pripyat	Belorussia	1969	60,767
Ramit	Tadzhik	1959	16,168
Repetek	Turkmen	1928	34,600
Ritsa	Georgia	1957, 1930	16,289
Saguram	Georgia	1957, 1946	5,247
Sary-Chelek	Kirgiz	1960	23,868
Satapliyskiy	Georgia	1958, 1935	354
Sayano-Shushenskoye	RSFSR	1976	389,570
Severo-Osetinsk	RSFSR	1967	67,538
Shirvan	Azerbaidzhan	1969	69,550
Sikhote-Alin	RSFSR	1935	340,000
Slitere	Latvia	1957, 1921	9,409

Zapovedniks of the Soviet Union (continued)

Name of Zapovednik	Location (republic)	Year(s) Established	Area (hectares)
Sokhonda	RSFSR	1974	210,988
Stolby	RSFSR	1925	47,154
Teberda	RSFSR	1936	83,122
Tigravaya balka	Tadzhik	1938	52,212
Tsentralno-Chernozem	RSFSR	1935	4,795
Tsentralno-Lesnoy	RSFSR	1960, 1931	21,348
Turianchai	Azerbaidzhan	1958	12,356
Ukrainian Steppe	Ukraine	1961	1,634
Ussuri (Suputinka)	RSFSR	1932, 1911	40,454
Vashlovani	Georgia	1957, 1935	4,868
Vilsandi (Vaika)	Estonia	1971, 1957	10,689
Visim	RSFSR	1971, 1946	13,319
Viydumyae	Estonia	1957	593
Volga-Kama	RSFSR	1960	9,393
Voronezh	RSFSR	1927	31,053
Vrangelya (Wrangel Island)	RSFSR	1976	795,650
Yalta	Ukraine	1973	14,173
Zaamin	Uzbek	1960	10,560
Zakataly	Azerbaidzhan	1930	25,200
Zavidov	RSFSR	1972	84,000
Zeravzhan	Uzbek	1975	2,518
Zeya	RSFSR	1963	82,300
Zhigulevsk	RSFSR	1966, 1927	19,130
Zhuvintas	Lithuania	1937	5,420
Total			8.96 million

APPENDIX F

Biographical Information

Chaplin B. Barnes directs the Office of International Activities of the National Audubon Society. He was vice-chairman and National Audubon's coordinator for the EARTHCARE Conference. Mr. Barnes is chairman of the International Committee of Natural Resources Council of America and is its representative on the Environment Liaison Board, Nairobi and Geneva. He is also a trustee of the Rare Animal Relief Effort, Inc.; a member of the advisory board of the Center for International Environment Information; and a member of the executive board of the American Committee for International Conservation.

Vladimir A. Borissoff is secretary general of the USSR Ministry of Agriculture's Central Laboratory on Nature Conservation. A humanist by education, Mr. Borissoff has been working in the field of nature conservation since 1955, when he joined the Commission for Nature Conservation of the USSR Academy of Science as an assistant scientific officer. He became secretary general of the Central Laboratory on Nature Conservation in 1967. Mr. Borissoff is the author of approximately thirty publications on nature conservation and protected natural areas.

F. Herbert Bormann is Oastler Professor of Forest Ecology in the School of Forestry and Environmental Studies at Yale University. Between 1956 and 1966, he taught botany at Dartmouth College. A member of the Ecology Advisory Committee of the U.S. Environmental Protection Agency, Dr. Bormann was president of the Ecological Society of America in 1970 and 1971. He currently (1973–1977) serves on the executive committee of the Assembly of Life Sciences of the National Research Council. Dr. Bormann was elected a member of the American Academy of Sciences in 1972 and of the National Academy of Sciences in 1973. He is author or coauthor of more than ninety scientific papers on forest ecology, conservation, and related subjects.

Arthur Bourne gave up a formal scientific career to join the conservation movement, becoming, in the early sixties, deputy director of the British Council for Nature Intelligence Unit. He was the founder and former chairman of the Ocean Resources Conservation Association, and for several years observer on the International Whaling Commission for the British section of the World Wildlife Fund. His work has ranged from studies on whaling pressures in the Atlantic, Arctic, and Antarctic Oceans, and sealing methods and pressures in the Gulf of St. Lawrence, to investigations into peaceful uses of the seabed and the effects of mineral extraction on oceanic life. He is the author of *Pollute and Be Damned* and *The Earth's Crust*, and editor, with F. Steele, of *The Man/Food Equation*.

David R. Brower, the founder and president of Friends of the Earth, has been a preeminent leader in the conservation movement for nearly four decades. His accomplishments are legion. He is a past executive director of the Sierra Club and the editor of many Sierra Club books. He was instrumental in keeping dams out of Dinosaur National Monument and the Grand Canyon, and was in the forefront of efforts to establish the North Cascades, Redwood, and Kings Canyon national parks. Mr. Brower was a cofounder and past director of the John Muir Institute for Environmental Studies and a founder of the League of Conservation Voters and the Trustees for Conservation. He is a member and former chairman of the Natural Resources Council of America and a director of the Rachel Carson Trust for the Living Environment.

Noel J. Brown has been director of the New York liaison office and assistant to the director of the United Nations Environment Programme since 1973. From 1963 until 1973, he served in the United Nations as special assistant to the director, Political Affairs Division, and as political affairs officer, Department of Political and Security

Council Affairs. Dr. Brown has served as visiting professor at Hunter College, New York, and the University of the West Indies. His professional interests are international humanitarian, environmental law, and governmental and public policy. He has conducted research on the United Nations and the Third World and on international responses to and responsibility toward national insurgencies. The author of *Morality and Modern Warfare: Biographical Essay* (1960), and of *OAS and the Newly Independent States of the Caribbean: Some Policy Questions* (1968), Dr. Brown is on the editorial board of *Worldview Magazine* and is a member of the International Council on Environmental Law.

Gerardo Budowski is head of the Forest Sciences Department of the Centro Agronómico Tropical de Investigación y Enseñanza in Turrialba, Costa Rica. From 1970 until 1976, Dr. Budowski was director-general of the International Union for Conservation of Nature and Natural Resources (IUCN) in Morges, Switzerland. An agronomist and forester by training, Dr. Budowski joined the staff of UNESCO in 1967 as program specialist for Ecology and Conservation. While with UNESCO, he organized the Intergovernmental Conference on the Rational Use and Conservation of the Biosphere, which took place in 1968, and subsequently organized the follow-up Man and the Biosphere Program. He is also author of more than ninety publications on ecology, conservation, tropical forestry, and related subjects.

Charles H. Callison is the executive vice-president of the National Audubon Society and, as a leading authority on national legislation in the field of natural resources, has assisted in the drafting of important conservation bills. In addition to his work on national conservation problems, Mr. Callison writes for *Audubon Magazine,* and in 1967, he edited the book *America's Natural Resources* for the Natural Resources Council of America. He is a past secretary and conservation director of the National Wildlife Federation and a past chairman of Natural Resources Council of America. In 1974, he received the Frances K. Hutchinson Medal from the Garden Club of America for national service in conservation.

Roy E. Cameron is a deputy director of Argonne National Laboratory's Land Reclamation Program in Argonne, Illinois. He was formerly a senior scientist and member of the technical staff at the Jet Propulsion Laboratory in Pasadena, California. A soil scientist and microbiologist by training, Dr. Cameron is an authority on desert ecology and has been the leader of numerous research teams to deserts and other harsh regions of the world, including seven expeditions to the Antarctic. He is the author and coauthor of more than 110 publications on the ecology, soils, and microbiology of arid lands. Currently, he is active in the reclamation and rehabilitation of strip-mined lands, especially those in arid and semiarid areas.

Eric Carlson is head of the United Nations Environment Programme's (UNEP's) program in human settlements and is also senior program officer for the United Nations Habitat and Human Settlements Foundation. Born in Sweden, but now a citizen of the United States, Mr. Carlson has dealt with the problems of financing and organizing habitat and human-settlement programs for more than twenty-five years. From 1964 to 1973, he was chief of the Housing Section of the United Nations Centre for Housing, Building, and Planning. Since 1973, he has served with UNEP as senior adviser and in other key roles, including work with the Preparatory Planning Group for HABITAT. In 1972, Mr. Carlson was appointed honorary research fellow in the School of Environmental Studies of University College, London, and Albert Schweitzer Research Fellow at the School of International Affairs, Columbia University.

Roland C. Clement is a vice president of the National Audubon Society, as well as secretary for the Americas and chairman of the United States Section of the International Council for Bird Preservation. An author, editor, and lecturer, Mr. Clement conducted investigations in ornithology and subarctic ecology in Central Ungava, Labrador, during World War II. A trustee of the Environmental Defense Fund, Inc., and director of the World Wildlife Fund's United States Appeal, Mr. Clements was a contributor to *Life Histories of North American Birds*, published by the United States National Museum, and W. E. Clyde Todd's *Birds of the Labrador Peninsula*.

Kai Curry-Lindahl is senior adviser in ecology and conservation to the United Nations Environment Programme. From 1953 to 1974, he was director of the Department of Natural History, Nordic

Museum and Skansen, Stockholm. During those years, he also served as visiting professor of conservation at the University of California, Berkeley, and advised governments, United Nations agencies, and private institutions. Formerly a regional expert in ecology and conservation in Africa for UNESCO and the United Nations Development Programme, he is currently active in many conservation organizations, including the International Council for Bird Preservation and IUCN. Dr. Curry-Lindahl was editor of *Sveriges Natur,* organ of the Swedish Society for Conservation of Nature, and of *Acta Vertebratica,* and is the author or editor of seventy books and over 500 scientific papers.

René J. Dubos, a bacteriologist and professor emeritus at The Rockefeller University in New York City, was assistant editor for the International Institute of Agronomy in Rome from 1922 to 1924. A citizen of the United States, Dr. Dubos served as a research assistant at the Rutgers University New Jersey Experiment Station from 1924 to 1927. The recipient of twenty-eight honorary degrees, he has an international reputation in the fields of microbiology, pathology, and chemotherapy and is a member of the National Academy of Sciences, the American Philosophical Society, and the National Research Council. He is the author of numerous scientific articles and popular books, including *So Human an Animal,* for which he received the Pulitzer Prize for Non-Fiction in 1969, and *Only One Earth,* which he wrote with Barbara Ward.

Lars Emmelin is a lecturer in environmental studies and course director of the Environmental Studies Programme at the University and Institute of Technology, Lund, Sweden. He is a consultant on environmental education to the Organisation for Economic Cooperation and Development (OECD), UNESCO, and the Council of Europe. Mr. Emmelin is also the author of the monthly publication, *Environmental Planning and Conservation in Sweden,* which is issued in five languages by the Swedish Institute. Trained as a diver at the Laboratory of Aviation and Naval Physiology at Lund, Mr. Emmelin is currently working on problems of marine nature reserves.

M. Taghi Farvar directs the Centre for Endogenous Development Studies at Tehran, Iran, and was formerly a research associate and coordinator of the Program on Ecology and International Development at the Center for the Biology of Natural Systems, Washington University, in St. Louis. For the past ten years he has worked on problems of environmental protection in the developing countries, and has done fieldwork in Central America, North Africa, and the Middle East. Dr. Farvar organized the Human Environment Division in the Government of Iran and served as senior adviser to the United Nations Environment Programme. He is the author and editor of numerous publications on environment and development, including *The Careless Technology: Ecology and International Development* and *International Development and the Human Environment: An Annotated Bibliography.*

Rimmon C. Fay serves as state representative on the California Coastal Zone Conservation Commission, South Coast Region. A biochemist, bacteriologist, and marine biologist by experience and training, Dr. Fay is the owner and manager of Pacific Bio-Marine Supply Company of Venice, California. Dr. Fay's research interests lie in the areas of energy-transfer mechanisms in organisms, the natural history of marine organisms, methods of controlling water pollution, and biochemical processes in bacteria and marine organisms. A leading conservationist on the West Coast of the United States, Dr. Fay has made more than 2,500 dives in the coastal waters of California.

Victor Ferkiss, a political scientist and author, is professor of government at Georgetown University. In 1958 and 1959 he was a Rockefeller Foundation fellow at the University of California, Berkeley. A contributor to many professional periodicals, Dr. Ferkiss has taught at The Johns Hopkins University, the University of California, and St. Mary's College (California), among others. In 1959 and 1960, he was field director of training for Africa in Boston University's African Studies Program. A member of the World Future Society and the American Political Science Association, Dr. Ferkiss is the author of several books, including *Africa's Search for Identity* (1966), *Technological Man* (1969), and *The Future of Technological Civilization* (1974).

William L. Finger is vice president of the International Executive Service Corps and practices law in Kew Gardens (Queens), New York.

Eskandar Firouz is the deputy prime minister of Iran and director of that country's Department of the Environment. He was formerly a member of Parliament and deputy prime minister of the Iran Ministry of Agriculture and Natural Resources. Mr. Firouz is a member of the High Council of the Society for the Protection of Natural Resources and the Human Environment in Iran, vice-president of the International Union for Conservation of Nature and Natural Resources, secretary-general of the Iran National Botanical Gardens Organization, and a member of the board of trustees of the World Wildlife Fund.

Arturo Gómez-Pompa is director of the National Research Institute of Biotic Resources (INIREB), Jalapa, Mexico, and professor of botany and ecology at the National University of Mexico. Formerly the head of the Department of Botany of that university's Institute of Biology, Dr. Gómez-Pompa is cochairman of the National Ecology Program of the National Council of Science and Technology (CONACYT) and of the executive council of the Man and the Biosphere (MAB) Program of UNESCO, and was a trustee of The Institute of Ecology. He was the recipient of a fellowship from the John Simon Guggenheim Foundation, honorary research associate at the Gray Herbarium, and Mercer Research Fellow at the Arnold Arboretum of Harvard University in 1964. Dr. Gómez-Pompa is the author of some one hundred papers on land use, tropical botany, and biological education.

François Gros is director of Grande Traversée des Alpes Françaises in Grenoble, France. An expert mountaineer, geologist, and national ski instructor, Monsieur Gros' responsibility involves the administration of trail management, preservation of natural refuges, and the organization of services related to the use, maintenance, and conservation of the French Alps. He is a former inspector of the Vanoise National Park in the Maurienne Valley.

Lawrence S. Hamilton is professor of forestry and conservation in the Department of Natural Resources, Cornell University, where he teaches courses in forest ecology, land-use planning, and international natural resources, and heads an interdisciplinary graduate program in the conservation of natural resources. He is also the director of the Sierra Club's Tropical Forests Task Force and is an active member of several conservation organizations, including the Sierra Club. In addition, Dr. Hamilton has been a visiting professor in Australia at the universities of New England and Queensland. At one time he was a zone forester for the Ontario Department of Lands and Forests and an extension forester for the New York State College of Agriculture. Dr. Hamilton is currently studying the ecological basis for planning land and water use through resource analysis.

Jay S. Hammond, governor of Alaska since 1974, was a member of the Alaska House of Representatives from 1960 to 1966 and of the Alaska Senate from 1968 to 1972, of which he was president in 1972 and 1973. In 1968 he received the Legislature Conservationist of the Year Award. He has also served as the mayor of Bristol Bay Borough. Formerly a trapper, a pilot for the U.S. Fish and Wildlife Service, a guide for fishermen and big-game hunters, and a commercial fisherman, Governor Hammond homesteaded in Alaska in 1952. He is the author of a book of verse and of articles on the outdoors.

Marian S. Heiskell is a newspaper executive and civic worker. She is director of special activities for *The New York Times* and serves on the advisory council for National Parks, Historic Sites, and Buildings and Monuments. She is on the board of managers of the New York Botanical Garden and the board of trustees of the Parks Council, and is a member of the Commission for the Gateway National Recreation Area and the Commission of the State Park and Recreation Commission for the City of New York. Ms. Heiskell is also co-chair (with the mayor) of the Council on the Environment of the City of New York.

Eleanor C. J. Horwitz is staff writer for the Society of American Foresters. An ecologist and environmental educator, she has served as coordinator of outdoor education for Lane County, Oregon, consultant to the New York State Department of Education, and was an instructor at the Massachusetts Audubon Society. Ms. Horwitz is the author of *Clearcutting: A View from the Top* (1974) and numerous articles and pamphlets.

Moustafa Imam is head of the Department of Natural Resources at Cairo University's Institute of African Research and Studies in Giza. A native of Egypt, Dr. Imam was a teaching assistant in botany at Cairo University from 1953 to 1960, and a Government Scholar at the University of California at Davis from 1960 to 1965. He was then a research officer of the Desert Institute in Cairo from 1965 to 1972.

Thomas B. Johansson is a member of the board and executive committee of the Environmental Studies Program at the University of Lund, Sweden. His work has focussed on a new method for detecting trace metals in environmental samples, especially particulate matter in the atmosphere. At present, Dr. Johansson is affiliated with the Ministry of Education's Secretariat for Studies of the Future in Stockholm, where he is studying the relationship between energy and society.

F. Wayne King has been director of zoology and conservation for the New York Zoological Society since 1973. Dr. King is a member of the board of governors of the American Society of Ichthyologists and Herpetologists, and serves on that organization's Committee on Environmental Quality. Past chairman of the American Association of Zoological Parks and Aquariums' Wildlife Conservation Committee, he is presently a member of the U.S. Department of the Interior's Fish and Wildlife and Parks National Sciences Advisory Committee, the Survival Service Commission, and the Crocodile Specialist Group of the International Union for Conservation of Nature and Natural Resources. Dr. King is the author of some forty professional papers on zoology and conservation.

Thomas E. Lovejoy is program director of the World Wildlife Fund, Washington, D.C., and has served as acting chairman and director of the Wildlife Preservation Trust International, treasurer of the International Council for Bird Preservation, Pan American Section, and coordinator of the J. Paul Getty Wildlife Conservation Prize. Dr. Lovejoy has made several trips to South and Central America to study bird populations, and he spent two months collecting birds in Kenya for the Peabody Museum of Yale University in 1964. In addition to his fieldwork, Dr. Lovejoy has studied the epidemiology of arboviruses associated with rain-forest birds.

Emile A. Malek is professor of parasitology and director of the Laboratory of Schistosomiasis and Medical Malacology at the Tulane University Medical Center. A former scientist with the World Health Organization, he has served as a consultant to the World Health Organization, the Peace Corps, and the U.S. Public Health Service. Dr. Malek teaches courses in malacology and human helminthology and is the author of many scientific papers on human parasitology, especially schistosomiasis and bilharziasis.

Arthur R. Marshall, ecologist and conservationist, worked in various capacities in the Bureau of Commercial Fisheries and the U.S. Department of the Interior between 1955 and 1970. In 1970 and 1971, he was professor of ecology and director of the University of Miami's Laboratory for Estuarine Research. From 1971 to 1973 he was director of the Division of Applied Ecology in the university's Center for Urban and Regional Studies. Mr. Marshall has served as a consultant to the University of Florida's Urban and Regional Development Center and as chairman of the governing board of the Saint Johns River Water Management District. He is author of numerous articles on the effects of development on wetlands and on the costs of growth in cities, and was chosen Conservationist of the Year by the Florida Audubon Society in both 1964 and 1972 and Outstanding Conservationist of the Year by the Florida Wildlife Federation in 1972. In 1970, the Governor's Conservation Award was conferred upon him by the state of Florida.

G.V.T. Matthews is the director of the International Waterfowl Research Bureau and director of research and general director of the Wildfowl Trust in Slimbridge, England. Dr. Matthews is the author of more than fifty papers on the navigation, orientation, general behavior, and conservation of birds. The second edition of his book, *Bird Navigation,* was published by Cambridge University Press in 1968.

J. Michael McCloskey, executive director of the Sierra Club since 1969, joined the club's staff in 1961. From 1961 until 1965, he represented the Sierra Club and the Federation of Western Outdoor Naturalists on conservation matters in the Pacific Northwest. Mr. McCloskey played a leading role in instituting a landscape-management policy in the national forests of the United States; was active in support of the Wilderness Preservation Act of 1964; developed the early stages of the successful campaign for a North Cascades National Park; and was the principal legislative advocate for Redwood National Park. He is a member of the American Committee for International Wildlife Protection and the planning board of the Environmental Management Institute at the University of Southern California. He received the California Conservation Council Award in 1969.

Edgardo Mondolfi, the former executive director of Consejo de Bienestar Rural (Rural Welfare Council) in Caracas, Venezuela, is a professor of vertebrate zoology and wildlife management in the Central University of Venezuela, president of the Venezuelan Association for the Protection of Nature, and president of the Committee for Protection of Native Fauna. Especially concerned with the wildlife of the tropical rain forest, Ingeniero Mondolfi is also a member of many other organizations, among them the executive board of the International Union for Conservation of Nature and Natural Resources, the Venezuelan Council of Renewable Natural Resources, the American Society of Mammalogists, and the Association for Tropical Biology.

Robert Muller is director and deputy to the under-secretary-general for Inter-Agency Affairs and Co-ordination of the United Nations, a post he has held since 1973. A member of the United Nations' staff since 1948, he has also served as senior adviser to the secretary-general of the United Nations Trade and Development Conference and associate director of the United Nations Division of Natural Resources. A former political adviser to United Nations forces in Cyprus and director of the United Nations Budget, Dr. Muller has given many speeches and written many articles on world affairs.

Norman Myers was the United Nations Food and Agriculture Organisation's wildlife and national parks officer for Africa in 1975. He has now resumed work as a consultant, specializing in the relationships between conservation and economic development. Dr. Myers has carried out surveys and research in Africa and southeastern Asia for the International Union for Conservation of Nature and Natural Resources, the World Wildlife Fund, the Smithsonian Institution, the

World Bank, and the United Nations Environment Programme. He is a writer, photographer, and lecturer on wildlife conservation in Africa, and has also conducted fieldwork in Brazil. Dr. Myers is the author of *The Long African Day* and more than sixty articles. At present, he is writing a book on threatened species and genetic reservoirs, considered as components of man's common global heritage.

Roderick Nash, professor of history and environmental studies and chairman of the Environmental Studies Program at the University of California, Santa Barbara, is currently engaged in a study, sponsored by the Rockefeller Foundation, of world parks and preserves. A member of the National Park Centennial Commission, Dr. Nash has developed courses in environmental history, and, in 1970, led in creating a multidisciplinary major in environmental studies at the University of California, Santa Barbara. A specialist in American social and intellectual history, he is also an expert in wilderness preservation, exploration, and interpretation. As a leader of Santa Barbara's response to the oil spill of 1969, Dr. Nash wrote the "Santa Barbara Declaration of Environmental Rights." He is the author of eight books, including *Wilderness and the American Mind*.

Frank X.J.C. Njenga is counsellor to the Permanent Mission of the Republic of Kenya to the United Nations and legal adviser to the Kenya Ministry of Foreign Affairs. A member of the Kenya delegation to the Security Council in 1973 and 1974, he was leader of the Kenya delegation to the first session of the Third United Nations Conference on the Law of the Sea and deputy leader of the delegation at the second session. From 1969 to 1972, he was leader of the Kenya delegation to the Committee on the Peaceful Uses of the Sea-Bed and Ocean Floor and Subsoil Thereof Beyond the Limits of National Jurisdiction.

Perez M. Olindo was director of Kenya National Parks from 1966 to 1976. After graduating from college he joined the Kenya Game Department first as a biologist, then as a divisional warden. He joined the National Parks Service in 1965 and was appointed its director a year later. The author or coauthor of papers on the impact of tourism and economic development on conservation efforts in developing countries, fire and the conservation of habitat in Kenya, and contamination of Lake Nakuru by metals and pesticides, Dr. Olindo was instrumental in promoting the Convention on International Trade in Endangered Species of Wild Flora and Fauna. In recognition of his leadership and accomplishments, Dr. Olindo was awarded an honorary doctoral degree by Michigan State University in 1975, the Bronze Centennial Medallion by the Second World Conference on National Parks, and the Conservationist of the Year Award by the Safari Club of Washington. He was also made Knight Commander of the African Redemption by the late President Tubman of Liberia.

Roger Tory Peterson is an ornithologist, author, artist, scientist, photographer, and lecturer. Dr. Peterson is a special consultant to and has been associated with the National Audubon Society for thirty-five years. He is art director for the National Wildlife Federation, and serves on the board of directors of the World Wildlife Fund. A frequent delegate to ornithological and bird-protection conferences, Dr. Peterson is a fellow of the American Association for the Advancement of Science and the American Ornithologists' Union, vice-president of the Society of Wildlife Artists (England), honorary member of the British Ornithologists' Union, and honorary fellow of the Zoological Society of London. He formulated the renowned "Peterson system" for the identification of birds in the field, is the author of *Field Guide to the Birds* and *The World of Birds*, and is editor of Houghton Mifflin Company's Field Guide Series.

Russell W. Peterson heads "New Directions," a new citizens' lobbying organization that focuses on international issues. He was formerly chairman of the President's Council on Environmental Quality and governor of the state of Delaware from 1969 to 1973. For some twenty-six years, Dr. Peterson was associated with E. I. DuPont de Nemours and Company of Wilmington, Delaware, in various research and development capacities. In 1971, he received the World Wildlife Fund's gold medal and was named Conservationist of the Year by the National Wildlife Federation.

Douglas H. Pimlott is professor of zoology and forestry at the University of Toronto. A wildlife ecologist and an authority on big-game and wolf ecology and population dynamics, he began his career as a biologist with Newfoundland's Department of Mines and Resources in 1950. From 1958 until 1962, Dr. Pimlott was with the Province of Ontario's Department of Lands and Mines and in 1962, he joined the University of Toronto. One of Canada's leading conservationists, Dr. Pimlott has served as president of the Canadian Nature Federation and was one of the leading forces behind the founding of the Canadian Arctic Resources Committee, an independent organization devoted to encouraging informed public debate on economic development in northern Canada.

Ghillean T. Prance, director of botany for the New York Botanical Garden, is an authority on the taxonomy of tropical plant families. From 1973 to 1975, he was director of the graduate program of the Instituto Nacional de Pesquisas da Amazônia, Manaus, Brazil. Born in Brandeston, England, Dr. Prance became associated with the New York Botanical Garden as a research assistant in 1963, and in 1968 became B. A. Krukoff Curator of Amazonian Botany. Since 1960, he has been active in botanical exploration in various parts of the tropics, and between 1964 and 1975 was the leader of a series of expeditions to Amazonian Brazil. The executive director of the Organization for Flora Neotropica, Dr. Prance is a fellow of the Linnean Society of London and a member of many botanical societies.

Philip R. Pryde, associate professor of geography at San Diego State University, is a specialist in natural resource management and conservation in the Soviet Union. Previously, he studied and taught at the University of Washington, and from 1966 to 1969, his research was aided by Foreign Area Fellowships, awarded jointly by the Social Science Research Council and the American Council of Learned Societies. During the summer of 1967, he travelled some 20,000 kilometers through European Russia, the Transcaucasus, and Siberia researching his doctoral dissertation. Best known for his classic work, *Conservation in the Soviet Union,* published in 1972 by Cambridge University Press, Dr. Pryde has written extensively for professional and popular periodicals on the geography, economics, and resources of the Soviet Union and the United States. He has travelled in western Europe and the United States gathering data on physical geography, environmental quality, and water resources.

Appendix F

Patricia Scharlin Rambach, lecturer, author, and consultant, is a trustee of the Population Reference Bureau, consultant to the United Nations Institute for Training and Research, and has been international program officer at the Sierra Club's Office of International Environment Affairs since 1972. Previously she was editor-in-chief of publications for the Carnegie Endowment for International Peace in New York. Ms. Rambach is a member of the IUCN American Committee on International Conservation and the task force on the Law of the Sea; the International Development Conference; and the advisory committee of the Center for International Environment Information.

Nathaniel P. Reed, assistant secretary for fish and wildlife and parks at the U.S. Department of the Interior since 1971, has served in environmental posts and as consultant for three Florida governors. His primary interests have been problems of the Everglades National Park and air and water pollution. A former member of the Florida Pollution Control Commission, he was appointed first chairman of the Florida Department of Air and Water Pollution Control in 1969. Mr. Reed has served frequently as federal-state interactor, influencing two nationally significant environmental decisions: abandonment of the across-Florida barge-canal project and the 1970 Jetport Pact. As assistant secretary, he played a major role in implementing the Endangered Species Act of 1973, fashioning the Nationwide Outdoor Recreation Plan, and formulating the administration's 33 million-hectare proposals for new national parks, wildlife refuges, forests, and scenic rivers in Alaska.

Nicholas A. Robinson, convenor and vice-chairman of the EARTHCARE Conference, is an attorney and counsellor at law associated with Marshall, Bratter, Greene, Allison and Tucker in New York. He is chairman of the Sierra Club's International Committee, member of the board of directors of the United Nations Association of the U.S.A., and member of the United States Committee for UNICEF. He was on the United States National Commission for UNESCO from 1968 to 1970, the legal advisory committee to the President's Council on Environmental Quality from 1970 to 1972, and was a member of the United States delegation to the USSR Bilateral Environmental Law Exchange in 1974. He was also a member of the governor of New York's Environment Advisory Task Force in 1975.

Jeanne Sauvé, Canada's minister of communications since 1975, was minister of the environment at the time of the EARTHCARE Conference. She was minister of state in charge of science and technology from 1972 to 1974, and had previously acted as a journalist, broadcaster, and lecturer in the United States and Canada. From 1966 to 1972, Madame Sauvé was the general secretary of the Fédération des Auteurs et des Artistes du Canada. In 1972, she was named one of seven founding members of the Institute of Political Research by Prime Minister Pierre Trudeau and was awarded an honorary doctor of science degree by New Brunswick University in 1974.

Gerald S. Schatz is a staff officer of the National Academy of Sciences, Washington, D.C., and editor of the academy's science-policy bulletin, *News Report.* Formerly, he was a staff member of the U.S. House of Representatives' Research and Technical Programs and Conservation and Natural Resources subcommittees of the Committee on Government Operations. Also a member of the Arctic Institute of North America and of the board of directors of the Antarctican Society, Mr. Schatz has travelled in the Arctic in connection with policy studies and coordinated the Antarctican Society's Colloquium on Science, Technology, and Sovereignty in the Polar Regions, the proceedings of which he also edited.

Raymond J. Sherwin is a judge of the Superior Court in Solano County, California, vice-president for the Sierra Club's International Program, chairman of the Save the John Muir Trail Association, and, with Dr. Elvis J. Stahr of the National Audubon Society, cochairman of the EARTHCARE Conference. Long active in the Sierra Club, he was elected to its board of directors in 1969, was president of the board in 1971 and 1972, and served as director of the Sierra Club of London and the Sierra Club Legal Defense Fund. A former member of the United States Delegation on Environmental Cooperation with the USSR, Judge Sherwin was in 1972 a member of the National Executive Advisory Committee on Natural Gas. He has since been a member of the National Executive Advisory Committee on Power, and is currently a member of the Conservation Law Society of America, the Planning and Conservation League, and the California Roadside Council.

William J. L. Sladen has been professor of microbiology and arian ecology at the School of Hygiene, Johns Hopkins University, since 1969. Between 1947 and 1955, he was medical officer and biologist and then senior research scientist with the Falkland Islands Dependencies Survey (later, the British Antarctic Survey). For his services in Antarctica, Dr. Sladen was made a Member of the Order of the British Empire in 1949, and was awarded the Polar Medal in 1953. In 1956 and 1957, he was a Rockefeller Foundation Fellow at Johns Hopkins Hospital and School of Hygiene. A prolific writer, he is the author of many articles on ornithology, ecology, physiology, biochemistry, microbiology, and pathology, and has made scientific motion pictures on Antarctica and animal behavior. Dr. Sladen is a fellow of the American Ornithologists' Union and of the London Zoological Society.

Elvis J. Stahr became president of the National Audubon Society in 1968, after a distinguished twenty-year career in government and education. At the time of the EARTHCARE Conference, he was a member of the National Commission for World Population Year, the National Petroleum Council, the International Energy Project Subcommittee of the Committee for Economic Development, and numerous other organizations. In 1972, he was a member of the United States delegation to the United Nations Conference on the Human Environment in Stockholm, and he has served on the Joint United States–USSR Committee on Cooperation for Protection of the Environment. Dr. Stahr has served in the administrations of five U.S. presidents, most notably as secretary of the army, a post he held in 1961 and 1962. Dr. Stahr served as president of West Virginia University from 1959 to 1961, and as president of Indiana University from 1962 through 1968. The recipient of twenty-three honorary degrees, he received the first doctor of environmental science degree ever awarded.

David H. Stansbery is director and curator of the Museum of Zoology at Ohio State University and professor in the university's department of zoology, where he teaches courses on field zoology and animal ecology. He is also curator of natural history for the Ohio Historical Society. Dr. Stansbery is a specialist in the zoogeography, ecology, evolution, and systematics of freshwater animals, especially bivalve mollusks and decapod crustaceans. He is a fellow of the American Association for the Advancement of Science and the Ohio Academy of Science, and is affiliated with numerous other professional associations and societies. Dr. Stansbery has served as the chairman of the Water-Use Committee of the Ohio Environmental Council and as president and permanent council member of the American Malacological Union. At the time of the EARTHCARE Conference, he was president of the Ohio State University Chapter of Sigma Xi. In 1974, the Ohio Department of Natural Resources conferred upon Dr. Stansbery the Ohio Conservation Achievement Award.

Margaret M. Stewart is professor of vertebrate biology at the State University of New York at Albany and consultant to the New York State Department of Environmental Conservation. An authority on amphibians, reptiles, and small mammals, she has served as visiting teacher of summer field courses and research for the City University of New York at several field centers, including the Cranberry Lake Biological Station and the Caribbean Biological Center, Jamaica, West Indies. Dr. Stewart has taught courses on vertebrate ecology and behavior, environmental biology, and "People, Resources in Ecological Perspective," and was recently awarded several fellowships, awards, and research grants to study Jamaican amphibians. Her publications include *Amphibians of Malawi: An Illustrated Handbook* and, with A. H. Benton, *Field Manual for Vertebrates of the Northeastern States*.

Lee M. Talbot, an internationally renowned ecologist, has held the post of assistant to the chairman for international and scientific affairs of the President's Council on Environmental Quality since 1975. He previously served as senior scientist for the council. He began his professional career in 1951 as a field biologist at the Naval Arctic Research Laboratory in Barrow, Alaska, and has conducted ecological research in some ninety countries. Dr. Talbot was instrumental in reversing a U.S. policy that favored commercial killing of whales, in securing the ban on poisons used to kill predators, and in helping to develop the U.S.–USSR agreement on environmental protection. The Convention on Control of Trade in Endangered Species owes its origin to Dr. Talbot as well. He is the author of five books and more than 135 professional articles and papers. He has served as consultant to or member of numerous conservation organizations, and received the Albert Schweitzer Medal of the Animal Welfare Institute in 1974, the centenary of Schweitzer's birth.

Oliver Thorold is an environmental lawyer whose headquarters are in London. At the time of the EARTHCARE Conference, he was involved in establishing a center to encourage the development of environmental law in the United Kingdom. Among his many other activities, Mr. Thorold is a member of the board of editors of the *International and Comparative Earth Law Journal* and vice-chairman of the Lawyers' Ecology Group, London.

Edward L. Towle is president of the Island Resources Foundation and former president of the Caribbean Conservation Association, Saint Thomas, United States Virgin Islands. He has served as associate curator of naval history at the Smithsonian Institution, as director of the Caribbean Research Institute at the College of the Virgin Islands, and as a consultant to several island governments and international organizations including UNESCO, UNDP, FAO, and IUCN. Dr. Towle is the author of some thirty articles and coauthor of a new book on ecological guidelines for the planning of island development.

Jan W. van Wagtendonk, research scientist in fire ecology at Yosemite National Park, California, has worked for the U.S. National Park Service since 1972. He currently is studying the recreational carrying capacities of wilderness areas and the role of fire in ecosystems of the Sierra Nevada. Dr. van Wagtendonk's doctoral research concerned the fire and fuel relationships in mixed-conifer ecosystems of Yosemite National Park. A member of the Society of American Foresters and of the Ecological Society of America, he is the author of *Refined Burning Prescriptions for Yosemite National Park*, published by the National Park Service in 1975.

John W. Winchester, professor and chairman of the Department of Oceanography at Florida State University, is a physical chemist who specializes in activation analysis and atmospheric and marine geochemistry. He was previously professor of oceanography at the University of Michigan and assistant director of the Great Lakes Research Division. Dr. Winchester held visiting Fulbright scholarships to the

Netherlands in 1955 and 1956, and was assistant and then associate professor of geochemistry at M.I.T. between 1956 and 1966. Dr. Winchester has taught or conducted research at many institutions, among them Columbia University, Oak Ridge National Laboratory, the University of Alaska, as well as Taipei and La Plata. He belongs to numerous professional societies, including the American Chemical Society, the Geological Society of America, and the American Geophysical Union.

R. Michael Wright has been active on legal issues related to the environment since his graduation from law school in 1970. He is currently director and general counsel for International Programs at The Nature Conservancy in Arlington, Virginia. A former fellow of the Rockefeller Foundation in Environmental Affairs, Mr. Wright has lectured at Stanford University, where he developed and taught a course on the problems of international environmental disruption. He is a member of the American Society of International Law and of the executive board of the American Committee for International Conservation.

Avraham Yoffe, chairman of the Nature Reserves Authority in Tel Aviv since 1964, is regarded as Israel's leading environmentalist. An avid outdoorsman, he has had a lifelong devotion to Middle-East wildlife. A member of the Knesset (Parliament) since 1973, he began his service in the Haganah during college and remained until the state of Israel was formed in 1948, at which time he entered the Israeli defense forces. He has continued to watch over Israel's 26,000 hectares of nature preserves, including the famous Hai-Bar Reserve, containing some 100 species of biblical animals, many of which are on the verge of extinction. General Yoffe has visited many foreign nations as representative of his government and as consultant and adviser on worldwide conservation problems. He has participated in international conferences sponsored by the International Union for Conservation of Nature and Natural Resources and the World Wildlife Fund, and is well-known in international circles involved in the conservation of nature and wildlife.

APPENDIX G

Conference Participants

Dr. Mohammed Reza Amini
Head, Aquatic Ecology and
  Fisheries Group
Department of Environmental
  Conservation
Tehran
Iran

Mr. Chaplin B. Barnes
Director of International Activities
National Audubon Society
950 Third Avenue
New York, New York 10022
U.S.A.

Mr. Vladimir A. Borissoff
Secretary-General
Central Laboratory on Nature Conservation
USSR Ministry of Agriculture
Znamenskoye-Sadki, 142790
P.O. Vilar
Moscow Region
USSR

Dr. F. Herbert Bormann
Oastler Professor of Forest Ecology
School of Forestry and Environmental Studies
Yale University
370 Prospect Street
New Haven, Connecticut 06511
U.S.A.

Dr. Arthur G. Bourne
17 Church Road
Flitwick, Bedfordshire
England

Dr. David R. Brower
President, Friends of the Earth
529 Commercial Street
San Francisco, California 94111
U.S.A.

Dr. Noel J. Brown
Senior Liaison Officer
United Nations Environment Programme
485 Lexington Avenue
New York, New York 10017
U.S.A.

Dr. Gerardo Budowski
Head, Forest Sciences Department
Centro Agronómico Tropical de
  Investigación y Enseñanza (CATIE)
Apartado 74
Turrialba
Costa Rica

Mr. Charles H. Callison
President, Public Lands Institute
1740 High Street
Denver, Colorado 80218
U.S.A.

Dr. Roy E. Cameron
Deputy Director
Land Reclamation Program
Argonne National Laboratory
9700 South Cass Avenue
Argonne, Illinois 60439
U.S.A.

Mr. Eric Carlson
Head, Human Settlements
United Nations Environment Programme
Post Office Box 30552
Nairobi
Kenya

Mr. Roland C. Clement
Weed Avenue
RR 2
Norwalk, Connecticut 06850
U.S.A.

Dr. Kai Curry-Lindahl
Professor
Royal Ministry for Foreign Affairs
Box 16121
S-103 23
Stockholm 16
Sweden

Dr. René J. Dubos
Professor Emeritus
The Rockefeller University
66th Street and York Avenue
New York, New York 10021
U.S.A.

Mr. Lars Emmelin
Course Director
Miljövårdsprogrammet
Lunds Universitet och Tekniska Högskola
Helgonavägen 5
S-223 62 Lund
Sweden

Dr. M. Taghi Farvar
Director, Centre for Endogenous
  Development Studies
46 Sahba Street
Amirabad Avenue
Tehran
Iran

Dr. Rimmon C. Fay
Commissioner, California Coastal Zone
  Conservation Commission
South Coast Region
Post Office Box 1450
Long Beach, California 90801
U.S.A.

Dr. Victor C. Ferkiss
Professor, Department of Government
Georgetown University
Washington, D.C. 20007
U.S.A.

Mr. William L. Finger
Vice-President
International Executive Service Corps
622 Third Avenue
New York, New York 10017
U.S.A.

Mr. Eskandar Firouz
Pasteur Avenue
Firouz Street No. 38
Tehran
Iran

Dr. Arturo Gómez-Pompa
Director, Instituto de Investigaciones sobre
  Recursos Bióticos, A. C.
Heroico Colegio Militar No. 7
Apartado Postal 63
Jalapa, Veracruz
Mexico

Appendix G

M. François Gros, Directeur
Grande Traversée des Alpes Françaises
16, rue Pierre-Dupont
38000 Grenoble
France

Dr. Lawrence S. Hamilton
Professor of Forestry and Conservation
Department of Natural Resources
Cornell University, Fernow Hall
Ithaca, New York 14853
U.S.A.

The Honorable Jay S. Hammond
Governor, State of Alaska
716 Calhoun
Juneau, Alaska 99801
U.S.A.

Ms. Marian S. Heiskell
870 United Nations Plaza
New York, New York 10017
U.S.A.

Ms. Eleanor C. J. Horwitz
Public Affairs
Society of American Foresters
Wild Acres
5400 Grosvenor Lane
Washington, D.C. 20014
U.S.A.

Dr. Moustafa Imam
Head, Department of Natural Resources
Institute of African Research and Studies
Cairo University
Giza
Egypt

Dr. Thomas B. Johansson
Environmental Studies Program
University of Lund
Helgonavägen 5
S-223 62 Lund
Sweden

Dr. F. Wayne King
Director, Zoology and Conservation
New York Zoological Society
185th Street and Southern Boulevard
Bronx, New York 10460
U.S.A.

Mr. Ismat T. Kittani
United Nations Secretariat
United Nations
New York, New York 10017
U.S.A.

Dr. Thomas E. Lovejoy
Program Director
World Wildlife Fund
1319 18th Street, N.W.
Washington, D.C. 20036
U.S.A.

Dr. Emile A. Malek
Director, Laboratory of Schistosomiasis
  and Medical Malacology
Department of Tropical Medicine
Tulane University Medical Center
1430 Tulane Avenue
New Orleans, Louisiana 70112
U.S.A.

Mr. Arthur R. Marshall
Post Office Box 613
Interlachen, Florida 32048
U.S.A.

Professor G.V.T. Matthews
Director, International Waterfowl
  Research Bureau
Slimbridge, Gloucestershire GL2 7BX
England

Mr. J. Michael McCloskey
Executive Director, Sierra Club
530 Bush Street
San Francisco, California 94108
U.S.A.

Ing. Edgardo Mondolfi
Quinta Masapo
Avenida Norte
Alta Florida
Caracas 105
Venezuela

Dr. Robert Muller
Deputy Under Secretary-General and Director
Inter-Agency Affairs and Coordination
United Nations
New York, New York 10017
U.S.A.

Dr. Norman Myers
6, Benson Place
Oxford
England

Dr. Roderick Nash
Professor of History and Environmental
   Studies
Department of History
University of California, Santa Barbara
Santa Barbara, California 93106
U.S.A.

Mr. Frank X.J.C. Njenga
Counsellor, Permanent Mission of the
   Republic of Kenya to the United Nations
866 United Nations Plaza
New York, New York 10017
U.S.A.

Dr. Perez M. Olindo
National Council of Science and
   Technology
P.O. Box 30007
64 Nairobi
Kenya

Dr. Roger Tory Peterson
Neck Road
Old Lyme, Connecticut 06371
U.S.A.

Dr. Russell W. Peterson
Director
Office of Technology Assessment
600 Pennsylvania Avenue, S.E.
Washington, D.C. 20510
U.S.A.

Dr. Douglas H. Pimlott
Ramsay Wright Zoological Laboratories
Department of Zoology
University of Toronto
25 Harbord Street
Toronto 5, Ontario
Canada

Dr. Ghillean T. Prance
Director of Botanical Research
The New York Botanical Garden
Bronx, New York 10458
U.S.A.

Dr. Philip R. Pryde
Professor, Department of Geography
San Diego State University
San Diego, California 92182
U.S.A.

The Honorable Nathaniel P. Reed
Jupiter Island
Post Office Box 375
Hobe Sound, Florida 33455
U.S.A.

Mr. Nicholas A. Robinson, Esquire
Marshall, Bratter, Greene, Allison, and
   Tucker
430 Park Avenue
New York, New York 10022
U.S.A.

The Honorable Jeanne Sauvé
Minister, Department of Communications
100 Metcalfe Street
Ottawa, Ontario K1A 0C8
Canada

Ms. Patricia Scharlin-Rambach
Director
Office of International Environment Affairs
Sierra Club
800 Second Avenue
New York, New York 10017
U.S.A.

Mr. Gerald S. Schatz
Editor, *News Report*
Office of Information
National Academy of Sciences
2101 Constitution Avenue, N.W.
Washington, D.C. 20418
U.S.A.

Appendix G

The Honorable Raymond J. Sherwin
Judge of Superior Court
Department 2
Superior Court
Hall of Justice
600 Union Street
Fairfield, California 94533
U.S.A.

Dr. William J. L. Sladen
Professor of Microbiology and Arian Ecology
Department of Pathobiology
School of Hygiene and Public Health
The Johns Hopkins University
615 North Wolfe Street
Baltimore, Maryland 21205
U.S.A.

Dr. Elvis J. Stahr
President, National Audubon Society
950 Third Avenue
New York, New York 10022
U.S.A.

Dr. David H. Stansbery
Professor of Zoology
Director, Museum of Zoology
The Ohio State University
Sullivant Hall
1813 North High Street
Columbus, Ohio 43210
U.S.A.

Dr. Margaret M. Stewart
Professor, Department of Biological Sciences
State University of New York at Albany
1400 Washington Avenue
Albany, New York 12222
U.S.A.

Dr. Lee M. Talbot
Assistant to the Chairman for International
  and Scientific Affairs
Council on Environmental Quality
722 Jackson Place, N.W.
Washington, D.C. 20006
U.S.A.

Mr. Oliver Thorold
Barrister-at-Law
3, Brunswick Mansions
Handel Street
London W.C. 1
England

Dr. Edward L. Towle
President, Island Resources Foundation, Inc.
Post Office Box 4187
Saint Thomas, U.S. Virgin Islands 00801
U.S.A.

Dr. Jan W. van Wagtendonk
Research Scientist
National Park Service
Post Office Box 577
Yosemite National Park, California 95389
U.S.A.

Dr. John W. Winchester
Chairman, Department of Oceanography
Florida State University
Tallahassee, Florida 32306
U.S.A.

Mr. R. Michael Wright
Director and General Counsel
International Program
The Nature Conservancy
1800 North Kent Street
Arlington, Virginia 22209
U.S.A.

General (Res.) Avraham Yoffe
Director, Nature Reserves Authority
16, Ha-Natziv Street
Tel-Aviv
Israel

CREDITS

Illustrations

Frontispiece: Courtesy National Aeronautics and Space Administration.
Page 74: Courtesy National Aeronautics and Space Administration.
Page 102: Courtesy National Center for Atmospheric Research.
Page 186: Courtesy National Aeronautics and Space Administration.
Page 302: Courtesy Allan Novick.
Page 464: Both courtesy William R. Curtsinger, copyright by the photographer.
Page 512: Courtesy William G. Damroth.
Page 616: Calligraphy by Charles Chang Chao.

Text

We also acknowledge with gratitude the permissions the following authors, publishers, and individuals have granted us to use previously published material:

Page iv: From Henry Beston's *The Outermost House: A Year of Life on the Great Beach of Cape Cod*, by permission of Doubleday & Company, Inc. Copyright 1928 by Doubleday, Doran and Company.

Page 74: From "Passage to India" from *Leaves of Grass* by Walt Whitman, by permission of Doubleday & Company, Inc. Copyright 1938 by Doubleday, Doran and Company.

Page 74: From an address to the National Institute of Social Science on 25 November 1969. Used with the permission of Col. Frank Borman.

Pages 130 and 336: From "El Gran Océano" and "Algunas Bestias" from *Canto General* by Pablo Neruda, by permission of Editorial Losada S.A., Buenos Aires.

Page 130: From Lü-shih ch'un-ch'iu in *Science and Civilisation in China,* volume 3, by Joseph Needham (New York, 1959), by permission of Cambridge University Press.

Page 152: From "The Marshes of Glynn" from Sidney Lanier's *Hymns of the Marshes* (New York, 1907), by permission of Charles Scribner's Sons.

Page 152: From *Poemas Completas* by Antonio Machado, by permission of Espasa-Calpe, S.A., Madrid. Copyright © 1940 by the publisher.

Page 186: From *Four Quartets* by T. S. Eliot, by permission of Harcourt, Brace & World, Inc., New York. Copyright 1943 by T. S. Eliot.

Pages 336 and 410: From "A Song of Nezahualcoyotl" and "Storm Song" from *The Magic World: American Indian Songs and Poems* (New York, 1971), edited by William Brandon, by permission of William Morrow & Company, Inc.

Page 358: From *The Sacred Pipe: Black Elk's Account of the Seven Rites of the Oglala Sioux,* recorded and edited by Joseph Epes Brown (The Civilization of the American Indian Series, volume 36), by permission of the University of Oklahoma Press. Copyright 1953 by the publishers.

Page 464: Reprinted, with minor changes, from "Song about the Reindeer, Musk Oxen, Women, and Men Who Want to Show Off," from *Eskimo Poems from Canada and Greenland* by Knud Rasmussen, translated by Tom Lowenstein, by permission of the University of Pittsburgh Press. Copyright © 1974 by Tom Lowenstein.